Introduction to

ECONOMIC STATISTICS

WILLIAM C. MERRILL

Associate Professor of Economics,
Iowa State University

KARL A. FOX

Distinguished Professor and Head of Department of Economics
Iowa State University

JOHN WILEY & SONS, INC

NEW YORK · LONDON · SYDNEY · TORONTO

10 9 8 7 6 5 4
Library of Congress Catalogue Card Number: 75–91154

SBN 471 59620 5

Printed in the United States of America

Preface

This book is designed for use by undergraduate students majoring in economics and business administration. Our goal is to provide the reader with an understanding of the basic principles and statistical techniques that are used most frequently in economic analysis and business decision making. We have relied heavily on the use of examples to accomplish this objective. Most of the examples are of the "work-along" type intended to illustrate a particular principle or technique. Many, however, have been drawn from quantitative studies in economics to illustrate the variety of ways that particular statistical techniques can be used for analyzing economic data. A more detailed discussion of many of these examples can be found in Fox's *Intermediate Economic Statistics.*

Some topics that are frequently covered in introductory courses have been excluded from this textbook. There is, for example, no discussion of the analysis of variance or simultaneous estimation techniques. Such subjects are certainly important but were excluded because we felt it advisable to emphasize basic probability.

The reader will find more material here than can be covered adequately during one semester. The core material is contained in Chapters 4 to 9, which cover probability, sampling, estimation, hypothesis testing, and simple regression. The principles of probability discussed in Chapters 4 to 6 are essential for full understanding of most of the other material. The techniques of estimation, hypothesis testing, and simple regression are important tools for business and economic analysts.

Chapter 2 is intended as a brief introduction to descriptive statistics and can be covered quickly. We feel that most undergraduates will be able to study this topic on their own once they are familiar with the basic principles of probability and the techniques of estimation.

The use of index numbers for measuring changes in economic aggregates is discussed in Chapter 3. An understanding of the strengths and weaknesses of different types of index numbers is important in applied economics but is not essential for understanding the material in Chapters 4 to 9. Thus, some readers may find it advisable to consider Chapter 3 as reference material and bypass it initially.

Our discussion of multiple regression (Chapter 10) is an introduction both to the subject itself and to the notation commonly used in econometrics. Readers who are unfamiliar with matrix algebra will find Chapter 10 difficult and may prefer to move directly to the discussion of time series analysis. Knowledge of multiple regression, while helpful, is not essential for an understanding of Chapter 11, which discusses standard procedures for dissecting individual time series. The limitations of these procedures are considered in a chapter appendix, which stresses the use of economic models in explaining movements in time series data.

Chapter 12 is drawn from Fox's *Intermediate Economic Statistics* and serves as an introduction to some of the problems encountered in the construction of large-scale economic models. The reader should be thoroughly familiar with multiple regression analysis before beginning this chapter; determinants are used but matrix notation is not employed. Emphasis is placed upon causal ordering as a basis for achieving identification and selecting appropriate estimation techniques in multiple equation models.

Considerable space is devoted to a mathematical appendix covering the elementary concepts of calculus and matrix algebra. We also present some basic rules of algebra and summations and a section on set theory in relation to probability and statistics.[1] Most of the mathematical concepts used in the text are explained in the appendix.

Economics at the graduate level has become a mathematical as well as a social science. Mathematical instruction in the elementary and secondary schools is improving rapidly. We believe the presentation here represents a reasonable aspiration level as of 1970; the appendix should help the reader with limited mathematical training to achieve this norm.

Textbooks have some generic limitations. As Thomas Kuhn says, "They address themselves to an already articulated body of problems, data, and theory, most often to the particular set of paradigms to which the scientific community is committed at the time they are written. . . . They have to be rewritten in the aftermath of each scientific revolution, and once rewritten,

[1] The set theory section is based on S. S. Wilks, *Mathematical Statistics* (New York: John Wiley and Sons, Inc.), 1962, pp. 1-7.

they inevitably disguise not only the role but the very existence of the revolutions that produced them."[2]

Economics has undergone a sweeping revolution since 1930. One has only to compare the articles on economic thought in the *International Encyclopedia of the Social Sciences* (1968) with those in the *Encyclopaedia of the Social Sciences* (1931), or to compare Robert H. Strotz's article on "Econometrics" (1968) with Oskar Morgenstern's article on "Mathematical Economics" (1931).

The word "econometrics" does not appear at all in Morgenstern's article or in the index to the 1931 *Encyclopaedia.* In 1968, Strotz could write:

"Areas of application of econometrics within economics have been steadily widened. There is now scarcely a field of applied economics into which mathematical and statistical theory has not penetrated, including economic history. . . . With its success as a major intellectual movement within economics, econometrics [broadly defined to include mathematical as well as statistical economics] is losing its identity and is disappearing as a special branch of the discipline, becoming now nearly conterminous with the entire field of economics. . . ."[3]

While Strotz adds appropriate qualifications to this statement, the present authors would agree that the 1955 concept of an econometrician has been incorporated (along with other skills) into the 1970 concept of an economist.

An undergraduate textbook in economic statistics can convey only a glimpse of the dynamism of the field's development, the excitement of new discoveries, and the satisfaction of successful applications. If the reader seriously intends to become an economist, he should begin to acquaint himself with some of the classic articles and monographs and with the current professional journals; these are extensions of the real scientific community in time and space.

In this connection, one of the authors (Karl A. Fox) would like to acknowledge some intellectual debts of long standing to Arthur Mowbray, Harry R. Wellman, George M. Kuznets, and Robert A. Gordon of the University of California, whose courses started him out in economic statistics and business cycle analysis in 1941–1942. They introduced him to the early work of Sewall Wright, R. A. Fisher, Mordecai Ezekiel, Henry Schultz, Jan Tinbergen, Ragnar Frisch, Tjalling Koopmans, and Herman Wold, whose pre-1942

[2] Thomas S. Kuhn, *The Structure of Scientific Revolutions* (Chicago: University of Chicago Press), 1962, pp. 135-136.

[3] Robert H. Strotz, "Econometrics," in *International Encyclopedia of the Social Sciences* (Macmillan Company and Free Press), 1968, Volume 4, p. 352.

classics contained or foreshadowed many of the ideas which came to dominate scientific practice in economics during the 1950's and 1960's. Students building foundations in econometrics today should become acquainted with the post-1942 classics by Tinbergen, Frisch, Koopmans, Wold, Richard Stone, Trygve Haavelmo, Herbert A. Simon, Lawrence R. Klein, Henri Theil, R. L. Basmann, James Tobin, Zvi Griliches, Dale Jorgenson, E. Malinvaud, and others.

The other author (William C. Merrill) wishes to acknowledge the indirect contributions of many former professors and colleagues who, in the classroom and in discussions, have stimulated interest in and increased his understanding of quantitative economic analysis. In this regard, special thanks are due Professors George Break, Roy Radner, Dale Jorgenson, Saul Hymans, and Gene Savin, all masters of the art of adding excitement to the mixture of economics and statistics.

Many others have contributed, directly or indirectly, to the development of this book. We are especially grateful to Helen M. Brown and Rita Bauman, whose fine typing and careful proofreading made our job much easier. Dr. Arnold Faden reviewed the material on hypothesis testing and assisted in several revisions. Wayne Lacy read a late version of the manuscript and recommended important changes. We wish to thank, without implicating, a number of reviewers who made suggestions for clarifying parts of the text, notably Franklin Fisher, Edward Greenberg, and R. W. Pfouts.

We are also indebted to the Literary Executor of the late Sir Ronald A. Fisher, F.R.S.; to Frank Yates, F.R.S.; and to Oliver and Boyd Ltd., Edinburgh, for permission to reprint Table III from their book Statistical Tables for Biological, Agricultural and Medical Research, and to the other authors and publishers who have given their kind permission to use various materials and tables presented in this book.

Ames, Iowa
May 19, 1969

William C. Merrill
Karl A. Fox

Contents

CHAPTER 1

Economics and Statistics

Statistical methods are used in almost every phase of human activity today. In business and economics, statistical methods are among the most important tools used for making decisions. As a result of developments taking place in the computer industry today, modern statistical methods will be even more important to decision makers in the future.

The term *economic statistics* can be given a variety of interpretations. In a narrow sense, the term refers to numerical data about such economic variables as national income, employment, and prices. In a broader sense, economic statistics refers to a body of statistical theory and methodology that has proven useful in the analysis of business and economic problems. We shall generally use the term economic statistics in this broader sense.

Economic statistics can be presented in varying degrees of mathematical refinement. Most of the mathematics used in this book is covered in modern college algebra although some calculus is used beginning in Chapter 5 and some matrix algebra in Chapter 10. The reader will need some familiarity with the concept and notation of functions, the use of mathematical tables, procedures for solving algebraic equations, and in Chapter 4, some knowledge of set theory. Brief reviews of these topics are presented in the mathematical appendix at the end of the book.

1.1 THE SCOPE OF ECONOMIC STATISTICS

Economics is the study of how people choose to use scarce resources to produce and allocate various commodities. It is one of a number of sciences that deal with man and society. The boundaries between these sciences are somewhat arbitrary and continually change over time. Specialists in different social sciences often emphasize different aspects of the same general problems or study the same aspects in different ways.

New theories and new information developed by researchers in other social sciences often change the thinking of economists. In a similar manner, new statistical techniques developed for use in other sciences are often rapidly adapted by economists. Thus, economic statistics also changes over time. The development of multidisciplinary fields such as management science and operations research can be expected to increase the rate of importation of new methods into economic statistics from other disciplines. The student who learns those parts of mathematics, probability theory, and statistics that are most relevant to applied economics will at the same time be acquiring a key to the understanding of important segments of several other disciplines. His work in economics will be broadened by his ability to recognize the relevance for economics of new techniques developed in other disciplines.

In the preface to his book, *Models of Man: Social and Rational*, Herbert A. Simon describes his attempt to lay the foundations for a "science of man" that would be broader than any of the existing social science disciplines. He believes that this science rests on two principal mechanisms, those of influence and of choice.

Mechanisms of influence are replicated in many of the social sciences, as in the power structure diagrams of sociologists and political scientists, and asymmetric relationships between individuals (parent and child, supervisor and employee, and the like, studied under such headings as small-group dynamics, transactional analysis, and other categories). National economic policy, for example, involves mechanisms of influence in the selection of instruments of economic policy to accomplish economic targets or objectives such as high employment, rapid economic growth, stable price levels, and balance of payments equilibrium. In the late 1960's, causal ordering diagrams and criteria for identification of individual relationships in complicated models are being used in political science by Hayward Alker and others, and by a number of research workers in sociology, social psychology, and political science.

Mechanisms of choice include the economist's models of firms trying to maximize profit and consumers trying to maximize the "utilities" of their incomes. But problems of choice pervade all phases of human life including the allocation of one's time, choice of friends, choice of occupations, choice of places to live; choice among political candidates; choice on the part of state highway commissions of the routes of new highways and whether they shall be two or four lanes in width; choice of the government in military and defense planning, and many other fields.

The concepts of operational research and systems analysis are also appearing in many areas of the social sciences. Examples are David Easton's *A Systems Analysis of Political Life* and the Brookings Institution quarterly econometric model of the United States.

Historians of science are also forming interesting hypotheses as to the nature of scientific discovery and the flow of concepts from one science to another. Thomas S. Kuhn's *The Structure of Scientific Revolutions* (1962) stresses the concept of a *paradigm* or model in describing scientific discovery. When scientists encounter situations that do not fit the existing paradigm, a considerable amount of scientific activity and speculation is set in motion to try to explain the difference. Often the difference can only be explained by a sweeping reformulation of the "true" basic nature of the phenomena, which in many specific cases seem to be satisfactorily explained by the existing paradigm or theory.

Arthur Koestler, in *The Act of Creation* (1964), explains scientific discovery as the juxtaposition of unlike matrices or conventional patterns of thought. Archimedes, according to tradition, was given the problem of deciding whether an elaborately carved crown was made of pure gold or whether it had been alloyed with a cheaper metal. While turning this problem over in his mind, Archimedes immersed himself in his daily bath, and the water level rose just as it had on countless other days. However, on this day Archimedes suddenly realized that the volume of his own body could be accurately measured by the volume of the water it displaced. Hence, he could measure the volume of the crown by putting it in a vessel of water; then, dividing the weight of the crown by the weight of the same volume of water, he could easily find whether the ratio corresponded with that of a sample of pure gold or with some combination of gold and lighter metals. According to history, Archimedes was so excited by the brilliant simplicity of his discovery that he leaped from the tub and ran out into the street shouting "Eureka!" (I have found it!)

Koestler's analysis of scientific discovery is in no way inconsistent with Kuhn's. In our present context, it is important to remember that sudden brilliant reformulations of scientific explanations may render obsolete any specific method, technique, or conceptual framework which currently defines a recognized specialty within a particular discipline such as economics. Frequently, scientific discovery has the effect of demonstrating that a number of problems that have hitherto occupied widely separated groups of specialists are simply special cases of a more general, and hence more powerful, paradigm or model.

It is important, therefore, to try to learn the theoretical core of any particular field, such as economic statistics, as a body of *principles* rather than as an array of individual cookbook recipes. Ideas of probability, of good or optimal decisions, and of estimating relationships among variables arise in such an enormous variety of contexts that a mastery of the underlying theories of choice, hypothesis testing, and estimation will open up areas of application far wider than economics itself.

Economic theory is often inspired by real economic problems. The theory expresses the essential features of the problems and generally provides suggestions as to how the problems can be solved. When possible, it is prudent to test the validity of economic theories before recommendations based on the theories are put into effect. For economists, economic theory is antecedent to and takes precedence over economic statistics. Once it has been rather fully developed, economic theory is the primary guide in deciding which economic magnitudes are worth measuring and how they should be measured. Economic statistics, then, enters the picture as a guide to measuring individual variables, estimating relationships between variables, and judging the validity of the theory.

Economic statistics is an especially important subject to those people concerned with the economic policies of governments. *Economic policy* involves the exercise of governmental powers to influence the course of economic development and current economic activity. The need for quantifying the favorable and unfavorable effects of alternative economic policies upon various groups presents a tremendous challenge to economic statisticians. If the economic statistician can demonstrate, within a moderate range of error, that a proposed policy will transfer $100 million of annual income from Group A to Group B, legislators are at least placed under pressure to decide consciously whether they wish to impose a penalty of this magnitude upon Group A in order to achieve a similar increase in income for Group B. The chains of causation and interaction which determine the ultimate effects of national economic policies are extremely complex and require a highly sophisticated combination of economic theory and statistical technique for their analysis.

Economic statistics is an important subject for the economist regardless of his area of specialization. Agricultural economics, labor economics, money and banking, public finance, national income analysis, monetary policy, fiscal policy, business cycle analysis, consumption economics, marketing, and international trade are just a few of the specialties within economics where economic statistics may be and have been applied. A great many of the applications in these fields involve statistical problems of measurement, estimation, and testing. The increased availability of electronic computers makes it possible for specialists in these fields to use more sophisticated statistical techniques than ever before. Therefore, the scope of economic statistics is rapidly expanding at the present time.

1.2 MEASURING ECONOMIC VARIABLES

The study of the behavior of individual economic units is commonly called *microeconomics*. In contrast, the study of economic relationships and interactions of large aggregates, such as national income, is called *macroeconomics*. Thus, variables reflecting the behavior of an individual economic unit are called *microvariables* and magnitudes reflecting the summation of all similar economic units in a region, a nation, or a sector of the economy are called *macrovariables*.

In concept, it would be quite possible to record every economic transaction made by the members of *a particular household*. The household members could, of course, keep such a record for themselves. Each item purchased could be precisely identified (with brand names, sizes, quality designations, and the like); the number of units of the commodity purchased at a given store on a given day could be noted, along with the price paid for it by the particular household member. At this level of detail, household records would involve many thousands of economic transactions in the course of a year. Some of these are conveniently aggregated. For example, a salaried worker's only money income may be in the form of a monthly paycheck. The household's use of electricity for many kinds of uses (lighting, washing clothes, cooking meals) may likewise be summarized in a single monthly bill from the power company. In this way, most of the economic activities of a household can be measured in the form of prices, quantities, amounts of money paid out, amounts of money received, and assets owned. Such quantitative data are the raw materials of the economic statistician.

Most *businesses* of any size keep detailed sets of accounts. Every economic transaction reflected in the family income and expenditures of a household will be reflected in the receipts of the firm that sold the commodity or the disbursements of the firm that paid the salary or wages to the household member. As a rule, firms purchase inputs (raw materials and factors of production) and sell outputs (goods or services). Thus, the economic activities of a firm can also be expressed, for the most part, in terms of prices, quantities, expenditures, and receipts.

Finally, the activities of government agencies and private nonprofit institutions can also be summarized in the form of microvariables such as numbers of employees, wage and salary disbursements, and purchases of goods and services.

"Social accounts," or national income and product accounts, are in concept simple summations of the accounts of individual households, business firms, nonprofit institutions, and governmental units. If every microeconomic unit kept accurate and detailed accounts according to some uniform format, we could simply process them all through a computer, arriving at the corresponding set of macrovariables by direct addition. In practice, of course, we do not have such detailed and uniform records covering all the economic transactions of all microunits. Hence, the construction of national income and product accounts for, say, the United States requires the matching, splicing, and reconciliation of economic records and estimates from many sources.

Microvariables reflect the decision-making activities of an identifiable household or business firm. The individual household actually observes the prices that we describe as microvariables and, in the light of these prices and of its income and assets, selects the quantities of each good and service purchased. The particular set of quantities that the household selects influences the level of satisfaction attained by members of the household. As a rule, its purchases are made without consultation with other households and its purchases have only a diffuse effect, if any, on the senses of well-being of other households.

In contrast, macrovariables such as average prices paid and total quantities purchased by millions of households cannot be related directly to the economic welfare of a corresponding "macrohousehold." None of the individual households is trying to optimize the economic position of such a mystical macrohousehold. Hence, we must use considerable care in attaching welfare connotations to changes in macrovariables or in assuming that relationships between macrovariables will have the same interpretation as relations between microvariables for a particular household. Nevertheless, comprehensive sets of macrovariables such as national income and product accounts can tell us a great deal about the overall performance of a national economy.

Since the human mind cannot encompass and interpret the totality of economic activity as reflected in every transaction of every micro-economic unit, we usually classify and summarize the basic transactions of economic units before they are used in economic analysis. There are a great many problems involved in the classification and aggregation of such transactions. For example, we may decide to preserve full commodity detail but to aggregate a household's purchases of each commodity over time into weekly, monthly, quarterly, or annual totals. Such totals are, in a sense, abstractions. The price of canned tomatoes, for example, may have differed from day to day within any of these time units, so the "the average price of tomatoes" during a given month or year is an abstraction that was never directly observed or experienced by members of the household.

Further summary of the accounts of an individual household would probably involve aggregating over commodities. First, we might lump together the different brands of canned tomatoes purchased during the year. If we then divide total expenditures for canned tomatoes by the number of cans purchased, we obtain an average price that is even more of an abstraction than before because the "average brand" that goes with the average price has no real existence.

In some cases, it may be necessary to proceed to higher levels of commodity aggregation such as all canned vegetables, all vegetables, all fruits and vegetables, and all foods. Further aggregation would lead to even more abstract groupings, such as those of consumer durable goods, consumer nondurable goods, and services. Such classifications are commonly used in national income and product accounts.

Concepts such as "average price" or "total quantity purchased" become increasingly abstract at each successive level of aggregation. This fact should be kept in mind when trying to interpret apparent relationships between aggregates. Suppose, for example, that the relationship between the "average price of canned tomatoes" and the "average quantity of canned tomatoes purchased per household" was estimated. We might find that a reduction in average price tends to be accompanied by an increase in average quantity purchased. This does not necessarily mean that a reduction in the price of Brand A tomatoes will lead to an increase in the quantity of Brand A tomatoes purchased. Nor can we conclude that the apparent relationship between "average price" and "average quantity purchased" will always hold. There could be significant differences in the kinds of consumers who buy from different stores or in the attitudes of consumers toward price reductions for different brands. In this case, a change in the distribution of prices may lead to a change in the average quantity of tomatoes purchased even though there may be no change in the average price.

In many cases, the type of classification or aggregation used will depend on the intended use of the data. At the manufacturing level, for example, a single large corporation may operate a number of plants, each of which produces quite a different product line. For purposes of constructing input-output tables and production indexes, government agencies may want to assign each *plant* to an appropriate industry classification. Such industry data may not be useful, for example, to the economist interested in industrial organization because, theoretically, it is the *firm* and not the individual plant that serves as the basic unit for decision making and profit maximization.

The form of the data on economic variables partly determines the statistical techniques that can be used to analyze the data. For example, the use of time series data often requires different statistical techniques than the use of cross-sectional data. Similarly, data based on small samples cannot always be analyzed in the same way as data based on large samples.

At first introduction, the problems of measuring economic variables and selecting appropriate statistical techniques may seem extremely complicated. In reality, acceptable solutions to many of these problems are extremely simple. The application of the basic principles of economic statistics to new problems of measurement and estimation often requires imagination and careful thought, however.

With these thoughts in mind, let us describe briefly the material to be covered in this book and then get underway.

1.3 THE SCOPE OF THIS BOOK

The goal of this book is to help you develop an *understanding* of the basic principles of economic statistics. Most of these principles are little more than refinements of everyday thinking. Even so, you undoubtedly will have to read and reread sections, carefully study the examples, write down the main assumptions, memorize key formulas, and work the exercises at the end of each chapter if you are to master this subject. Do not be too concerned if you do not completely understand everything when first reading a chapter. Be sure to return to the material from time to time, however, once you are several chapters beyond those things that were not clear at first. If you are making progress, you will be amazed at how easily a chapter reads once you are two chapters beyond it.

Although the emphasis in this book is on applying statistical methods to business and economic research, keep in mind that most of the statistical methods discussed can be used readily in other social sciences because the fundamental principles underlying these methods are identical regardless of the field of application.

A great deal of "shorthand" notation is used in modern statistics, and the symbols used have specific meanings that, by convention, are the same for most statisticians. You may find it advantageous at first to learn to read words where we write symbols. For example, the phrase "the distribution of $\bar{X}$ depends on the parameters μ and $\sigma/\sqrt{n}$" is read "the distribution of the sample mean depends on the population mean and the standard deviation of the population divided by the square root of the sample size." Even in words, however, the phrase has little meaning unless you are familiar with the definitions of the words distribution, sample mean, population, parameter, and standard deviation. Thus, a small amount of additional time spent on understanding definitions when they first appear will usually save a great deal of time later. Eventually, the symbols will automatically trigger a specific meaning in the mind and there will be no further need to translate them into words. Indeed, you will undoubtedly find economic statistics to be more enjoyable once you have learned "to think in the notation."

At times, it is helpful to substitute numbers for symbols when you are trying to understand a mathematical statement. For example, the first time you come across the formula

$$\overline{X} = \frac{1}{n}(X_1 + X_2 + \cdots + X_n)$$

you might let $n = 3$, $X_1 = 2$, $X_2 = 3$, $X_3 = 4$ and compute $\overline{X} = (1/3)(2 + 3 + 4) = 3$. This practice will be used less frequently as your understanding of the mathematical notation used in statistics develops. Always keep in mind, however, that symbols can denote other things besides numbers. The symbol θ, for example, could denote a number, a function, an action, or an idea, depending on the context in which it is used.

The material in this book is divided into four groups of chapters. The first three chapters serve as an introduction to economic statistics. They provide information on how to describe statistical data and measure changes in economic variables. In Chapters 4 to 6 the basic rules of probability are discussed and then applied to theoretical and sampling distributions. The main ideas of point estimation, interval estimation, and hypothesis testing are presented in Chapters 7 and 8. In Chapters 9 to 12, the emphasis is on regression and time series analysis. Economists use the statistical techniques discussed in the final chapters when measuring relationships between economic variables, testing economic theories, and making forecasts.

CHAPTER 2

Descriptive Statistics

The applied economist frequently must decide how to summarize and present economic data. Numerical tables, line graphs, and picture diagrams are just a few of the devices that can be used. He may also have to determine how the data should be classified or grouped or whether the information needs to be (or can be) summarized in one or several descriptive numbers. The particular method selected to present the information will depend on the type of data available and the audience for which the information is intended. For example, the governor of California may want a very detailed description of the income distribution of farmers in his state. The newspaper reader in San Francisco may prefer to have the data summarized into one number: the state's average farm income. The job of presenting economic data usually is not a difficult one once you are familiar with the commonly used descriptive tools. A few such tools will be discussed in this chapter.

2.1 GRAPHS

Raw statistical data usually consist of a series of measurements. A listing of such measurements seldom conveys much information to the reader, however. Therefore, it is frequently useful to graph the raw data so that their importance is more readily apparent. As an example, suppose we select 20 rice farms in Peru and estimate the cost per kilo of growing rice on each farm. The raw data are listed in Table 2.1.

Many readers will find a graphical summary of the rice costs to be more interesting and more useful than the detailed list in Table 2.1. Therefore, let us consider several ways by which these observations could be displayed.

Table 2.1 Production Cost of Rice for 20 Farms in Peru (Figures in Soles per Kilo)

1.20	1.75	1.60	1.65
1.55	2.00	1.40	1.55
1.35	2.10	1.65	1.50
1.10	1.65	2.20	1.60
1.05	1.75	1.30	1.90

2.1.1 Frequency Histograms and Polygons

One way to summarize the data on rice production costs is to use a *frequency histogram.* Such a histogram can be obtained by plotting the data in Columns (2) and (3) of Table 2.2. The resulting frequency histogram is presented in Figure 2.1.

Table 2.2 was obtained by counting the number of observations falling within each selected class interval. The choice of class intervals is arbitrary, but the use of equal class intervals will usually make the frequency distribution easier to read as well as easier to construct. Graphical presentation of frequency distributions is also somewhat easier when class intervals are equal. The number of classes used depends on the data available and the purpose for which the frequency distribution is intended. In general, the fewer observations available, the fewer classes necessary. Frequency distributions usually have at least five or six classes but seldom more than about 15.

Table 2.2 Frequency Distribution for Grouped Measurements on Rice Production Costs

(1) Class Limits[a]	(2) Class Midpoint[a]	(3) Frequency	(4) Relative Frequency	(5) Cumulative Frequency
1.01–1.20	1.105	3	0.15	3
1.21–1.40	1.305	3	0.15	6
1.41–1.60	1.505	5	0.25	11
1.61–1.80	1.705	5	0.25	16
1.81–2.00	1.905	2	0.10	18
2.01–2.20	2.105	2	0.10	20
Totals		20	1.00	

[a] Figures in soles per kilo.

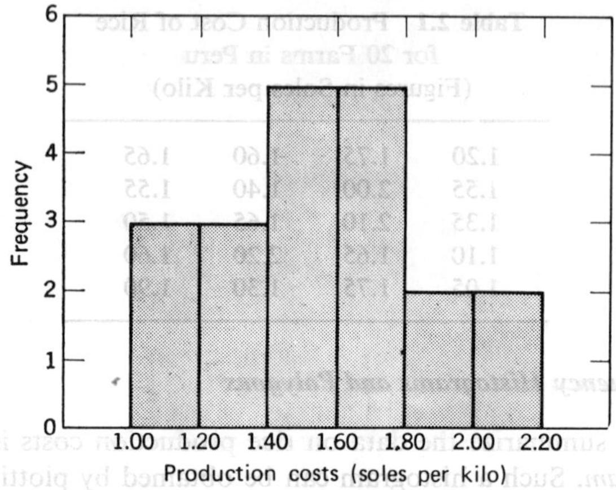

Figure 2.1 Histogram of rice production costs on 20 farms in Peru.

Before proceeding further, we need to introduce some terminology that is frequently used in discussions on the graphical presentation of data. The term *class boundaries* refers to the dividing lines between successive classes while *class limits* refers to the largest and smallest values that fall into a given class. This distinction is illustrated below using three classes.

Class	*Class Limits*	*Class Boundaries*
I	\$4.00–5.99	\$3.995–5.995
II	6.00–7.99	5.995–7.995
III	8.00–9.99	7.995–9.995

The class limits are generally used when tabling data. The class boundaries are useful when it is necessary to "fill the gaps" so that the classes used cover a continuous scale. The length of a *class interval* is the difference between the upper- and lower-class boundaries. Thus, in the illustration presented above, the length of the class interval for Class II is (\$7.995–5.995) = \$2.00 and *not* (\$7.99–6.00) = \$1.99. The *class midpoint* (or class mark) is the midpoint between the class boundaries (or the class limits) of a class. Thus, the class midpoint for Class II above is equal to

$$\frac{1}{2}(\$6.00 + 7.99) = \frac{1}{2}(\$5.995 + 7.995) = \$6.995$$

There are other ways to present the data in Table 2.2 graphically. *Frequency polygons* are often used in place of frequency histograms. A frequency polygon is constructed by plotting the class frequencies at their respective class midpoints and then connecting these points with straight lines. Such a frequency

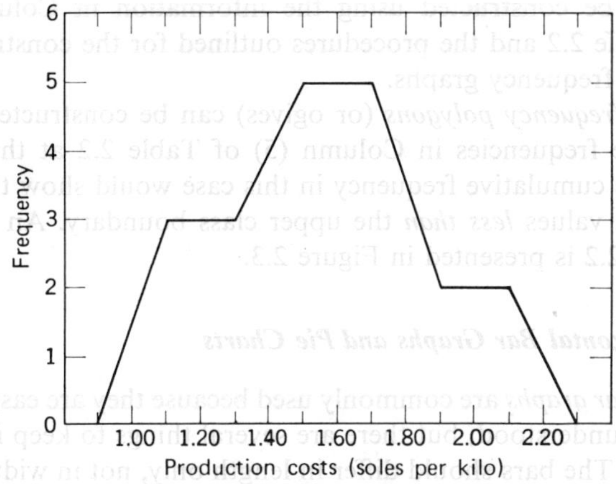

Figure 2.2 Frequency polygon of rice production costs on 20 farms in Peru.

polygon is drawn in Figure 2.2. An extra class is usually added at each end of the distribution so that both ends of the polygon can be brought down to the horizontal axis. The lines connecting the class frequencies at the class midpoints provide no information on the frequency of observations between the class midpoints but simply an idea of the shape of the underlying frequency distribution.

The relative frequencies presented in Column (4) of Table 2.2 were obtained by dividing each class frequency by the total number of observations. The relative frequencies constitute a *percentage distribution* because they tell us what percent of the total observations fall in each class. Relative frequency histograms, relative frequency polygons, and cumulative relative frequency

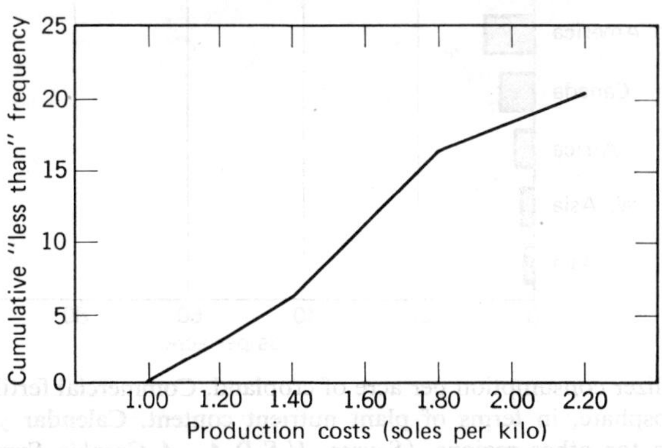

Figure 2.3 Cumulative frequency polygon of rice production costs on 20 farms in Peru.

polygons can be constructed using the information in Columns (1), (2), and (4) of Table 2.2 and the procedures outlined for the construction of the corresponding frequency graphs.

Cumulative frequency polygons (or ogives) can be constructed by plotting the cumulative frequencies in Column (5) of Table 2.2 at the upper class boundary. The cumulative frequency in this case would show the number of items that had values *less than* the upper class boundary. An ogive for the data in Table 2.2 is presented in Figure 2.3.

2.1.2 Horizontal Bar Graphs and Pie Charts

Horizontal bar graphs are commonly used because they are easy to construct and are easily understood, but there are several things to keep in mind when devising them. The bars should differ in length only, not in width. The graph should have a clearly defined zero line and an unbroken quantity scale. A space should be left between each bar, and the bars should be ranked in order of magnitude. If there is an "all others" category, it is usually shown as the bottom bar regardless of size. Scale markings should be shown above the

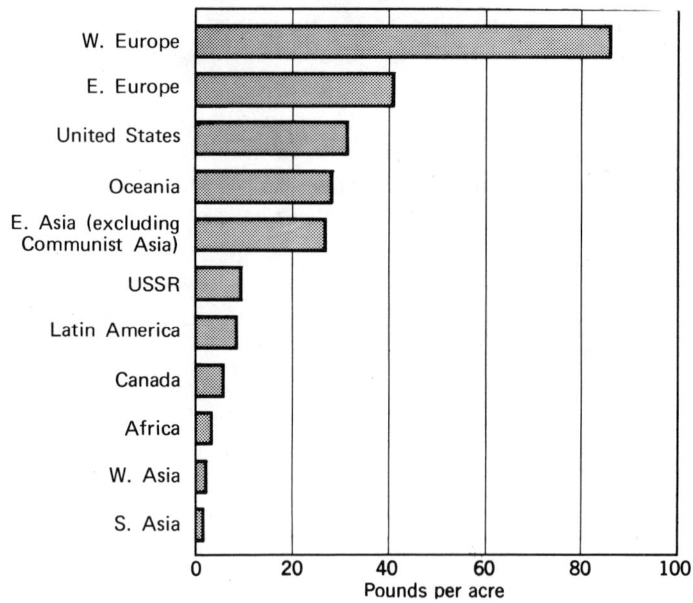

Figure 2.4 Fertilizer consumption per acre of cropland. Commercial fertilizers, excluding ground rock phosphate, in terms of plant nutrient content. Calendar year for USSR; year ending June for other regions. (*Source. U.S.D.A., A Graphic Summary of World Agriculture*, Misc. Pub. No. 705, 1964)

top bar and carried down through the graph by grid lines. Figure 2.4 provides an example of a horizontal bar graph.

Pie charts, such as those shown in Figure 2.5, are commonly used to show

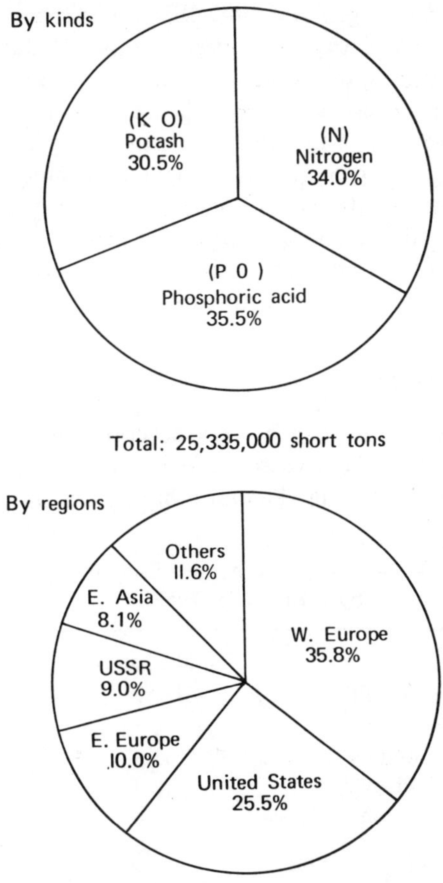

Figure 2.5 World consumption of fertilizer, average 1957–1961. Commercial fertilizers, excluding ground rock phosphate, in terms of plant nutrient content. Calendar year for USSR; year ending June for other regions. (*Source. U.S.D.A., A Graphic Summary of World Agriculture,* Misc. Pub. No. 705, 1964.)

the *relative* importance of components in an aggregate. They are not satisfactory for comparing differences in absolute magnitudes, however. When constructing pie charts, keep them as simple as possible. Do not include too many categories. Label the segments of the circle horizontally so that the labels are easily read. If there is no conventional order of categories, it is a good idea to arrange the components of the circle clockwise in order of magnitude. If there is an "all others" category, it is usually placed last regardless of magnitude. The "by kinds" pie graph in Figure 2.5 was not arranged in order of magnitude because fertilizer ingredients are usually listed in the order N-P-K, that is, nitrogen first, phosphoric acid second, and potash third.

There are many other ways to present statistical data graphically. Dot graphs, ratio-scale graphs, maps, vertical bar graphs, line graphs, and picture graphs are just a few. There is no single "best" way to present data. Always try to select a method that will allow the data to be presented as clearly and simply as possible. Examples of good graphical techniques can be found in many government publications, newspapers, and magazines. There are excellent references available on how to graph statistical data. Several of these are listed at the end of this chapter. In the following section, we shall illustrate why graphs must be constructed with care.

2.1.3 Misuse of Graphs

Graphs of statistical data can be very deceptive at times. You should be aware of the ease with which graphs can be misused and that any type of graph can be made misleading. The monthly figures on total money supply

Table 2.3 Total Money Supply by Months for 1963

Month	Total Money Supply[a]
January	151.8
February	148.4
March	147.6
April	149.8
May	147.5
June	148.3
July	149.5
August	149.2
September	150.6
October	152.5
November	154.8
December	157.2

[a] Averages of daily figures in billions of dollars.
Source. Board of Governors of Federal Reserve System.

in Table 2.3 will be used to illustrate how easily so-called "gee-whiz graphs" can be constructed.[1]

Inspection of the data in Table 2.3 reveals that the total supply of money increased by about 3.1 percent during 1963. Total gross national product increased by about 3.4 percent during the same period. If the data are presented in a graph with a zero line at the bottom, as in Figure 2.6, it is

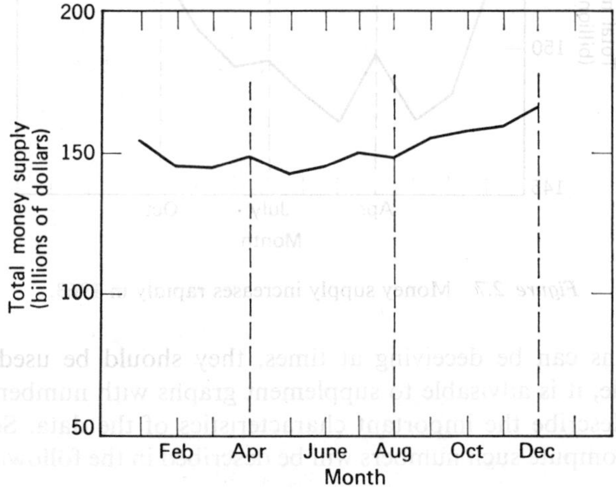

Figure 2.6 Money supply remains constant in 1963.

obvious that the supply of money did not change very much during 1963. The title of Figure 2.6 is not entirely justified, however, It would be better to label the figure "Total Money Supply in 1963" and let the reader draw his own conclusions as to whether or not the money supply remained constant.

Now suppose that, for some reason, we want to "emphasize" that the total supply of money increased in 1963 and to "suggest" that the increase was very rapid. This is easily done by chopping off the bottom of the graph in Figure 2.6, stretching the upper half, and shrinking the horizontal axis. The result is presented in Figure 2.7. The new graph is certainly impressive. Its title leaves nothing to chance. Anyone looking at this graph can just feel the money being poured into the economy. Even the careful reader is likely to be deceived. We have made a 3 percent annual increase appear as dynamic as a 100 percent increase—and it was not even necessary to change the data.

[1] For additional examples, see D. Huff, *How to Lie with Statistics* (New York: W. W. Norton and Company, Inc., 1954).

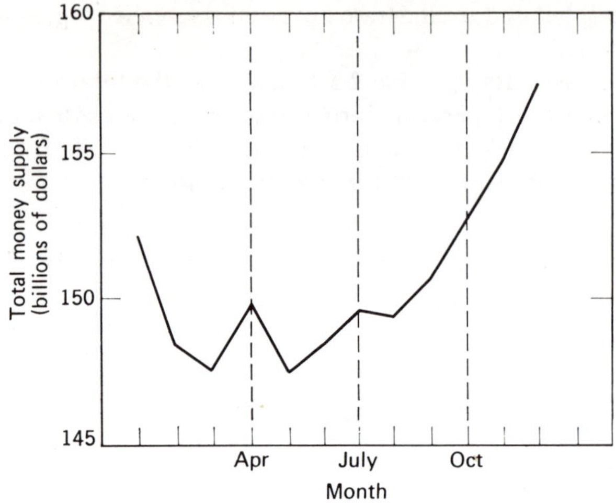

Figure 2.7 Money supply increases rapidly in 1963.

Since graphs can be deceiving at times, they should be used with care. When possible, it is advisable to supplement graphs with numbers that summarize and describe the important characteristics of the data. Several ways by which to compute such numbers will be described in the following sections.

2.2 MEASURES OF CENTRAL TENDENCY

The most commonly used measures of central location are the arithmetic mean, median, and mode. There are advantages and disadvantages to using each of these measures. The measure selected depends both on the nature of the statistical data and the purpose for which the measure is used.

2.2.1 The Mean

The *arithmetic mean* is sometimes called an *average* or simply a *mean*. We shall usually refer to it as the mean. Use of the full name "arithmetic mean" is sometimes necessary, however, to distinguish this mean from the geometric mean or the harmonic mean.

Let $X_1, X_2, \ldots, X_n$ be a set of real numbers. The mean of this set is defined as

$$\overline{X} = \frac{1}{n}\sum_{i=1}^{n} X_i \tag{2.1}$$

The mean is frequently used as a measure of central location and, as a result, most people are familiar with it. It takes account of each number in the set of observations with equal weight and is unique in the sense that a given set of numbers has only one mean. The mean can always be computed from numerical data and is easy to work with in many statistical applications.[2]

The main disadvantage of the mean is that it is affected by either extremely large or extremely small numbers. Thus, it is not always the most meaningful measure of central location. This disadvantage can be especially serious when working with small sets of numbers. To illustrate this, suppose we select six families living in a small midwestern city and estimate their annual incomes. Listing the results we have:

Family	*Income*
1	$6,000
2	$5,000
3	$6,000
4	$8,000
5	$10,000
6	$55,000

The mean annual income of the first five families is $7000 while the mean annual income of all six families is $15,000. A single observation, the $55,000 income, had a very large effect on the mean annual income figure. The median would provide a somewhat better estimate of the "average" annual income in this case.

2.2.2 *The Median*

The *median* is the central value of a set of observations that have been arranged in order of magnitude. If there is an odd number of values, the median is found by ordering the values in the set from smallest to largest and then selecting the middle value. If there is an even number of values, the median is equal to the mean of the two middle values of the ordered set.

The median is easy to compute when the set of observations is small. For instance, if we wanted to estimate the median cost of producing rice in Peru, we would begin by ordering the numbers in Table 2.1 as follows:

1.05, 1.10, 1.20, 1.30, 1.35, 1.40, 1.50, 1.55,
1.55, 1.60, 1.60, 1.65, 1.65, 1.65, 1.75, 1.75,
1.90, 2.00, 2.10, 2.20

[2] Readers not familiar with the notation of the definition (2.1) should review Part I of the mathematical notes in Appendix A.

There is an even number of observations in this set (20). Therefore, the median is equal to the mean of the two middle numbers, or

$$\frac{1.60 + 1.60}{2} = 1.60 = \text{the median}$$

The median is not affected by extreme values as in the case of the mean, but it is difficult to compute when there is a large number of observations. The median does not use much of the information contained in the set of observations, but it is a useful measure of central tendency when dealing with small sets of numbers, especially if the sets contain one or more extreme values. The median does not have as many statistical uses as the mean because the median is difficult to handle algebraically. In a later chapter, we shall show that the median also lacks several desirable statistical properties possessed by the mean.

2.2.3 The Mode

The *mode* is defined as the value which occurs most frequently in a set of observations. For example, the modal cost in Table 2.1 is 1.65. Unfortunately, in some sets of observations there is no mode. If each value in a set of measurements is different, it makes little sense to say that each value is a mode. In such a case, we would say that the mode does not exist. The mode is not necessarily unique; a set of numbers may have two or three different modes. The mode does not lend itself to other statistical purposes because, like the median, it is difficult to handle algebraically.

The concept of a modal value can be used with either numerical or nonnumerical data. Suppose that a marketing research firm wants to determine the best color for the box of a new brand of soap. It selects 100 housewives and asks them if they prefer a red, blue, green, or yellow soap box. It finds that 40 prefer red, 25 blue, 15 green, and 20 yellow. In this case, red is the modal color.

The disadvantages of the mode usually tend to outweight its advantages except when working with nonnumerical data such as in the example above. For this reason, it is less commonly used than either the mean or the median.

2.2.4 The Weighted Mean

There are many situations in applied economics in which we want to take account of the relative importance of different quantities when computing an average. Suppose, for example, that we wanted to compute the average price of steel in 1966. We could obtain the price per ton of steel plate, bars, tubes, wire rod, etc., add all of these prices up, and divide by the number of prices in the set. This would not provide a good estimate of the "average" price of steel, however, because steel mills sell different quantities of these steel products. A better estimate of average steel prices could be obtained by using a weighted mean. The prices of the various steel products would be multiplied by the respective quantities sold, the results added up, and the sum divided by the total quantity of steel products sold. In other words, the "average" price of steel is equal to the total revenue from steel products divided by the total quantity of steel products sold. Similarly, the average hourly wage of automobile workers can be estimated by dividing total wage payments to auto workers by total hours worked. The resulting average hourly wage is a weighted mean.

Let $X_1, X_2, \ldots, X_n$ be a set of numbers and $w_1, w_2, \ldots, w_n$ be a second set of numbers called weights. The weighted mean is defined as

$$\bar{X}_w = \frac{\sum_{i=1}^{n} w_i X_i}{\sum_{i=1}^{n} w_i} \tag{2.2}$$

The weighted mean will be used later in the discussion of index numbers. The arithmetic mean discussed in Section 2.2.1 is simply a special case of the weighted mean where all of the weights are equal to one.

2.3 MEASURES OF DISPERSION

There are many cases in which more information is needed about a set of observations than is contained in a measure of central location. A feed manufacturer, for example, may be interested in knowing both the average price of fish meal last year and the extent to which the price varied during the year. If he finds considerable variation in the price of fish meal, he may then be interested in investigating ways by which his company can buy a larger proportion of its fish meal requirements when prices are low.

The most commonly used measures of dispersion are the range, mean absolute deviation, and standard deviation.

2.3.1 The Range

By definition, the *range* of a set of numbers is found by subtracting the smallest number in the set from the largest number. Consider the list of interest rates on new homes given in Table 2.4. The range of these interest rates is equal to

$$R = (6.32 - 5.81) = 0.51$$

Table 2.4 Contract Interest Rates on First Mortgages for New Homes During the First 10 Months of 1966

Month	Interest Rate (Percent)
January	5.81
February	5.85
March	5.90
April	5.99
May	6.02
June	6.07
July	6.12
August	6.18
September	6.22
October	6.32

Source. Federal Reserve Bulletin, December, 1966.

The range tells us about the dispersion of the extreme values in a set of observations but does not use any information about the dispersion of the observations between these two values. Thus, in a sense, it uses very little of the information contained in the data. As a result, it is not a very reliable measure of dispersion. Its main advantage is that it is easy to compute. This is an important advantage, however, in some quality control situations.

2.3.2 The Mean Absolute Deviation

The *mean absolute deviation* of a set of numbers is defined as

$$MD = \left(\frac{1}{n}\right) \sum_{i=1}^{n} |X_i - \bar{X}| \tag{2.3}$$

where $\bar{X}$ is the arithmetic mean of the set of numbers and $|X_i - \bar{X}|$ signifies that all differences between the individual numbers and the mean are to be considered positive numbers. The mean absolute deviation, therefore, takes into account only the *magnitude* of the deviations from the mean and not the *sign* of the deviations. The mean absolute deviation uses more of the information contained in a set of numbers than does the range but is more difficult to compute. Absolute values are also rather difficult to work with in most statistical applications.

Table 2.5

Month	Excess Reserves[a] (in millions of dollars)	Absolute Value of the Deviations from the Mean $\lvert X_i - \bar{X} \rvert$
January	−4	7.5
February	4	0.5
March	3	0.5
April	5	1.5
May	5	1.5
June	−2	5.5
July	12	8.5
August	4	0.5
September	5	1.5
October	3	0.5
Totals	35	28.0

[a] *Source.* Federal Reserve Bulletin, December, 1966.

The data in Table 2.5 on excess reserves held by the reserve city banks in Chicago during the first 10 months of 1966 will be employed to illustrate the use of the mean absolute deviation.

To calculate the mean absolute deviation, first compute the mean, then the absolute values of the deviations from the mean, and finally divide the sum of absolute values by the number of observations. The mean absolute deviation in this case is

$$\text{MD} = (28/10) = 2.8$$

This value tells us that, on the average, the monthly excess reserves of the reserve city banks in Chicago during the first 10 months of 1966 deviated by 2.8 million dollars from the monthly average of 3.5 million dollars.

Had the range been used to measure the dispersion of the excess reserves, we would have obtained

$$R = 12 - (-4) = 16 \text{ million dollars}$$

The large difference in the values of the mean absolute deviation and the range is due to the fact that they measure different aspects of the variation in the excess reserves. The range measures the extent of the total variation in excess reserves during the year. The mean absolute deviation is an average of the monthly variation of excess reserves from the annual average of excess reserves.

2.3.3 The Standard Deviation

The *standard deviation* is the most commonly used measure of dispersion. It is easy to compute and has a number of desirable statistical properties that will be discussed in later chapters.

By definition, the standard deviation of a set of observations is equal to:

$$s = \sqrt{\frac{\sum_{i=1}^{n}(X_i - \bar{X})^2}{n}} \tag{2.4}$$

where n is the number of observations in the sample and $\bar{X}$ is the sample mean.[3]

The standard deviation is harder to compute than the range, but it uses more of the information in a set of numbers and is easier to work with than either the range or the mean deviation. Squaring the difference between each observation and the mean tends to give more weight to those observations which are further away from the mean. A deviation of 2, when squared, is 4. A deviation of 6, when squared, is 36. Although the *absolute difference* between 6 and 2 is only 4, the deviation of 6, is, in a sense, given 9 times as much "weight" as the deviation of 2. The result is an extremely useful measure of dispersion, however.

The data presented earlier on excess reserves held by reserve banks in Chicago can be used to illustrate the use of the standard deviation (Table 2.6). To calculate the standard deviation, first compute the mean, then the deviations from the mean, square these deviations, divide the sum of squared deviations by the number of observations, and take the square root of the result. The standard deviation obtained is

$$s = \sqrt{\frac{166.50}{10}} = \sqrt{16.55} = 4.07$$

When working with large sets of data, the standard deviation is usually computed using the formula

$$s = \sqrt{\frac{n\sum_{i=1}^{n} X_i^2 - \left(\sum_{i=1}^{n} X_i\right)^2}{n^2}} \tag{2.5}$$

[3] The formula

$$s' = \sqrt{\frac{\sum_{i=1}^{n}(X_i - \bar{X})^2}{n-1}}$$

is also used to compute the standard deviation. For large n, the difference between s and s' is small. The statistical properties of s and s' will be considered in a later chapter.

Table 2.6

Month	Excess Reserves (in millions of dollars)	Deviations from the Mean	Squared Deviations
January	−4	−7.5	56.25
February	4	0.5	0.25
March	3	−0.5	0.25
April	5	1.5	2.25
May	5	1.5	2.25
June	−2	−5.5	30.25
July	12	8.5	72.25
August	4	0.5	0.25
September	5	1.5	2.25
October	3	−0.5	0.25
Totals	35	0	166.50

The computing formula (2.5) is obtained directly from the definition (2.4) by multiplying out $\sum_{i=1}^{n} (X_i - \bar{X})^2$ and simplifying the result. The use of (2.5) saves a good deal of time when working with large samples, since it is not necessary to compute each deviation from the mean. The sum of squares, $\sum_{i=1}^{n} X_i^2$, and the sum of the X's, $\sum_{i=1}^{n} X_i$, can be obtained simultaneously on most desk calculators. Square roots can be obtained easily from tables of square roots or by using logarithms.

2.3.4 The Coefficient of Variation

The standard deviation is a measure of the *absolute* dispersion in a set of observations but is not always useful for comparison purposes. The standard deviation of family expenditures on food, for example, cannot be compared directly with the standard deviation of family size because the units differ. If the units are the same, a comparison of standard deviations provides information about differences in *absolute* variations but tells nothing about differences in *relative* variation.

The *coefficient of variation* is a commonly used measure of relative variation. It relates the standard deviation to the mean and presents the result as a percentage that can be used for comparison purposes. The coefficient of variation for a set of observations is

$$V = \frac{s}{\bar{X}} \cdot 100 \tag{2.6}$$

where s is the standard deviation and $\overline{X}$ is the arithmetic mean.

To illustrate the use of the coefficient of variation, suppose that a feed manufacturer keeps records on monthly sales of his best livestock feed, SuperGrow, and the price at which the feed sells each month. At the end of the year he summarizes these records and finds:

Average price of SuperGrow	\$150 per ton
Standard deviation of prices	\$ 30 per ton
Coefficient of variation for prices	20%
Average sales	5000 tons
Standard deviation of sales	500 tons
Coefficient of variation for sales	10%

The coefficients of variation indicate that there was a greater relative variation in the price of SuperGrow than in the tons of SuperGrow sold each month. This suggests that sales of SuperGrow may not be very responsive to price changes.

2.4 FURTHER REMARKS AND EXAMPLES

We have already pointed out that the techniques selected to summarize the information in a set of data depend partly on what we intend to do with the summary once it has been obtained. Another factor that needs to be considered when selecting descriptive techniques is the type of data available.

Statistical observations on economic variables can be classified into two general types: (1) quantitative and (2) qualitative. By quantitative observations, we mean numerical measurements made on a set of objects. For example, a set of prices per pound of bananas in selected stores in Cleveland are quantitative observations.

The term qualitative observations refers to observations that can be classified into one of several mutually exclusive classes which may be nonnumerical. For example, an agricultural economist takes a survey to determine how many farmers in Ohio use hybrid seed corn. In his questionnaire he may use the question, "Did you use hybrid seed corn on your farm last year? Yes or No?" The farmer's answer (yes or no) is a qualitative statistical observation. For recording purposes, a "yes" may be denoted by a one and a "no" by a zero, but the numbers one and zero are not measurements and therefore are not quantitative observations.

All of the graphical techniques presented in Section 2.1 may be used with either quantitative or qualitative observations. Bar graphs and pie charts are used more frequently to present qualitative data than are histograms or frequency polygons. Quantitative data, on the other hand, are probably most frequently presented using histograms or frequency polygons.

Except for the mode, the measures of central tendency and dispersion considered in Sections 2.2 and 2.3 are useful when working with quantitative data but provide very little, if any, information when dealing with qualitative observations. The concept of modal class of a set of qualitative observations is a useful one at times, but the modal class itself usually would not be interpreted as a measure of central tendency.

Economists frequently assume that the available observations were drawn from some basic population. In such cases we are usually more interested in attempting to describe or to make inferences about the basic population than in merely summarizing and presenting the available data. The distinction between a *sample* and the *population* from which the sample is drawn is an important one in statistics. Whether a set of data constitutes a population or a sample depends on how we intend to use it. If, for example, we want to estimate the average annual income of workers in the automobile industry, then figures on the incomes of all employees of the Ford Motor Company represent a sample. If we want to know the average annual income of Ford employees, then income figures for all employees of the company constitute a population.

When used with sample data, the formula (2.1) for the mean provides an *estimator* of the population mean which is a *parameter* of the population from which the sample was drawn. We shall use different notation to distinguish estimators from parameters. For example, the arithmetic mean of a sample will usually be denoted by $\bar{X}$ and the arithmetic mean of a population by μ. The sample mean $\bar{X}$ serves as an estimator of the population parameter μ. The sample mean is only one of several estimators of μ, however. The median and mode of a sample may also be used as estimates of μ, for example.

If our main interest is to obtain an idea of the form of the population underlying the sample observations or if the data constitute a population, then graphical techniques such as histograms or frequency polygons are particularly helpful. If, on the other hand, our objective is to draw inferences about the population from the sample data, we would probably turn directly to measures of central tendency and dispersion such as those discussed in Sections 2.2 and 2.3. We shall discuss deductive statistics at length in later chapters.

The following examples will serve to illustrate and expand on several of the ideas presented so far in this chapter.

Example 1: Suppose that the Institute of Life Insurance wants to illustrate that the average premium payment per $1000 of life insurance in force has generally been declining over the years. The set of observations available consists of average premium payments per $1000 of life insurance in force during, say, the past 30 years. A simple line graph such as is shown in Figure 2.8 will serve quite nicely to illustrate the downward trend. The line

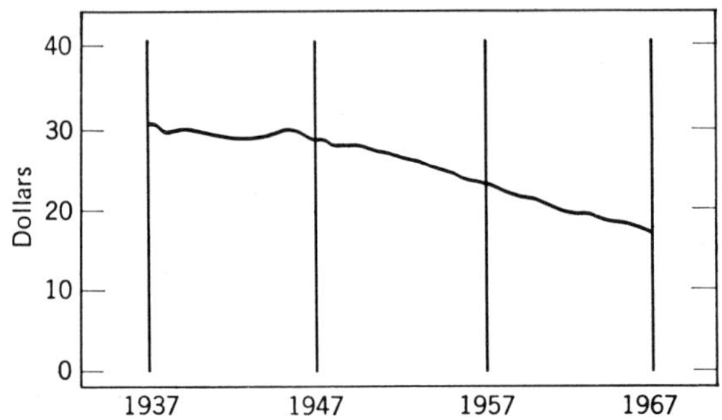

Figure 2.8 Average premium payment per thousand dollars of life insurance in force (United States life insurance companies). (*Source. Life Insurance Fact Book*, 1968, p. 52.)

graph provides very little information but leaves the reader with the general impression that insurance premiums have fallen during the past 30 years. The graph as such provides no information as to the type of average being used or whether the average used is based on the premiums of all (or only some) United States life insurance companies. The average premium rate for 1937 represents something very different than the average premium rate in 1967. The decline in premium rates is explained partly by the growth of group life insurance and the increased sales of family income plans with term insurance policies which have low premiums because they carry no cash values. The line graph provides no information as to what is happening to the average premium rate of a particular type of life insurance policy.

Example 2: A light bulb manufacturer has a set of measurements on how long a set of 1000 light bulbs produced by a new process continued to work under test conditions. If the manufacturer is primarily interested in drawing inferences about the average life of all light bulbs manufactured by the new process, then the arithmetic mean and standard deviation of the observations would provide the basic information needed.

If the manufacturer is interested in comparing the "new" bulbs with "standard" bulbs as part of a sales campaign, then a bar graph or picture graph would be a more appropriate way to summarize the information.

Picture graphs can be deceptive at times. The area of the "new" bulb in Figure 2.9 exceeds the area of the "standard" bulb by more than the 73 percent difference in height. Consumers who look at the pictures rather than the words are left with the impression that the new bulbs will last somewhat

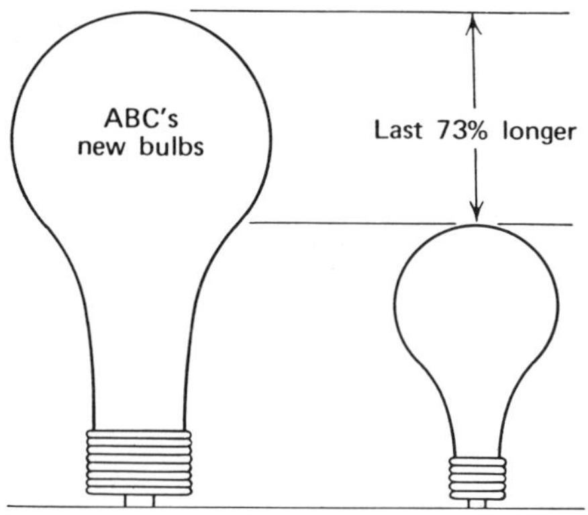

Figure 2.9 New bulbs last longer.

longer than they actually do. This situation is easily corrected by drawing the "new" bulb with 73 percent more volume rather than 73 percent more height.

Example 3: An instructor interested in illustrating the wide difference in scores on a particular test would probably use the range of the test scores. If his main interest is comparing the variability of scores on two different tests, he would probably compute the coefficient of variation for the two tests. If the test is scored on the basis of, say, 100 points, the set of test scores represents a set of quantitative observations. The letter grades assigned to the test papers are qualitative observations.

Example 4: A survey of 600 Iowa farmers was taken to determine how farmers obtained and used market news. The survey revealed that nearly 95 percent of the farmers selling hogs listened to market news on the radio while less than 10 percent of the farmers selling cream or whole milk listened to the radio for market news on these products. Such percentages were also computed for farmers selling cattle, soybeans, corn and eggs. A bar graph of the survey results is presented in Figure 2.10. The survey results could also be presented as a set of six circle diagrams; one diagram per product. Line graphs or frequency polygons would not be particularly meaningful in this case. One could compute the average number of farmers that listen to market news on the radio by summing over products and dividing by six but the result would have little, if any, significance.

Example 5: An economist has data on the profit rates of 1000 small supermarkets and wants to obtain a general picture of the distribution of profit rates of all small supermarkets. A histogram or frequency polygon would provide such a picture.

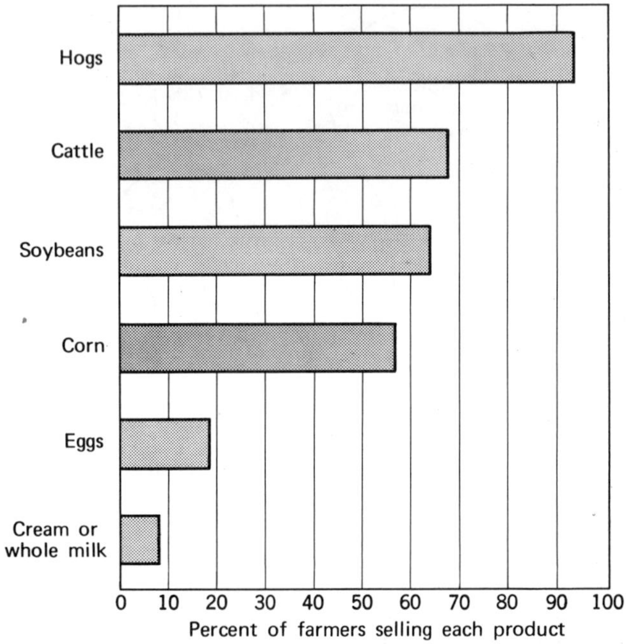

Figure 2.10 Radio listening according to product. (*Source. Research Bulletin 413*, Agricultural Experiment Station, Iowa State College, Ames, Iowa, August 1954).

The arithmetic mean and standard deviation would be needed if the intent was to draw inferences about the average profit rate of all small supermarkets. The coefficient of variation would be useful when comparing the variability of profit rates of small supermarkets with the variability of profit rates of large supermarkets.

* 2.5 MEASURES OF DISTRIBUTION SHAPE

Two frequency distributions can have identical means and standard deviations and yet look very different. One distribution may be very symmetrical while the second may be skewed either to the right or to the left. Sometimes we need to describe a distribution more completely than is possible by using only measures of central location and dispersion. In such situations, there is often a need for numbers that describe the *shape* of frequency distributions. Two such measures, which are especially useful in problems of curve fitting, will be discussed in this section.

* This section may be omitted without loss of continuity.

2.5.1 A Measure of Skewness

The mean, median, and mode are all equal when a frequency distribution is symmetrical. A frequency distribution is said to be "skewed" if it is not symmetrical about the mean. The skewness is said to be *positive* (or "to the right") if the mean exceeds the median and *negative* (or "to the left") if the median exceeds the mean. Some examples of skewed frequency distributions are shown in Figure 2.11. The frequency distributions in Figure 2.11 can be thought of as distributions obtained from a very large set of observations.

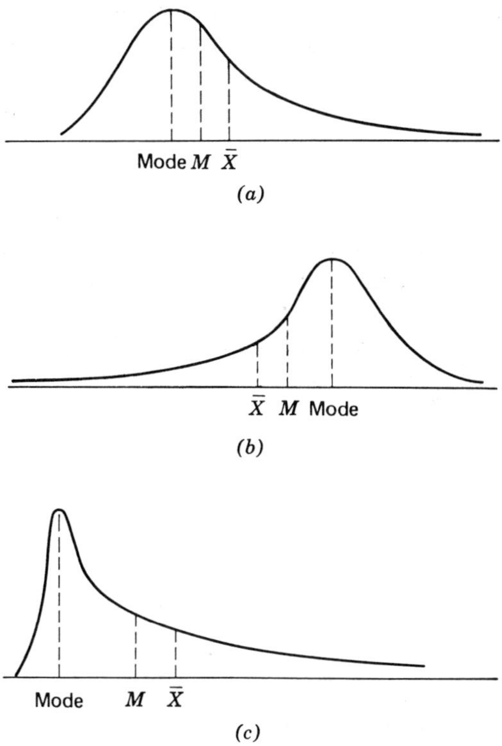

Figure 2.11 Examples of skewed distribution. (*a*) Slight positive skewness. (*b*) Slight negative skewness. (*c*) Extreme positive skewness.

The *coefficient of skewness* α_3 (alpha three) is the most widely used measure of skewness. It is defined as:

$$\alpha_3 = \frac{\frac{1}{n}\sum_{i=1}^{n}(X_i - \bar{X})^3}{s^3} \qquad (2.7)$$

where s^3 is the cube of the standard deviation and $\overline{X}$ is the arithmetic mean. This measure of skewness is a pure number and can therefore be used for comparison purposes.

To calculate α_3, first compute the sum of cubed deviations from the mean, divide this sum by the number of observations, and divide the result by the cube of the standard deviation. A computing formula for α_3 can be obtained from the definition formula (2.7) by using the rules of summation discussed in Section A of the Mathematical Appendix at the end of the book. Computing formulas are especially useful when dealing with large samples.

For a symmetrical distribution, α_3 is zero. Positive values for α_3 indicate that the distribution is positively skewed; in other words, the distribution has a long "tail" to the right. Negative values for α_3 indicate the distribution has a long "tail" to the left. The larger the value of α_3, the more skewed the distribution.

2.5.2 A Measure of Peakedness (Kurtosis)

Two distributions may have the same mean, the same standard deviation, be perfectly symmetrical, and still look very different if one is "peaked" and the other is "flat." A distribution is said to be *leptokurtic* if it is very peaked and has relatively wide tails while distributions which are rather flat and have relatively thin tails are called *platykurtic*. This is illustrated in Figure 2.12.

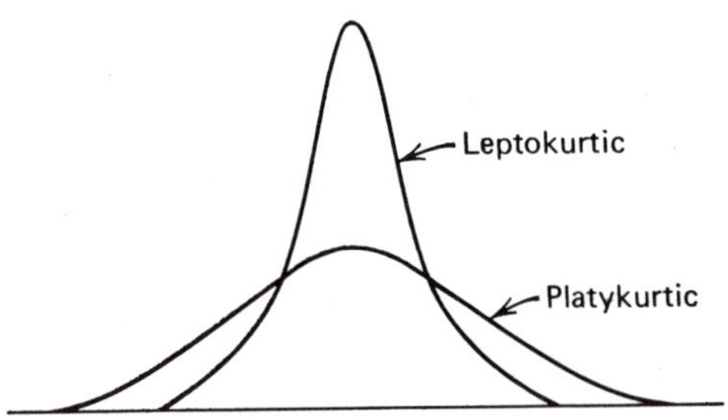

Figure 2.12 Leptokurtic and platykurtic distributions.

The *coefficient of peakedness,* α_4 (alpha four), is the most commonly used measure of peakedness. It is defined as

$$\alpha_4 = \frac{\frac{1}{n}\sum_{i=1}^{n}(X_i - \overline{X})^4}{s^4} \tag{2.8}$$

where s^4 is the fourth power of the standard deviation and $\overline{X}$ is the arithmetic mean. This measure of peakedness does not depend on the location or variance of the distribution and can, therefore, be used for comparison purposes.

The bell-shaped *normal distribution* is commonly used as the standard by which to measure peakedness. For the normal distribution, α_4 is equal to 3. If α_4 exceeds 3, then the distribution is considered to be leptokurtic (peaked). If α_4 is less than 3, then the distribution is considered to be platykurtic (flat.)

EXERCISES

Section A, Mathematical Appendix

1. Let $n = 4$, $a = 2$, $b = 3$ and $k = 10$. Compute the values of the summations presented below using

$X_1 = 2,\ X_2 = 4,\ X_3 = 5,\ X_4 = 3$
$Y_1 = 3,\ Y_2 = 5,\ Y_3 = 6,\ Y_4 = 2$

(a) $\sum_{i=1}^{n}(X_i - Y_i)^2$

(b) $\sum_{i=1}^{3}(k + aX_i - bY_i)$

(c) $\sum_{i=1}^{2}(2X_i^{\,2} - 3X_i Y_i)$

(d) $\sum_{i=1}^{n}(Y_i - \overline{Y})$ where $\overline{Y}$ is the arithmetic mean

(e) $\sum_{i=1}^{n}(Y_i - \overline{Y})^2$

(Readers having difficulty with this problem should study Section A of the Mathematical Appendix before proceeding to Chapter 3.)

Section 2.1

2. The man-on-the-street from radio station WOW in Coolwater, Iowa, stops 30 people at random on the street and asks them their ages. The result are listed below

Ages of 30 people selected at random

30	32	21	15	26	43
18	37	23	14	27	44
16	32	39	21	28	39
42	40	39	22	21	40
35	26	39	25	39	22

(a) Summarize this information in the form of a frequency distribution table.
(b) Present the data in the form of a frequency histogram.
(c) Use the data to draw a relative frequency polygon.
(d) Graph a cumulative frequency polygon.

3. In 1966 the home offices of 3047 property-liability insurance companies were scattered throughout the United States. The number of home offices located in Midwestern states is listed below.

State	Number of Home Offices
Illinois	302
Iowa	196
Kansas	38
Minnesota	193
Missouri	71
Nebraska	76
North Dakota	39
South Dakota	54

Source. Insurance Facts, 1966.

(a) Present the data in the form of a horizontal bar graph.
(b) In what other ways could the data be presented?

4. The life insurance in force in the United States in 1965 can be classified as follows.

Type	Amount (in Millions of Dollars)
Ordinary	$497,630
Group	306,113
Industrial	39,818
Credit	56,993
Total	$900,554

(a) Present the data in the form of a pie chart.
(b) In what other ways could the data be presented?

Section 2.2

5. Compute the average number of insurance company home offices per state in the Midwest from the information given in Exercise 3.

6. Using the data from Exercise 2, compute the median age of the 30 people interviewed by the man-on-the-street from radio station WOW.

7. Using the data in Exercise 2, compute the modal age.

8. Let

$$\begin{aligned} w_1 &= 0.2 & x_1 &= 12 \\ w_2 &= 0.3 & x_2 &= 4 \\ w_3 &= 0.3 & x_3 &= 6 \\ w_4 &= 0.2 & x_4 &= 18 \end{aligned}$$

Compute the weighted mean of the X's.

9. During a weekend sale on meat, the Silver Cloud Supermarket sold the following amounts of meat.

Type of Meat	Price (Dollars per Pound)	Quantity (Pounds)
Beef	$0.80	2000
Pork	0.60	900
Chicken	0.30	1200
Turkey	0.50	700
Fish	0.40	500

What was the average price per pound of meat sold?

Section 2.3

10. Using the data in Exercise 2, what is the range of the ages of the people interviewed?

11. The number of hurricanes reaching the continental United States per year during the 1955–1965 period are presented below.

Year	Number of Hurricanes
1955	3
1956	1
1957	1
1958	0
1959	3
1960	2
1961	2
1962	0
1963	1
1964	4
1965	1
Total	18

(a) Compute the mean absolute deviation of the number of hurricanes reaching the United States during the 1955–1965 period.
(b) Compute the standard deviation for the same period.
(c) Compute the range for the 1955–1965 period.
(d) Could a number of hurricanes equal to the arithmetic mean ever be observed in any given year?

12. A cardboard box manufacturer produces three types of small boxes. The strength of each type of box is tested by taking a sample of 100 boxes and determining the pressure required to burst each box. The results of the tests are summarized below.

Type of Box	*A*	*B*	*C*
Average bursting pressure (pounds per square inch)	150	200	300
Standard deviation of the bursting pressures	40	50	60

(a) Which type of box has the smallest *absolute* variation in bursting pressure?
(b) Which type of box has the largest *relative* variation in bursting pressure?

Section 2.4

13. The average life of light bulbs produced by the standard process is known to be 2000 hours. Light bulbs produced by a new process are found to have an average life of 3000 hours.
(a) Let the diameter of a circle represent the average life of a light bulb, and draw two circles representing the average life of bulbs produced by the "standard" and "new" processes.
(b) Let the area of a circle represent the average life of a light bulb, and draw two circles representing the average life of bulbs produced by the "new" and "standard" processes. (The area of a circle is equal to πr^2.)

14. Explain under what circumstances the following sets of data may be considered to constitute a sample.
(a) The test scores of all the members in a statistics class.
(b) Data on total national income for all years during the 1960 to 1970 period.
(c) The weight of all bags of cement produced by the Great Cement Company on July 29, 1968.

15. Explain under what circumstances each of the sets of data presented in Exercise 14 may be considered to constitute a population.

Section 2.5

16. Is the distribution of ages obtained from the data in Exercise 2 skewed to the right (positive) or to the left (negative)?

SELECTED REFERENCES

Huff, D., *How to Lie With Statistics*, W. W. Norton & Company, Inc., New York, 1954.

Neter, J. and W. Wassermann, *Fundamental Statistics for Business and Economics*, Allyn and Bacon, Inc., Boston, 1961. Explains the use of graphical devices in greater detail with special emphasis on business and economic applications.

Waugh, F. V., *Graphic Analysis in Agricultural Economics*, Agricultural Handbook No. 128. United States Department of Agriculture, Washington, D.C. Explains a wide variety of ways in which graphs can be used in analysis of data.

CHAPTER 3

Measuring Changes in Economic Aggregates

An *index number* is a statistical device for measuring differences between groups of data. There are hundreds of ways to construct index numbers. The particular method selected will depend, to a large extent, on the purpose of the index number.

Index numbers are in wide use today. They are used by economists to measure changes in prices, psychologists to measure differences in IQ, historians to compare changes in populations, businessmen to measure changes in sales, and labor unions to assure that wages increase with the general price level. Index numbers will be in even wider use tomorrow. There is an increasing demand for "tailor-made" index numbers, and the availability of electronic computers makes it easier to supply them. It is important, therefore, to understand how index numbers can be used and the problems involved in their construction.

Economists frequently use index numbers to measure changes in economic aggregates. In this chapter we shall discuss some of the basic problems involved in measuring such changes and a few of the index numbers that are commonly used.

*3.1 SOME HISTORICAL BACKGROUND

Economic statisticians have devoted a great deal of effort to improving the coverage of particular index numbers, the accuracy and speed of collection of the individual price and quantity series, the weighting formulas, and other

* This section may be omitted without loss of continuity.

technical aspects of index number construction. From time to time, some of the best-known economists in the United States and other countries have been asked to advise their governments on methods of improving index numbers that had taken on great political importance.

Successive waves of interest in index numbers have generally been stimulated by problems of major political and economic importance. According to Irving Fisher, the price changes caused by the Napoleonic Wars and the effects of paper money led to proposals for a "standard from materials" or a "tabular standard of value" which amounted to the correction by means of an index number of contracts to pay sums of money in the future. Jevons (1863) in England used a particular form of index number to estimate the "fall in the value of gold" caused by the outpourings of newly opened gold mines in the United States and Australia beginning in 1849 and 1851. In 1864 Laspeyres worked out price index numbers for Hamburg. He objected to Jevon's use of a simple geometric mean and proposed the index which now bears his name in the index number literature. In 1874 Paasche, reacting to Laspeyres' proposal, suggested the alternative formula which now bears his name. Both Laspeyres' and Paasche's index numbers are still used today.

One of the earliest index number problems to receive serious attention in the United States was that of measuring the *purchasing power of money*. One practical motivation for this concern was the opposition of interest between debtors and creditors with respect to mortgages and other long-term loans involving fixed dollar amounts. As settlement expanded westward after the Civil War, many farmers went into debt to acquire land. Subsequent declines in prices of farm products created hardships for them. A farmer who had borrowed the money equivalent of 1000 bushels of wheat was obligated to pay back the equivalent of 2000 bushels of wheat if the price of wheat had declined 50 percent by the time the mortgage was due. From the early 1870's until the 1890's, the general drift of the price level for primary products was downward and the consequent pressure on debtors (mostly farmers) had important political repercussions. Many farmers became intensely interested in monetary policies that might relieve their situation by raising the general level of commodity prices—if necessary, by redefining the monetary unit itself. The idea of a "commodity dollar" that would buy the same fixed bundle of agricultural and other raw or primary products at different points in time became attractive to some economists. As farmers sold their products at wholesale prices, and as the only published price series of national significance were wholesale prices, the most relevant and measurable bundle of commodities consisted of primary products.

Against this background, some very able economists became interested in "the measurement of general exchange value" (Walsh, 1901) and "the purchasing power of money" (Fisher, 1911). Their interests converged to some extent with lines of thought initiated by European economists concerned with monetary policy and with measuring changes in the terms of trade—i.e., the relative prices of commodities exported and commodities imported by their countries—which also required aggregative measures of changes in wholesale prices.

When the official United States index number of wholesale prices was initiated, it was evidently intended to measure an aggregate of great economic importance and high political interest. Sweeping changes in the United States economy and society since 1902 have changed the nature of the questions which require answers in terms of wholesale price indexes.

The growth of industry and labor unions created interest in measuring changes in the "cost of living" and in the real wages or purchasing power of urban workers. In the United States this interest led to development of the Bureau of Labor Statistics' Cost of Living Index, which was subsequently renamed the Consumer Price Index. The rationale for such an index is that a wage worker is exchanging his labor for consumer goods. For example, if the price of an hour of labor increased at an annual rate of 4 percent while the average price of an appropriate bundle of consumer goods remained constant, *real hourly wages* would be increasing by 4 percent a year. As of 1964, the Consumer Price Index was weighted to reflect the expenditure patterns of "city wage earner and clerical worker families" in a recent base period.

Prices received by farmers in the United States trended upward from about 1896 to 1910, and political concern about farmers as a debtor class dwindled. However, the agricultural depression of the 1920's, which followed much speculative buying of land during 1915–1920 and a precipitous drop in prices of farm products during 1920–1921, reestablished farmers as a debtor class at a time when most nonfarm people were prospering. This led to a special interest in measuring changes in the economic well-being of farmers over time and comparing the well-being of farmers with that of other groups.

If we looked at the farmer exclusively as the proprietor of a business, it would seem appropriate to compare the prices he received for his farm products with the prices of the goods and services he used in their production. If, on the other hand, we looked upon farm families as consumers, the appropriate comparison would involve prices of goods and services used in farm family living and the net incomes of farm residents from all sources. However, political interest focused on the relationship between prices received for all farm products and prices paid for goods and services used both in production and in family living, and the ratio between these two indexes was christened the "parity ratio."

"Parity" was a concept of great political potency in the United States for

40 years, beginning in the 1920's. It was the focus of most work on price index numbers related to agriculture and also gave impetus to the development of estimates of gross and net farm income.

Index numbers of physical output developed much later than index numbers of prices. It was a relatively easy matter to collect wholesale prices at a few commodity exchanges and central markets, and these prices had at least a barometric significance to buyers and sellers throughout the nation. It is a much more arduous and expensive undertaking to collect information about production, which may be accurately known for a given firm only by the proprietor of that firm. In the United States a few quantity indexes of limited scope were published in the early and middle 1920's. The first comprehensive index of output for the manufacturing sector, the Federal Reserve Board's Index of Industrial Production, was published in 1927.

The implementation in the 1930's of national income and product accounts for the United States opened the way for a more comprehensive view of the role of price and quantity indexes. The "macroeconomic revolution," which followed the publication of J. M. Keynes' *General Theory of Employment, Interest and Money* (1936), brought a new level of need, and hence of aspiration, for comprehensive and consistent measurements of all of the major macroeconomic variables involved in business cycles. As a result, both price and quantity indexes became more common, more widely used, and more important.

3.2 UNWEIGHTED INDEX NUMBERS

When constructing index numbers, businessmen and economists usually work with prices (p) and quantities (q). The base-year prices are denoted by p_0, given-year prices by p_t, base-year quantities by q_0, and given-year quantities by q_t. For some purposes, it is necessary to attach a commodity subscript as well as a time period subscript to the symbols for prices and quantities. In such cases, the symbols p_{it} and q_{it} are used to denote the price and quantity of commodity i during period t. If there are n commodities, the total value of the commodities in period t would be

$$\text{total value in period } t = \sum_{i=1}^{n} p_{it} q_{it}$$

If the summation runs over all n commodities, the subscript i is dropped and $\sum_{i=1}^{n} p_{it} q_{it}$ is written as $\sum p_t q_t$. This greatly simplifies the notation used in index number formulas.

Unweighted index numbers are very easy to compute and very useful for some purposes. They must be constructed and used with special care, however.

The simplest type of unweighted index number is a percentage that shows how something differs from a given standard or base period. Suppose, for example, that eggs sold for 50 cents a dozen in January and 60 cents a dozen in February. If January is designated as the base period, then the "base period price" is $p_0 = \$0.50$ and the price of eggs in February (period 1) is denoted by $p_1 = \$0.60$. The ratio (p_1/p_0) is called a *price relative*. An unweighted index number is obtained by multiplying the price relative times 100. Doing this, we find the unweighted price index number for eggs in January to be

$$I_0 = \frac{p_0}{p_0} \cdot 100 = \frac{0.50}{0.50} \cdot 100 = 100$$

while in February the price index is

$$I_1 = \frac{p_1}{p_0} \cdot 100 = \frac{0.60}{0.50} \cdot 100 = 120$$

Comparing the index numbers I_0 and I_1, we find that the price of eggs was 20 percent higher in February than in January. (If the price of eggs dropped to 35 cents a dozen in March, what is the value of the March price index for eggs?)

As a second example, suppose that we want to measure the performance of a new car salesman, Mr. Jones, during his first month on the job. One way to measure performance would be in terms of number of cars sold each week. This information is presented in Table 3.1.

Table 3.1 Sales Performance of Mr. Jones

	Number of cars sold		Indexes of Sales		
	t	q_t	I_s	I_y	I_n
1st week	0	5	100	50	41.7
2nd week	1	8	160	80	66.7
3rd week	2	6	120	60	50.0
4th week	3	12	240	120	100.0

Using the first week as the base period, we can compute an index of sales by using the formula

$$I_s = (q_t/q_0) \cdot 100$$

where q_0 is the number of cars sold during the first week (the base period) and q_t is the number of cars sold in period t. This index of sales allows us to compare the salesman's performance during each week of the month on the basis of his own performance during his first week with the company. The index number for the second week tells us that he sold 60 percent more cars that week than he did the first week. The ratio (q_t/q_0) is called a *quantity relative*. The index numbers I_s are shown in the I_s column of Table 3.1.

We could also compare Mr. Jones' performance with that of another salesman or with some standard level of sales. One standard that could be used is the average weekly sales of other salesmen during the previous year. Suppose this average is ten cars a week. A new index can be computed using the formula

$$I_y = (q_t/10) \cdot 100$$

The results are presented in the I_y column of Table 3.1 and show that Mr. Jones' sales during his second week on the job are 20 percent below last year's average.

Notice that the particular base period or standard used can make a big difference in the value of the index number. In this case, Mr. Jones would probably prefer that his performance be measured by index I_s. His boss may feel that the index I_y is a better measure of performance. Judging Mr. Jones' performance by last year's average weekly sales may not be justified, however, if car sales were unusually high last year. A better standard for comparison may be the average weekly sales of the company's other car salesmen during the month. The new index would measure performance differently than either I_s or I_y.

It is easy to misinterpret index numbers. Suppose an index of Mr. Jones' sales was computed using the fourth week as the base period. The formula in this case is

$$I_n = (q_t/12) \cdot 100$$

The results are presented in Column I_n of Table 3.1. It is tempting (but not correct) to look at Column I_n and conclude that Mr. Jones' sales increased by (100 – 41.7) or 58.3 percent during the month. In reality, sales increased by (100 – 41.7)/41.7 = (240 – 100) or 140 percent as shown by the I_s column of Table 3.1. People sometimes tend to think of the base period value, 100, as an "average" or "normal" value and then to consider index numbers less than 100 to be below "average" and those above 100 to be above "average." Such interpretation is seldom justified. Still, it would not be surprising if Mr. Jones strongly preferred index I_s to I_n in this particular case.

The way in which an index number is used will also affect people's opinions as to how that index number should be constructed. Suppose Mr. Jones receives a bonus each week which is computed by multiplying the index I_s times one dollar. In this case, his bonus for the third week is \$120. If his boss informed him that the index I_y would be used in the future, Mr. Jones would be an unhappy man because his future bonus would always be only one-half as much as it would be otherwise. There surely would be some discussion as to which index was the "best."

In some cases a *simple aggregate index* can be used to summarize and measure changes taking place in several economic variables. The price data in Columns (1) and (2) of Table 3.2 will be used to illustrate the construction of such indexes.

Table 3.2 Grain Prices

	(1) Price (Cents per Pound)	(2)	(3) Price Relative
Product	1960	1964	(p_t/p_0)
Rice	12	16	1.33
Corn	9	6	0.67
Wheat	10	20	2.00
Totals	31	42	4.00
Index based on formula (3.1)	100	135	—
Index based on formula (3.2)	—	—	133

The formula for a simple aggregate index is

(3.1)
$$I = \frac{\sum p_t}{\sum p_0} \cdot 100$$

where $\sum p_t$ is the sum of the given-year prices and $\sum p_0$ is the sum of the base-year prices. Using 1960 as the base year, the price index for 1964 turns out to be

$$I = \frac{100 \cdot (16 + 6 + 20)}{(12 + 9 + 10)} = 135$$

The main weakness of a simple aggregate price index is that it can produce different results depending on the units for which the prices are quoted. Suppose that the price of corn was given in dollars per hundred pounds while the prices of rice and wheat were given in cents per pound. Using formula (3.1) with the new units, we obtain

$$I = \frac{100 \cdot (0.16 + 6.00 + 0.20)}{(0.12 + 9.00 + 0.10)} = 69$$

When the price of corn is quoted in dollars per hundred pounds, it dominates the price index. As a result, the simple aggregate price index for 1964 indicates that grain prices fell during the period from 1960 to 1964. In actual fact, only the price of corn fell. The price of corn dominates because it has an implicit weight 100 times greater than the implicit weights given the prices of rice and wheat. The unconscious weighting process employed in the construction of simple aggregate indexes can yield misleading results in index number calculations.

The problem of units is not too serious in the grain price example because all prices can be quoted on a per pound basis and the prices per pound are not extremely different. If we used a simple aggregate price index to measure changes in the prices of household appliances, however, the units problem would be difficult to solve in many cases. The simplest solution, in such a case, would be to construct a price index using price relatives.

A *price relative* is the given-year price of a commodity divided by its base-year price. The resulting price relative is a pure (or unitless) number because the units in which the prices are quoted cancel out in the division process. An index number can be obtained by taking an average of the price relatives. The average is usually obtained by using either the arithmetic mean or the geometric mean. The *arithmetic mean of price relatives* is defined by

$$I = \frac{\sum \left(\frac{p_t}{p_0}\right)}{n} \cdot 100 \tag{3.2}$$

where (p_t/p_0) are the price relatives and n is the number of price relatives in the sum.[1]

Using the definition (3.2) and the grain price data in Table 3.2, we obtain a 1964 price index equal to 133. This can be verified by adding up the price relatives listed in Column (3) of Table 3.2, multiplying the sum by 100, and dividing the result by the number of price relatives which were added up. The price index based on price relatives is the same even if the price of corn is quoted in dollars per 100 pounds.

Indexes (3.1) and (3.2) can be converted from price indexes to quantity indexes by replacing the p's with q's.

Very few of the indexes in wide use today are computed from formulas as simple as (3.1) and (3.2). Neither of these simple aggregate indexes explicitly takes into account the relative importance of the economic variables that enter the indexes. For example, if twice as much rice is consumed as wheat,

[1] Keep in mind that $\left(\sum \frac{p_t}{p_0}\right)$ is a shorthand notation of $\left(\sum_{i=1}^{n} \frac{p_{it}}{p_{i0}}\right)$ where i indexes the items included in the price index.

then it seems reasonable to consider a change in the price of rice to be more important than a change in the price of wheat. This can be done by attaching a weight to the price of rice. The changes needed to convert indexes (3.1) and (3.2) to weighted indexes are small—but important.

3.3 WEIGHTED INDEX NUMBERS

Although the index numbers discussed in the previous section are commonly referred to as unweighted, they are in reality weighted in the sense that the weights are implicit rather than explicit. In the arithmetic mean of price relatives index (3.2), for example, an implicit weight of 1 is assigned to each price relative. Thus, if this index were used to measure changes in food prices, a change in the price of salt would be given the same "weight" as a change in the price of beef. It is obvious that such an implicit weighting is far from realistic in many cases.

The construction of useful index numbers often requires a conscious effort to assign weights to each item included. In many practical problems, the appropriate weights seem obvious. For example, if a monthly index of retail food prices is being computed, the number of quarts of milk purchased during the month would seem to be an appropriate weight for the price of milk in the index. But even when the appropriate weights seem obvious, there is usually a certain degree of arbitrariness in the way they are defined. In a monthly index of retail food prices, for example, we may decide to define "quantity of milk purchased" as "the average number of quarts of milk purchased each month by an urban family of four with an annual income in the range \$6000 to \$10,000." But why a "family of four" instead of a "family of five?" Why not specify that the annual income should be in the range \$5000 to \$8000? There may be many good reasons for defining the "quantity of milk purchased" in a certain way, but there is still a degree of arbitrariness in the final definition.

Once guidelines are established for the definition of one weight, those guidelines should be followed, to the greatest possible extent, when defining other weights. The definition of the "quantity of bread purchased," for example, should be consistent with the definition used for milk. Thus, it would *not* be appropriate in this case to define the "quantity of bread purchased" as the "total loaves of bread sold in an urban area with a population of 10,000 people."

It is impossible to make a final selection of weights before selecting the index number formula to be used. There are a great number of weighted index numbers that could be used. We shall discuss several of the more commonly used formulas in this section.

3.3.1 The Paasche Price Index

The *Paasche index* is one of the most widely known *weighted aggregate indexes.* A Paasche price index can be obtained from the simple aggregate index (3.1) by multiplying the prices in the numerator and denominator by the given-year quantities. Paasche price index is defined by

$$P = \frac{\sum p_t q_t}{\sum p_0 q_t} \cdot 100 \tag{3.3}$$

To calculate a Paasche price index, first obtain the sum of given-year prices multiplied by given-year quantities, divide this sum by the sum of base-period prices multiplied by given-year quantities, and multiply the result by 100.

Table 3.3 Grain Prices and Quantities

	Price (Dollars per ton)		Quantity (Tons)	
Product	1960	1964	1960	1964
	p_0	p_t	q_0	q_t
Rice	24.00	32.00	100	80
Corn	18.00	12.00	40	40
Wheat	20.00	40.00	50	70

The price and quantity data in Table 3.3 will be used to show these calculations. The value of given-year quantities at given-year prices is equal to

$$\sum p_t q_t = 32(80) + 12(40) + 40(70) = 5840$$

The value of given-year quantities at base-year prices is equal to

$$\sum p_0 q_t = 24(80) + 18(40) + 20(70) = 4040$$

The Paasche price index is therefore equal to

$$P = 100(5840/4040) = 144$$

This index number tells us that the value of the *given-year quantities* increased by 44 percent as a result of price changes between 1960 and 1964. The interpretation commonly given the index number is to say that "prices increased by about 44 percent between 1960 and 1964."

A serious disadvantage of the Paasche index, in many applications, is that the weights (q_t) change each period. In the case of price indexes, this is a costly disadvantage if it is difficult to estimate the given-year quantities that serve as weights. The problem of changing weights becomes a very serious one when there is a need to compute a weighted index on a daily, weekly, or monthly basis. In such cases, the cost of obtaining new quantity weights each period usually prohibits the use of a Paasche index.

3.3.2 The Laspeyres Price Index

The *Laspeyres price index* uses base-year quantities as fixed weights for the given and base-year prices. It is equal to

$$L = \frac{\sum p_t q_0}{\sum p_0 q_0} \cdot 100 \tag{3.4}$$

The data in Table 3.3 are used to illustrate the calculations necessary to obtain a Laspeyres index number. The value of the base-year quantities at given-year prices is equal to:

$$\sum p_t q_0 = 32(100) + 12(40) + 40(50) = 5680$$

The value of base-year quantities at base-year prices is equal to

$$\sum p_0 q_0 = 24(100) + 18(40) + 20(50) = 4120$$

The Laspeyres price index is therefore equal to

$$L = 100 \left(\frac{5680}{4120}\right) = 138$$

This index number tells us that the value of the *base-year quantities* increased by 38 percent as a result of price increases between 1960 and 1964. The usual interpretation would be that "prices increased by about 38 percent between 1960 and 1964."

The Laspeyres price index is equivalent to a weighted aggregate index of price relatives in which the weights are equal to the base-period expenditures $p_0 q_0$. The weighted aggregate index of price relatives is equal to

$$I = \frac{\sum \frac{p_t}{p_0} \cdot w}{\sum w} \cdot 100 \tag{3.5}$$

where w denotes the weights. If we let $w = p_0 q_0$, the index (3.5) becomes

$$I = \frac{100 \cdot \sum \frac{p_t}{p_0} \cdot p_0 q_0}{\sum p_0 q_0} = \frac{100 \cdot \sum p_t q_0}{\sum p_0 q_0} = L$$

because the p_0's in the numerator cancel out.

Both the Paasche and Laspeyres price indexes are special cases of the more general index number:

$$I = \frac{100 \cdot \sum_{i=1}^{n} p_{it} w_i}{\sum_{i=1}^{n} p_{i0} w_i} = \frac{\sum p_t w}{\sum p_0 w} \cdot 100 \tag{3.6}$$

where w represents the weights assigned to the given-year and base-year prices. The weights can be either fixed or variable. To obtain the Paasche price index, let $w = q_t$. The Laspeyres price index is obtained by letting $w = q_0$. A new price index could be obtained by letting the weights w be fixed quantities, say, q_a, which differ from either the base or given-year quantities.

All price indexes discussed so far can be converted to quantity indexes by replacing the p's with q's and q's with p's. For example, the Laspeyres quantity index is equal to

$$L = \frac{\sum q_t p_0}{\sum q_0 p_0} \cdot 100 \tag{3.7}$$

This quantity index will be used in Section 3.4.

3.4 THE CONCEPT OF AN IMPLICIT PRICE DEFLATOR

Price indexes obtained from national income and product accounts data are generally called *implicit price deflators*. They are based on index number formulas which we have already discussed, but they are more comprehensive than index numbers obtained from relatively small samples.

A consumer price index, for example, may include several hundred price relatives. An attempt is made to include commodities which are important in their own right and which are believed to be fairly representative of other similar commodities. In other words, a (judgment) sampling problem is involved in the selection of particular commodities to represent even a small subgroup such as fresh vegetables. If we defined the population of all commodities in an advanced economy in such detail as to recognize, for example, all different sizes, colors, styles, and brands of men's shoes, the total number of commodities would no doubt exceed a million.

While in concept it would be possible to design a data-processing system which would record prices and quantities of each of a million commodities, in practice our estimates of gross national product and its components are pieced together from more aggregative data from a variety of sources.

Practically all business firms keep track of their total receipts and expenditure, so it takes little extra effort for proprietors to send in to a public agency or to a trade association research unit a dollar figure on their last week's or last month's sales. Implicitly, such dollar sales figures from a sample of food supermarkets would include several thousand commodities whereas the official Consumer Price Index includes only about 100 food items.

An exact figure on the total dollar sales of a given store during 1965 would implicitly contain every microtransaction rung up in its cash registers during the year. This dollar value can be related to the comparable dollar figure for another year, say, 1960, but it cannot be used to obtain measures of both changes in physical volumes of commodities sold and changes in prices. If we wish to partition a change in money values into a quantity component and a price component, we must *either* make a direct estimate of the quantity component and derive the price component by dividing the quantity index into the dollar value series *or* we must make a direct estimate of the price component and derive the quantity component by dividing the price index into the dollar value series.

Whichever component is estimated directly may be measured with various degrees of accuracy. But its precision will affect not only its own usefulness but also the usefulness of the derived estimate of the other component. If the quantity component is estimated directly, the price component derived by dividing quantity into value is known as an *implicit price deflator*.

In the United States, implicit price deflators are an integral part of the gross national product and income tables. As of 1964, for example, direct estimates were made of the 1963 gross national product and its component in constant (1954) dollars. Corresponding estimates were, of course, made in current dollars—that is, in the prices actually existing in each quarter or year. Then implicit price deflators for the gross national product and its major components were calculated by dividing the constant (1954) dollar series into the current dollar series.

These calculations may be illustrated for the total gross national product for the year 1954 and 1963.

	1954	*1963*
	(*All GNP figures in billions of dollars*)	
1. Gross national product in current dollars	363.1	583.9
2. Gross national product in constant (1954) dollars	363.1	492.6
3. Implicit price deflator for the gross national product	1.000	1.185
4. Index of quantity change	1.000	1.357

The implicit price deflator for the gross national product turns out to be a Paasche index number. Thus, for 1963 we have

$$\text{Paasche price index} = \frac{\sum p_t q_t}{\sum p_0 q_t} = \frac{583.9}{492.6} = 1.185$$

The index of quantity change given in Row 4 is obtained by dividing each number in Row 2 by 363.1. This index is the Laspeyres index of quantity change. Thus, for 1963 we have

$$\text{Laspeyres quantity index} = \frac{\sum q_t p_0}{\sum q_0 p_0} = \frac{492.6}{363.1} = 1.357$$

Implicit price deflators are also published for the following components of the gross national product: (1) personal consumption expenditures (total), and separately for (a) durable goods, (b) nondurable goods, and (c) services; (2) new construction (total), and separately for (a) residential nonfarm construction and (b) other construction; (3) producers' durable equipment; (4a) gross exports and (4b) gross imports of goods and services; and (5) government purchases of goods and services (total), and separately for (a) federal and (b) state and local.

The chief advantage of implicit price deflators is their exact arithmetical consistency with the corresponding quantity and value series. Given the existence of discretionary elements in the choice of any standard index number formula, the consistency aspect of implicit price deflators and the two related series is a telling argument in their favor.

* 3.5 MATHEMATICAL TESTS FOR INDEX NUMBERS

There are many different criteria that can be used when trying to decide which index number formula to use. We have already pointed out a few such criteria. Unweighted index numbers, for example, may be preferred at times because they are easy to compute. Simple aggregate indexes may not be desirable because of the units problem. The Laspeyres index may be preferred to the Paasche index when it is difficult to obtain new weights each period.

There are also several mathematical "tests" available by which to compare index numbers. The time reversal test and the factor reversal test will be considered in this section. These tests are more easily presented if we drop the convention of multiplying index numbers by 100. In other words, when referring to index number formulas (3.1) to (3.7), we will assume that the multiplier 100 has been crossed out.

* This section may be omitted without loss of continuity.

3.5.1 *The Time Reversal Test*

The *time reversal test* requires that, if the subscripts 0 and t are interchanged in an index number formula, the resulting index must be the reciprocal of the first. Very few index numbers satisfy the time reversal test.

The Laspeyres price index does not satisfy the time reversal test because

$$L^* = \frac{\sum p_0 q_t}{\sum p_t q_t} \neq \left(\frac{1}{L}\right) = \frac{\sum p_0 q_0}{\sum p_t q_0}$$

The Paasche price index also fails the time reversal test.

The time reversal test expresses the belief that the relative changes from year to year in a series of index numbers should be the same regardless of the base year selected. For example, if an index with a base at 1960 has a value of 4.00, then changing the base period to 1965 should result in a 1960 value of 0.25. Thus, if an index formula satisfies the time reversal test, we will have:

	Value of the Index Number for		
	1960		*1965*
Index A, base year = 1960	1.00	⟶	4.00
Index B, base year = 1965	0.25	⟶	1.00

The time reversal test requires that the ratios (1.00/4.00) and (0.25/1.00) be equal.

The simple aggregate index (3.1) is an example of an index number that satisfies the time reversal test. To verify this, change the subscripts in formula (3.1) and compare the result with the reciprocal of (3.1). We find that

$$I^* = \frac{\sum p_0}{\sum p_t} = (1/I) = \frac{\sum p_0}{\sum p_t}$$

3.5.2 *The Factor Reversal Test*

The *factor reversal test* requires that a price index multiplied by its corresponding quantity index be equal to the value index. By definition, the value index is

$$V = \frac{\sum p_t q_t}{\sum p_0 q_0} \tag{3.8}$$

Price relatives for single commodities satisfy the factor reversal test. In other words, for a commodity i, the price relative times the quantity relative is equal to the value relative. Mathematically we have

$$(p_{it}/p_{io})(q_{it}/q_{io}) = (p_{it}q_{it}/p_{io}q_{io})$$

The Laspeyres and Paasche indexes do not satisfy the factor reversal test, but their geometric mean does. Irving Fisher, in his extensive study of index number formulas in 1922, referred to the geometric mean of the Laspeyres and Paasche indexes as the "ideal index number."[2] This terminology is still used. The "ideal" price index is equal to

$$I_p^* = \sqrt{\frac{\sum p_t q_0}{\sum p_0 q_0} \cdot \frac{\sum p_t q_t}{\sum p_0 q_t}} \tag{3.9}$$

and the "ideal" quantity index is

$$I_q^* = \sqrt{\frac{\sum q_t p_0}{\sum q_0 p_0} \cdot \frac{\sum q_t p_t}{\sum q_0 p_t}} \tag{3.10}$$

The "ideal" price index satisfies the factor reversal test as can be verified by showing that:

$$(I_p^*)(I_q^*) = V$$

This "ideal" index is still far from being a "perfect" index. The need to change weights each period in the Paasche portion of the "ideal" index is a serious disadvantage in many applications.

3.6 SOME SPECIAL PROBLEMS IN MEASURING AGGREGATIVE CHANGES

In actual practice, mathematical "tests" of the "correctness" of index number formulas cannot replace sound judgment in the selection and construction of index numbers. Such judgment requires an understanding of the special problems that arise in actual construction of index numbers over a period of time as well as a knowledge of the mathematical characteristics of index number formulas. Several special problems in the construction of index numbers from economic data will be discussed in this section.

3.6.1 The Problem of Changing Quality

The measurement of changes in quality is partly a philosophical problem and partly a pragmatic one. At one time, the consumer's price index may

[2] Irving Fisher, *The Making of Index Numbers: A Study of Their Varieties, Tests, and Reliability*, Third Edition (Boston and New York: Houghton Mifflin Co., 1927).

reasonably have given considerable weight to commodities such as flour in 50-pound sacks, and to cracked wheat and cornmeal used as home cooked cereals. Today the Consumer Price Index gives much less weight to such "old-fashioned" commodities. A great number of commodities are now sold by the retailer prepackaged into small and presumably convenient units.

On the one hand, it may appear that the modern consumer is required to lay out a considerable amount of money for cardboard packages which he would formerly have spent for the contents themselves! However, over the past 40 years the whole food-marketing system has reoriented itself to a pattern of self-service and consumer packages, and modern kitchens and appliances have also been adapted to the storage and preparation of modern commodities.

This kind of shift in quality and in the meaning of "price per pound" of a food product is primarily of interest in connection with the measurement of long-time changes in levels of living; it is not particularly serious in measuring short-time price changes for commodities that retain their identities intact during the period concerned.

Very complex commodities, such as new automobiles, present a more serious problem. Do changes in styling, affecting only the contour of the "outer skin" of the automobile, constitute changes in quality? This can *perhaps* be measured in a short-term market sense if we are permitted to plan appropriate experiments and have access to sales records of automobile manufacturers and dealers. In the late 1930's Andrew Court suggested the calculation of "hedonic indexes" to quantify the more fundamental structural and mechanical improvements in automobile quality over a long period of years.

Griliches, in the early 1960's, suggested methods of adjusting automobile prices for the more dramatic changes in quality. By means of multiple regression techniques, the net differences that consumers were evidently willing to pay for features such as automatic transmissions were estimated.[3] The estimates had to be based on time periods in which the quality improvement was an optional feature in new models or in which sales of used automobiles both with and without the particular quality improvement were taking place. This opportunity for comparison is lost after the quality improvement has become standard equipment even on the lower priced models. However, if consumers were willing to pay a premium, say, of $100 for a new feature while it was still optional, the implication is that the now-universal quality feature continues to add $100 to the intrinsic value of the current generation of new cars.

Quality changes may also be highly important in producers goods, par-

[3] Multiple regression techniques are discussed in Chapter 10.

ticularly during the early years of the development of a new technology. Can we measure changes in the output of electrical power generators by means of changes in numbers of generators alone? If the generator itself is identified in terms of output units, so that we are dealing with a price series for generators of 1000 kilowatt capacity, does this automatically eliminate the effects of quality change? Or, if we shift the comparison to diesel engine, is an engine which delivers 100 horsepower from 100 units of fuel the same quality as an engine which delivers 100 horsepower from 70 units of fuel? As part of a total production process, it would appear that the generator with the greater fuel conversion efficiency would be worth a higher price.

The chances are that many problems of this type can be given a quantitative solution with the use of more sophisticated economic models (such as process analysis and linear or quadratic programming), and these might provide us with some guidance in improving our measures of price and production change. The basic problem of quality measurement is, in some respects, a very important one. If all the price relatives in our index numbers reflect the average quality of whatever generation of new products is sold in each year, the corresponding quantity series must reflect only changes in the number of physical units of the commodity produced in some fairly naive and gross sense. If production measured in this latter sense is increasing at a rate of only 2 percent a year *per worker* and if the quality of an average unit in the sense of (a) its capacity to satisfy consumer wants or (b) its capacity to facilitate production of other goods and services is also increasing at the rate of 2 percent a year, then our present estimate of the rate of increase in production per worker is only half as high as it should be. Or, if one economy is characterized by rapid improvements in quality and another is primarily engaged in increasing outputs of simple standardized commodities, failure of official output measures in the first country to reflect the quality factor is equivalent to using an index-weighting formula which is appropriate to the conditions of the second country but not to those of the first.

Obviously, if the dollar values of deliveries to final demand are correctly measured and if the measures of quantity change do not take proper account of improvements in quality, the result is that the implicit price deflators overestimate the rate of increase in prices. If policy makers responsible for economic stabilization interpret these increases in the implicit price deflators as indicators of inflationary pressure, they may be led to pursue deflationary fiscal and monetary policies that are not justified by the facts.

The preceding paragraphs are not intended to prejudge the facts concerning the rate of quality change in the United States and other advanced economies or the degree to which our quantity indexes may be biased downward and our price indexes upward as a result of existing practices in index number construction. They are only intended to indicate that the measurement of quality change is an extremely important one for research and investigation along with other problems involved in the construction and interpretation of economic aggregates and index numbers.

3.6.2 *The Problem of Changing Commodities*

At a formal level, the problem of entry of new commodities into the consumption pattern and hence into index numbers can be waived by saying that the new commodity really was in the old index after all but with a zero weight! Television sets justified only a minor weight in 1946; A Paasche index of quantity change would have led to a moderate increase in this weight in each successive year, but the cumulative impact on the index over a ten-year period would be substantial. Similarly, a commodity that is going out of fashion, and eventually out of production, would be assigned progressively smaller weights in the existing index number formula so that its exit from the index (or its assignment of a zero weight!) may simply represent the end point of a gradual trend which has been affecting the index number over a decade or more.

The problem can thus be faced on a common-sense basis. The population of relevant commodities in an advanced economy is a continually changing one. We are acutely aware of this when we step into a museum exhibit containing an authentic replica of a kitchen, a grocery store, or a drugstore as it existed in 1900 or 1910.

Any particular index number formula (weighting pattern) will be most valuable for comparisons between two adjacent years; it will probably be quite good for comparisons over five or six years; tolerable for comparisons over eight or ten years; and seriously misleading for longer-term comparisons. As the content of our pattern of consumption changes, the yardstick with which we measure further changes must itself be adjusted.

3.6.3 *The Problem of Interregional Comparisons*

Comparisons of price level differences or consumption level differences between regions and nations involve the same formal difficulties as do comparisons for a given country or region over a period of years. Within a country such as the United States, the expenditure patterns of households of the same type and with the same money incomes may differ because of differences in the relative prices of commodities or services between areas and because of differences in climate which affect the materials and practices used in construction of suitable houses, costs of heating and air conditioning, clothing needs, and recreational opportunities. There may also be differences based in cultural (historical) factors.

There are also differences in city size. For example, the largest cities in the United States have extensive rapid transit systems which enable many people to commute from home to work without automobiles. At the same time, the downtown areas of these cities are so congested that automobile ownership may be something of a liability.

Hence, an attempt to compare "the" costs of living in New York City and in Santa Barbara, California raises the same problems of choice between Laspeyres and Paasche (and perhaps other) formulas as do comparisons over time. The New Yorker will probably find that the particular bundle of goods and services he consumes in New York would cost him more in Santa Barbara. On the other hand, the particular set of goods and services that is typical of his income group in Santa Barbara would cost more money in New York City. If the New Yorker is spending $6000 a year, we could perhaps show him what $6000 a year will buy in Santa Barbara *if* he allocates it according to the pattern which is typical for Santa Barbara households at that expenditure level. He might then be able to respond to the question as to whether he would feel better off spending $6000 Santa Barbara style in Santa Barbara or spending $6000 New York style in New York.

3.7 SHIFTING THE BASE OF AN INDEX NUMBER

People working with index numbers often find it necessary to shift the base of an index number series from one time period to another. This may be necessary, for example, if index numbers of one series are to be easily compared with those of another. Most people work with index numbers that other people have computed and, therefore, seldom have access to the original data from which the index numbers were obtained. Even if the original data can be obtained, the complete reconstruction of an index number series may require more time and money than is available. In such cases, it is necessary to use a short-cut method to shift the base of the index numbers.

The short-cut method commonly used is to divide all of the index numbers in the original series by the original value of the index number for the new base period and multiply by 100. This can be illustrated using the data in Table 3.4.

To obtain the new price index (with base 1955) for 1955, divide the original price index (with base 1960) for 1955 by 80 and multiply by 100. This gives us $(80/80) \cdot 100 = 100$. The new price index for 1956 is equal to $(60/80) \cdot 100 = 75.0$. Continue to divide each index number in the old series by 80 and multiply by 100 until the complete new series is obtained.

The short-cut method for shifting the base of an index number is also useful when it is necessary to splice two or more index number series together.

Table 3.4 Changing the Base of an Index Number

Year	Old Price Index (1960 = 100)	New Price Index (1955 = 100)
1955	80	100.0
1956	60	75.0
1957	70	87.5
1958	90	112.5
1959	98	122.5
1960	100	125.0
1961	112	140.0
1962	116	145.0
1963	110	137.5

Table 3.5 Method for Splicing Two Index Number Series

Year	(1) Old Index (1955 = 100)	(2) New Index (1960 = 100)	(3) Old Index Shifted to 1960 Base	(4) Spliced Index Series (1960 = 100)
1955	100		125	125
1956	96		120	120
1957	88		110	110
1958	76		95	95
1959	60		75	75
1960	80	100	100	100
1961		108		108
1962		95		95
1963		98		98
1964		103		103
1965		107		107

For example, suppose we have two price index series for coffee, one which has a 1955 base and runs from 1955 to 1960, and another which has a 1960 base and runs from 1960 to 1965. This situation is presented in Table 3.5. If we want to splice these two series together, they will have to have a common base period. In this case, we would shift the base of the first series from 1955 to 1960 and then combine the two series. There is no need to alter the index number series for the 1960 to 1965 period because it already has 1960 as the base year.

The short-cut method for shifting the base of an index number series assures that the relative changes from year to year in the series are not affected. In other words, that

$$\frac{\text{Period } t \text{ index on old base}}{\text{Period } t+1 \text{ index on old base}} = \frac{\text{Period } t \text{ index on new base}}{\text{Period } t+1 \text{ index on new base}} \tag{3.11}$$

The 1956 and 1957 data in Table 3.4 can be used to illustrate this. We require that

$$\frac{\text{1956 index on 1960 base}}{\text{1957 index on 1960 base}} = \frac{\text{1956 index on 1955 base}}{\text{1957 index on 1955 base}}$$

Putting in the appropriate numbers, we find that

$$(60/70) = (75/87.5) = 0.857$$

which verifies that the relative changes were maintained in this case. (*Suggestion.* Verify that (3.11) is true for the 1961 and 1962 data in Table 3.4.)

In some cases, the results of the short-cut method are identical to those which would have been obtained if a new index series had been completely reconstructed on the new basis. This would be true, for example, if the original index numbers were obtained from the simple aggregate index formula (3.1). To prove this, let t represent period t, o the original base period, and a the new base period. Then the short-cut method requires that the original index number for the new base period, which is $\sum p_a/\sum p_0$, be divided into each of the index numbers, $\sum p_t/\sum p_0$, in the original series. This gives us

$$I^* = \frac{\sum p_t/\sum p_0}{\sum p_a/\sum p_0} = \frac{\sum p_t}{\sum p_a}$$

which is exactly the index formula we would have used if we had completely reconstructed the index number series on a new base.

If the original index number series had been obtained using the Laspeyres price index, then the short-cut method results in a new index number equal to

$$I^* = \frac{\sum p_t q_0 / \sum p_0 q_0}{\sum p_a q_0 / \sum p_0 q_0} = \frac{\sum p_t q_0}{\sum p_a q_0}$$

This new index is not equal to the Laspeyres price index because the original base period quantities q_0 act as weights rather than the new base period quantities q_a.

For most index number formulas, the short-cut method of shifting the base from one period to another will not give exactly the same results as those obtained if the index were recomputed on the new base. In many practical applications, the short-cut method is the only practical alternative, however. The short-cut method is adequate for many uses—as long as you keep in mind its limitations.

* 3.8 SOME PRACTICAL INDEX NUMBER CALCULATIONS AND COMPARISONS

Calculations for the two most commonly used index number formulas are illustrated here using basic data on the retail prices and per capita consumption of four kinds of meat. The original price and quantity series are shown in Table 3.6 and the price *relatives* and quantity *relatives* in Table 3.7. The base year chosen for the relatives is 1950. The relatives are multiplied by 100 for better comparison with the aggregative index numbers in Table 3.8 which accord with the almost universal practice on the United States of publishing official index numbers on a base of 100 rather than 1.00.

In Table 3.8, Column 1 shows total expenditures for the four meats in current dollars. The subscript t in the formula in the column heading indicates that the variable, to which it is attached, changes values from year to year. Column 7 shows the total value of current year quantities of the four meats measured at constant 1950 prices. Here, the subscript ($_0$) implies that the variable in question remains fixed during the entire span of years. Conversely, Column 8 shows the total value of the four meats at current prices, but with quantities fixed at their 1950 levels. This is obviously the cost of a "market basket" or composite commodity consisting of fixed quantities of each meat. Hence, the year-to-year changes in Column 7 can arise only from

* This section may be omitted without loss of continuity.

Table 3.6 Meats: Basic Series on Per Capita Consumption and Retail Prices, United States, 1950–1960

	Per Capita Consumption				Retail Price			
Year	q_1 Beef and Veal	q_2 Lamb and Mutton	q_3 Pork	q_4 Chicken and Turkey	p_1 Beef and Veal	p_2 Lamb and Mutton	p_3 Pork	p_4 Chicken and Turkey
		(pounds)				(cents per pound)		
1950	57.1	3.6	64.4	24.7	67.9	67.1	49.6	57.0
1951	50.2	3.0	66.9	26.1	79.4	74.6	53.3	59.7
1952	55.5	3.7	67.3	26.8	77.9	72.6	51.8	60.0
1953	69.7	4.2	59.1	26.7	62.2	61.9	57.2	58.5
1954	72.1	4.1	55.8	28.1	61.6	63.2	58.3	52.8
1955	73.1	4.1	62.1	26.4	60.8	61.2	49.3	54.8
1956	75.9	4.0	62.6	29.8	59.4	61.5	46.9	47.8
1957	74.7	3.7	56.8	31.4	63.5	65.1	54.2	46.7
1958	69.8	3.6	56.0	34.1	72.9	70.9	58.3	46.1
1959	69.7	4.0	62.9	35.2	74.4	67.0	51.5	42.3
1960	73.1	4.1	60.7	34.6	72.6	66.3	50.8	42.7

Source. United States Department of Agriculture, Economic Research Service.

Table 3.7 Meats: Retail Price and Per Capita Consumption Relatives, United States, 1950–1960 (1950 = 100)

Year	Beef and Veal	Lamb and Mutton	Pork	Chicken and Turkey	Beef and Veal	Lamb and Mutton	Pork	Chicken and Turkey
	Quantity Relatives (times 100)				Price Relatives (times 100)			
1950	100.0	100.0	100.0	100.0	100.0	100.0	100.0	100.0
1951	87.9	83.3	103.9	105.7	116.9	111.2	107.5	104.7
1952	97.2	102.8	104.5	108.5	114.7	108.2	104.4	105.3
1953	122.1	116.7	91.8	108.1	91.6	92.3	115.3	102.6
1954	126.3	113.9	86.6	113.8	90.7	94.2	117.5	92.6
1955	128.0	113.9	96.4	106.9	89.5	91.2	99.4	96.1
1956	132.9	111.1	97.2	120.6	87.5	91.7	94.6	83.9
1957	130.8	102.8	88.2	127.1	93.5	97.0	109.3	81.9
1958	122.2	100.0	87.0	138.1	107.4	105.7	117.5	80.9
1959	122.1	111.1	97.7	142.5	109.6	99.9	103.8	74.2
1960	128.0	113.9	94.3	140.1	106.9	98.8	102.4	74.9

Table 3.8 Meats: Comparisons of Price and Per Capita Consumption Indexes Using Laspeyres and Paasche Formulas, United States, 1950–1960 (1950 = 100)

	(1)	(2)	(3)	(4)	(5)	(6)	(7)	(8)
	Expenditures in Current Dollars		Price Indexes		Quantity Indexes		Work Columns	
Year	$\sum p_t q_t$ Dollars	Value index, 1950 = 100[a]	Laspeyres 1950 = 100	Paasche 1950 = 100	Laspeyres 1950 = 100	Paasche 1950 = 100	$\sum p_0 q_t$ Dollars	$\sum q_0 p_t$ Dollars
1950	$87.2079	100.000	100.000	100.000	100.000	100.000	$87.2079	$87.2079
1951	93.3362	107.027	111.336	110.906	96.503	96.130	84.1582	97.0941
1952	96.8621	111.070	109.249	109.049	101.853	101.667	88.8240	95.2737
1953	95.3779	109.368	102.090	100.740	108.565	107.129	94.6771	89.0309
1954	94.3730	108.216	100.949	98.923	109.395	107.199	95.4008	88.0356
1955	92.0365	105.537	94.263	93.690	112.645	111.960	98.2356	82.2048
1956	91.1484	104.519	89.604	89.138	177.255	116.645	102.2557	78.1416
1957	95.2926	109.271	97.516	95.989	113.837	112.054	99.2748	85.0418
1958	101.8047	116.738	106.768	104.929	111.255	109.338	97.0228	93.1102
1959	101.8199	116.755	101.491	100.540	116.128	115.040	101.2727	88.5085
1960	101.3987	116.272	99.880	99.201	117.209	116.412	102.2152	87.1035

[a] Column (1) divided by $87.2079 and multiplied by 100. Formula is $\left(\frac{\sum p_t q_t}{\sum p_0 q_0}\right)$ times 100.

changes in quantities and the year-to-year changes in Column 8 can arise only from changes in prices.

In this notation, $\sum p_0 q_0$ represents the dollar value of expenditures in 1950, which we shall use as our index number base in all comparisons.

The expenditures in 1950 were \$87.2079. We divide this figure by itself and multiply by 100 to establish the base value for 1950 at the convenient figure of 100. Then, the 1951 value of the Laspeyres price index is obtained by dividing \$97.0941 by \$87.2079, which (times 100) equals 111.336. The 1951 value of the Paasche index is obtained by dividing \$93.3362 by \$84.1582, which (times 100) equals 110.906. The difference is about 0.4 index points.

The maximum difference between the two price indexes occurs in 1958 at 1.8 index points. The Laspeyres index indicates that meat prices were nearly 6.8 percent higher in 1958 than in 1950; the Paasche index indicates a price difference of only 4.9 percent. The Laspeyres and Paasche quantity indexes show a maximum difference of 2.2 index points in 1954.

It is interesting to note in Table 3.8 that the Laspeyres price and Laspeyres quantity indexes always change in opposite directions. The same is true for the Paasche price and Paasche quantity indexes, although the price change between 1953 and 1954 is only 0.1 index points. This inverse relationship between price and quantity is consistent with the conclusion of the theory of consumption that demand curves slope downward to the right.

Also note that the value index in Column (2) of Table 3.8 can be obtained by multiplying either a Laspeyres quantity and a Paasche price index or a Paasche quantity and a Laspeyres price index.

EXERCISES

Section 3.1

1. What is meant by the term "parity ratio?"
2. What factors probably account for the fact that index numbers of physical output developed much later than index numbers of prices?

Section 3.2

3. What are the main disadvantages of simple unweighted aggregate price indexes?

4. Use the following data to compute:
(a) A simple unweighted aggregate price index with base 1960.
(b) A price index using the arithmetic mean of price relatives with base 1960.

Average Price
(in Dollars)

Commodity	1960	1962
Wheat	1.74	2.04 per bushel
Rice	4.55	5.03 per bag
Corn	1.00	1.10 per bushel
Soybeans	2.17	2.54 per bushel
Barley	0.84	0.91 per bushel

5. Using the data in Exercise 4, compute an index number based on the *median* of the price relatives. How does this index compare in value to the index based on the *mean* of the price relatives?

Section 3.3

6. Using the following data, compute:
(a) The Paasche price index.
(b) The Laspeyres price index.
(c) The Laspeyres quantity index.

	Price		Quantity	
Product	1960	1967	1960	1967
X	9	10	40	50
Y	6	5	30	40
Z	12	10	20	20

Section 3.4

7. Compute the implicit price deflators for gross national product using the data given below.

Total Gross National Product
(in Billions of Dollars)

Year	Current Prices	1958 Prices
1955	398.0	438.0
1956	419.2	446.1
1957	441.1	452.5
1958	447.3	447.3
1959	483.7	475.9
1960	503.7	487.7
1961	520.1	497.2
1962	560.3	529.8
1963	590.5	551.0
1964	631.7	580.0
1965	681.2	614.4
1966	739.5	647.7

Source. Economic Report of the President, 1967.

8. Use the implicit price deflators obtained in Exercise 7 to estimate *real* personal consumption expenditures from the following money figures.

Year	Personal Consumption Expenditures (in Billions of Current Dollars)
1958	290.1
1959	311.2
1960	325.2
1961	335.2
1962	355.1
1963	375.0
1964	401.4
1965	431.5
1966	465.0

Source. Economic Report of the President, 1967.

Section 3.5

9. Show that the Paasche price index does not satisfy the time reversal test for index numbers.

10. Show that an index number of the form $I = \sum p_t w_k / \sum p_0 w_k$ where the w_k are fixed weights, does (or does not) satisfy the time reversal test.

Section 3.6

11. Assume that you are in charge of starting a new price index series to measure changes in the wholesale prices of farm products in five large cities on the East coast. Discuss the problems you would expect to face in the short-run and in the long-run. What price index formula would you use? Why?

Section 3.7

12. Splice the following index number series together using 1958 as the base period.

Price Index of Latin American Exports

Year	Old Series	New Series
1953	118	
1954	115	
1955	105	
1956	100	
1957	96	
1958	90	100
1959		95
1960		95
1961		93
1962		91
1963		94

13. Shift the spliced index series obtained in Exercise 12 from a 1958 base to a 1963 base.

SELECTED REFERENCES ON INDEX NUMBERS

Allen, R. G. D., *Mathematical Economics*, Second Edition, Macmillan and Co. Ltd., London and New York, 1959, 812 pp. Chapter 20 (pages 694–724) outlines the principle of " perfect aggregation " based on the pioneering work by H. Theil (1954). Allen's references on pages 723–724 are a good introduction to the literature on *aggregation* as distinct from the narrower topic of *index numbers*. The aggregation literature was largely stimulated by problems encountered in constructing econometric models of national economies. Allen's earliest reference is to Tinbergen (1939).

Economic Report of the President: Together with the Annual Report of the Council of Economic Advisers (January 1964 and subsequent years). Each annual issue of the *President's Economic Report* includes an appendix on "statistical tables relating to income, employment, and production." In the January 1964 issue, this appendix covers more than 100 pages (pages 201–304). It includes most of the official measures of economic aggregates in the United States which are used by policy makers and their economic advisers for appraising the current state of the economy. The major categories of tables include national income or expenditure; employment, wages, and productivity; production and business activity; prices; money supply, credit, and finance; government finance; corporate profits and finance; agriculture; and international statistics. Many of the annual series extend back to 1929. This annual appendix is an extremely useful, compact, and well-selected handbook for professional economists and students.

Fisher, Irving, *The Making of Index Numbers: A Study of Their Varieties, Tests, and Reliability*, Third Edition, Houghton Mifflin Co., Boston and New York, 1927, 538 pp. The first edition was published in 1922. The third edition, revised, contains interesting supplementary materials connected with the many book reviews and widespread controversy elicited by the 1922 edition. The book contains discussions and calculations involving a total of more than 100 alternative index number formulas. In addition to extensive technical appendixes, his brief appendix (pages 458–460) on "landmarks in the history of index numbers" and his selected bibliography (pages 519–520) provide an entry into the classical literature on index numbers from 1863 to 1921. Fisher's book marked the culmination of the first 60 years of interest in index number formulas by professional economic statisticians.

Haberler, Gottfried, *Der Sinn Der Indexzahlen*, Verlag von J. C. D. Mohr (Paul Siebeck), Tubingen, 1927, 134 pp. A selected bibliography (pages 129–132) contains many references to the early literature on index numbers in addition to those mentioned in Fisher's bibliography. For example, references are given to the original papers by Laspeyres (1864 and 1871) and by Paasche (1874), describing and justifying the two index number formulas that are still in almost universal use. The fact that the crossproduct of a Laspeyres price index and a Paasche quantity index or of a Paasche price index and a Laspeyres quantity index gives a mathematically exact measure of changes in the current dollar value of the "composite commodity" involved is likely to insure the survival of the Laspeyres and Paasche formulas for a long time to come. Although the phrase *perfect aggregation* entered the economic literature only in the 1950's, the preoccupation of Irving Fisher and others with the mathematical consistency of index numbers of price, quantity, and value may be regarded as a concern with "perfect aggregation." The relevant aspects of index number theory may well be assimilated into the broader topic of measurement of economic aggregates during the 1960's.

Joint Economic Committee, United States Congress, *1964 Supplement to Economic Indicators: Historical and Descriptive Background*, (prepared for the Joint Economic Committee by the Committee Staff and the Office of Statistical Standards, Bureau of the Budget), United States Government Printing Office ,Washington, 1964 pp. This supplement gives brief but accurate descriptions of the principal index numbers and economic aggregates which are published monthly in *Economic-Indicators* by the Joint Economic Committee of the United States Congress. The monthly publication, *Economic Indicators*, is prepared by the staff of the President's Council of Economic Advisers. The *Supplement* also contains references to the basic official descriptions of the major aggregates and index numbers—descriptions published by the agencies directly responsible for the series.

Theil, H., *Linear Aggregation of Economic Relations*, North-Holland Publishing Co., Amsterdam, 1954, 205 pp. This is the first systematic treatment of aggregation problems in the modern sense.

United States Department of Commerce, Office of Business Economics, *Survey of Current Business*, July 1964, pp. 7–40. These pages contain a complete set of estimates of components of the national income and product accounts in 1963 and for a number of earlier years. Page 7 also refers to the official description of the most recent major revision in the United States social accounts, *United States Income and Output* (1957). The same tables are published each year in the July issue of the *Survey of Current Business*.

CHAPTER 4

Basic Probability

The word *probability* is used very loosely in everyday conversation to suggest that there is uncertainty about what has happened in the past, what will happen in the future, or what is happening at the present. The businessman may decide to market a new product because the "probability" is high that the product will be successful. The sports fan may bet against his favorite football team because the "probability" of it winning is very small. The farmer may not cut his hay because it is "likely" to rain. Neither the businessman, the sports fan, nor the farmer may know how to define or measure probability; yet each of them found the idea of "probability" useful when making a decision. But what is probability? How can it be measured? How can it be used? These are a few of the questions to be answered in this chapter.

The idea of probability plays an important role in many decision-making situations. Suppose a radio tube manufacturer wants to produce tubes that will last at least 2000 hours. He could thoroughly test all of the tubes, but it would be expensive and reduce the expected life of the good tubes. In such cases, a quality control program based on sampling would be used. The manufacturer would decide what type of sample to take and the basis for accepting or rejecting a batch of tubes. As another example, suppose an economist estimates the price elasticity of demand for a product and wants to know if this elasticity is significantly greater than zero. The economist needs a procedure that is reasonable to him, acceptable to others, and that will allow him to arrive at a decision. He will turn to testing procedures which are built on the foundations of probability theory.

Probability models can be very useful for making predictions. Suppose that public health authorities know that an infectious disease is spread by contact but that not everyone is susceptible to the disease. Probabilistic models can be developed to describe the likely progress of such epidemics. These models can help health authorities predict how long the epidemic will last, how many people will be infected, and how rapidly the epidemic will move through a community. As another example, suppose a plant specialist wants to develop a new type of soybean. The basic rules of probability can help him predict the outcome of crossing different known varieties.

The word probability is interpreted and used in several different ways in statistics. There is little disagreement about the mathematical foundations of probability, but there are some differences of opinion as to when the rules of probability can be applied. At the extremes there are two schools of thought.

The *objective* (or *classical*) *school* holds that the rules of probability apply only to events that can be repeated under the same conditions. The word objective is used to imply that, if any two people accurately and separately computed the probability of an event, they would always arrive at the same answer. Thus there is a "probability" associated with receiving ten face-cards in a game of bridge or winning a lottery in which 10,000 people hold tickets because the basic "experiments" could be repeated under the same conditions and different people would assign the same values to the probabilities. Members of the classical school would not assign a "probability" to the possibility that your brother will become a millionaire within five years or that John Doe will be the first man on the moon. Such events are not outcomes of an experiment that could be repeated under the same conditions. Thus attempts to assign "probabilities" to the outcomes may result in widely different results.

The *personalistic* (or *subjective*) *school* regards probability as a measure of personal belief that an event has happened, will happen, or is happening. The "personalist" is willing to use probability techniques in many applications where the "objectivist" would not. The personalist realizes that personal probabilities may be different and that two people using the same probability techniques may, therefore, arrive at different answers. He believes, however, that decision makers can make better decisions by using personal probabilities systematically. If he is correct, then the personalistic school has a strong argument in its favor.

More people have been trained in the objective school and its position is still the more popular. The personalistic school has been growing rapidly and is likely to be more popular in the future. We will use the concept of *objective* probability in most of this chapter.

At first reading, you cannot expect full understanding of all material in this chapter. The ideas presented here will have greater significance after you have seen them in action in later chapters. We recommend, therefore, that this chapter be reread from time to time.

4.1 THE BASIC RULES OF PROBABILITY

4.1.1 The Definition of Probability

In this section, we shall define probability and introduce some terminology to be used later. We begin with the terminology.

The word *experiment* is used in statistics to describe any act that can be repeated under the same conditions. Consider an experiment of tossing a single coin. The outcome of tossing the coin is not known prior to the performance of the experiment, but the set of possible outcomes can be listed. The coin may land either "heads" or "tails." Each of these outcomes is called an *elementary* (or simple) *event*. An elementary event is an outcome of an experiment that cannot be decomposed into a combination of other events. A set S listing all possible outcomes of an experiment is called a *sample space*. More precisely, a sample space is a set of elements such that any outcome of the experiment corresponds exactly to one element in the set.

Elementary events will be represented by either letters or numbers, depending on which is most convenient. In the case of the coin-tossing experiment, it is sometimes convenient to let H stand for the outcome heads and T for tails. The sample space would then be written as $S = \{H, T\}$. At other times, it may be more convenient to denote heads by the number 0 and tails by 1. In this case, the sample space becomes $S = \{0, 1\}$.

The outcomes of an experiment can often be described in several different ways. Thus an experiment may have several sample spaces. Consider the experiment of simultaneously tossing a dime and a penny. Suppose we are interested in whether each coin falls heads (H) or tails (T). If we let the first letter in a pair designate the outcome for the dime and the second letter that for the penny, the sample space can be written as $S_1 = \{HH, HT, TH, TT\}$. Alternatively, we may be interested only in the number of heads that appear. When tossing two coins, we may observe 0, 1, or 2 heads. The sample space could therefore be written as $S_2 = \{0, 1, 2\}$. Both S_1 and S_2 satisfy the definition of a sample space because every outcome of the experiment corresponds to exactly one element in each set.

In general, any subset of a sample space S is called an event. If the subset consists of a single element of S, it is called an elementary event or a *sample point*. If the subset consists of more than one element of S, it is called a compound event.

We often want to know the probabilities of various events associated with an experiment. In many cases, such probabilities can be computed by setting up a sample space of equally likely outcomes and then counting the sample points corresponding to the events we are interested in. For example, suppose that a die is tossed and we want to compute the probability that a 3 will occur.

The sample space S is equal to $S = \{1, 2, 3, 4, 5, 6\}$. If each of the 6 possible outcomes is equally likely, the probability that a 3 will occur is 1/6. In other words, there is one outcome, of the 6 equally likely outcomes, that corresponds to the die landing 3.

In reality, the assumption that each of the possible outcomes is equally likely may not be justified. This assumption is based on abstract reasoning rather than experience. Unless we have sufficient reason to believe that the outcomes are not equally likely, the assumption of equally likely outcomes is used in order to simplify the process of computing probabilities. We shall use this abstract approach to explain the basic rules of probability.

In general terms, a probability is a number assigned to an event. Suppose an experiment can result in N different, equally likely outcomes and N_c of these outcomes corresponds to the event C. The probability of the event C is defined as

$$P(C) = \frac{N_c}{N} \tag{4.1}$$

The number N_c must be greater than or equal to zero and less than or equal to the total number of possible outcomes N. Therefore the probability of the event C must be a number greater than or equal to zero but less than or equal to one. That is,

$$0 \leq P(C) \leq 1$$

All probabilities must satisfy this condition.

If $P(\bar{C})$ denotes the probability that an item selected at random does *not* have the characteristic C, then $P(\bar{C}) = 1 - P(C)$. In words, the probability that the event C does *not* occur is equal to one minus the probability that the event C occurs. To prove this, let the number of items not having characteristic C be denoted by $N_{\bar{c}}$. Since $N_{\bar{c}} = N - N_c$, the probability that C does not occur is equal to

$$P(\bar{C}) = \frac{N - N_c}{N} = \frac{N}{N} - \frac{N_c}{N} = 1 - P(C)$$

This is a very useful property to understand.

This idea can also be presented graphically using a Venn (or circle) diagram such as Figure 4.1.[1] Suppose the sample space is represented by a circle. The outcomes in the sample space can be divided into two groups: (1) those with characteristic C and (2) those without characteristic C.

[1] The idea of using diagrams to represent relationships between sets comes from the English mathematician, John Venn.

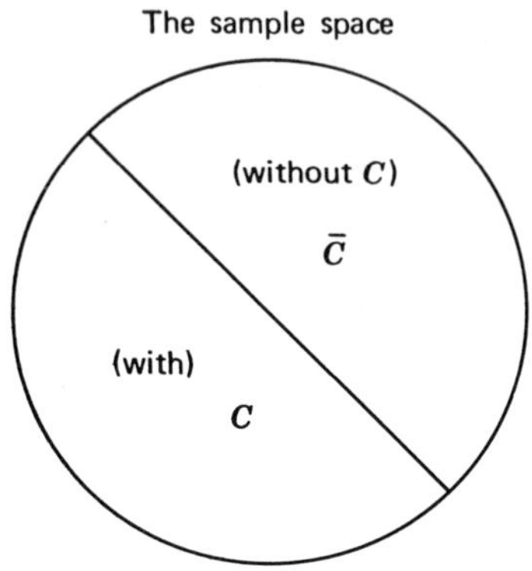

Figure 4.1 A Venn diagram.

If an item is selected from this sample space, it either has the characteristic C or it does not. Stated symbolically, we have

(4.2) $$P(C) + P(\bar{C}) = 1$$

and it follows directly that $P(\bar{C}) = 1 - P(C)$.

The main ideas presented so far can be illustrated with the help of a pair of dice. Suppose an experiment E consists of throwing a pair of dice. The sample space of the experiment consists of 36 different elementary events. These are shown in Table 4.1. The outcome of the experiment consists of a

Table 4.1 The Elementary Events for one Throw of Two Dice

		Number of Spots on DIE I (White)					
	b \ a	1	2	3	4	5	6
Number of Spots on DIE II (Red)	1	(1, 1)	(1, 2)	(1, 3)	(1, 4)	(1, 5)	(1, 6)
	2	(2, 1)	(2, 2)	(2, 3)	(2, 4)	(2, 5)	(2, 6)
	3	(3, 1)	(3, 2)	(3, 3)	(3, 4)	(3, 5)	(3, 6)
	4	(4, 1)	(4, 2)	(4, 3)	(4, 4)	(4, 5)	(4, 6)
	5	(5, 1)	(5, 2)	(5, 3)	(5, 4)	(5, 5)	(5, 6)
	6	(6, 1)	(6, 2)	(6, 3)	(6, 4)	(6, 5)	(6, 6)

pair of numbers (a, b). Suppose the event B is defined as "throwing a total of 9 points." The event B is a composite event consisting of the four elementary events (6, 3), (5, 4), (4, 5), and (3, 6). If the dice are "fair" dice, then all elementary events in the sample space are equally probable. In this case, as will be shown later, we can add up the probabilities of the elementary events to obtain

$$P(B) = P(6, 3) + P(5, 4) + P(4, 5) + P(3, 6)$$
$$= \left(\frac{1}{36}\right) + \left(\frac{1}{36}\right) + \left(\frac{1}{36}\right) + \left(\frac{1}{36}\right) = \left(\frac{1}{9}\right)$$

This probability could also be obtained by counting the number of elementary events "favorable" to "throwing a total of 9 points" and dividing by the total number of elementary events. This gives

$$P(B) = \left(\frac{4}{36}\right) = \left(\frac{1}{9}\right)$$

The probability of "not throwing a total of 9 points" is equal to one minus the probability of "throwing a total of 9 points." Thus, P (not throwing a total of 9 points) $= 1 - P(B) = 1 - (1/9) \equiv 8/9$. In other words, out of the 36 equally likely outcomes, there are 32 outcomes corresponding to "not throwing a total of 9 points." Thus the probability of "not throwing a total of 9 points" is 32/36 or 8/9.

It is important that the event of interest be defined precisely. Suppose, for example, that one die is red and the other white. The probability that the red die lands 3 and the white die 4 is 1/36 because only the sample point (3, 4) corresponds to this event. The event "one die lands 3 and the other 4" is an entirely different event. This event occurs when the red die lands 3 and the white die 4 and also when the red die lands 4 and the white die 3. Its probability is therefore equal to

$$P(3, 4) + P(4, 3) = \frac{2}{36}$$

The importance of precise definitions of events will become increasingly evident in later samples.

The following discussion of several of the basic rules of probability is limited to random experiments having a finite number of possible outcomes. We shall therefore be dealing with the special case of finite sample spaces.

4.1.2 The Addition Rule of Probability

Suppose a person is selected at random from a group of 100 persons, of whom 40 have brown hair, 30 blue eyes, and 20 both brown hair and blue eyes. Let the event "having brown hair" be denoted by the letter B and the event "having blue eyes" by the letter C. The term "selected at random" implies that each of the 100 persons is equally likely to be the one selected. Since 40 of the group have brown hair, the probability of selecting a person with brown hair is $P(B) = 40/100 = 0.40$. The probability of selecting a person with blue eyes, on the other hand, is $P(C) = 30/100 = 0.30$.

Suppose now that we wish to determine the probability that the person selected will have either brown hair or blue eyes (or both). This probability is denoted by $P(B \cup C)$ and is easily determined by inspection of the Venn diagram in Figure 4.2.[2] The rectangle represents the entire set of 100 persons.

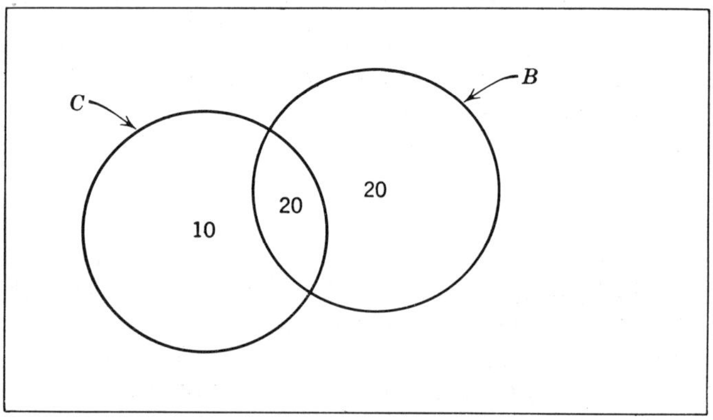

Figure 4.2 Sample space representing 100 persons.

The circles B and C represent the persons having brown hair and blue eyes, respectively. The overlap of the circles reflects the fact that some persons have both brown hair and blue eyes. In total there are 50 persons who have either brown hair or blue eyes or both. Thus $P(B \cup C) = 50/100 = 0.50$. Notice that the probability that the person selected has both brown hair and blue eyes is equal to $P(B \cap C) = 20/100 = 0.20$.[3]

The probability $P(B \cup C)$ can also be obtained using the so-called *addition rule of probability* which states that the probability that either event B or C (or both) will occur on a given trial of an experiment is equal to

$$P(B \cup C) = P(B) + P(C) - P(B \cap C) \tag{4.3}$$

The probability $P(B \cap C)$ that both B and C will occur is subtracted out once because the outcomes favorable to both B and C have been counted twice: once when $P(B)$ was computed and again when $P(C)$ was computed.

[2] The notation used here indicates that we are dealing with the *union* of the sets of elementary events favorable to B or C.

[3] The notation $P(B \cap C)$ indicates that we are dealing with the *intersection* of B and C. That is, the set of elementary events common to both B and C.

If the events B and C cannot occur together, they are called mutually exclusive events. In this case the probability $P(B \cap C)$ that both B and C will occur is zero. The *addition rule for mutually-exclusive events* is, therefore,

$$P(B \cup C) = P(B) + P(C) \tag{4.4}$$

This basic rule was used earlier to compute the probability of throwing 9 points when rolling two fair die.

The addition rule for mutually exclusive events follows directly from the fact that if a set of events (say, A, B, C, and D) are mutually exclusive and exhaust all possible outcomes of an experiment, then

$$P(A \cup B \cup C \cup D) = P(A) + P(B) + P(C) + P(D) = 1 \tag{4.5}$$

In other words, if none of the possible outcomes of an experiment are favorable to events other than A, B, C, or D, and the events A, B, C, and D are mutually exclusive, then the sum of their probabilities is equal to 1. A special case of this rule was discussed earlier when the event "not C" was denoted by $\bar{C}$ and it was shown that $P(C) = 1 - P(\bar{C})$. The events C and $\bar{C}$ are called *complementary events.*

These basic rules are readily illustrated with the help of a standard deck of cards. Consider an experiment that consists of drawing a card at random from a deck of 52 cards. Let event A be "drawing an ace" and event B be "drawing a king." The probability of drawing an ace *or* a king is equal to

$$P(A \cup B) = \left(\frac{4}{52}\right) + \left(\frac{4}{52}\right) = \left(\frac{8}{52}\right) = 0.154$$

The probability of *not* drawing an ace is equal to

$$P(\bar{A}) = 1 - P(A) = 1 - \left(\frac{4}{52}\right) = 0.923$$

Suppose, however, that we want to determine the probability of drawing an ace (A) or a spade (S). The events S and A are not mutually exclusive events because the probability of drawing the ace of spades is equal to $P(A \cap S) = 1/52$. Thus the probability of drawing an ace or a spade is equal to

$$P(A \cup S) = P(S) + P(A) - P(A \cap S) = \frac{13}{52} + \frac{4}{52} - \frac{1}{52} = \frac{4}{13}$$

The word "or" in probability is always used in the inclusive sense. Thus, the phrase "A or B" is always read "A or B or both."

If one card is drawn from a well-shuffled deck of 52 cards, what is the probability that the card is a black card or a queen? Let B represent the event "drawing a black card" and Q the event "drawing a queen." There are 26 elementary events favorable to event B and each has a probability of 1/52. Thus $P(B) = 26/52$. Letting Q_s, Q_d, Q_h, and Q_c represent the queen of spades, diamonds, hearts, and clubs, respectively, the probability of drawing a queen is seen to be equal to

$$P(Q) = P(Q_s) + P(Q_d) + P(Q_h) + P(Q_c) = \frac{4}{52}$$

Since there are two black queens, the probability that the card drawn is both a queen and black is $P(B \cap Q) = 2/52$. Using the addition rule of probability, we can now compute

$$P(B \cup Q) = P(B) + P(Q) - P(B \cap Q) = \frac{28}{52}$$

4.1.3 The Multiplication Rule of Probability

It is frequently necessary to deal with probabilities for part rather than all of a sample space. The probability that a person randomly selected from a group of persons 20 years old will live to be 65 differs from the probability that a person randomly selected from the entire population will live to be 65. Similarly, the probability that a student selected at random from a group of straight-A students will pass a given statistics course would undoubtedly be higher than the probability for a student selected at random from all students at the university. In each of these examples, we are concerned with the probability of an event in a given subset of the general sample space and we note that the probability of the event in the subset may differ from that in the sample space. The probabilities associated with events in a given subset of the sample space are called *conditional probabilities.*

We shall use the sample space representing a toss of a pair of dice given in Table 4.1 to illustrate the idea of conditional probability. Let w represent the number of spots on Die I and r the number of spots on Die II. Suppose we restrict our attention to the subset B of events defined by $w + r \leq 4$. There are 6 equally likely elementary events in this subset. They are (1, 1), (1, 2), (1, 3), (2, 1), (2, 2), and (3, 1). The probability that the sum of the dots on the dice is less than or equal to 4 is equal to $P(B) = 6/36$.

Next let C represent the event defined by $w + r \leq 3$. The elementary events favorable to C are (1, 1), (1, 2), and (2, 1). Thus the probability of C is $P(C) = P(C \cap B) = 3/36$. But what if we want to determine the probability of C given that $w + r$ is less than or equal to 4? The usual notation for the probability of an event C given B is $P(C \mid B)$ where the vertical bar is read "given." Since 3 of the 6 equally likely elementary events in B are favorable to C, we find that $P(C \mid B) = 3/6$.

Consider now the event A defined by $r = 2$. The elementary events of the entire sample space favorable to A are (2, 1), (2, 2), (2, 3), (2, 4), (2, 5), and (2, 6). The probability of A is obviously $P(A) = 6/36$. What is the probability of A given B? The elementary events favorable to both A and B are (2, 1) and (2, 2). Thus $P(A \cap B) = 2/36$ and $P(A \mid B) = 2/6$.

Let us now proceed to develop a general procedure for finding conditional probabilities. The type of situation we have been dealing with is pictured in general terms in the Venn diagram in Figure 4.3.

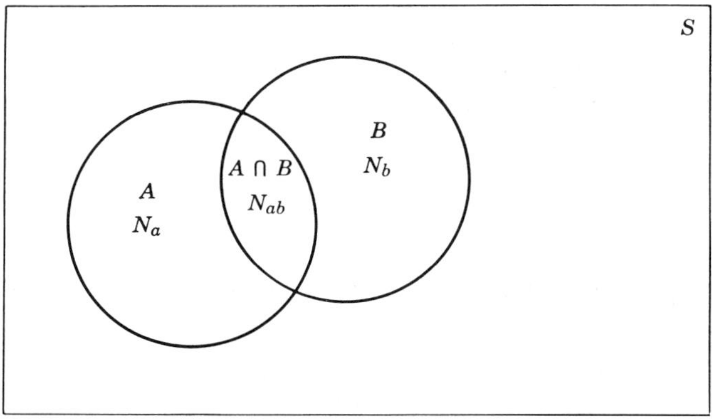

Figure 4.3 Venn diagram.

Suppose there are N equally likely outcomes in the sample space S, of which N_a are favorable to event A, N_b favorable to B and N_{ab} favorable to both A and B. Then

$$P(A) = \frac{N_a}{N} \qquad P(B) = \frac{N_b}{N}$$

$$P(A \cap B) = \frac{N_{ab}}{N} \qquad P(A \mid B) = \frac{N_{ab}}{N_b}$$

$$P(B \mid A) = \frac{N_{ab}}{N_a}$$

If $P(B)$ and $P(A \mid B)$ are known, we can obtain $P(A \cap B)$ directly by noting that

$$P(A \cap B) = P(B)P(A \mid B) = \frac{N_b}{N}\frac{N_{ab}}{N_b} = \frac{N_{ab}}{N} \tag{4.6}$$

This result is frequently referred to as the *multiplication rule* of probability. You may readily verify that $P(A \cap B)$ is also equal to the product $P(A)P(B \mid A)$.

The multiplication rule of probability is frequently used to obtain conditional probabilities. If $P(B)$ is not equal to zero, we may divide both sides of (4.6) by $P(B)$ to obtain

$$P(A \mid B) = \frac{P(A \cap B)}{P(B)} \tag{4.7}$$

Similarly, the probability of B given A is found to be

$$P(B \mid A) = \frac{P(A \cap B)}{P(A)}$$

provided that $P(A)$ is not equal to zero.

Applying the multiplication rule of probability to the dice problems presented previously, we can verify that

$$P(C \mid B) = \frac{3/36}{6/36} = \frac{3}{6} \qquad \text{and} \qquad P(A \mid B) = \frac{2/36}{6/36} = \frac{2}{6}$$

as was shown.

The general multiplication rule of probability when extended to three events A, B, and C becomes

$$P(A \cap B \cap C) = P(A)P(B \mid A)P(C \mid A \cap B) \tag{4.8}$$

Thus, if there are N equally probable elementary events, N_a of which are favorable to A, N_{ab} favorable to $A \cap B$, and N_{abc} favorable to $A \cap B \cap C$, then

$$P(A \cap B \cap C) = \frac{N_{abc}}{N} = \left(\frac{N_a}{N}\right)\left(\frac{N_{ab}}{N_a}\right)\left(\frac{N_{abc}}{N_{ab}}\right)$$

The probability $P(A \cap B \cap C)$ takes six different forms, one of which is given by (4.8), another of which is

$$P(A \cap B \cap C) = P(A)P(C \mid A)P(B \mid A \cap C)$$

It is left to the reader to derive the other four forms.

To illustrate the use of (4.8), suppose a card is drawn at random from a regular deck of 52 cards. What is the probability that the card is a heart given that the card is a red face-card? (Jacks, queens, and kings are referred to as face-cards.) Let H denote the event "heart," R the event "red," and F the event "face-card." Since $P(R)$ and $P(F \mid R)$ are both positive numbers, we have

$$P(H \mid F \cap R) = \frac{P(H \cap R \cap F)}{P(R)P(F \mid R)}$$

The probability of drawing a red card is $P(R) = 26/52$. The probability that the card is a face-card, given it is a red card, is

$$P(F \mid R) = \frac{P(F \cap R)}{P(R)} = \frac{6/52}{26/52} = \frac{6}{26}$$

The probability that the card is a heart and a red face-card is

$$P(H \cap R \cap F) = \frac{3}{52}$$

Thus

$$P(H \mid F \cap R) = \frac{3/52}{(26/52)(6/26)} = \frac{3}{6}$$

This result can be verified directly by computing the probability that a card drawn at random from the set of six red face-cards is a heart.

If the occurrence or nonoccurrence of an event A has no effect on the probability of occurrence of event B and vice versa, the events A and B are referred to as *independent events*. Suppose, for example, that a fair coin is to be flipped twice and you are trying to decide whether or not to bet on heads or tails on the second toss. Would information on the outcome of the first toss be of any value to you in making your decision? The outcome of the first toss certainly does not alter the probability that the coin will land heads on the second toss. In other words, the outcomes of each toss are independent of the outcome of the other. In this case, knowing the outcome of the first toss would be of no value when predicting the outcome of the second toss.

Stated in more rigorous terms, two events A and B are said to be independent events if

$$P(A \cap B) = P(A)P(B) \tag{4.9}$$

This definition is sometimes referred to as the multiplication rule for independent events. Comparing (4.9) with the general multiplication rule (4.6), it is obvious that, if A and B are independent events, then

$$P(A) = P(A \mid B) \qquad \text{and} \qquad P(B) = P(B \mid A)$$

If A and B are independent events with probability greater than zero, then A and B must have at least one point in common. In other words, if

$$\frac{N_a}{N} = \frac{N_{ab}}{N_b} > 0 \qquad \text{and} \qquad \frac{N_b}{N} = \frac{N_{ab}}{N_a} > 0$$

then N_{ab} must be greater than zero.

Suppose that a pair of dice is thrown. We want to show that the event B, "two spots on die I," and the event A, "two spots on die II," are independent events. The sample space of the experiment consists of the 36 points listed in Table 4.1. The points favorable to A are

$$\{(2, 1), (2, 2), (2, 3), (2, 4), (2, 5), (2, 6)\}$$

and those favorable to B are

$$\{(1, 2), (2, 2), (3, 2), (4, 2), (5, 2), (6, 2)\}$$

These events have the point (2, 2) in common, and this point has a probability of (1/36). If A and B are independent events, then $P(A)P(B)$ must be (1/36). We find that $P(A) = (6/36)$, $P(B) = (6/36)$, and $P(A)P(B) = (1/6)(1/6) = (1/36) = P(A \cap B)$ as expected.

There is a danger of confusing independent and mutually exclusive events. We have just seen that two events which are independent cannot be mutually exclusive. The condition for mutually exclusive events is that $N_{ab} = 0$. But when A and B are independent events with nonzero probability, we must have N_{ab} greater than zero. The concept of independence is not limited to two events. In the case of three events A, B, and C, the events are said to be independent if and only if:

(a) $P(A \cap B \cap C) = P(A)P(B)P(C)$.
(b) All events are independent in pairs.

Both conditions are important. It is possible that one condition may be met while the other is not. Consider, for example, the following three events defined with respect to the throw of a 10-sided die which has one of the numbers from 1 to 10 on each side:

A: An odd number
B: A number greater than 4 but less than 10
C: A number divisible without remainder by 3 or 10

In this case:

$$A = \{1, 3, 5, 7, 9\}$$
$$B = \{5, 6, 7, 8, 9\}$$
$$C = \{3, 6, 9, 10\}$$

If each number is equally likely to land up, then $P(A) = 1/2$, $P(B) = 1/2$, and $P(C) = 2/5$. The elementary event 9 is the only event common to A, B, and C; hence $P(A \cap B \cap C) = 1/10$. In this case, we also have $P(A)P(B)P(C) = (1/2)(1/2)(2/5) = 1/10$ and thus condition (a) is satisfied.

If we consider the independence of A, B, and C pairwise, we find:

$$A \cap C = \{3, 9\}, \quad P(A \cap C) = \frac{2}{10} = P(A)P(C)$$

$$B \cap C = \{6, 9\}, \quad P(B \cap C) = \frac{2}{10} = P(B)P(C)$$

$$A \cap B = \{5, 7, 9\}, \quad P(A \cap B) = \frac{3}{10} \text{ which is not equal to } P(A)P(B)$$

The events A, B, and C are therefore not independent events, since condition (b) is not satisfied.

It is also possible that condition (b) is met and condition (a) is not. Suppose, for example, that we write the numbers 1 to 12 on a set of 12 cards and proceed to draw a card at random.

Consider the events A, B, and C defined as:

$$A = \{7, 8, 9, 10, 11, 12\}$$
$$B = \{4, 9\}$$
$$C = \{1, 3, 6, 7, 9, 12\}$$

In this case $P(A) = 1/2$, $P(B) = 1/6$ and $P(C) = 1/2$. Checking for pairwise independence, we find that:

$$A \cap B = \{9\}, \quad P(A \cap B) = \frac{1}{12} = P(A)P(B)$$

$$A \cap C = \{7, 9, 12\}, \quad P(A \cap C) = \frac{3}{12} = P(A)P(C)$$

$$B \cap C = \{9\}, \quad P(B \cap C) = \frac{1}{12} = P(B)P(C)$$

Thus, A, B, and C are pairwise independent and condition (b) is satisfied. Condition (a), on the other hand, is not met, since $A \cap B \cap C = \{9\}$ and $P(A \cap B \cap C) = 1/12$ which is not equal to $P(A)P(B)P(C) = (1/2)(1/6)(1/2) = 1/24$. Thus A, B, and C are not independent events.

4.1.4 Some Probability Problems

The probability examples discussed so far have been extremely simple. In practical problems, the probability models describing very simple situations can be extremely complicated. Full consideration of such models is beyond the scope of this book, but we are ready to consider more difficult problems than those discussed previously.

Example 1: Suppose a tomato is selected at random from a truck in the New York wholesale market. We classify the tomato as good (G) or bad (B), put it back, select a second tomato, and continue the process until three tomatoes have been inspected. In this sample of three, we find two good tomatoes and one bad. What is the probability of this result, if 10 percent of the tomatoes in the truck are bad?

Let the random variable X measure the number of bad tomatoes.[4] Thus X can be 0, 1, 2, or 3. The observed outcome was $X = 1$. The probability of selecting a bad tomato on a particular draw is $p = 0.10$. The probability of a good tomato is therefore $(1 - p) = 0.90$. There are three different ways in which one bad tomato could occur in a sample of three. Either the first or second or third tomato could have been bad. The events $B \cap G \cap G$, $G \cap B \cap G$, and $G \cap G \cap B$ are mutually exclusive events. Therefore, if we know the probability of each of these events, the probabilities can be added up to find that of "one bad tomato in a sample of size three." In other words, the probability is that $X = 1$. The multiplication rule for independent events can be used to find the probabilities of $B \cap G \cap G$, $G \cap B \cap G$, and $G \cap G \cap B$ because the outcome of the first draw does not affect the outcome of the second draw, nor does the outcome of the first two draws affect the outcome of the third. Therefore, we have

$$\begin{aligned} P(B \cap G \cap G) &= p(1 - p)(1 - p) = (0.10)(0.90)(0.90) &&= 0.081 \\ P(G \cap B \cap G) &= (1 - p)p(1 - p) = (0.90)(0.10)(0.90) &&= 0.081 \\ P(G \cap G \cap B) &= (1 - p)(1 - p)p = (0.90)(0.90)(0.10) &&= 0.081 \end{aligned}$$

and

$$P(X = 1) = P(B \cap G \cap G) + P(G \cap B \cap G) + P(G \cap G \cap B) = 0.243$$

What is the probability of one good and two bad tomatoes in a sample of size three?

Example 2: In a product-testing procedure, each radio on an assembly line must pass two inspection points before being packaged for shipment. The probability is $p_1 = 0.6$ that a defective radio is detected at the first inspection point and $p_2 = 0.7$ that a defective radio is detected at the second inspection point. What is the probability that a defective radio will be packaged for shipment?

A defective radio has to pass both inspection points before it is packed. The probability that such a radio will pass the first inspection point is $P(A) = (1 - p_1) = 0.40$. The probability of passing the second inspection point, given that the defective radio passed the first point, is $P(B \mid A) = (1 - p_2) = 0.30$. The probability that a defective radio will pass both inspection points is, therefore,

$$P(A \cap B) = (1 - p_1)(1 - p_2) = (0.40)(0.30) = 0.12$$

[4] The term random variable will be defined precisely later. For the moment, X can be thought of as representing the outcome of the experiment which consists of selecting three tomatoes from the truck.

We conclude that about 12 percent of the defective radios will be packaged for shipment under the present inspection plan.

Example 3: In an experiment with an insecticide for flies, the Black Cloud Fly Spray Company finds that 80 percent are killed on the initial application but that those which survive develop a resistance. The percentage of the survivors killed in further applications is one-half as large as the percentage killed on the immediately preceding application. What is the probability that a fly will survive four applications given that it survived the first two?

Let K_i denote that the fly was killed on the ith application of the fly spray and S_i denote survival on the ith application; then

$$\begin{aligned} P(K_1) &= 0.8 & P(S_1) &= 0.2 \\ P(K_2 \mid S_1) &= 0.4 & P(S_2 \mid S_1) &= 0.6 \\ P(K_3 \mid S_2) &= 0.2 & P(S_3 \mid S_2) &= 0.8 \\ P(K_4 \mid S_3) &= 0.1 & P(S_4 \mid S_3) &= 0.9 \end{aligned}$$

and

$$P(S_3 \cap S_4) = P(S_4 \mid S_3)P(S_3 \mid S_2)P(S_2) = (0.9)(0.8)(1.0) = 0.72$$

This is an extension of the multiplication rule for independent events.

What is the probability that a fly will survive the first four applications? (*Answer.* 0.0864.)

4.2 PERMUTATIONS AND COMBINATIONS

Many experiments have a large number of possible outcomes. In such cases, it is usually difficult to list and count the sample points. For example, suppose an experiment consists of drawing five cards without replacement from a deck of 52 cards. What is the probability of drawing four aces? There are 48 different five-card hands with four aces and 2,598,960 different ways to draw five cards out of a deck of 52. Thus the probability of four aces is 1/54,145. In such problems there is often a need for convenient, general methods to deal with the large numbers involved. The principles of permutations and combinations are commonly used to provide the needed shortcuts in counting.

4.2.1 Permutations

Suppose a salesman wishes to go by plane from New York to Chicago and then on to Denver. He finds there are two flights from New York to Chicago and three from Chicago to Denver that fit his schedule. How many different ways can he fly to Denver?

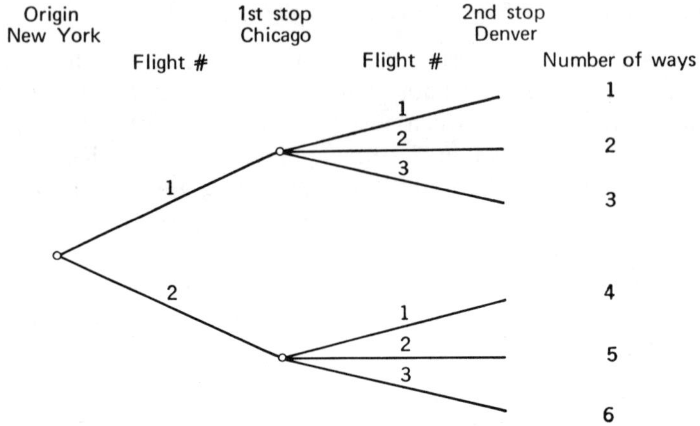

Figure 4.4 Tree diagram of different ways to Denver.

The possibilities are shown as a *tree diagram* in Figure 4.4. We find there are six different ways to fly to Denver. (If there are four flights from Denver to San Francisco, how many different ways can the salesman fly from New York to San Francisco?)

When the salesman arrives in Denver, he finds there are two other salesmen waiting to see his client, Mr. Jones. In how many different sequences could Mr. Jones meet separately with these three salesmen? The possibilities are again shown as a tree diagram in Figure 4.5.

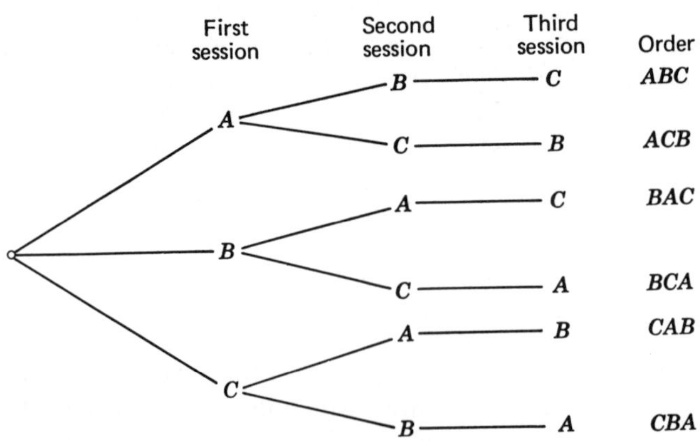

Figure 4.5 Tree diagram of different sequences in which to meet salesmen.

There are six different ways in which Mr. Jones could meet the salesmen. He can select any of the three to begin with; he then has two to select from for the second session, and one left to talk with during the third meeting. Note that the tree diagram takes order into account.

In both of these cases, we were interested in the number of *different* ways to do something. More precisely we were interested in the number of permutations.

A *permutation* is an arrangement of objects in a definite order. The same objects arranged in a different order constitute a different permutation.

Tree diagrams can be used to show the various definite orders, but they are inconvenient when there is a large number of objects. Fortunately, a shortcut multiplication method is available that is especially useful when dealing with large numbers.

The multiplication method states that if an operation can be performed in n_1 ways, after which a second operation can be performed in n_2 ways, after which a third operation can be performed in n_3 ways, and so on for k operations, then the k operations can be performed together in

(4.10) $$(n_1)(n_2)(n_3) \cdots (n_k) \text{ different ways}$$

In the case of the salesman flying to Denver, there were $n_1 = 2$ ways to fly from New York to Chicago and $n_2 = 3$ ways to go from Chicago to Denver. Thus there were $(n_1)(n_2) = 6$ ways to fly from New York to Denver. In the case of Mr. Jones meeting the three salesmen, there were $n_1 = 3$ salesmen to select from for the first meeting, $n_2 = 2$ for the second, and $n_3 = 1$ for the last interview. Thus there were $(n_1)(n_2)(n_3) = 6$ different ways to meet the salesmen.

As a third example, suppose a telephone company decides the first three numbers in a telephone number are to be 232. How many seven-number phone numbers of the form 232-_ _ _ _ are possible? If zero is allowed in any of the last four places, there are ten ways to fill each space or (10)(10)(10)(10) = 10,000 different numbers possible.

In many probability problems it is convenient to use a special form of the multiplication method to compute the number of permutations. This special form uses factorial notation.

The product of all whole numbers from 1 to n is called *n factorial*, and is denoted by $n!$. Thus,

$$n! = (n)(n-1)(n-2)(n-3) \cdots (3)(2)(1)$$

For example, five factorial is equal to $5! = (5)(4)(3)(2)(1) = 120$.

Factorials increase in size very rapidly. Seven factorial, for example, is equal to 5040 while ten factorial is 3,628,800. In many probability problems, one factorial is divided by another. In such cases, the number of required multiplications can usually be reduced by canceling. Suppose that the probability of an event B is equal to

$$P(B) = \frac{8!}{9!}$$

Noting that $(n + 1)! = (n + 1)(n!)$, this is quickly reduced to

$$P(B) = \frac{8!}{(9)(8!)} = \frac{1}{9}$$

by canceling.

The number of permutations of a set of n different objects, taken all together, is equal to $n!$ and is denoted by ${}_nP_n$. Symbolically, we have

$${}_nP_n = n!$$

This result is simply a special case of the multiplication method summarized by equation (4.10). To illustrate this, imagine n empty spaces in a parking lot and that there are n cars waiting to be parked. Any of the n cars can be parked in the first space. There are then $(n - 1)$ cars that can be parked in the second space. Since each way the first space can be filled can be combined with each way the second space can be filled, there are $(n)(n - 1)$ ways to fill the first two spaces. There are $n - 2$ ways to fill the third space. Thus, there are $(n)(n - 1)(n - 2)$ ways to fill the first three spaces. This line of reasoning is continued down to the last parking space. We thereby conclude that there are

$$(n)(n - 1)(n - 2) \cdots (2)(1) = n!$$

different ways to fill n spaces with n objects.

Suppose, however, that there are only r parking spaces, where r is a number less than or equal to the number of cars n. We want to fill these r spaces with r cars chosen from all n cars. The first space can be filled in n ways, the second in $(n - 1)$ ways, and so on. Since there are r spaces, we have r consecutive factors starting with n and going down to $(n - r + 1)$. In other words, the number of permutations of a set of n different objects, taken r at a time, is equal to

$${}_np_r = (n)(n - 1)(n - 2) \cdots (n - r + 1)$$

If we multiply ${}_np_r$ by $(n - r)!/(n - r)!$, we obtain

$$(4.11) \qquad {}_np_r = \frac{n!}{(n - r)!}$$

where r is less than or equal to n. The formula (4.11) applies for $r = n$ because, by convention, we define $0! = 1$. For example, if $n = 3$ and $r = 3$, we have ${}_3P_3 = 3!/(3 - 3)! = 3!/0! = 3!/1 = 6$.

Suppose that six new men have been hired by the Mighty Radio Corporation. The production manager decides to station three of these men at three different locations on the assembly line. In how many different ways can the new men be assigned to the three locations? In other words, how many permutations are there of six men, taken three at a time?

In this case,

$$n! = (6)(5)(4)(3)(2)(1) = 720$$
$$(n-r)! = (6-3)! = 3! = (3)(2)(1) = 6$$

and

$${}_6P_3 = (720/6) = 120$$

Notice that the computations can be shortened considerably if, where possible, numbers are cancelled before beginning to multiply. Thus,

$${}_6P_3 = \frac{(6)(5)(4)(\not{3})(\not{2})(\not{1})}{(\not{3})(\not{2})(\not{1})} = (6)(5)(4) = 120$$

Let the six men be denoted by the letters *A*, *B*, *C*, *D*, *E*, and *F*, and the three locations on the assembly line by the Roman numerals I, II, and III. What is the probability of the event, say, *Z*, that location *I* is filled by *A*, II by *B*, and III by *C*?

There are ${}_6P_3 = 120$ different ways in which the three locations can be filled, only one of which corresponds to event *Z*. Therefore, we have

$$P(Z) = \left(\frac{1}{120}\right)$$

In other words, there is one chance in 120 that the event *Z* will occur.

The multiplication rule of probability could also be used to find the probability of *Z*. We know that:

1. $p_1 = P(A \text{ at I}) = (1/6)$ because *A* is 1 of 6 men.
2. $p_2 = P(B \text{ at II} \mid A \text{ at I}) = (1/5)$ because *B* is 1 of 5 men left after *A* is located at I.
3. $p_3 = P(C \text{ at III} \mid B \text{ at II and } A \text{ at I}) = (1/4)$.

Therefore the probability of event *Z* is

$$P(Z) = (p_1)(p_2)(p_3) = \left(\frac{1}{6}\right)\left(\frac{1}{5}\right)\left(\frac{1}{4}\right) = \left(\frac{1}{120}\right)$$

The general rule for finding the number of permutations may be stated as follows:

If there are n objects, of which n_1 are alike, another n_2 are alike, another

n_3 are alike, and so on, for k kinds of objects; then the number of permutations of n objects, taken all together, is

$$(4.12) \qquad {}_nP(n_1, n_2, n_3, \ldots, n_k) = \frac{n!}{n_1!\,n_2!\,n_3!\ldots n_k!}$$

where $n_1 + n_2 + \cdots + n_k = n$.

For example, if 5 coins are tossed, in how many ways can we get 3 heads? In this case, $n = 5$ and we have $n_1 = 3$ coins that are alike in the sense that they all have the characteristic heads (H). Thus there are $n_2 = 2$ *tails* (T) and

$${}_5P(3, 2) = \frac{5!}{3!\,2!} = 10$$

ways in which we can obtain 3 heads and 2 tails when tossing 5 coins. Listing the possible ways, we obtain

HHHTT, HHTHT, HTHHT, THHHT, HHTTH, HTHTH, THHTH, HTTHH, THTHH, TTHHH

(If 4 coins are tossed, in how many ways can we get 2 heads?)

If all n objects are different, we have $n_1 = n_2 = \cdots = n_k = 1$, in which case

$${}_nP(1, 1, 1, \ldots, 1) = {}_nP_n = n!$$

If, on the other hand, only r of the objects are different while $(n - r)$ objects are alike, we have

$${}_nP(\underbrace{1, 1, \ldots, 1}_{r \text{ one's}}, n - r) = {}_nP_r = \frac{n!}{(n - r)!}$$

Note that the $(n - r)$ objects may be alike in the sense that they are not among the r objects that are being arranged.

4.2.2 Combinations

Suppose we have n different objects and want to select r of them without regard to the order in which the r objects are selected. A selection without regard to order is called a *combination*. For example, consider the 3 letters A, B, and C. In how many ways can we arrange these letters 2 at a time? If we take order into account, there are ${}_3P_2 = (3!/1!) = 6$ different ways to arrange the 3 letters taken 2 at a time. These are AB, AC, BA, BC, CA, and CB. If the order in which the objects are arranged is ignored, there are only 3 ways to select 2 letters from A, B, C. The three combinations are AB, AC, and BC. In other words, when order is ignored, AB is identical to BA, AC identical to CA, and BC identical to CB.

The formula for finding the number of combinations of a set of n different objects, taken r at a time, is derived from the rule for finding the number of permutations. A set of r objects gives rise to $r!$ permutations. The total number of permutations of n objects taken r at a time is therefore divided by $r!$ to determine the number of possible selections in which the order of the objects is not considered. For example, there are 120 ways to select three books from a set of six if the order is taken into account. Each group of three books gives rise to six permutations. Thus, there are $(120/6) = 20$ different sets of three books that can be obtained from a set of six. Using general notation, the number of combinations of n different objects, taken r at a time, is

$$ {}_nC_r = \frac{n!}{(r!)(n-r)!} \tag{4.13} $$

Other common abbreviation for $_n C_r$ are $\binom{n}{r}$ and C_r^n.
In the case of the six books, taken three at a time, we have

$$ \binom{n}{r} = \binom{6}{3} = {}_6C_3 = \frac{6!}{3!3!} = 20 $$

As a second example, suppose a committee of two is selected from a group of four people. How many different committees are possible? Using the combinations formula, we obtain ${}_4C_2 = 6$. If we denote the four people by the letters A, B, C, and D, the six possible committees are AB, AC, AD, BC, BD, and CD. The 12 ways in which these committees could be selected are listed below.

Committee	*Permutations Giving Rise to the Committee*	
AB	AB	BA
AC	AC	CA
AD	AD	DA
BC	BC	CB
BD	BD	DB
CD	CD	DC

(Construct a similar table for committees of three persons selected from the group of four.)

Before turning to several probability problems involving permutations and combinations, let us note that

$$ {}_nC_{(n-r)} = {}_nC_r $$

In words, the number of combinations of n things taken $(n-r)$ at a time is the same as the number taken r at a time.

4.2.3 Some Probability Problems

Example 1: Suppose four coins are thrown and we want to compute the probability of observing two heads and two tails. There are $2^4 = 16$ different ways in which the coins could fall. The number of cases favorable to two heads is

$$_4P(2, 2) = \frac{4!}{2!2!} = 6$$

Therefore the probability of two heads is $P(2 \text{ heads}) = (6/16) = (3/8)$.

The general rule for coin-tossing probabilities can be stated: if a fair coin is thrown n times, the probability of r heads is

$$P(r \text{ heads}) = {}_nC_r\left(\frac{1}{2}\right)^n$$

Example 2: A box of ten radio tubes contains five bad tubes and five good ones. A radio serviceman randomly selects four of the tubes and goes out to repair a radio. What is the probability that all four of the tubes are bad?

There are $_{10}C_4$ different ways in which the serviceman can select four tubes without regard to the order in which they are selected. There are $_5C_4$ ways to select four bad tubes from the set of five bad tubes. The probability of four bad tubes is therefore, equal to

$$P(4 \text{ bad tubes}) = \left(\frac{_5C_4}{_{10}C_4}\right) = \left(\frac{1}{42}\right)$$

This could also be found by using the multiplication rule of probability. The serviceman selects the four tubes without replacement. The probability that the first tube selected is bad is $p_1 = (5/10)$. If the first tube is bad, there are four bad tubes and five good tubes left. The probability that the second tube is bad is therefore $p_2 = (4/9)$. Given the first two tubes are bad, the probability that the third tube is bad is $p_3 = (3/8)$, and given the first three are bad, the probability that the fourth one is bad is $p_4 = (2/7)$. The probability of four consecutive bad tubes is therefore

$$P(4 \text{ bad tubes}) = p_1 p_2 p_3 p_4 = \left(\frac{1}{42}\right)$$

Keep in mind that p_2, p_3, and p_4 are conditional probabilities.

What is the probability that the serviceman will select exactly one good and three bad tubes? Again, there are $_{10}C_4 = 210$ different ways to select four tubes without regard to order. There are $_5C_3 = 10$ ways to select three bad tubes from a set of five bad ones and ${}_5C_1 = 5$ ways to select one good tube from a set of five good ones. Any set of three bad tubes can be selected with any of the five ways to select a good tube; thus there are

$$({}_5C_3)({}_5C_1) = 50 \text{ different ways}$$

to select three good tubes and one bad tube. The probability of selecting exactly one good tube in a set of four tubes is therefore equal to

$$P(1 \text{ good and } 3 \text{ bad tubes}) = \left(\frac{50}{210}\right) = \left(\frac{5}{21}\right)$$

The multiplication and addition rules of probability could also be used to find the probability of one good (G) and three bad (B) tubes. The multiplication rule would be used to find the probabilities $P(G \cap B \cap B \cap B)$, $P(B \cap G \cap B \cap B)$, $P(B \cap B \cap G \cap B)$, and $P(B \cap B \cap B \cap G)$. The sum of these probabilities is (5/21).

In general, if a population has n elements of which n_1 have characteristic A and $n_2 = n - n_1$ have characteristic B, then the probability of selecting exactly k elements with characteristic A in a random selection of r elements is equal to

$$P(k \text{ of } A) = \frac{[{}_{n_1}C_k][{}_{n_2}C_{r-k}]}{{}_nC_r} \tag{4.14}$$

The set of probabilties obtained from $P(k$ of $A)$ as k ranges from zero to n_1 or r, whichever is smaller, is called the hypergeometric distribution.

In the radio tube example, we wanted the probability of three good tubes in a random selection of four. Thus,

$$n_1 = 5 \qquad n_2 = 5 \qquad k = 3$$
$$n = 10 \qquad r = 4 \qquad (n - k) = 1$$

Example 3: The five employees of XYZ Corporation all drive to work and park their cars in one of the three parking lots near their office. It is assumed that the employees have no particular preference for any of the three lots. What is the probability that, on any given day, there will be two cars in one lot, two cars in another, and one car in the third?

Each of the five cars could be parked in any of the three different lots. Using the multiplication method (4.10), we find there are $n = 3^5 = 243$ different ways in which the cars could be parked. (An operation consists of parking a car, and there are three different "ways" to perform this operation. Each "way" corresponds to one of the parking lots.)

For convenience, number the parking lots I, II, and III and consider the event A that 2 cars are in I, 2 cars and in II, and 1 car is in III. Using the general rule for the number of permutations (4.12), we find there are

$$m = \frac{5!}{2!2!1!} = 30 \text{ different ways}$$

in which event A could occur. (A car is an object, and two cars are "alike" if they are located in the same parking lot.)

If the parking lots are selected at random by the drivers, then each of these 30 ways is equally likely and has a probability of $(1/n) = (1/243)$. The probability of event A is therefore $P(A) = (30/243) = (10/81)$.

Define the event B as 2 cars in I, 1 car in II, and 2 cars in III. The event C can be defined as 1 car in I, 2 cars in II, and 2 cars in III. Events B and C have the same probability as event A.

The events A, B, and C are mutually exclusive and exhaust the set of events consisting of 2 cars in one lot, 2 cars in another, and 1 car in the third (without regard to order). The addition rule of probability (4.4) therefore applies. We have

$$P(A) + P(B) + P(C) = (3)\left(\frac{10}{81}\right) = \left(\frac{10}{27}\right)$$

* 4.3 ALTERNATIVE APPROACHES TO PROBABILITY

The definition (4.1) of probability as a ratio of two known numbers is a useful one in many situations but is difficult to apply at times. To illustrate why, consider an experiment that consists of throwing a coin and observing the occurrence of an event H defined as the coin landing heads. In this case, it seems reasonable to assume there are $N = 2$ different, equally likely outcomes, one of which corresponds to the event H. The probability of a head is therefore $P(H) = (1/2)$. But what if a thumbtack is thrown instead of a coin? Again there are only two possible outcomes: (1) the thumbtack can either be point up (U), or (2) point down (D). (The thumbtack is down if the edge of the head and the tip of the thumbtack both touch the table.) It is known that $P(\mathrm{U}) = 1 - P(D)$ but neither $P(U)$ nor $P(D)$ is known because there is no reason to believe that "up" and "down" are equally likely events. There are several alternative approaches to probability that could be used in such cases.

4.3.1 The Relative Frequency Approach

Suppose that an experiment E can be repeated over and over again under uniform conditions. Furthermore, let us assume that the experiment has an

* This section may be omitted without loss of continuity.

outcome A that has an unknown probability $P(A)$. Suppose that the experiment is repeated n times and that the outcome A is observed m times. The relative frequency m/n could be used as an estimate of the true probability $P(A)$. In many cases it seems reasonable to assume that the relative frequence m/n will become stable and tend to approach the true probability $P(A)$ as the number of trials n becomes large. In such cases, we could define the probability of the event A as a number $P(A)$ which the relative frequency m/n tends to approach as n becomes large. The true probability $P(A)$ is unknown but can be estimated by the relative frequency m/n. The estimated value of $P(A)$ in this case is based on experience rather than abstract reasoning. For example, suppose an experiment consists of tossing a thumbtack. This experiment has two outcomes; the tack either lands up (U) or down (D). If the experiment is performed $n = 10{,}000$ times and the thumbtack lands up $m = 1500$ times, then our estimate of the probability $P(U)$ is $m/n = 1500/10{,}000 = 0.15$. Since n is large, this estimate is expected to be very close to the true probability that the thumbtack will land up.

This experimental approach could also be used to estimate the probability that a coin will land heads. Such an estimate would allow us to check the assumption that heads and tails are equally likely. The main difficulty with using the experimental approach to estimate probabilities is that it is very time-consuming and often costly. Probabilities are generally computed by using definition (4.1) when it is possible to do so.

4.3.2 The Subjective Approach

The subjective (or personalistic) approach to probability is very flexible and may be applied in a wide variety of situations by those who accept the view that "probability measures the confidence that a particular individual has in the truth of a particular proposition."[5] This is a rather nebulous definition of probability in the sense that different members of the "subjective school," when faced with the same evidence, may assign different values to the probability of an event. The probabilities they use are not necessarily based on abstract reasoning or the outcomes of repeated experiments. A "personalist" may talk about the probability of a steel strike next month or the probability that it will rain tomorrow. Such probabilities are not based on abstract reasoning and, given the irreversible nature of time, repeated experiments involving such events are impossible. Indeed, the same individual, at a different point in time, may assign a new probability to an event after new evidence is taken into account. (Provided, of course, that he is a "personalist.")

[5] L. J. Savage, *Foundations of Statistics* (New York: John Wiley and Sons, Inc., 1954), p. 3.

The subjective approach to probability has become increasingly popular in recent years. Investigators using this approach are often able to provide more realistic solutions to problems than would otherwise be possible. Keep in mind that the basic rules of probability apply to both "objective" and "subjective" probabilities. The flexibility of the subjective approach is due primarily to the fact that the "personalist" is willing to assign "probabilities" to many events that, from the classical viewpoint, do not have probabilities associated with them.

In some cases, it is useful to have a systematic procedure by which to revise probabilities as additional information accumulates. One such procedure will be presented below.

Suppose we have three identical urns. Urn E_1 contains two black balls, urn E_2 two red balls, and urn E_3 one red and one black ball. An urn is selected at random and a ball is then taken from the urn. After selecting an urn, but before selecting a ball, we could ask, "What is the probability that urn E_1 (or E_2 or E_3) has been selected?" In the strict classical sense, there is no probability associated with an event that has already occurred; either urn E_1 has been selected or it has not. The identity of the urn is still unknown, however, and a personal probability could be attached to the possibility that urn E_1 was selected. The probability that urn E_1 (or E_2 or E_3) would be selected prior to the actual selection is (1/3). It seems reasonable, then, to say that the probability remains equal to (1/3) until the urn's identity is actually known (reasonable, that is, if you believe in personal probabilities). We therefore have

$$P_1(E_1) = P_1(E_2) = P_1(E_3) = \left(\frac{1}{3}\right)$$

These are called *prior probabilities.*

A ball is drawn at random from the chosen urn and found to be red. The prior probabilities can now be adjusted to take this new information into account. A red ball could not have been drawn from urn E_1. Therefore the probability is now zero that urn E_1 was selected. The probability of drawing a red ball (event A) from urn E_2 is $P(A \mid E_2) = 1$. The probability of drawing a red ball from urn E_3 is $P(A \mid E_3) = (1/2)$. Thus it seems reasonable, on the basis of the information obtained by drawing a red ball, to consider the probability that urn E_2 was selected to be twice as large as the probability for urn E_3. The new probabilities are

$$P_2(E_1) = 0,\ P_2(E_2) = \left(\frac{2}{3}\right), \qquad \text{and} \qquad P_2(E_3) = \left(\frac{1}{3}\right)$$

these probabilities are called *posterior probabilities.*

Assume that the original ball is replaced and a second ball is drawn which is also red. The probability of two consecutive red balls from urn E_2 is equal to 1 while the probability of two consecutive red balls from urn E_3 is equal to (1/4). The new posterior probabilities are therefore

$$P_3(E_1) = 0,\ P_3(E_2) = \left(\frac{4}{5}\right), \qquad \text{and} \qquad P_3(E_3) = \left(\frac{1}{5}\right)$$

The ball is replaced and a third selection made which turns out to be black (event B). The probability that urn E_2 was selected now becomes 0, and the new posterior probabilities are

$$P_4(E_1) = 0,\ P_4(E_2) = 0, \qquad \text{and} \qquad P_4(E_3) = 1$$

We have been dealing with a situation in which the probabilities change as actions are taken and new information is obtained. The procedure used to obtain the posterior probabilities is based on Bayes' theorem.

Bayes' theorem. Let $E_1, E_2, \ldots E_k$ be mutually exclusive events such that $P(E_1) + P(E_2) + \cdots + P(E_k) = 1$.

Let A be an arbitrary event which is known to occur in conjunction with or as a consequence of one of the events E_i. The probability of the occurrence of event E_i, given the occurrence of A, is

$$P(E_i \mid A) = \frac{P(E_i \cap A)}{P(A)} = \frac{P(E_i)P(A \mid E_i)}{P(A)} \tag{4.15}$$

where $P(A)$ is equal to $\sum_{i=1}^{k} P(E_i \cap A)$.

This theorem was used as follows.

Let E_1 be the event "selecting urn E_1"
E_2 be the event "selecting urn E_2"
E_3 be the event "selecting urn E_3"
A be the event "selecting a red ball"
B be the event "selecting a black ball"

Initially, we have

$$P(E_1 \cap A) = P_1(E_1)P(A \mid E_1) = \left(\frac{1}{3}\right)(0) = 0$$

$$P(E_2 \cap A) = P_1(E_2)P(A \mid E_2) = \left(\frac{1}{3}\right)(1) = \frac{1}{3}$$

$$P(E_3 \cap A) = P_1(E_3)P(A \mid E_3) = \left(\frac{1}{3}\right)\left(\frac{1}{2}\right) = \frac{1}{6}$$

$$P(A) = \frac{1}{2}$$

Therefore,

$$P_2(E_1) = P(E_1 \mid A) = \frac{P(E_1 \cap A)}{P(A)} = 0$$

$$P_2(E_2) = P(E_2 \mid A) = \frac{P(E_2 \cap A)}{P(A)} = \frac{(1/3)}{(1/2)} = \frac{2}{3}$$

$$P_2(E_3) = P(E_3 \mid A) = \frac{P(E_3 \cap A)}{P(A)} = \frac{(1/6)}{(1/2)} = \frac{1}{3}$$

After the second draw, the probabilities are recomputed and we have

$$P(E_1 \cap A) = P_2(E_1)P(A \mid E_1) = 0$$

$$P(E_2 \cap A) = P_2(E_2)P(A \mid E_2) = \left(\frac{2}{3}\right)(1) = \frac{2}{3}$$

$$P(E_3 \cap A) = P_2(E_3)P(A \mid E_3) = \left(\frac{1}{3}\right)\left(\frac{1}{2}\right) = \frac{1}{6}$$

$$P(A) \qquad = \left(\frac{5}{6}\right)$$

which gives us

$$P_3(E_1) = P(E_1 \mid A) = \frac{P(E_1 \cap A)}{P(A)} = 0$$

$$P_3(E_2) = P(E_2 \mid A) = \frac{P(E_2 \cap A)}{P(A)} = \frac{(2/3)}{(5/6)} = \frac{4}{5}$$

$$P_3(E_3) = P(E_3 \mid A) = \frac{P(E_3 \cap A)}{P(A)} = \frac{(1/6)}{(5/6)} = \frac{1}{5}$$

After the third draw, the new probabilities are computed

$$P(E_1 \cap B) = P_3(E_1)P(B \mid E_1) = (0)(1) = 0$$

$$P(E_2 \cap B) = P_3(E_2)P(B \mid E_2) = \left(\frac{4}{5}\right)(0) = 0$$

$$P(E_3 \cap B) = P_3(E_3)P(B \mid E_3) = \left(\frac{1}{5}\right)\left(\frac{1}{2}\right) = \frac{1}{10}$$

$$P(B) = \left(\frac{1}{10}\right)$$

and the posterior probabilities become

$$P_4(E_1) = P(E_1 \mid B) = \frac{P(E_1 \cap B)}{P(B)} = 0$$

$$P_4(E_2) = P(E_2 \mid B) = \frac{P(E_2 \cap B)}{P(B)} = 0$$

$$P_4(E_3) = P(E_3 \mid B) = \frac{P(E_3 \cap B)}{P(B)} = \frac{(1/10)}{(1/10)} = 1$$

The preceding example illustrates a method by which probabilities can be continually adjusted as new information is obtained. If a red ball had been obtained on the third draw, the posterior probability $P_4(E_2)$ would have been equal to (8/9) and $P_4(E_3)$ equal to (1/9). If a fourth draw also produced a red ball, then

$$P_5(E_2) = \left(\frac{16}{17}\right) \quad \text{and} \quad P_5(E_3) = \left(\frac{1}{17}\right)$$

After drawing four straight red balls, we would be fairly certain that the urn was E_2. After drawing ten straight red balls, we would be *almost* completely certain. But the possibility that the urn is urn E_3 can never be completely ruled out as long as red balls continued to be drawn. Notice, on the other hand, that once a black ball was drawn (after a sequence of red balls), we were able to conclude that the urn was definitely not E_2. Further selections would provide no new information about the identity of the urns once a black ball was drawn.

A better understanding of the preceding example can be obtained by studying the following, shorter, somewhat simpler illustration of Bayes' Theorem.

A radio assembly plant has two assembly lines. Line I produces 80 percent of the radios and line II produces 20 percent. On the average, 5 percent of the radios assembled on line I and 10 percent of those assembled on line II are defective. A radio selected at random from those ready to be shipped is defective. What is the probability that it was asembled on line I? On line II?

Using the notation of Bayes' theorem, the events A, E_1, and E_2 are:

A: the radio is defective
E_1: the radio was produced on line I
E_2: the radio was produced on line II

The probability that a radio was produced on line I is $P(E_1) = 0.80$. The probability that the radio is defective *given* that it was produced on line I is $P(A \mid E_1) = 0.05$. The probabilities for line II are $P(E_2) = 0.20$ and $P(A \mid E_2) = 0.10$.

The probability that a radio was assembled on line I *and* is defective is

$$P(E_1 \cap A) = P(E_1)P(A \mid E_1) = (0.8)(0.05) = 0.04$$

while the similar probability for line II is

$$P(E_2 \cap A) = P(E_2)P(A \mid E_2) = (0.2)(0.10) = 0.02$$

The probability of a defective item from either line I or line II is

$$P(A) = P(E_1 \cap A) + P(E_2 \cap A) = 0.06$$

Prior to knowing that the selected radio is defective, the probabilities that it was produced on lines I and II, respectively, are $P(E_1) = 0.80$ and $P(E_2) = 0.20$. Bayes' theorem can be used to obtain posterior probabilities from these prior probabilities once is it known that the radio is defective. The new probabilities are

$$P(E_1 \mid A) = \frac{P(E_1 \cap A)}{P(A)} = \left(\frac{0.04}{0.06}\right) = 0.67$$

$$P(E_2 \mid A) = \frac{P(E_2 \cap A)}{P(A)} = \left(\frac{0.02}{0.06}\right) = 0.33$$

EXERCISES

Section 4.1

1. The following mortality table shows the number of survivors to various ages of 100,000 new-born white American males.

Age x	*Survivors to Age x*
0	100,000
10	93,601
20	92,293
30	90,092
40	86,880
50	80,521
60	67,787
70	46,739
80	19,860
90	2,812
100	65

Using these figures:
(a) Estimate the probability of a new-born infant living to be 60 years old.
(b) Estimate the probability of a 20-year-old living until he is 50 years old.
(c) How would you estimate the probability of living to be 65?

2. We have two urns. Urn A contains two black balls and one white ball. Urn B contains two white balls and one black ball. An urn is chosen at random and two balls are drawn successively from the urn chosen.
   (a) What is the probability of getting two white balls if the balls are drawn *without* replacement?
   (b) What is the probability of getting two black balls if the balls are drawn *with* replacement?
3. Four men match coins simultaneously to see who pays for coffee. The odd man pays. What is the probability that one trial will suffice to determine the odd man?
4. A bowl contains 200 poker chips. 100 are blue, 60 are red, and 40 are white. A series of five chips are drawn with replacement.
   (a) What is the probability that the chips will be in the order RNRNR, where R is red and N is "not red?"
   (b) What is the probability that all of the chips will be blue?
   (c) What is the probability of two red and three white chips?
5. The average duck hunter has a probability of 0.30 of hitting a duck on any given shot. Four average duck hunters surround a small duck pond. Each hunter has one shell. Only one duck comes within shooting range during the morning. What is the probability that the hunters have duck for lunch if each hunter shoots at the duck?

Section 4.2

6. A group of eight Republicans and four Democrats attend a city council meeting. A committee of four persons is selected at random from this group. What is the probability that the committee will consist of exactly two Republicans and two Democrats?
7. In how many ways can five different books be arranged on a shelf?
8. Suppose you have a 1¢, 5¢, 10¢, and 50¢ piece. How many different sums of money can you form from these four coins?
9. A marketing research company wants to determine customer reactions to three different types of packages, four package colors, and five package sizes. For each market test, it selects a package of a certain type, color, and size. How many market tests will be necessary in order to test all possible package variations?
10. A bowl contains 15 poker chips. 8 are blue, 4 are red, and 3 are white. If 6 poker chips are selected at random, what is the probability of getting exactly 3 blue, 1 red, and 2 white chips?

Section 4.3

11. A dairy company has two bottle washers. Washer A washes 20 percent of all bottles used by the company each day and chips 4 percent of the bottles it washes. Washer B washes the remaining bottles and chips 2 percent of those it washes.

(a) What is the probability that a washed bottle selected at random is chipped?
(b) A washed bottle selected at random is chipped. What is the probability that it was washed by bottle washer A? By washer B?

12. A taxi company classifies its drivers as class G (great), class S (so-so), and class B (bad), depending on the driver's past accident record. The probability that a class G driver will have at least one accident a month is 0.02, for a class S driver the probability is 0.04, and for a class B driver it is 0.08. The company knows that 80 percent of its drivers are class G, 10 percent are class S, and 10 percent are class B. One of its drivers calls in to report his recent accident. What is the probability that the driver is a class G driver? Class S? Class B?

13. Let E_1, E_2, and E_3 be mutually-exclusive events such that

$$P(E_1) + P(E_2) + P(E_3) = 1$$

Let A be an arbitrary event such that

$$P(A) = P(E_1 \cap A) + P(E_2 \cap A) + P(E_3 \cap A)$$

Draw a circle diagram representing the same space, and then partition the sample space into areas representing E_1, E_2, E_3, and A.

SELECTED REFERENCES

Feller, W., *An Introduction to Probability Theory and Its Applications*, John Wiley & Sons, Inc., New York, 1950. An algebraic approach to probability. Provides detailed explanations of the use of permutations and combinations in probability problems.

Fellner, William, *Probability and Profit*, Richard D. Irwin, Inc., Homewood, Ill., 1965. An excellent study of economic behavior along Bayesian lines.

Hymans, Saul H., *Probability Theory with Applications to Econometrics and Decision-Making*, Prentice-Hall, Inc., Englewood Cliffs, N. J., 1967.

Mosteller, F., R. E. K. Rourke, and G. B. Thomas, Jr., *Probability With Statistical Applications*, Addison-Wesley Publishing Company, Inc., Reading, Mass., 1961. A set theory approach to probability with a wide variety of illustrations and applications.

CHAPTER 5

Probability Distributions

In many probability problems, we are interested in the probabilities associated with every possible outcome of an experiment. Such a set of numbers is called a *probability distribution*. Suppose, for example, that three coins are tossed. This experiment has eight, equally likely outcomes: *HHH*, *HHT*, *HTH*, *THH*, *HTT*, *THT*, *TTH*, and *TTT*. If we let X represent the number of heads, then X may take on the values 0, 1, 2, or 3. The probability distribution of X consists of the set of possible numbers that X can take on and their associated probabilities. The probability distribution of X is, therefore, equal to:

Number of heads, X	0	1	2	3
Probability of X heads	$\frac{1}{8}$	$\frac{3}{8}$	$\frac{3}{8}$	$\frac{1}{8}$

The number X is called a random variable.

The term *random variable* is used frequently in statistics and is, by definition, a variable (or function) whose value is a real number determined by the outcome of an experiment. For example, the number of spots that turn up on the roll of a die is a random variable that has the values 1, 2, 3, 4, 5, or 6. If each of these numbers is multiplied by two, we obtain a new random variable that takes on the values 2, 4, 6, 8, 10, 12. (Write down the probability distribution of this random variable.)

If a random variable X can take on only certain isolated values in an interval rather than all values in an interval, it is called a *discrete random variable*. For example, if in the interval 10 to 20, the random variable X takes on only the discrete values 11, 12.35, 15, and 19.89, then X is a discrete random variable. Probability distributions associated with discrete random variables are referred to as *discrete probability distributions. A continuous random variable* is one that can take on any value in an interval. Thus, a continuous random variable could take on any value in the interval 10 to 20, for example. Probability distributions associated with continuous random variables are called *continuous probability distributions.*

A probability distribution can be regarded as a relative frequency distribution for a population. Theoretical probability distributions are useful as benchmarks against which to compare the relative frequency distributions obtained from samples. They can also tell us what to expect before samples are actually taken. Furthermore, they provide criteria by which to judge estimators such as the mean and standard deviation and can serve as useful approximations when it is impossible, or too costly, to obtain sufficient information about actual frequency distributions. This section is intended as a brief introduction to probability distributions. The next chapter is devoted to a full discussion of several of the most important theoretical probability distributions.

5.1 DISCRETE PROBABILITY DISTRIBUTIONS

The probability distribution for an experiment that consists of throwing a pair of fair dice is presented in Table 5.1. The random variable X is the

Table 5.1 A Discrete Probability Distribution

X	2	3	4	5	6	7	8	9	10	11	12
$f(x)$	$\frac{1}{36}$	$\frac{2}{36}$	$\frac{3}{36}$	$\frac{4}{36}$	$\frac{5}{36}$	$\frac{6}{36}$	$\frac{5}{36}$	$\frac{4}{36}$	$\frac{3}{36}$	$\frac{2}{36}$	$\frac{1}{36}$

sum of spots that turn up on the two dice and takes on the integer values from 2 to 12. The probability distribution is therefore a discrete probability distribution. It is presented graphically in the form of a histogram in Figure 5.1. The rectangles of the histogram have been spread over a continuous scale and centered on the values of the random variable X even though there is no probability associated with the numbers between the integers. (In other words, the probability of sum 3 is 2/36 but the probability of sum 3.2 is zero.) The area of a rectangle measures the probability of the corresponding value of X because the base of the rectangle is equal to one and its height is equal to the probability.

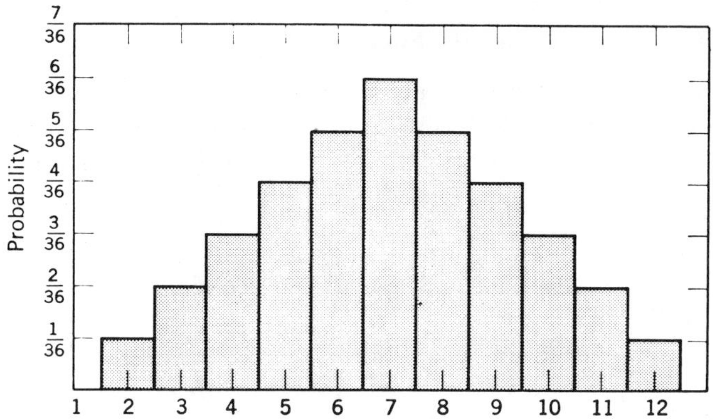

Figure 5.1 A discrete probability distribution.

In the following discussion, it is often necessary to distinguish between a random variable X and one of its values. The small letter x is used to denote a particular value of the random variable X. Subscripts are attached to x when it is necessary to distinguish between particular values of the random variable. Thus, in Table 5.1, $x_1 = 2$, $x_2 = 3$, $x_3 = 4$, etc.

Suppose that X is a discrete random variable that takes on the values $x_1, x_2, \ldots, x_m$. Furthermore, let $f(x)$ be a function such that $f(x_i) \geq 0$ for all i and

$$\sum_{i=1}^{m} f(x_i) = 1$$

The function $f(x)$ is called the *probability density function* (abbreviated p.d.f.) of the random variable X.[1]

The *distribution function* (or cumulative distribution function) $F(x)$ of a random variable X is defined as the probability that X is less than or equal to some particular value x. That is,

$$F(x) = P(X \leq x)$$

Since F(x) represents a probability, it is necessary that $0 \leq F(x) \leq 1$ for all values of X. Every random variable has a distribution function. If the distribution function is known for all values of x, the probability distribution of X will be completely known.

The cumulative probability distribution of a discrete random variable X consists of the set of values x_i that X can take on together with the set of probabilities $P(X \leq x_i)$.

[1] Some authors prefer to call $f(x)$ a probability function in the case of discrete random variables and a probability density function in the case of continuous variables.

The following examples will illustrate the concepts of a probability density function and a distribution function.

Example 1: Consider an experiment that consists of one toss of a coin. Let $x = 0$ if the coin lands tails and $x = 1$ if it lands heads. If p is the probability that the coin lands heads and $(1 - p)$ the probability that it lands tails, then

$$f(x) = p^x(1 - p)^{1-x}$$

The probability distribution and cumulative probability distribution can be presented as follows:

	Tails	*Heads*
x	0	1
$f(x)$	$(1 - p)$	p
$F(x)$	$(1 - p)$	1

Note that $f(0) = (1 - p) \geq 0$, $f(1) = p \geq 0$, and $f(0) + f(1) = (1 - p) + p = 1$ as required by the definition of a probability density function.

Example 2: Let X be a random variable that takes on the values 0, 1, 2, 3, 4, with probability given by

$$f(x) = \frac{3}{2}\frac{1}{x!(4 - x)!}$$

The probability distribution and cumulative probability distribution are presented below.

x	0	1	2	3	4
$f(x)$	$\frac{1}{16}$	$\frac{4}{16}$	$\frac{6}{16}$	$\frac{4}{16}$	$\frac{1}{16}$
$F(x)$	$\frac{1}{16}$	$\frac{5}{16}$	$\frac{11}{16}$	$\frac{15}{16}$	$\frac{16}{16}$

Example 3: Let X denote the number of spots that turn up on a single toss of a true die. The probability density function of X can be written as

$$f(x) = \begin{cases} \frac{1}{6} & \text{for } x = 1, 2, 3, 4, 5, 6 \\ 0 & \text{otherwise} \end{cases}$$

The probability distribution and cumulative probability distribution are therefore:

x	1	2	3	4	5	6
$f(x)$	$\frac{1}{6}$	$\frac{1}{6}$	$\frac{1}{6}$	$\frac{1}{6}$	$\frac{1}{6}$	$\frac{1}{6}$
$F(x)$	$\frac{1}{6}$	$\frac{2}{6}$	$\frac{3}{6}$	$\frac{4}{6}$	$\frac{5}{6}$	$\frac{6}{6}$

5.2 CONTINUOUS PROBABILITY DISTRIBUTIONS[2]

To illustrate the concept of a continuous random variable, assume we have a circular scale one unit long with a pointer pivoted at the center such as is shown in Figure 5.2. Suppose that when the pointer is whirled, it is

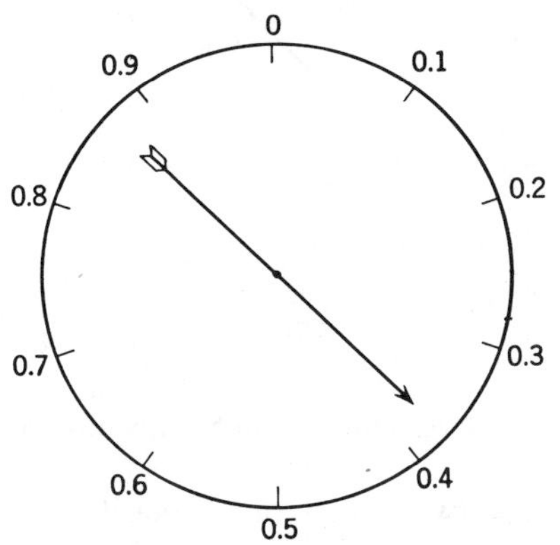

Figure 5.2 A random pointer.

equally likely to stop anywhere. The probability of the pointer stopping in any interval is equal to the length of that interval. For example, the probability of the pointer stopping between 0.3 and 0.5 is $(0.5 - 0.3) = 0.2$. If X is the distance from the zero-point to the point at which the pointer stops, then X

[2] Those unfamiliar with calculus should read the notes on calculus in Appendix II before reading this section. If the reader will keep in mind that integrals are like sums, it will be obvious that most of the statements made about continuous variables are really quite the same as those made about discrete variables.

is a continuous random variable. Since X is a random variable, it must have a distribution function $F(x)$. We know that X cannot be less than zero or greater than one. Thus,

$$F(x) = 0 \textit{ for } x \leq 0 \qquad \text{and} \qquad F(x) = 1 \text{ for } x > 1$$

For values between zero and one, the probability that X is less than or equal to a particular value x is x itself. Thus,

$$F(x) = x \text{ for } 0 < x \leq 1$$

The distribution function $F(x)$ is shown in Figure 5.3.

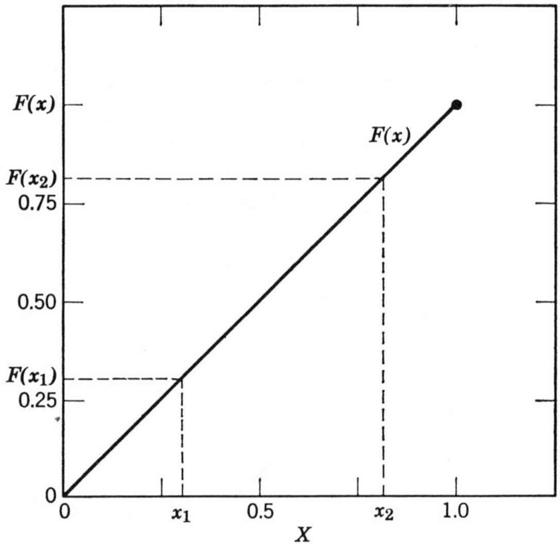

Figure 5.3 Cumulative probability distribution $F(x)$.

If we let x_1 and x_2 be two values of X such that $x_1 < x_2$, then it can be shown that

$$(5.1) \qquad P(x_1 < X \leq x_2) = P(X \leq x_2) - P(X \leq x_1) = F(x_2) - F(x_1)$$

In the case of the pointer, for example, if $x_1 = 0.3$ and $x_2 = 0.8$, then

$$P(0.3 < X \leq 0.8) = F(0.8) - F(0.3) = 0.5$$

This can be seen directly by inspection of Figure 5.3.

The discussion so far has been concerned with the probability that X lies between two values. Let us turn now to the problem of finding the probability of a single value of X. When X is a continuous random variable, the problem is easily solved because the probability that X takes on any particular value

x is equal to zero. This is easily verified with the help of the distribution function. Since $P(X = x_1) = P(x_1 < X \leq x_1)$, we have

$$P(X = x_1) = F(x_1) - F(x_1) = 0$$

The probability density function of a continuous random variable X is a function $f(x)$ such that

$$P(x_1 < X \leq x_2) = \int_{x_1}^{x_2} f(x)\,dx$$

where $f(x) > 0$ for all values of x and $\int_{-\infty}^{\infty} f(x)\,dx = 1$. The graph of the probability density function furnishes a graphical representation of the probability distribution of a continuous random variable X. The integral

$$\int_{x_1}^{x_2} f(x)\,dx$$

represents the area under the graph of $f(x)$ between the values x_1 and x_2. Thus, the area under the graph of $f(x)$ between x_1 and x_2 represents the probability that X will assume a value between x_1 and x_2. This is represented by the shaded area in Figure 5.4. The probability density function $f(x)$ can be

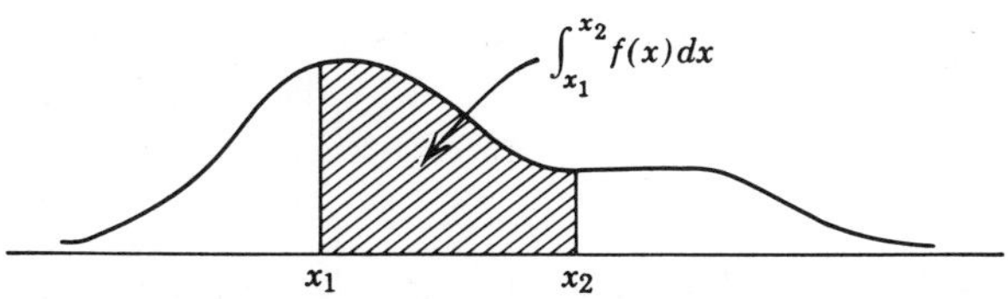

Figure 5.4 A continuous probability distribution.

obtained from the distribution function $F(x)$ by differentiation. That is,

$$f(x) = \frac{dF(x)}{dx} \tag{5.2}$$

Returning to the pointer problem, we have $F(x) = x$ for $0 < x \leq 1$ and, therefore,

$$f(x) = \frac{dF(x)}{dx} = 1$$

The graph of this probability density function is a horizontal line at $f(x) = 1$ over the interval $0 < x \leq 1$. It is presented in Figure 5.5. For this particular case,

$$F(x) = \int_0^x 1\,dx = x \qquad \text{for } 0 < x \leq 1$$

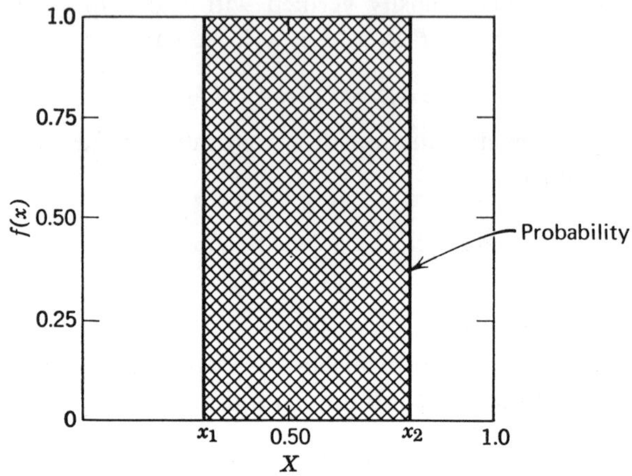

Figure 5.5 Probability density function for $f(x) = 1$.

and, from Figure 5.5, we see that

$$P(x_1 < X \leq x_2) = F(x_2) - F(x_1) = \int_{x_1}^{x_2} 1\,dx = x_2 - x_1$$

The probability that the pointer will stop between the values 0.3 and 0.8 is equal to 0.5. This probability is represented by the shaded area in Figure 5.5.

The following examples will illustrate further the concepts discussed in this section.

Example 1: Let X be a random variable with a probability density function defined by

$$f(x) = \begin{cases} \dfrac{1}{b-a} & \text{for } a < x < b \\ 0 & \text{otherwise} \end{cases}$$

The corresponding distribution function is given by

$$F(x) = \begin{cases} 0 & \text{for } x \leq a \\ \dfrac{(x-a)}{(b-a)} & \text{for } a < x < b \\ 1 & \text{for } x \geq b \end{cases}$$

This result is obtained by noting that

$$F(x) = \int_a^x \frac{1}{b-a}\,dx = \left[\frac{x}{b-a}\right]_{a<x<b} - \left[\frac{x}{b-a}\right]_{x=a} = \frac{x-a}{b-a}$$

Thus, if $x = a$, then $F(x) = 0$, and if $x = b$, we have $F(x) = 1$. The random variable X is said to have a *rectangular distribution* in the interval between a and b, or to be *uniformly distributed* between a and b. In the pointer problem, we were working with a rectangular distribution in the interval $a = 0$ to $b = 1$.

Example 2: Let X be a continuous random variable with a probability density function equal to

$$f(x) = \begin{cases} 3x^2 & \text{for } 0 < x < 1 \\ 0 & \text{otherwise} \end{cases}$$

The corresponding distribution function is given by

$$F(x) = \begin{cases} 0 & \text{for } x \leq 0 \\ x^3 & \text{for } 0 < x < 1 \\ 1 & \text{for } x \geq 1 \end{cases}$$

The probability density function and the distribution function of X are presented in Figures 5.6 and 5.7. As an illustration, suppose we want to find

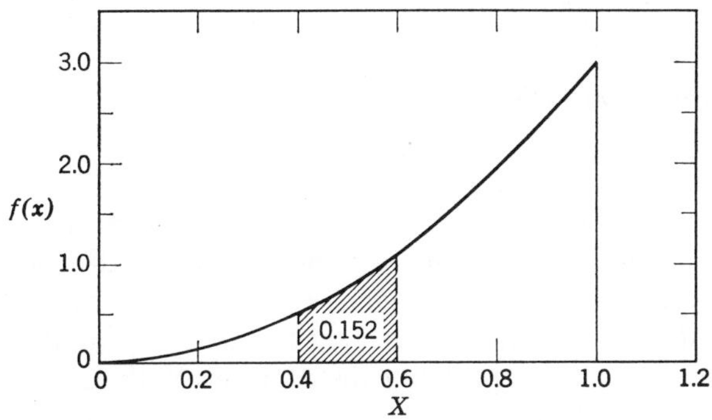

Figure 5.6 Probability density function $f(x) = 3x^2$.

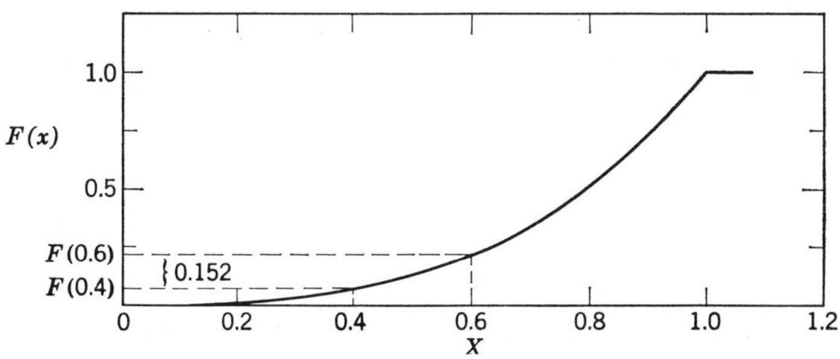

Figure 5.7 Distribution function $F(x) = x^3$.

the probability that X is between 0.4 and 0.6. This probability is equal to

$$P(0.4 < X \leq 0.6) = F(0.6) - F(0.4) = (0.6)^3 - (0.4)^3 = 0.152$$

This case is illustrated in Figures 5.6 and 5.7. (*Problem.* Prove that $f(x)$ is a probability density function.)

5.3 JOINT PROBABILITY DISTRIBUTIONS

In many cases, the outcome of a random experiment will be expressed by two or more simultaneously observed quantities. For example, we may measure the income (X) and expenditures (Y) of a randomly selected family. There are several ways to describe the possible relationship between the variables X and Y. A detailed discussion of statistical techniques commonly used by economists in this connection is postponed until Chapter 9. In this section, our discussion will be limited to explaining certain fundamental notions about the joint probability distribution of two random variables.

5.3.1 Discrete Random Variables

Suppose that a box contains an equal number of each of three types of cards that are marked with the numbers 1, 2, or 3. A certain card game consists of drawing a card at random from the box. If the card is a 1 or a 3, a second card is drawn; otherwise the game ends. Let X be the number on the first card drawn and Y be the number on the second or zero if no second card is drawn. The joint probability distribution of the random variables X and Y is given in Table 5.2.

In Table 5.2, the probability that X is equal to 1 and Y is equal to 2 is $P(X = 1,\ Y = 2) = 1/9$, while the probability that X is equal to 1 without regard to the value assumed by Y is equal to $P(X = 1) = 1/3$.

In general, suppose that X and Y are discrete random variables that can take on the values $x_1, x_2, \ldots, x_m$ and $y_1, y_2, \ldots, y_m$, respectively. The *joint distribution function* of X and Y is defined as

(5.3) $$F(x, y) = P(X \leq x,\ Y \leq y)$$

The joint *cumulative probability distribution* of X and Y consists of the numbers x, y, and $F(x, y)$ for all values of X and Y.

The joint cumulative probability distribution for the card game is presented numerically in Table 5.3 and graphically in Figure 5.8. The variables X and

Table 5.2 Joint Probability Distribution of a Card Game

	X			
Y	1	2	3	$g(y)$
0	0	$\frac{1}{3}$	0	$\frac{3}{9}$
1	$\frac{1}{9}$	0	$\frac{1}{9}$	$\frac{2}{9}$
2	$\frac{1}{9}$	0	$\frac{1}{9}$	$\frac{2}{9}$
3	$\frac{1}{9}$	0	$\frac{1}{9}$	$\frac{2}{9}$
$f(x)$	$\frac{1}{3}$	$\frac{1}{3}$	$\frac{1}{3}$	

Table 5.3 Joint Cumulative Probability Distribution for a Card Game

	X		
Y	1	2	3
0	0	$\frac{3}{9}$	$\frac{3}{9}$
1	$\frac{1}{9}$	$\frac{4}{9}$	$\frac{5}{9}$
2	$\frac{2}{9}$	$\frac{5}{9}$	$\frac{7}{9}$
3	$\frac{3}{9}$	$\frac{6}{9}$	$\frac{9}{9}$

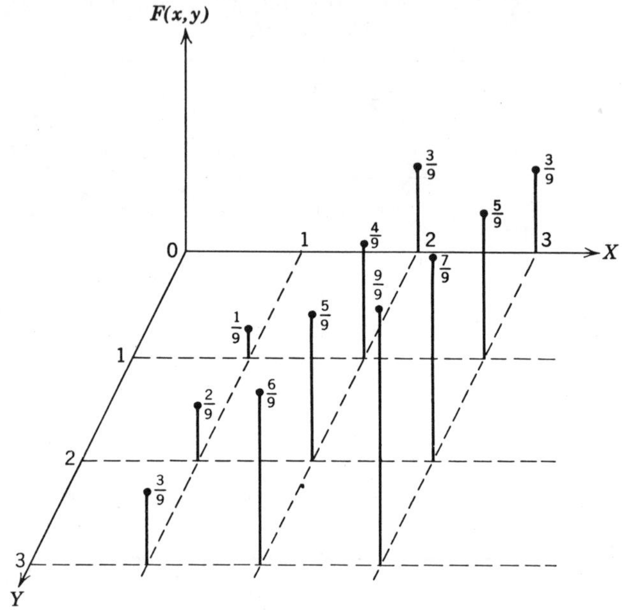

Figure 5.8 A joint cumulative probability distribution.

Y are plotted on the horizontal plane and $F(x, y)$ on the vertical axis in Figure 5.8. The height of the vertical lines over the horizontal plane represents the cumulative probability associated with the point at which the vertical lines are drawn.

It can be shown[3] that

$$(5.4)\quad P(x_1 < X \leq x_2, y_1 < Y \leq y_2) = F(x_2, y_2) - F(x_1, y_2) - F(x_2, y_1) + F(x_1, y_1)$$

Returning to the card game and letting $x_1 = 1, x_2 = 3, y_1 = 1$, and $y_2 = 2$, we find

$$P(1 < X \leq 3, 1 < Y \leq 2) = F(3, 2) - F(1, 2) - F(3, 1) + F(1, 1)$$

which is equal to

$$P(X = 2 \text{ or } 3 \text{ and } Y = 2) = \frac{7}{9} - \frac{2}{9} - \frac{5}{9} + \frac{1}{9} = \frac{1}{9}$$

[3] This result also applies to continuous random variables.

Inspection of Table 5.3 will verify this result.

If X and Y are discrete random variables and $f(x, y)$ is a function such that

$$\begin{aligned} & f(x, y) = P(X = x, Y = y) \\ (5.5) \qquad & f(x, y) \geq 0 \text{ for all values of } X \text{ and } Y \\ & \sum_x \sum_y f(x, y) = 1 \text{ (where the summations are all over values of } X \text{ and } Y) \end{aligned}$$

then $f(x, y)$ is the *joint probability density function* of X and Y.

The *marginal distribution of* X is a one-dimensional probability distribution associated with the joint distribution of X and Y and has a probability density function given by

$$(5.6) \quad f(x) = \sum_y f(x, y) \qquad \text{(where the summation is over all values of } Y)$$

The function $f(x)$ will be referred to as the *marginal density function* when we wish to emphasize that it is obtained from the joint probability density function of X and Y.

Returning again to the card game, the marginal density function of X is given by

$$f(x) = \begin{cases} \frac{1}{3} & \text{for } X = 1, 2, 3 \\ 0 & \text{otherwise} \end{cases}$$

This distribution is presented in the last row of Table 5.2. The marginal density function of Y is defined as

$$g(y) = \sum_x f(x, y)$$

and is presented in the far right column of Table 5.2.

The *conditional probability of* X for a given value of Y is given by

$$(5.7) \qquad P(X = x \mid Y = y) = f(x \mid y) = \frac{f(x, y)}{g(y)}$$

provided that $g(y)$ is greater than zero. When x varies, y being held fixed, $f(x \mid y)$ is a function of x and is called the *conditional probability function of* X for the given value y of Y. The variable X has as many conditional probability functions as there are values of Y.

In case of the card game, let the fixed (or given) value of Y be $y = 2$; then the conditional probability distribution of X is

$$f(1\mid 2)=\frac{f(1,2)}{g(2)}=\frac{(1/9)}{(2/9)}=\frac{1}{2}$$

$$f(2\mid 2)=\frac{f(2,2)}{g(2)}=\frac{(0)}{(2/9)}=0$$

$$f(3\mid 2)=\frac{f(3,2)}{g(2)}=\frac{(1/9)}{(2/9)}=\frac{1}{2}$$

Since Y has four values, X has four conditional probability distributions. These are given by $f(x\mid 0)$, $f(x\mid 1)$, $f(x\mid 2)$, and $f(x\mid 3)$. Similarly, Y has a separate conditional probability distribution for each value of X. The conditional probability function of Y is equal to

$$g(y\mid x)=\frac{f(x,y)}{f(x)}$$

Most of the concepts that have been discussed in this section are obvious from the discussions of probability in Chapter 4. The notation is somewhat different, but the ideas are basically the same. In summary, we have defined four types of probability distributions.

1. The joint probability distribution.
2. The joint cumulative probability distribution.
3. The marginal probability distribution.
4. The conditional probability distribution.

An understanding of the differences between these four types of probability distributions is required at several points in the following chapters.

5.3.2 Continuous Random Variables

Suppose that X and Y are continuous random variables with joint distribution function defined by

$$F(x,y)=P(X\leq x,\ Y\leq y)$$

The joint probability density function of X and Y can be obtained by differentiating $F(x, y)$ with respect to x and y. More precisely, the joint probability density function of X and Y is equal to

$$f(x,y)=\frac{\partial^2 F(x,y)}{\partial X\,\partial Y}$$

and is such that

$$P(x_1 < X \le x_2, y_1 < Y \le y_2) = \int_{x_1}^{x_2} \int_{y_1}^{y_2} f(x, y)\, dx\, dy$$

and

$$\int_{-\infty}^{\infty} \int_{-\infty}^{\infty} f(x, y)\, dx\, dy = 1$$

The marginal density function of X, denoted by $f(x)$, is obtained from $f(x, y)$ by integrating over y. In other words,

$$f(x) = \int_{-\infty}^{\infty} f(x, y)\, dy$$

Similarly, the marginal density function of Y is defined as

$$g(y) = \int_{-\infty}^{\infty} f(x, y)\, dx$$

Note the similarity between the definition of $f(x)$ and the definition (5.6) of the marginal density function for a discrete random variable. The summation in (5.6) is replaced by the integral sign when x is a continuous random variable. Both the summation in (5.6) and the integration above are over all values of Y.

The marginal density functions allow us to determine the probability that the value of one random variable lies in a given interval regardless of the value of the other variable. For example, the probability that X is greater than x_1 but less than or equal to x_2, regardless of the value of Y, is given by

$$P(x_1 < X \le x_2) = \int_{x_1}^{x_2} f(x)\, dx$$

The conditional probability of the event $(y_1 < Y \le y_2)$, given that $x < X \le x + \Delta x$, where Δx is a small positive number, is equal to

$$P(y_1 < Y \le y_2 \mid x < X \le x + \Delta x) = \frac{P(y_1 < Y \le y_2, x < X \le x + \Delta x)}{P(x < X \le x + \Delta x)}$$

It can be shown that, in the limit as Δx approaches zero, the probability $P(y_1 < Y \le y_2 \mid x < X \le x + \Delta x)$ is equal to

$$\int_{y_1}^{y_2} \frac{f(x, y)}{f(x)}\, dy$$

provided that $f(x)$ is not equal to zero.

The function $g(y \mid x) = f(x, y)/f(x)$ is called the *conditional density function of Y*. Similarly, the conditional density function of X is equal to

$$f(x \mid y) = \frac{f(x, y)}{g(y)}$$

In many problems of interest, the conditional density function $f(x \mid y)$ is known for each value of y along with the marginal density function $g(y)$. The joint density function $f(x, y)$ is then obtained directly from the relationship:

$$f(x, y) = f(x \mid y)g(y)$$

The following examples should provide added meaning to the above concepts and notation.

Example 1: Suppose that X and Y are continuous random variables having a joint probability density function

$$f(x, y) = \frac{1}{(b-a)(d-c)}$$

for $a < x < b$, $c < y < d$, and $f(x, y) = 0$ otherwise. The mass in this distribution is distributed with a constant density over the area of the rectangle $a < x < b$, $c < y < d$. This is illustrated in Figure 5.9. The total volume of

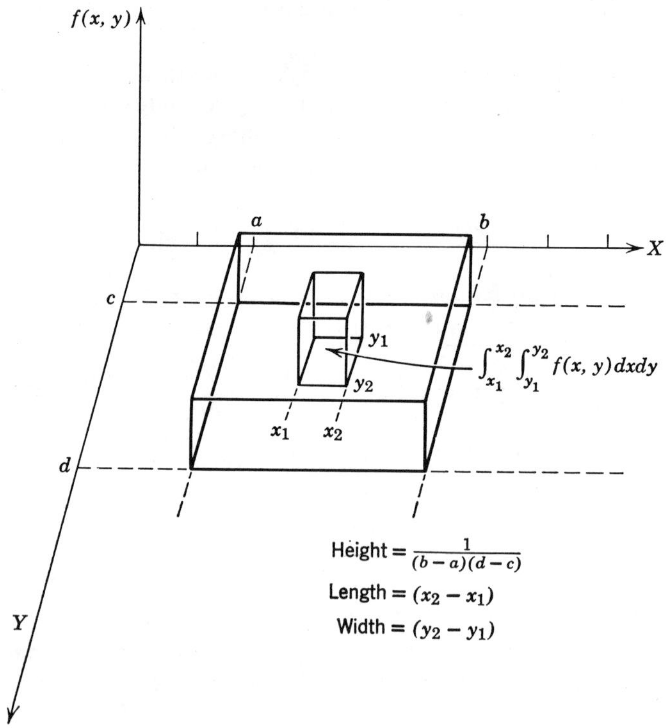

Figure 5.9 Probability density function $f(x, y) = 1/(b-a)(d-c)$.

the rectangular box shown in Figure 5.9 is equal to one. The probability that X lies in the interval from x_1 to x_2 and Y lies in the interval from y_1 to y_2 is given by

$$P(x_1 < X \leq x_2, y_1 < Y \leq y_2) = \int_{x_1}^{x_2} \int_{y_2}^{y_2} \frac{1}{(b-a)(d-c)}\, dx\, dy$$

$$= \frac{(x_2 - x_1)(y_2 - y_1)}{(b-a)(d-c)}$$

This probability is equal to the volume of the smaller rectangular box shown in Figure 5.9.

The distribution function associated with $f(x, y)$ is given by

$$F(x, y) = P(X \leq x, Y \leq y) = \int_a^x \int_c^y \frac{1}{(b-a)(d-c)}\, dx\, dy = \frac{(x-a)(y-c)}{(b-a)(d-c)}$$

The distribution function in this example can be thought of as measuring the volume of a rectangular box that has

$$\text{height} = \frac{1}{(b-a)(d-c)}$$

$$\text{length} = (x-a) \qquad \text{(distance on the } X \text{ axis)}$$

$$\text{width} = (y-c) \qquad \text{(distance on the } Y \text{ axis)}$$

Thus, in this particular case, $F(x, y)$ can be obtained simply by multiplying the height, length, and width of the probability cube over the area of the rectangle that has one corner at the origin of the probability distribution of X and Y.

The marginal density function of X is equal to

$$f(x) = \int_c^d \frac{1}{(b-a)(d-c)}\, dy = \frac{1}{(b-a)}$$

over the interval $a < x < b$. This is identical to the rectangular distribution discussed in Example 1 of Section 5.2.

The marginal density function of Y is equal to

$$g(y) = \int_a^b \frac{1}{(b-a)(d-c)}\, dx = \frac{1}{(d-c)}$$

over the interval $c < y < d$.

The conditional density function of X is equal to

$$f(x \mid y) = \frac{f(x, y)}{g(y)} = \frac{1}{(b-a)} = f(x)$$

Similarly, the conditional density function of Y is equal to

$$g(y \mid x) = \frac{f(x, y)}{g(x)} = \frac{1}{(d - c)} = g(y)$$

Suppose now that $a = 2$, $b = 6$, $c = 1$, and $d = 4$. Then

$$f(x, y) = \frac{1}{12} \qquad \text{and} \qquad F(x, y) = \frac{(x - 2)(y - 1)}{12}$$

$$\text{for } 2 < x < 6 \qquad \text{and} \qquad 1 < y < 4$$

The values of the distribution function $F(x, y)$, for several values of X and Y, are given in Table 5.4. For example, $F(4, 3) = 4/12$. Since X and Y are continuous, it is impossible to table $F(x, y)$ for all values x and y. Geometrically, $F(x, y)$ has the form of the distribution shown in Figure 5.10.

Table 5.4 Values of $F(x, y) = \dfrac{(x - 2)(y - 1)}{12}$

	x				
y	2	3	4	5	6
1	0	0	0	0	0
2	0	$\frac{1}{12}$	$\frac{2}{12}$	$\frac{3}{12}$	$\frac{4}{12}$
3	0	$\frac{2}{12}$	$\frac{4}{12}$	$\frac{6}{12}$	$\frac{8}{12}$
4	0	$\frac{3}{12}$	$\frac{6}{12}$	$\frac{9}{12}$	$\frac{12}{12}$

The probability that X lies in the interval 4 to 5 and Y lies in the interval 2 to 3, is equal to

$$P(4 < x \leq 5, 2 < Y \leq 3) = \int_4^5 \int_2^3 \frac{1}{12}\, dx\, dy = \frac{1}{12}$$

This is equivalent to

$$\begin{aligned} P(4 < x \leq 5, 2 < Y \leq 3) &= F(5, 3) - F(4, 3) - F(5, 2) + F(4, 2) \\ &= \frac{6}{12} - \frac{4}{12} - \frac{3}{12} + \frac{2}{12} = \frac{1}{12} \end{aligned}$$

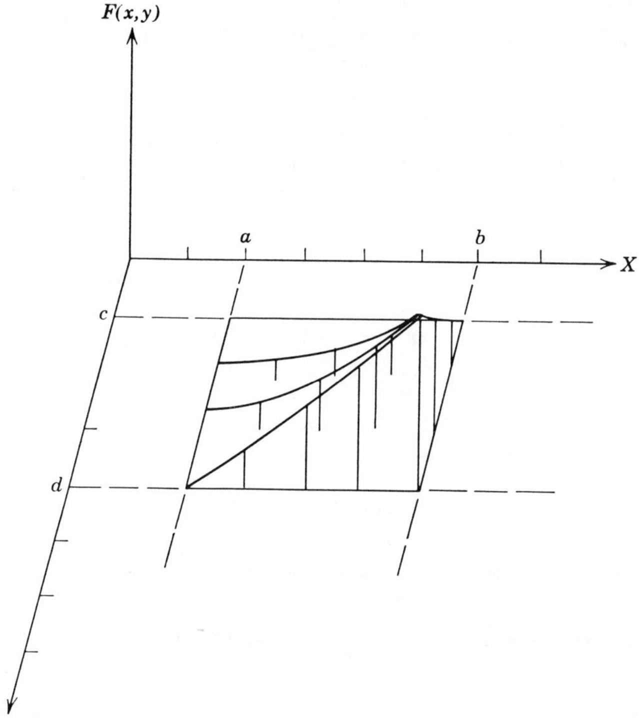

Figure 5.10 Distribution function $F(x, y) = (x - a)(y - c)/(b - a)(d - c)$.

The probability that X lies in the interval 4 to 5, regardless of the value of Y, is equal to

$$P(4 < x \leq 5) = \int_4^5 \frac{1}{4}\,dx = F(5) - F(4) = \frac{3}{4} - \frac{2}{4} = \frac{1}{4}$$

where

$$F(x) = \frac{x - a}{b - a} = \frac{x - 2}{4}$$

The probability that Y is in the interval from 2 to 3, given $X = 4$, is given by

$$P(2 < y \leq 3 \mid x = 4) = \int_2^3 g(y \mid x)\,dy = \int_2^3 \frac{1}{3}\,dy = \frac{1}{3}$$

Example 2: Suppose that X and Y are continuous random variables having a joint probability density function

$$f(x, y) = x + y$$

for $0 < x < 1, 0 < y < 1$, and $f(x, y) = 0$ otherwise. The values of $f(x, y)$ for several values of x and y are given in Table 5.5. A geometric representation of $f(x, y)$ is shown in Figure 5.11.

Table 5.5 Values of $f(x, y) = x + y$ for Selected Values of X and Y

	x			
y	0	0.4	0.8	1.0
0	0	0.4	0.8	1.0
0.4	0.4	0.8	1.2	1.4
0.8	0.8	1.2	1.6	1.8
1.0	1.0	1.4	1.8	2.0

The distribution function of X and Y is equal to

$$F(x, y) = \int_0^x \int_0^y (x + y)\, dx\, dy = \frac{x^2y + xy^2}{2}$$

for $0 < x < 1$, $0 < y < 1$, and $F(x, y) = 0$ otherwise. This function is used to obtain

$$P(x_1 < X \le x_2, y_1 < Y < y_2) = F(x_2, y_2) - F(x_1, y_2) - F(x_2, y_1) + F(x_1, y_1)$$

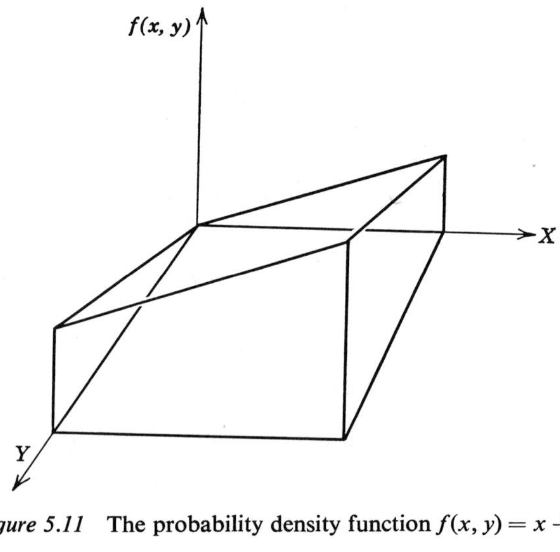

Figure 5.11 The probability density function $f(x, y) = x + y$.

The marginal density function of X is equal to

$$f(x) = \int_0^1 (x + y)\, dy = x + \frac{1}{2}$$

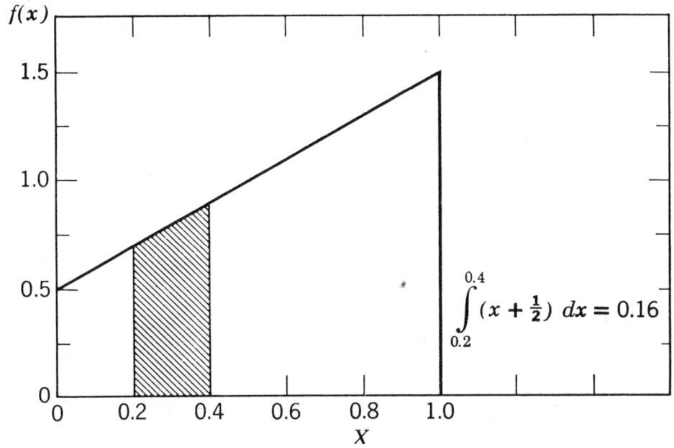

Figure 5.12 The marginal density function $f(x) = x + 1/2$.

over the interval $0 < x < 1$. The function $f(x)$ is shown in Figure 5.12. The marginal density function of Y is equal to

$$g(y) = \int_0^1 (x + y)\, dx = y + \frac{1}{2}$$

over the interval $0 < y < 1$.

The conditional density function of X is equal to

$$f(x \mid y) = \frac{f(x, y)}{g(y)} = \frac{2x + 2y}{2y + 1}$$

over the interval $0 < x < 1$ given a value of y in the interval $0 < y < 1$. The function $f(x \mid y)$ is shown in Figure 5.13.

The function $F(x, y)$ is used to compute *cumulative probabilities*. For example,

$$P(x \le 0.4, y \le 0.2) = F(0.4, 0.2) = \frac{(0.4)^2(0.2) + (0.4)(0.2)^2}{2}$$

$$= \frac{0.048}{2} = 0.024$$

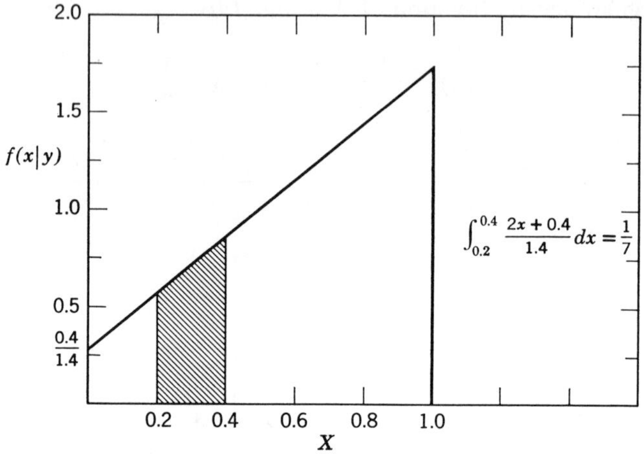

Figure 5.13 The conditional density function $f(x \mid y = 0.2) = (2x + 0.4)/1.4$.

The function $f(x, y)$ is used to obtain *joint probabilities*. Thus,

$$P(0.2 < x \leq 0.4, 0.1 < y \leq 0.4) = \int_{0.2}^{0.4} \int_{0.1}^{0.4} (x + y)\, dx\, dy$$

$$= F(0.4, 0.4) - F(0.2, 0.4) - F(0.4, 0.1) + F(0.2, 0.1)$$

$$= 0.064 - 0.024 - 0.010 + 0.003 = 0.033$$

The function $f(x)$ is used to compute *marginal probabilities*. In general,

$$P(x_1 < x < x_2) = \int_{x_1}^{x_2} \left(x + \frac{1}{2}\right) dx = \frac{1}{2}(x_2{}^2 - x_1{}^2 + x_2 - x_1)$$

For example,

$$P(0.2 < x \leq 0.4) = \int_{0.2}^{0.4} \left(x + \frac{1}{2}\right) dx = 0.16$$

This probability is represented geometrically in Figure 5.12 by the area under the graph of $f(x)$.

The function $f(x \mid y)$ is used to obtain *conditional probabilities*. In general,

$$P(x_1 < x < x_2 \mid y) = \int_{x_1}^{x_2} \frac{2x + 2y}{2y + 1} dx = \frac{x_2{}^2 - x_1{}^2 + 2y(x_2 - x_1)}{2y + 1}$$

Thus,

$$P(0.2 < x \leq 0.4 \mid y = 0.2) = \int_{0.2}^{0.4} \frac{2x + 2y}{2y + 1} dx = \int_{0.2}^{0.4} \frac{2x + 0.4}{1.4} dx = \frac{1}{7}$$

This probability is represented by the area under the graph of $f(x \mid y)$ in Figure 5.13.

5.4 INDEPENDENT RANDOM VARIABLES

The concept of independence of random variables is one of the most important concepts in statistics and will be used frequently in the following chapters.

5.4.1 *Discrete Random Variables*

Let X and Y be discrete random variables with a joint probability density function $f(x, y)$ and the marginal density functions $f(x)$ and $g(y)$, respectively. The random variables X and Y are said to be independent (or stochastically independent) if and only if

$$f(x, y) = f(x)g(y) \tag{5.8}$$

or

$$P(X = x,\ Y = y) = P(X = x)P(Y = y)$$

for every pair of values (x, y).[4]

We recall that the conditional density function of Y for a given value of X is a probability function defined for all x for which the marginal density function $f(x)$ is not equal to zero. That is,

$$g(y \mid x) = \frac{f(x, y)}{f(x)}$$

if $f(x) \neq 0$. If X and Y are independent, the $f(x, y) = f(x)g(y)$ and

$$g(y \mid x) = \frac{f(x)g(y)}{f(x)} = g(y)$$

Similarly, if X and Y are independent, the conditional density function of $X, f(x \mid y)$, is equal to the marginal density function $f(x)$. Thus, if X and Y are independent, the probability that X takes on any value does not depend on what the value of Y happens to be. This independence between the probabilities associated with one variable and the values of the other variable is the reason for referring to X and Y as independent random variables.

Example 1: The joint probability distribution of a certain card game is shown in Table 5.2. The random variables X and Y are *not* independent, however. For example, if $x = 2$ and $y = 0$, we have $f(2, 0) = 1/3$ which is not equal to $f(2)g(0) = (1/3)(1/3) = 1/9$. In other words, the joint probabilities are not equal to the product of the marginal probabilities.

[4] The word "stochastic" means "connected with random experiments and probability."

Example 2: Suppose that X and Y are discrete random variables with the joint probability distribution given in Table 5.6.

Table 5.6 The Joint Probability Distribution of X and Y

	x		
y	2	4	$g(y)$
1	$\frac{1}{8}$	$\frac{1}{8}$	$\frac{1}{4}$
3	$\frac{3}{8}$	$\frac{3}{8}$	$\frac{3}{4}$
$f(x)$	$\frac{1}{2}$	$\frac{1}{2}$	

To check for independence, we compare $f(x, y)$ and $f(x)g(y)$ for all values of x and y. We find that

$$f(2)g(1) = \left(\frac{1}{2}\right)\left(\frac{1}{4}\right) = \frac{1}{8} = f(2, 1)$$

$$f(4)g(1) = \left(\frac{1}{2}\right)\left(\frac{1}{4}\right) = \frac{1}{8} = f(4, 1)$$

$$f(2)g(3) = \left(\frac{1}{2}\right)\left(\frac{3}{4}\right) = \frac{3}{8} = f(2, 3)$$

$$f(4)g(3) = \left(\frac{1}{2}\right)\left(\frac{3}{4}\right) = \frac{3}{8} = f(4, 3)$$

and, therefore, X and Y are independent. In this case, the conditional probabilities should be equal to the marginal probabilities. To check this in one instance, let $x = 4$, $y = 3$, and compute

$$f(4 \mid 3) = \frac{f(4, 3)}{g(3)} = \frac{3/8}{3/4} = \frac{1}{2} = f(4)$$

The result is exactly as expected.

5.4.2 *Continuous Random Variables*

Let X and Y be continuous random variables with a joint probability density function $f(x, y)$ and the marginal probability density functions $f(x)$ and $g(y)$, respectively. The random variables X and Y are said to be independent if, and only if,

$$f(x, y) = f(x)g(y)$$

for all values of x and y.[5]

If X and Y are independent random variables, then

$$P(x_1 < X \leq x_2, y_1 < Y \leq y_2) = P(x_1 < X \leq x_2)P(y_1 < Y \leq y_2)$$

for every $x_1 < x_2$ and $y_1 < y_2$. The proof of this statement follows directly from the fact that the joint density function of X and Y is equal to $f(x)g(y)$ if X and Y are independent. Thus,

$$\begin{aligned} P(x_1 < X \leq x_2, y_1 < Y \leq y_2) &= \int_{x_1}^{x_2} \int_{y_1}^{y_2} f(x, y)\, dx\, dy \\ &= \int_{x_1}^{x_2} \int_{y_1}^{y_2} f(x)g(y)\, dx\, dy \\ &= \left[\int_{x_1}^{x_2} f(x)\, dx\right]\left[\int_{y_1}^{y_2} g(y)\, dy\right] \\ &= P(x_1 < X \leq x_2)P(y_1 < Y \leq y_2) \end{aligned}$$

as was to be shown.

As in the case of discrete random variables, the conditional density functions are equal to the marginal density functions when X and Y are independent.

In many practical problems, it is necessary to work with functions of random variables. For example, the arithmetic mean of a sample is a function of random variables. It can be shown that, if X and Y are independent random variables, then any random variable that is a function of X and any random variable that is a function of Y are independent. This statement is true for both discrete and continuous random variables.

The following examples will serve to illustrate the relationship of statistical independence of continuous random variables.

Example 1: Suppose that X and Y are continuous random variables having a joint probability density function

$$f(x, y) = x + y$$

[5] In general, the random variables $X_1, X_2, \ldots, X_m$ are independent if, and only if, their joint density function $f(x_1, x_2, \ldots, x_m)$ is the product of their marginal density functions $f_1(x_1), f_2(x_2), \ldots, f_m(x_m)$.

for $0 < x < 1$, $0 < y < 1$ and $f(x, y) = 0$ otherwise. We have seen that the marginal density functions of X and Y are

$$f(x) = x + \frac{1}{2} \qquad \text{and} \qquad g(y) = y + \frac{1}{2}$$

In this case, X and Y are not independent because

$$f(x)g(y) = \frac{1}{4} + \frac{x}{2} + \frac{y}{2} + xy$$

is not equal to $f(x, y)$ for all values of x and y.

Example 2: Suppose that X and Y are continuous random variables with a joint probability density function

$$f(x, y) = \frac{1}{(b - a)(d - c)}$$

for $a < x < b$, $c < y < d$, and $f(x, y) = 0$ otherwise. We recall from the previous discussion of this distribution that the marginal density functions of X and Y are

$$f(x) = \frac{1}{(b - a)} \qquad \text{and} \qquad g(y) = \frac{1}{(d - c)}$$

Hence, $f(x, y) = f(x)g(y)$ and, therefore, X and Y are independent. It was also shown previously that $f(x) = f(x \mid y)$ and $g(y) = g(y \mid x)$ as expected when X and Y are independent.

Suppose that we let the random variable W be a function of X and the random variable Z be a function of Y. More specifically, let $W = 2X$ and $Z = 3Y$. The joint probability density function of W and Z is equal to

$$f(w, z) = \frac{1}{6(b - a)(d - c)}$$

for $2a < w < 2b$, $3c < z < 3d$, and $f(w, z) = 0$ otherwise. The marginal density functions of W and Z are

$$f(w) = \int_{3c}^{3d} \frac{1}{6(b - a)(d - c)}\, dz = \frac{1}{2(b - a)}$$

and

$$g(z) = \int_{2a}^{2b} \frac{1}{6(b - a)(d - c)}\, dw = \frac{1}{3(d - c)}$$

We find that $f(w, z) = f(w)g(z)$ and, hence, W and Z are independent random variables.

5.5 MATHEMATICAL EXPECTATION

One of the most useful concepts in problems involving distributions of random variables is that of mathematical expectation. A number of rules for operating with expectations are presented in this section. These rules will be used extensively in the following chapters. Although the basic rules apply to both discrete and continuous random variables, the proofs usually will be carried out only for discrete random variables.

To illustrate the idea of a mathematical expectation, suppose that a large bowl contains $n_1 = 400$ \$5 bills and $n_2 = 100$ \$10 bills. You are offered an opportunity to pay \$7 for the privilege of selecting a bill at random from the bowl. The expected winnings can be estimated as follows. Let $n_1 + n_2 = n$; the expected amount won per draw is

$$\frac{5n_1 + 10n_2}{n_1 + n_2} = 5\left[\frac{n_1}{n}\right] + 10\left[\frac{n_2}{n}\right] = 5(0.8) + 10(0.2)$$
$$= 6 \text{ dollars}$$

The term on the far left is a weighted mean. The weights (n_1/n) and (n_2/n) are probabilities. The (n_1/n), for example, is the probability of drawing a five-dollar bill. Unless you like to gamble, you would probably pass up the opportunity to select a bill from the bowl because the cost is \$1 more than the expected return.

In general, if X is a discrete random variable that takes on the values $x_1, x_2, \ldots, x_n$ and has a probability density function $f(x)$, then the mathematical expectation of X is defined as

$$E(X) = x_1 f(x_1) + x_2 f(x_2) + \cdots + x_n f(x_n) = \sum x_i f(x_i) \tag{5.9}$$

In other words, to compute the mathematical expectation of X, multiply every possible value of X by its probability and add the results. $E(X)$ is also referred to as the *mean of X*, the *population mean*, or the *expected value.*

If X is a continuous random variable with probability density function $f(x)$, then

$$E(X) = \int_{-\infty}^{\infty} x f(x)\, dx \tag{5.10}$$

The term expected value will be used frequently in later chapters. Keep in mind that this term has a special meaning, and does *not* imply that the value occurs frequently or has a high probability. In many cases, it is not even possible for the random variable to be equal to its expected value. For example, on any particular draw from the bowl with \$5 and \$10 bills, we could never observe its expected value of \$6. Assuming that the bowl always contains 400 fives and 100 tens, the \$6 can be thought of as an "average" amount won per draw over a long series of draws. The expected value is merely a weighted mean of the possible values of a random variable with probabilities as weights.

The notion of expected value is analogous to the concept of a center of mass in physics. Consider the probability density function presented in Figure 5.6. Think of the area outlined by $f(x) = 3x^2$ as a block of lead of uniform density placed on a one-foot board, and try to determine the point at which a fulcrum would have to be placed in order for the board to remain balanced. The center of mass in this case is located at

$$\int_0^1 3x^3\, dx = 0.75$$

which corresponds to the expected value of X. Thus the fulcrum would have to be placed (0.75) (12 inches) = 9 inches.

Before presenting several important rules of expected values, let us note that the expected value of a constant c is equal to the value of the constant. That is, $E(c) = c$. This is somewhat like saying that if all cars have four wheels, the mean number of wheels per car is equal to four. We will make frequent use of the fact that $E(c) = c$, however.

If X is a random variable with expectation $E(X)$ and c is a constant, then

$$E(X + c) = E(X) + c \tag{5.11}$$

This result is easily obtained. By definition of the mathematical expectation of X, we have $E(X) = \sum x_i f(x_i)$. We know that for a discrete random variable the probability density function $f(x)$ is such that $\sum_i f(x_i) = 1$. Hence,

$$\begin{aligned} E(X + c) &= \sum (x_i + c) f(x_i) = \sum x_i f(x_i) + \sum c f(x_i) \\ &= E(X) + c \sum f(x_i) = E(X) + c \end{aligned}$$

The result (5.11) applies to continuous as well as to discrete random variables.

If X and Y are random variables with expectations $E(X)$ and $E(Y)$, then

$$E(X + Y) = E(X) + E(Y) \tag{5.12}$$

This result applies to both dependent and independent random variables. The result (5.12) is obtained as follows. Let $f(x, y)$ be the joint probability density function of X and Y. If X and Y are discrete random variables, then

$\sum_x f(x, y) = g(y)$ is the marginal distribution of y

$\sum_y f(x, y) = f(x)$ is the marginal distribution of x

By definition

$$E(X + Y) = \sum_x \sum_y (x + y)f(x, y)$$

and hence

$$\begin{aligned} E(X + Y) &= \sum_x \sum_y xf(x, y) + \sum_x \sum_y yf(x, y) \\ &= \sum_x xf(x) + \sum_y yg(y) \\ &= E(X) + E(Y) \end{aligned}$$

This result can be extended to a finite sum of random variables. Thus, if $X_1, X_2, \ldots, X_m$ are m random variables with expectations $E(X_1)$, $E(X_2)$, $\ldots$, $E(X_m)$, then

$$E(X_1 + X_2 + \cdots + X_m) = E(X_1) + E(X_2) + \cdots + E(X_m) \tag{5.13}$$

In other words, the expected value of the sum of random variables is equal to the sum of the individual expected values.

If X and Y are *independent* random variables with expectations $E(X)$ and $E(Y)$, then

$$E(XY) = E(X)E(Y) \tag{5.14}$$

That is, the expected value of the product of two *independent* random variables is equal to the product of the individual expected values. The proof of this result begins with the definition

$$E(XY) = \sum_x \sum_y xyf(x, y)$$

Since X and Y are independent, we have $f(x, y) = f(x)g(y)$ and, therefore,

$$E(XY) = \sum_x \sum_y xyf(x)g(y) = \sum_x xf(x) \sum_y yg(y) = E(X)E(Y)$$

This result can be extended to the product of a finite number of random variables. Thus, if $X_1, X_2, \ldots, X_m$ are m *independent* random variables with expectations $E(X_1)$, $E(X_2)$, $\ldots$, $E(X_m)$, then

$$E(X_1 X_2 \cdots X_3) = E(X_1)E(X_2) \cdots E(X_3) \tag{5.15}$$

The joint probability distribution presented in Table 5.7 will provide numerical examples for the above rules. To begin with, we have

$$E(X) = 2(0.4) + 3(0.4) + 4(0.2) = 2.8$$

and

$$E(Y) = 0(0.2) + 1(0.2) + 2(0.2) + 3(0.2) + 4(0.2) = 2.0$$

Table 5.7 Joint Probability Distribution of X and Y

	X			
Y	2	3	4	$g(y)$
0	0	0.1	0.1	0.2
1	0.1	0	0.1	0.2
2	0.1	0.1	0	0.2
3	0.2	0	0	0.2
4	0	0.2	0	0.2
$f(x)$	0.4	0.4	0.2	

Suppose a constant $c = 4$ is added to each value of X in Table 5.7. We then have a new random variable $W = X + c = X + 4$ that has the following probability distribution:

W	6	7	8
$h(w) = f(x)$	0.4	0.4	0.2

The expected value of W is

$$\begin{aligned} E(W) &= 6(0.4) + 7(0.4) + 8(0.2) = 6.8 \\ &= E(X) + 4 = 6.8 \end{aligned}$$

as it should be.

The values of the random variable $X + Y$ are presented in Table 5.8 along with the corresponding probabilities.

We find that

$$\begin{aligned} E(X + Y) = \sum\sum (x + y)f(x, y) &= 2(0) + 3(0.1) + 4(0.1) + 3(0.1) \\ &+ 4(0) + 5(0.1) + 4(0.1) + 5(0.1) + 6(0) + 5(0.2) \\ &+ 6(0) + 7(0) + 6(0) + 7(0.2) + 8(0) \\ = 4.8 \end{aligned}$$

which is identical to

$$E(X) + E(Y) = 2.8 + 2 = 4.8$$

Since $f(x, y)$ is not equal to $f(x)g(y)$ for the probability distribution presented in Table 5.7, we do not expect $E(XY)$ to be equal to $E(X)E(Y)$. It is

Table 5.8 Values of $X+Y$ and Their Probabilities

	X		
Y	2	3	4
0	2 (0)	3 (0.1)	4 (0.1)
1	3 (0.1)	4 (0)	5 (0.1)
2	4 (0.1)	5 (0.1)	6 (0)
3	5 (0.2)	6 (0)	7 (0)
4	6 (0)	7 (0.2)	8 (0)

left to the reader to verify that $E(XY) = 5.2$ while $E(X)E(Y) = 5.6$ in this example.

As an illustration of the case in which $E(XY) = E(X)E(Y)$, consider the independent random variables X and Y whose joint probability distribution is presented in Table 5.6. The expected values of the product XY in this case is

$$E(XY) = \sum\sum xyf(x, y) = 2\left(\frac{1}{8}\right) + 4\left(\frac{1}{8}\right) + 6\left(\frac{3}{8}\right) + 12\left(\frac{3}{8}\right) = 7.5$$

$$= E(X)E(Y) = (3)(2.5) = 7.5$$

as expected.[6]

[6] To illustrate the use of mathematical expectation in the case of a continuous distribution, suppose that X is a continuous random variable with p.d.f:

$$f(x) = \frac{1}{(b-a)} \text{ for } a < x < b$$

This is the rectangular distribution discussed in Section 5.4. The expected value of X is

$$E(X) = \int_a^b x\left(\frac{1}{(b-a)}\right) dx = \left(\frac{1}{(b-a)}\right)\int_a^b x\,dx = \frac{(b+a)}{2}$$

Furthermore, if c is a constant, then

$$E(X+c) = \int_a^b (x+c)\frac{1}{b-a}dx = \frac{1}{b-a}\int_a^b (x+c)\,dx = \frac{1}{b-a}\int_a^b x\,dx$$

$$+\frac{c}{b-a}\int_a^b 1\,dx = \frac{b+a}{2} + c$$

In general, if $\phi(x)$ is a function of x and if X is a random variable with probability density function $f(x)$, then

$$E[\phi(x)] = \sum_x \phi(x)f(x) \qquad \text{if } X \text{ is discrete}$$

$$E[\phi(x)] = \int_{-\infty}^{\infty} \phi(x)f(x)\,dx \qquad \text{if } X \text{ is continuous}$$

provided that $E[\phi(x)]$ exists and $\phi(x)$ is continuous if X is continuous.

Furthermore, if $\phi(x, y)$ is a function of the real variables x and y while X and Y are random variables with joint density function $f(x, y)$, then

$$E[\phi(x, y)] = \sum_x \sum_y \phi(x, y)f(x, y) \qquad \text{if } X \text{ and } Y \text{ are discrete}$$

$$E[\phi(x, y)] = \int_{-\infty}^{\infty} \int_{-\infty}^{\infty} \phi(x, y)f(x, y)\,dx\,dy \qquad \text{if } X \text{ and } Y \text{ are continuous}$$

provided that $E[\phi(x, y)]$ exists and $\phi(x, y)$ is continuous if X and Y are continuous.

By letting $\phi(x) = x$, $\phi(x) = x + c$, $\phi(x, y) = x + y$, or $\phi(x, y) = xy$, we see that several special forms of the general statements have already been discussed.

Suppose that $\phi(x) = cX$ where X is a random variable with probability density function $f(x)$ and c is a constant. The expected value of cX is equal to c times the expected value of X. That is, $E(cX) = cE(X)$. If, for example, X is a discrete random variable, we have

$$E(cX) = \sum cxf(x) = c \sum xf(x) = cE(X) \tag{5.16}$$

Still another special mathematical expectation is obtained by setting $\phi(x)$ equal to $(x - \mu)^2$ where μ is the expected value (the mean) of the random variable X. If X is a discrete random variable with probability density function $f(x)$, then

$$E[(x - \mu)^2] = \sum_i (x_i - \mu)^2 f(x_i)$$

This sum of products is a weighted average of the squares of the deviations of the values of X from the mean value μ. The "weight" associated with each $(x_i - \mu)^2$ is the probability $f(x_i)$. The quantity $E[(X - \mu)^2]$ is called the *variance of X*.[7] The variance of X is commonly denoted by Var(X), $V(X)$, or σ^2. The positive square root of the variance σ^2 is called the *standard deviation of X* and is denoted by σ.

[7] If X is a continuous random variable, then the variance of X is defined as

$$E[(X - \mu)^2] = \int_{-\infty}^{\infty} (x - \mu)^2 f(x)\,dx$$

There are several rules for operating with variances that will be used frequently in the following chapters. To begin with, we observe that

$$\begin{aligned}
\mathrm{Var}(X) = E[(X-\mu)^2] &= E[X^2 - 2\mu X + \mu^2] && \text{by squaring } (X-\mu) \\
&= E(X^2) - 2\mu E(X) + \mu^2 && \text{since } E \text{ is a linear operator} \\
&= E(X^2) - 2\mu^2 + \mu^2 && \text{since } E(X) = \mu \\
&= E(X^2) - \mu^2
\end{aligned}$$

where $E(X^2) = \sum x^2 f(x)$ if X is a discrete random variable.[8] The phrase "E is a linear operator" is used as a reminder that the expected value of a sum of random variables is equal to the sum of the expected values. Since $\mu^2 = [E(X)]^2$, this result is often written as

$$\sigma^2 = E(X^2) - [E(X)]^2 \tag{5.17}$$

By rearrangement of the terms in (5.17), we obtain $E(X^2) = \sigma^2 - [E(X)]^2$. This relationship often provides an easy way to find $E(X^2)$.

If c is a constant and X is a random variable with variance $\mathrm{Var}(X)$, then

$$\mathrm{Var}(X + c) = \mathrm{Var}(X) \tag{5.18}$$

This result is obtained as follows. We have

$$\begin{aligned}
\mathrm{Var}(X + c) &= E[X + c - E(X + c)]^2 \\
&= E[X + c - E(X) - c]^2 \\
&= E[X - E(X)]^2 = \mathrm{Var}(X)
\end{aligned}$$

Furthermore,

$$\mathrm{Var}(cX) = c^2\,\mathrm{Var}(X) \tag{5.19}$$

To verify this result, let $W = cX$. The variance of W is equal to

$$\mathrm{Var}(W) = E(W^2) - [E(W)]^2$$

Substituting cX for W, we obtain

$$\begin{aligned}
\mathrm{Var}(cX) &= E(c^2X^2) - [E(cX)]^2 \\
&= c^2E(X^2) - c^2[E(X)]^2 \\
&= c^2[E(X^2) - [E(X)]^2] = c^2\,\mathrm{Var}(X)
\end{aligned}$$

as was to be shown.

[8] If X is a continuous random variable, then

$$E(X^2) = \int_{-\infty}^{\infty} x^2 f(x)\,dx$$

5.6 COVARIANCE AND CORRELATION

The *covariance* of two random variables X and Y is defined as the expected value of the product of their paired deviations from their own means and is denoted by Cov(X, Y). In symbols:

$$\text{Cov}(X, Y) = E[(X - \mu_x)(Y - \mu_y)] \tag{5.20}$$

The covariance formula can be thought of as a generalization of the variance formula presented earlier. As will be shown later, the covariance is extremely useful when measuring the degree of dependence between the variables X and Y.

By definition, the variance of the sum of two random variables X and Y is equal to

$$\text{Var}(X + Y) = E[(X + Y - \mu_{x+y})^2]$$

It was shown earlier that the mean of $X + Y$ is equal to the mean of X plus the mean of Y regardless of whether X and Y are independent or not. That is,

$$\mu_{x+y} = E(X + Y) = E(X) + E(Y) = \mu_x + \mu_y$$

Substituting $\mu_x + \mu_y$ for μ_{x+y} and squaring the term $(X + Y - \mu_x - \mu_y)$, we obtain

$$\text{Var}(X + Y) = E[(X - \mu_x)^2 + 2(X - \mu_x)(Y - \mu_y) + (Y - \mu_y)^2]$$

Since the expected value of a sum is equal to the sum of the expected values, we have

$$\begin{aligned}\text{Var}(X + Y) &= E[(X - \mu_x)^2] + E[(Y - \mu_y)^2] + 2E[(X - \mu_x)(Y - \mu_y)] \\ &= \text{Var}(X) + \text{Var}(Y) + 2\,\text{Cov}(X, Y)\end{aligned}$$

In other words, if X and Y are random variables with finite variances, then the variance of their sum is equal to the sum of their variances plus two times their covariance.

A convenient computational formula for the covariance is easily derived from the definitional formula (5.20). Multiplying out the product $(X - \mu_x)(Y - \mu_y)$, we obtain

$$\text{Cov}(X, Y) = E[XY - \mu_x Y - \mu_y X + \mu_x \mu_y]$$

Recalling that the expected value of a sum is the sum of the expected values, we obtain

$$\text{Cov}(X, Y) = E(XY) - E(\mu_x Y) - E(\mu_y X) + E(\mu_x \mu_y)$$

But μ_x and μ_y are constants, and thus the rules $E(c) = c$ and $E(cX) = cE(X)$ apply. Therefore, we have

$$\text{Cov}(X, Y) = E(XY) - \mu_x E(Y) - \mu_y E(X) + \mu_x \mu_y$$

which is equivalent to

$$\text{Cov}(X, Y) = E(XY) - \mu_x \mu_y \tag{5.21}$$

because $E(Y) = \mu_y$ and $E(X) = \mu_x$.

As a numerical illustration, suppose that X and Y have the joint probability distribution presented in Table 5.9. We have

$$E(X) = \mu_x = 1\left(\frac{5}{9}\right) + 2\left(\frac{4}{9}\right) = \frac{13}{9}$$

$$E(Y) = \mu_y = -1\left(\frac{4}{9}\right) + 2\left(\frac{5}{9}\right) = \frac{6}{9}$$

$$E(X^2) = 1\left(\frac{5}{9}\right) + 4\left(\frac{4}{9}\right) = \frac{21}{9}$$

$$E(Y^2) = 1\left(\frac{4}{9}\right) + 4\left(\frac{5}{9}\right) = \frac{24}{9}$$

$$E(XY) = -1\left(\frac{2}{9}\right) - 2\left(\frac{2}{9}\right) + 2\left(\frac{3}{9}\right) + 4\left(\frac{2}{9}\right) = \frac{8}{9}$$

$$\text{Var}(X) = E(X^2) - \mu_x{}^2 = \left(\frac{189}{81}\right) - \left(\frac{169}{81}\right) = \frac{20}{81}$$

$$\text{Var}(Y) = E(Y^2) - \mu_y{}^2 = \left(\frac{216}{81}\right) - \left(\frac{36}{81}\right) = \frac{180}{81}$$

$$\text{Cov}(X, Y) = E(XY) - \mu_x \mu_y = \left(\frac{72}{81}\right) - \left(\frac{78}{81}\right) = -\left(\frac{6}{81}\right)$$

Thus

$$\begin{aligned}\text{Var}(X + Y) &= \text{Var}(X) + \text{Var}(Y) + 2\,\text{Cov}(X, Y)\\ &= \left(\frac{20}{81}\right) + \left(\frac{180}{81}\right) - \left(\frac{12}{81}\right) = \frac{188}{81} = 2.32\end{aligned}$$

Notice that the covariance is negative in this example. Although the variance of a random variable is always positive, the covariance of a pair of random variables can be either positive or negative.

Table 5.9 A Joint Probability Distribution

	X		
Y	1	2	g(y)
−1	$\frac{2}{9}$	$\frac{2}{9}$	$\frac{4}{9}$
2	$\frac{3}{9}$	$\frac{2}{9}$	$\frac{5}{9}$
f(x)	$\frac{5}{9}$	$\frac{4}{9}$	

We have seen that when X and Y are independent random variables, the expected value of their product is the product of their expected values. That is,

$$E(XY) = E(X)E(Y) = \mu_x \mu_y$$

Therefore, if X and Y are independent, their covariance must be zero because then

$$\text{Cov}(X, Y) = E(XY) - \mu_x \mu_y = \mu_x \mu_y - \mu_x \mu_y = 0$$

From this it follows that, if X and Y are independent, then the variance of the sum $X + Y$ is equal to the sum of the variance of X and the variance of Y. In symbols, if X and Y are independent random variables with finite variances, then

$$\text{Var}(X + Y) = \text{Var}(X) + \text{Var}(Y) \tag{5.22}$$

This result can be extended to the sum of a finite number of random variables. Thus, if $X_1, X_2, \ldots, X_m$ are m independent random variables with variances $\text{Var}(X_1), \text{Var}(X_2), \ldots, \text{Var}(X_m)$, then

$$\text{Var}(X_1 + X_2 + \cdots + X_m) = \text{Var}(X_1) + \text{Var}(X_2) + \cdots + \text{Var}(X_m) \tag{5.23}$$

A zero covariance does not necessarily imply that two random variables are independent, however. For example, we know that the variables X and Y having the joint probability distribution presented in Table 5.2 are not independent because $f(x, y) \neq f(x)g(y)$. Yet, if we compute the covariance of X and Y, we find

$$E(XY) = 1\left(\frac{1}{9}\right) + 3\left(\frac{1}{9}\right) + 2\left(\frac{1}{9}\right) + 6\left(\frac{1}{9}\right) + 3\left(\frac{1}{9}\right) + 9\left(\frac{1}{9}\right) = \frac{24}{9}$$

$$E(X) = 1\left(\frac{1}{3}\right) + 2\left(\frac{1}{3}\right) + 3\left(\frac{1}{3}\right) = 2$$

$$E(Y) = 0\left(\frac{3}{9}\right) + 1\left(\frac{2}{9}\right) + 2\left(\frac{2}{9}\right) + 3\left(\frac{2}{9}\right) = \frac{4}{3}$$

and

$$\text{Cov}(X, Y) = E(XY) + E(X)E(Y) = \left(\frac{24}{9}\right) - \left(\frac{24}{9}\right) = 0$$

Thus, in this particular case, the variance of the sum $X + Y$ would be equal to the sum of their variances even though they are not independent.

The units assigned to the covariance depend on the units in which X and Y are measured. For example, if X and Y are both measured in feet, then the covariance is expressed in terms of square feet. If, however, X was expressed in yards and Y in feet, the units of the covariance would be yards times feet. In some cases, it is convenient to have a measure of the degree of linear relationship between X and Y that does not depend upon their units of measurement. Such a measure can be obtained by transforming X and Y into unitless (or standardized) variables and then using the covariance of the new variables as a unitless measure of the degree of linear relationship between X and Y. To do this, we first form the variables

$$X^* = \frac{X - \mu_x}{\sigma_x} \qquad \text{and} \qquad Y^* = \frac{Y - \mu_y}{\sigma_y}$$

where σ_x and σ_y are the standard deviations of X and Y, respectively. The expected value of X^* is equal to

$$E(X^*) = E\left[\frac{X - \mu_x}{\sigma_x}\right] = \frac{1}{\sigma_x}[E(X - \mu_x)] = \frac{1}{\sigma_x}[E(X) - \mu_x] = 0$$

Letting $1/\sigma_x = c$ and using the rule that $\text{Var}(cX) = c^2\,\text{Var}(X)$, we find that

$$\text{Var}(X^*) = \text{Var}\left[\frac{X - \mu_x}{\sigma_x}\right] = \frac{1}{\sigma_x{}^2}\text{Var}(X - \mu_x)$$

But μ_x is a constant and therefore has a variance of zero. Thus, $\text{Var}(X - \mu_x)$ is equal to $\text{Var}(X) = \sigma_x{}^2$, from which it follows that $\text{Var}(X^*)$ is equal to one.

Similar calculations would show that the expected value of Y^* is equal to zero also and its variance is equal to one.

The variables X^* and Y^* are unitless and are called standardized variables. For example, if X is measured in pounds, then $(X - \mu_x)$ and σ_x are also measured in pounds. In this case, the variable X^* is obtained by dividing pounds into pounds and, therefore, the units cancel out leaving X^* unitless.

Since the expected values of X^* and Y^* are equal to zero, their covariance is equal to

$$\text{Cov}(X^*, Y^*) = E(X^*Y^*)$$

Making the appropriate substitutions and simplifying the result, we obtain

$$\text{(5.24)} \quad \text{Cov}(X^*, Y^*) = E\left[\frac{X - \mu_x}{\sigma_x} \cdot \frac{Y - \mu_y}{\sigma_y}\right] = \frac{1}{\sigma_x \sigma_y} E[(X - \mu_x)(Y - \mu_y)]$$
$$= \frac{\text{Cov}(X, Y)}{\sigma_x \sigma_y}$$

Since X^* and Y^* are unitless, their covariance is also unitless. For example, if X is measured in feet and Y in pounds, then Cov(X, Y) is measured in feet times pounds while σ_x and σ_y are measured in feet and pounds, respectively. The units cancel, however, when the covariance of X and Y is divided by the product $\sigma_x \sigma_y$ to obtain the covariance of X^* and Y^*.

The covariance of X^* and Y^* is commonly denoted by ρ and referred to as the *correlation coefficient*. The correlation coefficient is either positive or negative depending on the sign of Cov(X, Y). In a later chapter, it is shown that the correlation coefficient cannot be less than -1 or greater than $+1$. If $\rho = 1$, we say that there is a perfect positive correlation between X and Y. If $\rho = -1$, there is a perfect negative correlation between them. When X and Y are independent, Cov(X, Y) is equal to zero and, therefore, the correlation coefficient is also equal to zero. We have seen, however, that the covariance of X and Y can be zero even though X and Y are dependent. Therefore, when $\rho = 0$, we say X and Y are *uncorrelated* or that there is *no linear relationship* between X and Y.

As a numerical illustration, let us compute the correlation coefficient for the joint probability distribution presented in Table 5.9. We obtain

$$\rho = \frac{\text{Cov}(X, Y)}{\sigma_x \sigma_y} = \frac{-(6/81)}{\sqrt{20/81}\sqrt{180/81}} = \frac{-6}{\sqrt{3600}} = -0.10$$

Since the correlation coefficient is negative and close to zero, we say there is a weak negative linear relationship between X and Y.

5.7 SOME PROBABILITY PROBLEMS

The following examples will allow us to bring together a few of the many concepts presented in this chapter.

Example 1: A filling station receives gasoline once a week. Past sales records suggest that the probability distribution of weekly sales X measured in thousands of gallons is approximated very well by an isosceles triangle with a base extending from one to three thousand gallons per week and a vertex at two thousand gallons. The probability density function $f(x)$ is shown in Figure 5.14.

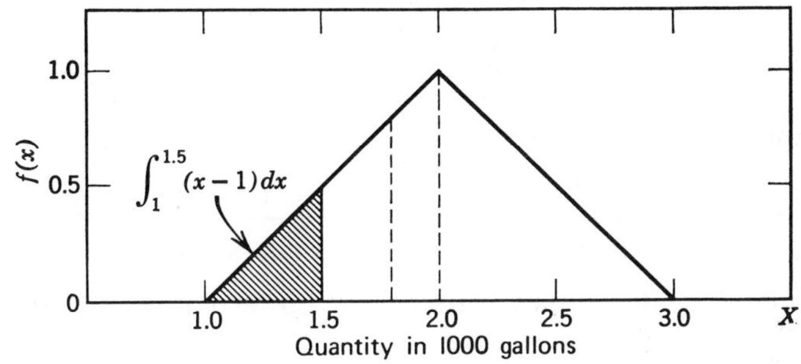

Figure 5.14 Probability density function for gasoline sales.

The total area of an isosceles triangle is equal to the length of the base b times the height h of the triangle at the vertex divided by 2. That is,

$$\text{Area} = \frac{bh}{2}$$

The length of the base in this example is equal to $b = (3 - 1) = 2$. If the triangle represents a probability density function, the total area of the triangle is equal to 1. Upon solving the equation,

$$\frac{2h}{2} = 1$$

we find the height of the triangle to be $h = 1$.

The probability density function in this case is equal to

$$f(x) = \begin{cases} (x-1) & \text{for} \quad 1 \le x \le 2 \\ (3-x) & \text{for} \quad 2 \le x \le 3 \\ 0 & \text{otherwise} \end{cases} \tag{5.25}$$

Thus, if $x = 1.5$, $f(1.5) = (1.5 - 1) = 0.5$, and if $x = 2.8$, then $f(2.8) = (3.0 - 2.8) = 0.2$.

The probability that x is less than a given number c for $1 \le c \le 2$ is equal to

$$P(X < c) = \int_1^c f(x)\,dx = \int_1^c (x-1)\,dx = F(c)$$

This probability is represented by the shaded area in the left-hand tail of the probability distribution in Figure 5.14. Since we are working with a right triangle, the area is easily found without using calculus. The area of a right triangle is equal to one-half times the base b times the height h. The base of the triangle b, in this case, is equal to $(c - 1)$ and the height is equal to $h = f(c) = (c - 1)$, provided that $1 \leq c \leq 2$. Therefore,

$$P(X < c) = \frac{(c-1)^2}{2}$$

For example, if $c = 1.5$, then

$$P(X < 1.5) = \int_1^{1.5} (x - 1)\, dx = \frac{(1.5-1)^2}{2} = \frac{0.250}{2} = 0.125$$

The probability that x lies between two given numbers c and d where $c < d$, $1 \leq c \leq 2$, and $1 \leq d \leq 2$ is equal to

$$P(c < X < d) = P(X < d) - P(X < c) = F(d) - F(c) = \frac{(d-1)^2}{2} - \frac{(c-1)^2}{2}$$

This probability is represented by the shaded area in Figure 5.15. The shaded

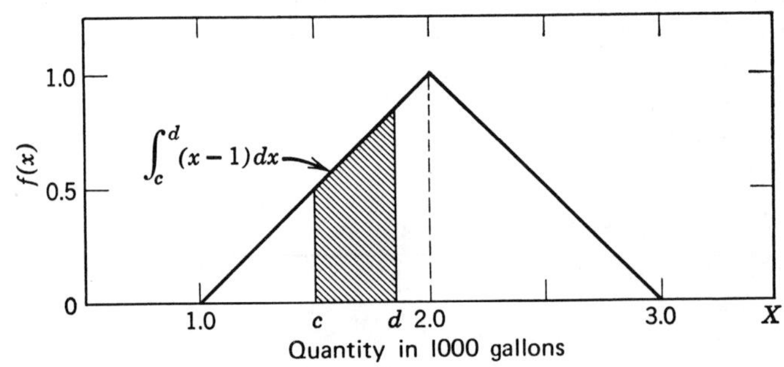

Figure 5.15 Probability density function for gasoline sales.

area is equal to the area under the triangle with base $(c - 1)$ subtracted from the area under the larger triangle with base $(d - 1)$. As example, let $c = 1.5$ and $d = 1.8$; then

$$P(1.5 < X < 1.8) = \frac{(1.8-1)^2 - (1.5-1)^2}{2} = \frac{0.64 - 0.25}{2} = 0.195$$

[Find $P(1.2 < X < 1.6)$.]

The probability density function (5.25) is symmetrical around $X = 2$.

Therefore,

$$P(X < 2) = P(X > 2) = 0.500$$

and

$$P(2 - k < X < 2) = P(2 < X < 2 + k)$$

where k is any given number between 0 and 1. Suppose, for example, that $k = 0.2$; then

$$P(1.8 < X < 2) = P(X < 2) - P(X < 1.8) = 0.500 - 0.320 = 0.180$$

and it follows immediately from the symmetry of $f(x)$ that

$$P(2 < X < 2.2) = 0.180$$

[Draw a graph of $f(x)$ and shade in the area corresponding to $P(2 < X < 2.2)$ and $P(1.8 < X < 2)$.]

Probabilities in the right-hand tail of $f(x)$ can be found directly by noting that the base of a right triangle located in the right-hand tail of $f(x)$ is equal to $(3 - x)$. Thus, for $2 \leq X \leq 3$, we have

$$P(r < X < 3) = \frac{(3 - r)^2}{2}$$

where r is a given number between 2 and 3. If $r = 2.6$, for example, then

$$P(2.6 < X < 3.0) = \frac{(3.0 - 2.6)^2}{2} = \frac{0.16}{2} = 0.08$$

[Find $P(2.8 < X < 3.0)$.]

If the probability distribution of X is to be used frequently, it would be helpful to have a table of the distribution function $F(m)$ where

$$F(m) = P(X < m) \quad \text{for} \quad 1 \leq m \leq 3$$

Table 5.10 is such a table in which only a few values of m are presented, but additional values could be added easily. As an illustration of the use of Table 5.10, we find that the probability that X lies between 1.3 and 1.9 is equal to

$$P(1.3 < X < 1.9) = F(1.9) - F(1.3) = 0.405 - 0.045 = 0.360$$

Similarly,

$$P(1.6 < X < 2.3) = 0.755 - 0.180 = 0.575$$

and

$$P(2.2 < X < 2.7) = 0.955 - 0.680 = 0.275$$

[Use Table 5.10 to find $P(X > 2.4)$.]

Table 5.10 The Distribution Function $F(m)$

m	$F(m)$	m	$F(m)$
1.0	0.000	2.1	0.595
1.1	0.005	2.2	0.680
1.2	0.020	2.3	0.755
1.3	0.045	2.4	0.820
1.4	0.080	2.5	0.875
1.5	0.125	2.6	0.920
1.6	0.180	2.7	0.955
1.7	0.245	2.8	0.980
1.8	0.320	2.9	0.955
1.9	0.405	3.0	1.000
2.0	0.500		

The expected value of X is equal to

$$E(X) = \int_1^2 x(x-1)\,dx + \int_2^3 x(3-x)\,dx = \frac{5}{6} + \frac{7}{6} = 2$$

as is obvious from an inspection of Figure 5.14.

Suppose the owner of the filling station does not want the station to be out of gas more than 8 percent of the time when the weekly tank truck comes to fill the station's storage tanks. What is the minimum amount of gas storage capacity the station should have? To answer this question, we must find that quantity of gas s for which $P(X > s) = 0.08$. The quantity s is obviously between two and three thousand gallons. We can calculate s by solving the equation $(3 - s)^2/2 = 0.08$ for the value of s that lies between 2 and 3. Doing so, we obtain $s = 2.6$. This can be obtained directly from Table 5.10 by noting that

$$P(X > s) = 1.00 - F(s) = 0.08$$

when $s = 2.6$.

Example 2: If a discrete variable X is such that

$$P(X = x) = \left(\frac{1}{m+1}\right)\left(\frac{m}{m+1}\right)^x = f(x) \tag{5.26}$$

for $m > 0$ and $x = 0, 1, 2, 3, \ldots$, then X is said to possess a *Pascal distribution.*

Part of the probability distribution of X for $m = 3$ is presented in Table 5.11 and is shown in Figure 5.16. The probabilities in Table 5.11 are equal to

$$P(X = x) = \left(\frac{1}{4}\right)\left(\frac{3}{4}\right)^x$$

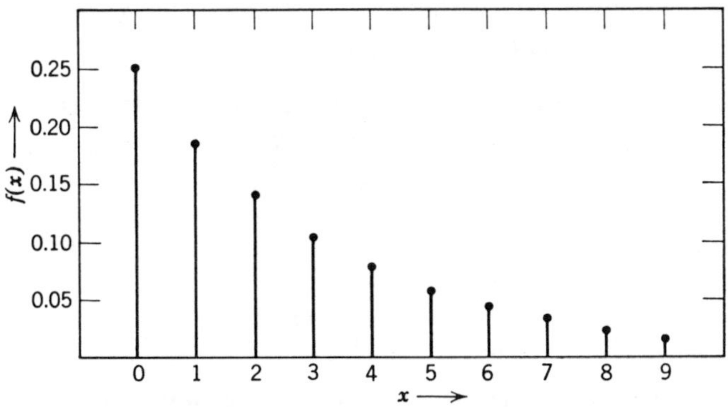

Figure 5.16 Pascal's distribution for $m = 3$.

Notice that as x becomes large, $P(X = x)$ becomes small. With $m = 3$, the probability $P(X = x)$ is effectively equal to zero by the time x reaches 20.

Table 5.11 Pascal's Distribution for $m = 3$

x	$f(x)$	x	$f(x)$
0	0.250	5	0.059
1	0.187	6	0.044
2	0.141	7	0.033
3	0.105	8	0.025
4	0.079	9	0.019

To verify that $f(x)$ is actually a probability density function, we must show that $f(x) \geq 0$ for all possible values of x and $\sum_x f(x) = 1$. That $f(x) \geq 0$ for $x = 0, 1, 2, 3, \ldots,$ follows directly from the restriction that m is greater than zero. We next need to show that $\sum_x f(x) = 1$. To simplify the notation, let

$$r = \frac{m}{m+1} \qquad \text{and} \qquad \phi = \sum_x \left(\frac{m}{m+1}\right)^x$$

Then

$$\sum{}' f(x) = \sum_x \frac{1}{m+1}\left(\frac{m}{m+1}\right)^x = \frac{1}{m+1}\sum_x \left(\frac{m}{m+1}\right)^x = \frac{1}{m+1}(\phi)$$

Note that

$$\phi = 1 + r + r^2 + r^3 + r^4 + \cdots$$

regardless of the value of m because $r^0 = 1$. Multiplying ϕ by r gives

$$r\phi = r + r^2 + r^3 + r^4 + \cdots$$

The summation ϕ is an infinite geometric sequence, the last term of which is equal to zero. Subtracting $r\phi$ from ϕ, we obtain

$$\phi - r\phi = (1 - r)\phi = 1$$

Solving for ϕ, we find

$$\phi = \frac{1}{(1 - r)} = m + 1$$

Therefore, $\sum_x f(x) = (m + 1)/(m + 1) = 1$ and we have shown that $f(x)$ is, indeed, a probability density function.

The expected value of X is equal to

$$E(X) = \sum_x x f(x) = \sum_x x\left(\frac{1}{m + 1}\right)\left(\frac{m}{m + 1}\right)^x = \frac{1}{m + 1}\sum_x xr^x$$

Let

$$S = \sum_x x\left(\frac{m}{m + 1}\right)^x = \sum_x xr^x$$

Then

$$S = r + 2r^2 + 3r^3 + 4r^4 + \cdots$$

and

$$rS = r^2 + 2r^3 + 3r^4 + \cdots$$

Subtracting rS from S, gives

$$S - rS = (1 - r)S = r + r^2 + r^3 + r^4 + \cdots$$

We note that

$$\phi = 1 + r + r^2 + r^3 + r^4 + \cdots$$

and therefore

$$(1 - r)S = \phi - 1 = m \cdot$$

from which it follows that $S = (m + 1)(m)$. Thus, the expected value of X is equal to

$$E(X) = \frac{(m + 1)(m)}{(m + 1)} = m$$

[Show that $\text{Var}(X) = m^2 + m$.]

Past records of a telegraph office suggest that the probability distribution of telegrams received between 4 and 5 A.M. is approximated very well by a Pascal's distribution with $m = 3$. What is the probability that the office will receive more than three telegrams tomorrow morning between 4 and 5 A.M.? This probability is quickly obtained by subtracting from one the probability that three or less telegrams are received. In this way, we obtain

$$P(X > 3) = 1 - P(X \leq 3) = 1.000 - 0.250 - 0.187 - 0.141 - 0.105$$
$$= 0.327$$

The information needed to compute $P(X > 3)$ is contained in Table 5.11.

What is the probability that the office will receive more than three telegrams, given that at least two telegrams have been received? We are interested here in finding the conditional probability

$$P(X > 3 \mid X \geq 2)$$

Let $X \geq 2$ be the event A and $X > 3$ be the event B; then

$$P(B \mid A) = \frac{P(AB)}{P(A)}$$

If $X > 3$, then the event $X \geq 2$ has occurred; thus $P(AB) = P(B)$. Since

$$P(B) = P(X > 3) = 1 - P(X \leq 3) = 0.327$$

and

$$P(A) = P(X \geq 2) = 1 - P(X < 2) = 0.563$$

it follows that

$$P(X > 3 \mid X \geq 2) = \frac{P(X > 3)}{P(X \geq 2)} = \frac{0.327}{0.563} = 0.582$$

The Pascal probability distribution used here provides a simple description of the arrival of telegrams during a given hour. Such a description may be adequate for making decisions as to the number of people to have in the office between 4 and 5 A.M. or determining the amount of equipment that is likely to be in operation at that time. The Pascal distribution, by itself, provides an inadequate description of telegram arrivals if it is necessary to answer questions such as:

(a) What is the probability of receiving five telegrams during a given 10-minute period between 4 and 5 A.M.?

(b) What is the probability of receiving more than five telegrams between 4 and 5 A.M., given that only two telegrams had been received by 4:50 A.M.?

Example 3: Let X and Y be discrete random variables which have the joint probability distribution presented in Table 5.12.

Table 5.12 The Joint Probability Distribution of X and Y

	X			
Y	1	2	3	$g(y)$
2	0.1	0.2	0.1	0.4
4	0.1	0.2	0.3	0.6
$f(x)$	0.2	0.4	0.4	

The expected values of X and Y are equal to

$$E(X) = \sum_i x_i f(x_i) = 1(0.2) + 2(0.4) + 3(0.4) = 2.2$$

and

$$E(Y) = \sum_j y_j g(y_j) = 2(0.4) + 4(0.6) = 3.2$$

[See equation (5.9).]

The expected value of $X + Y$ is equal to

$$\begin{aligned} E(X+Y) &= 3(0.1) + 4(0.2) + 5(0.1) + 5(0.1) + 6(0.2) + 7(0.3) \\ &= E(X) + E(Y) = 5.4 \end{aligned}$$

[See equation (5.12).]

It is easy to verify that $f(x, y)$ is not equal to the product $f(x)g(y)$ for all values of X and Y in Table 5.12. For example, if $x = 2$ and $y = 4$, then $f(2, 4) = 0.2$ while $f(2)g(4) = (0.4)(0.6) = 0.24$. Thus X and Y are not independent random variables, and we do not expect $E(X)E(Y)$ to be equal to $E(XY)$ in this case. The expected value of the product XY is, therefore, obtained directly from the definition $E(XY) = \sum \sum xyf(x, y)$. Carrying out the arithmetic, we find $E(XY) = 2(0.1) + 4(0.2) + 6(0.1) + 4(0.1) + 8(0.2) + 12(0.3) = 7.2$.

The covariance of X and Y is equal to

$$\text{Cov}(X, Y) = E(XY) - E(X)E(Y) = 7.2 - (2.2)(3.2) = 0.16$$

Thus, there is a weak, positive linear relationship between X and Y.

The conditional density function of X, given $Y = 2$, is equal to

$$f(x \mid Y = 2) = \frac{f(x, 2)}{g(2)} = \frac{f(x, 2)}{0.4}$$

and therefore

$$f(1 \mid Y = 2) = \left(\frac{0.1}{0.4}\right) = 0.25$$

$$f(2 \mid Y = 2) = \left(\frac{0.2}{0.4}\right) = 0.50$$

$$f(3 \mid Y = 2) = \left(\frac{0.1}{0.4}\right) = 0.25$$

[Find $f(x \mid Y = 4)$.]

The conditional mean of X, given $Y = 2$, is equal to

$$E(X \mid Y = 2) = \sum_i x_i f(x_i \mid Y = 2) = 1(0.25) + 2(0.50) + 3(0.25) = 2.0$$

[Find $E(X \mid Y = 4)$.]

If each value of X is multiplied by a constant $c = 2$, the expected value of the new random variable $2X$ is equal to $2E(X)$. Checking this numerically, we find

$$E(2X) = \sum_i 2x_i\, f(x_i) = 2(0.2) + 4(0.4) + 6(0.4) = 4.4 = 2E(X)$$

[See equation (5.16). Show that $E(3Y) = 3E(Y)$.]

If a constant $c = 2$ is added to each value of X, then

$$E(X + 2) = \sum_i (x_i + 2) f(x_i) = 3(0.2) + 4(0.4) + 5(0.4) = 4.2 = E(X) + 2$$

as expected. [See equation (5.11). Show that $E(Y + 3) = E(Y) + 3$.]

The variance of X is equal to

$$\text{Var}(X) = E(X^2) - (E(X))^2 = 1(0.2) + 4(0.4) + 9(0.4) - (2.2)^2 = 0.56$$

If each value of X is multiplied by a constant $c = 2$, the variance of the new variable $2X$ should be equal to $c^2 = 4$ times $\text{Var}(X)$. Numerically, we have

$$\begin{aligned}\text{Var}(2X) &= 4(0.2) + 16(0.4) + 36(0.4) - (4.4)^2 = 21.60 - 19.36 = 2.24 \\ &= 4\,\text{Var}(X)\end{aligned}$$

[See equation (5.19). Show that $\text{Var}(3Y) = 9\,\text{Var}(Y)$.]

If the constant $c = 2$ is added to each value of X, then $\text{Var}(X + 2)$ is equal to $\text{Var}(X)$. That is,

$$\begin{aligned}\text{Var}(X + 2) &= \sum_i (x_i + 2)^2 f(x_i) - \left(\sum_i (x_i + 2) f(x_i)\right)^2 = 9(0.2) + 16(0.4) \\ &\quad + 25(0.4) - (4.2)^2 = 18.20 - 17.64 = 0.56 = \text{Var}(X)\end{aligned}$$

[See equation (5.18). Show that $\mathrm{Var}(Y+3) = \mathrm{Var}(Y)$.]

A new random variable W is defined as $W = [X - E(X)]/\sigma_x$. The random variable W takes on the values

$$w_1 = \frac{1.0 - 2.2}{0.748} = -1.604$$

$$w_2 = \frac{2.0 - 2.2}{0.748} = -0.267$$

and

$$w_3 = \frac{3.0 - 2.2}{0.748} = 1.069$$

The expected value of W is equal to

$$E(W) = -1.604(0.2) - 0.267(0.4) + 1.069(0.4) = 0$$

[Show that $\mathrm{Var}(W) = 1$.]

The variance of $X + Y$ is equal to

$$\begin{aligned} \mathrm{Var}(X+Y) &= E[(X+Y-E(X+Y))^2] = E[(X+Y-5.4)^2] \\ &= \sum_i \sum_j (x_i + y_j - 5.4)^2 f(x_i, y_j) \\ &= (-2.4)^2(0.1) + (-1.4)^2(0.2) + (-0.4)^2(0.1) \\ &\quad + (-0.4)^2(0.1) + (0.6)^2(0.2) + (1.6)^2(0.3) = 1.84 \end{aligned}$$

To check this result, recall that

$$\begin{aligned} \mathrm{Var}(X+Y) &= \mathrm{Var}(X) + \mathrm{Var}(Y) + 2\,\mathrm{Cov}(X, Y) \\ &= 0.56 + 0.96 + 2(0.16) = 1.84 \end{aligned}$$

[See Section 5.6. Show that $\mathrm{Var}(X+Y+c) = \mathrm{Var}(X+Y)$ when c is a constant equal to 3.]

* 5.8 DECISION MAKING

There are several ways to define "rational" behavior. Generally, the concept of rationality involves doing the best you can to get what you want under the given circumstances. This concept can be useful in theory but is often too vague in practice. It gives no hint as to what is "best," what is "wanted," or what constitutes the "circumstances." Therefore, it is not surprising that many alternative definitions of rationality have been suggested.

* This section may be omitted without loss of continuity.

Economists often assume that rational behavior consists of trying to maximize mathematical expectations. This simplification of reality is useful for theoretical purposes and is often very helpful both in applied economics and in business administration. The maximization of mathematical expectations usually requires a great deal of information which in practice may not be available or may be very costly to obtain.

The "rational" decision maker may try to assure some minimum gain when there is very little information on the probabilities of some events or when the payoff from the decision depends on an unknown action taken by someone else.

In many cases, the "rational" decision maker may be willing to accept a lower expected profit if, by doing so, he can reduce the variation of his profits. For example, an investor may prefer to buy a number of different stocks even though the expected return on his investment is considerably less than could be obtained from investing all of his money in one particular stock. Many people would argue that there is a great deal of wisdom in the old saying, "Don't put all of your eggs in one basket." Attempting to maximize expected profits may be a very "irrational" thing to do at times.

Regardless of how rational behavior is defined, there are still many questions to answer before the definition can be used in actual decision-making situations. How can expectations be described? What do they depend on? How can they be translated into decision rules? How can the necessary probabilities be estimated? How large does the probability of an event have to be before the event can be classified as "reasonably certain?" These questions are difficult to answer in general terms even though the answers may be obvious in particular cases. In the following examples, we shall deal with situations in which most such questions can be answered.

Example 1: A farm implement company has set the date for its annual equipment show and has to decide if the show is to be held indoors or outdoors. It estimates that, if the show is held outdoors and it does not rain, the company will net \$6000. If it rains, however, the outdoor show will net only \$1000. If the show is held indoors, the company will net \$2000 if it rains and \$3000 if it does not. In summary, we have:

	Weather	
Type of Show	*Rain*	*No Rain*
Outdoors	\$1000	\$6000
Indoors	\$2000	\$3000

The company is sure to earn \$2000 if the show is held inside and could make \$6000 on an outdoors show—provided it did not rain. If the company can estimate the probability of rain, it could compute the expected profits for both types of shows and select the one with the maximum expected profits. Let us assume that the company feels there is a "50-50" chance of rain. The probability of rain is, therefore, equal to 0.50 and the expected profits are

$$E_0(\pi) = 0.5(1000) + 0.5(6000) = \$3500$$

for an outdoor show, and

$$E_I(\pi) = 0.5(2000) + 0.5(3000) = \$2500$$

for an indoor show. In this case, the rational decision for an expected profit maximizer would be to hold the show outdoors.

New information may cause the company to modify this decision. Suppose the company checks with the weather bureau and learns that the probability of rain is 0.90. The expected profits then become

$$E_0(\pi) = 0.9(1000) + 0.1(6000) = \$1500$$

for an outdoor show, and

$$E_I(\pi) = 0.9(2000) + 0.1(3000) = \$2100$$

for the indoor show.

The general decision rule is to hold the show indoors as long as the probability of rain is greater than 0.75, and outdoors otherwise. This decision rule is derived by equating the expected profits from an outdoor show to those for an indoor show and solving for the probability of rain p. If p is unknown, the expected profit for the outdoor show is

$$E_0(\pi) = 1000p + 6000(1 - p) = 6000 - 5000p$$

where $(1 - p)$ is the probability of no rain. The expected profit for the indoor show is

$$E_I(\pi) = 2000p + 3000(1 - p) = 3000 - 1000p$$

Equating $E_0(\pi)$ and $E_I(\pi)$, we have

$$6000 - 5000p = 3000 - 1000p$$

or

$$4000p = 3000$$

Solving for p, we obtain $p = 0.75$. If p is greater than 0.75, the expected profit of holding the show inside is greater than the expected profit of holding the show outside [i.e., $E_I(\pi) > E_0(\pi)$ for $p > 0.75$].

In order to make good decisions, the payoffs of different actions must be accurately estimated. Suppose the company learns that a debt of $2000 must be paid the day after the show or the company will be forced out of business. Furthermore, suppose that the only way the company can pay the debt is from the profits earned at the show and that it estimates the value of remaining in business to be $20,000. The payoff matrix then becomes:

	Weather	
Type of Show	*Rain*	*No Rain*
Outdoors	−$19,000	$26,000
Indoors	$22,000	$23,000

An amount equal to $20,000 has been added to the payoffs in the original table except for when the show is held outdoors and it rains in which case the company is forced out of business and "loses" $20,000. The decision rule becomes to hold the show indoors as long as the probability of rain is greater than 0.068, and outdoors otherwise.

Example 2: Tony's Pizza Parlor decides to advertise "special prices" on its giant pizzas. To make eating at Tony's more fun, Tony makes up a game. When you order a giant pizza, Tony gives you a plastic chip which is red on one side and blue on the other. You cover the chip with your hand and place it on the table. Tony then guesses which color is facing up. If you place red up and Tony guesses red, you pay the regular price of $4 for your giant pizza. If you place red up and Tony guesses blue, the pizza is free. If you place blue up, you pay $2 if Tony's guess is red and $3 if it is blue. The payoff matrix is therefore equal to:

	Your Decision	
Tony's Guess	*Red*	*Blue*
Red	$4	$2
Blue	0	$3

Tony's problem is to decide whether to say "red" or "blue." If he always says red, then the best strategy for his customers is always to select blue. In that way, they never have to pay more than $2 for a giant pizza. If Tony always says blue, his customers would soon catch on and always select red, thereby receiving one of Tony's fine pizzas free. Tony would like to select a strategy which would insure him of some minimum gain. He can do this as follows: Suppose he says red (R) with relative frequency p and blue (B) with relative frequency $(1-p)$. Then, if the customers always select red, Tony's gain will be

$$g(R) = p(4) + (1-p)(0) = 4p$$

If the customers always select blue, Tony's gain is

$$g(B) = p(2) + (1-p)(3) = 3 - p$$

It can be shown mathematically that if Tony chooses p, so that $g(R) = g(B)$, this will be the best strategy for him in the sense that no other strategy will provide a larger minimum gain in the long run. Equating $g(R)$ and $g(B)$ and solving for p, we obtain

$$4p = 3 - p$$

and, therefore, $p = 3/5$. Tony's minimum gain is equal to

$$\left(\frac{3}{5}\right)(4) = \$2.40$$

Tony could make his decisions by placing 30 red and 20 blue marbles in his top hat, selecting a marble at random, and letting the color of the marble represent his guess. This would be called *randomized* decision making.

Is it rational to base your decisions on some random event? In this type of situation, many people would say yes.

Example 3: A small store orders several birthday cakes each day from the local bakery. If the cakes are sold on the day they are baked, the store earns \$2 per cake, otherwise the remaining cakes are sold the following day at a loss of \$1 per cake. On the basis of past sales, the store obtains the following probability distribution:

Number of cakes that could be sold, x	0	1	2	3	4	5	6
Probability of selling x cakes, $f(x)$	0.10	0.10	0.30	0.20	0.20	0.10	00

Let

$S =$ the number of cakes ordered

$x =$ the number of cakes which could be sold

C_1 = the profit from selling a cake on the day it is ordered
C_2 = the loss from selling a cake the day after it is ordered

The profit, C_1, can be considered to be the "cost" of not having enough cakes, while the loss, C_2, is a holding cost.

The "cost" associated with having S cakes in stock is either:

1. $(x - S)C_1$ if the store could sell more cakes than it has (i.e., $x > S$), or
2. $(S - x)C_2$ if the store has more cakes than it can sell (i.e., $S \leq x$).

The expected total cost is therefore equal to

$$E(TC) = C_2 \sum_{x=0}^{S} (S - x)f(x) + C_1 \sum_{x=S+1}^{\infty} (x - S)f(x) \tag{5.27}$$

The store can maximize its expected profits by minimizing expected total costs.

The store's decision variable is the number of cakes ordered, S. Using formula (5.27), the store can compute expected total costs, $E(TC)$, for different values of S and then order the number of cakes which minimizes $E(TC)$. Using $C_1 = \$2$ and $C_2 = \$1$, we obtain:

Number of cakes ordered, S	0	1	2	3	4	5	6
Expected total cost, $E(TC)$	5.20	3.50	2.10	1.60	1.70	2.40	3.40

The store can minimize its cost and thereby maximize its profits by ordering three cakes each day.

Summary. These three examples illustrate that there is no one, natural or compelling model of rational decision making which the intelligent person can safely be assumed to use. The technique selected depends on the information available, the cost of additional information, and the time available in which to consider the alternatives. When the degree of uncertainty is small or the consequences of making a wrong decision are unimportant, there is little need for sophisticated decision models. When making important decisions, however, such models can be extremely useful. In the next chapter, we shall discuss additional tools that can be used in the construction of such models. This information will allow us to discuss more realistic decision-making situations later.

EXERCISES

Section 5.1

1. A whole number is chosen at random between 1 and 8, inclusive.
(a) Define a random variable for this experiment.
(b) Sketch the probability distribution of the random variable defined in part (a).

2. Three dice are rolled. Let X denote the number of dots obtained.
(a) Compute the probability distribution of X.
(b) Graph the probability distribution of X.

3. A bag contains six black and four white balls. Four balls are drawn *without replacement*. Let X denote the number of black balls drawn.
(a) What is the probability distribution of X?
(b) Graph the cumulative probability distribution.

4. A certain function $f(x)$ is equal to

$$f(x) = \begin{cases} \dfrac{3}{2(4-x)!\,x!} & \text{for } x = 0, 1, 2, 3, 4 \\ 0 & \text{otherwise} \end{cases}$$

Explain why $f(x)$ is (or is not) a probability density function.

5. The uniform distribution is

$$f(x) = \frac{1}{n} \quad \text{for } x = 1, 2, 3, \ldots, n$$

(a) Give an example of this distribution.
(b) If $n = 9$, what is the probability that x is greater than 7 given that x is greater than 4?

Section 5.2

6. Let T denote the time at which a standard 12-hour clock stops.
(a) Find the cumulative probability distribution $F(t) = P(T \leq t)$.
(b) What assumptions were made in order to find $F(t)$ in part (a)?
(c) Graph $F(t)$.
(d) What is the probability that the clock will stop between 4 and 6?

7. If $f(x) = x/2$ when $0 < x < 2$, and zero otherwise, find the probability that:
(a) $x < 1$

(b) $\dfrac{1}{2} < x < \dfrac{4}{3}$

(c) $x > 1$ given $x > \dfrac{1}{2}$

8. If $f(x) = 2x$ when $0 < x < 1$, and zero otherwise, find a number m such that $P(X \geq m) = P(X \leq m)$.

9. Let X be a number chosen at random from the numbers that lie between the numbers A and B, where A is less than B.
(a) Find the probability density function and cumulative probability distribution function of X.
(b) Graph these functions.
(c) Find $P(A < X \leq 0.5B)$.

10. Graph the rectangular probability distribution that has a probability density function $f(x) = 0.2$ for $0 < x < 5$. What is the probability that x is greater than 2 but less than 3?

Section 5.3

11. If $f(x, y) = 1$ when $0 < x < 1, 0 < y < 1$, and zero otherwise, find the probability that:
(a) $x < 0.25$ and $y < 0.25$.
(b) $x + y < 0.5$
(c) $x > 1$ given $y > 1$

12. The following table contains $P(X = x,\ Y = y)$.

	X			
Y	1	2	3	4
1	0.10	0.10	0.05	0.05
2	0.05	0.10	0.05	0.10
3	0.05	0.05	0.20	0.10

(a) What is the probability that $X < 2$ and $Y < 3$?
(b) Find the marginal probability distribution for X. (For Y).
(c) Find the conditional probability distribution of X given $Y = 3$.
(d) What is the probability that $X < 3$ given $Y < 3$?
(e) Let $W = 2X$. Find the joint probability distribution of W and Y.
(f) If $W = 2X$, find the probability that $W < 4$ and $Y < 2$.

Section 5.4

13. Explain why (or why not) the random variables in Exercise 12 are (or are not) statistically independent.

14. Suppose that $f(x, y) = 1$ when $0 < x < 1$, $0 < y < 1$, and zero otherwise. Are x and y statistically independent?

Section 5.5

15. Let X, Y, and Z be independent random variables with:

$E(X^2) = 100 \quad E(X) = 6$
$E(Y^2) = 80 \quad E(Y) = 6$
$E(Z^2) = 120 \quad E(Z) = 6$

Find:
(a) $\text{Var}(X + Y + Z)$
(b) $\text{Var}(X + Y)$
(c) $\text{Var}(3X + Y)$
(d) $\text{Var}(4 + 2X + 2Z)$

16. Find (1) $E(X)$, (2) $E(X^2)$, and (3) $E[X - E(X)]^2$ for the following probability distribution:

X	-2	4	6
$f(x)$	0.2	0.5	0.3

17. A bowl contains 80 red and 20 blue poker chips.
(a) Three chips are drawn successively at random with replacement. What is the mathematical expectation of the number of red chips? Of the number of blue chips?
(b) If you pay \$1 each time a red chip is drawn and receive \$5 each time a blue chip is drawn, would you expect to win or lose money in the long run?

18. Let X denote the number of spots obtained in a throw with an ordinary die.
(a) Compute the expected value of X.
(b) What assumptions were necessary in order to obtain this result?
(c) Explain the meaning of the result.
(d) Compute the variance of X.
(e) Let $W = X^2$. Is W a random variable? What is the probability that W is greater than zero but less than 17?

19. A radio repairman wants to replace a defective tube in a radio. He has four tubes in the bottom of his repair kit, but only one of them will work. He selects the tubes at random one after another without replacement. What is the expected number of tubes he will try before he gets the radio fixed?

20. Let W, X, and Y be independent random variables with $E(W) = 1$, $E(X) = 2$, $E(Y) = 4$, $\text{Var}(W) = 1$, $\text{Var}(X) = 2$, and $\text{Var}(Y) = 2$.
(a) Find: $E(W + X + Y)$, $E(2X)$, $E(3X + Y)$, and $E[W(X + Y)]$.
(b) Find: $\text{Var}(X + Y + W)$, $\text{Var}(2X)$, and $\text{Var}(3X + Y)$.
(c) Which of the values in parts (a) and (b) could not be computed without the assumption that W, X, and Y are independent?

21. Suppose that the joint probability distribution of the random variables X and Y is:

	X	
Y	2	4
-1	$\frac{1}{5}$	$\frac{1}{5}$
2	$\frac{1}{5}$	$\frac{2}{5}$

(a) What is the marginal distribution of X? (Of Y?)
(b) What is the conditional probability distribution of X, given Y is equal to 2?
(c) Are X and Y independent random variables?
(d) Find $E(X)$, $E(X^2)$, $\mathrm{Var}(X)$, $E(XY)$, and $E(4X)$.
(e) If $c = 2$, show that $E(X + c) = E(X) + c$.
(f) Find the correlation coefficient.
(g) Let $W = (X - \mu_x)/\sigma_x$. Show that $E(W) = 0$ and $\mathrm{Var}(W) = 1$ in this case.

22. A fair coin is thrown three times.
(a) What is the probability distribution for the number of heads?
(b) What is the mathematical expectation of the number of heads?
(c) What is the variance of the probability distribution?
(d) If you wanted to bet on the number of times the coin would fall heads in three throws, would you bet on the *expected value*?

23. A coin is tossed. If the coin lands heads, a die is tossed. If the coin lands tails, it is tossed again. Assign the number 1 to heads and the number 0 to tails. Let X denote the number to come up on the first toss (i.e.; 0, 1) and Y the number to come up on the second toss (i.e.; 0, 1, ..., 6).
(a) What is the joint probability distribution of X and Y?
(b) What is the marginal probability distribution of X?
(c) What is the conditional probability distribution of Y if X is equal to one?
(d) Find $P(X + Y < 4)$.
(e) Compute $E(X + Y)$.
(f) Are X and Y independent?
(g) Compute $\mathrm{Var}(X + Y)$.

Section 5.6

24. Using the joint probability distribution given in Exercise 12, compute $E(X)$, $E(Y)$, $E(XY)$, $\mathrm{Var}(X)$, $\mathrm{Var}(Y)$, $\mathrm{Cov}(X, Y)$, and $\mathrm{Var}(X + Y)$.

25. Let X be a random variable with mean μ and variance σ^2.
Let $W = 2(X - \mu)/\sigma$.

(a) Show that the expected value of W is equal to zero.
(b) What is the variance of W?
(c) What are the units of W?
(d) Find $\text{Cov}(X, W)$.

26. Suppose that X and Y are discrete random variables with the following probability distribution:

	X		
Y	1	3	5
2	0.2	0.2	0.2
6	0.2	0.1	0.1

Find: $\text{Var}(X)$, $\text{Var}(Y)$, and $\text{Cov}(X, Y)$.

Section 5.7

27. A random variable X has a probability density function $f(x) = 0.5x$ when $0 < x < 2$, and zero otherwise. Table the distribution function $F(x)$ for $x = 0$, 0.1, 0.2, ..., 1.9, 2.0. Using the table of $F(x)$, find:
(a) $P(X < 0.5)$
(b) $P(0.8 < X < 1.8)$
(c) $P(X > 0.8 \mid X > 0.5)$

28. A discrete random variable X is such that $P(X = x) = 0.5(0.5)^x$ for $x = 0$, 1, 2, 3, ...
(a) Calculate $E(X)$.
(b) Table $P(X = x)$ for $x = 0, 1, 2, \ldots, 10$.
(c) Find $P(X > 3)$, $P(4 \leq X \leq 6)$, and $P(X > 3 \mid X > 1)$.

Section 5.8

29. By inspection of its records, a trucking company finds that 20 percent of its trucks sent out have a mechanical failure if they are not checked beforehand while 10 per cent have a mechanical failure even though they are checked beforehand.
(a) What is the probability that at least two out of the next five trucks sent out will have a mechanical failure if they are not checked before going on the road?
(b) If the cost of checking each truck before it is sent out is \$5 and the cost of a breakdown on the road is \$100, would you recommend that every truck be checked before each trip? Explain your decision.

30. An automobile company has to ship a heavy piece of equipment to its assembly plant in Peru. The machine can be sent by air or by sea. The cost by air is generally higher, but there is the possibility of a dock strike which would delay the machine's arrival in Peru. The cost matrix is estimated to be:

Decision	Dock Strike	No Dock Strike
Ship by sea	\$5000	\$1000
Ship by air	\$2000	\$2000

(a) If the probability of a dock strike is 0.4, which method of transportation would minimize expected costs?

(b) How large can the probability of a dock strike be before it becomes profitable to ship by air?

31. An ice cream salesman earns \$20 a day on the average when the sun shines. If it rains he earns only \$4 a day. He can always earn \$12 a day as a painter regardless of the type of weather.

(a) If the TV weatherman on the 7 A.M. news program mentions that there is a 60 percent chance of rain, should the ice cream salesman decide to sell ice cream or to become a painter for the day?

(b) How large does the probability of rain have to be before he will decide not to sell ice cream?

SELECTED REFERENCES

Brunk, H. D., *An Introduction to Mathematical Statistics*, Ginn and Company, New York, 1960. Provides additional material on probability distributions, mathematical expectation, covariance, and correlation analysis. Level of mathematics slightly higher than that used in this chapter.

Cramer, Harald, *The Elements of Probability Theory*, John Wiley & Sons, New York, 1955. Provides a wide variety of problems and examples involving probability distributions. Level of mathematics slightly higher than that used in this chapter.

Wilks, S. S., *Elementary Statistical Analysis*, Princeton University Press, Princeton, N. J., 1951. Provides an excellent elementary introduction to both discrete and continuous probability distributions.

CHAPTER 6

Theoretical Distributions

The more important theoretical distributions are easy to work with, are well known, and have proven useful in a wide variety of situations. They tell us what to expect if the world (or random variable) behaves as we assume it does. They serve as benchmarks against which to compare observed distributions and act as substitutes for actual distributions when the latter are costly to obtain or cannot be obtained at all. They provide decision makers with a logical basis for making decisions, and are useful in making predictions on the basis of limited information or theoretical considerations. In short, theoretical distributions play many important roles in statistical theory.

An understanding of only a few of the main theoretical distributions will greatly increase your capacity to read and comprehend scientific work in nearly all fields of knowledge. Plant specialists find the *binomial distribution* to be useful when predicting the outcome of crossing different varieties of plants. Quality control specialists use the *Poisson distribution* as a basis for deciding if a production process is working correctly. Anthropologists use the *normal distribution* when comparing different populations. Marketing research firms use the *chi-square distribution* to determine if there are significant differences in consumers' reactions to a product change.

Agronomists use the *F distribution* to determine if changes in cultivation techniques have any significant effect on crop yields. Economists use the *t distribution* to decide if price changes have any significant effect on consumer demand for a product. All of these distributions are used at times by specialists in every field of knowledge, and each will be discussed in this chapter.

6.1 THE BINOMIAL DISTRIBUTION

The binomial probability distribution is a discrete probability distribution that is useful in describing an enormous variety of real-life events. A few examples of such events will serve to illustrate this.

A quality control inspector wants to know the probability of obtaining 4 bad light bulbs in a random sample of 20 bulbs if 10 percent of the bulbs are defective. He can quickly obtain the answer from tables of the binomial probability distribution.

A pharmaceutical company wants to test the effectiveness of a new vaccine in preventing a certain disease. In preliminary tests, 100 people who have been exposed to the disease are vaccinated. Under certain conditions, this is a binomial experiment and the company could use the binomial distribution when analyzing the results of the experiment.

A baseball fan wants to estimate the probability that his favorite player will get five hits during his next ten times at bat. The fan can use the player's batting average as an estimator of a probability and obtain the answer with the help of the binomial probability distribution.

6.1.1 Binomial Experiments

In general, an experiment is classified as a *binomial experiment* if it has four properties:

1. There is a fixed number of trials,
2. Each trial has only two possible outcomes that traditionally are called "success" or "failure" but are given appropriate names in specific applications,
3. The probability of a "success" is the same on each trial,
4. The trials are statistically independent.

In short, a binomial experiment consists of n independent ($n \geq 1$) binomial (or Bernoulli) trials, all of which have the same probability of success. With a little practice, you will find binomial experiments very easy to identify.

Suppose a baseball fan wants to determine the probability that his favorite player will get five hits in ten times at bat. The "experiment," in this case, consists of ten trials. The outcome of each trial can be classified as a hit (success) or nonhit (failure). The baseball fan may believe that it is reasonable to assume that the probability of a hit is known and remains the same each time the player is at bat and that the trials are independent of each other. If so, we have a binomial experiment and the binomial probability distribution can be used to obtain the probability of five hits in ten times at bat. (In what situations would the assumptions of constant and independent probabilities not be justified?) If the fan is interested in how many times his favorite player has to be up to bat before he gets five hits, then the number of trials (times at bat) becomes a random variable and we no longer have a binomial experiment.

A bowl contains red, white, and blue poker chips. Ten chips are selected at random from the bowl with replacement. Is this a binomial experiment? The answer depends on the type of probability question being asked. It is a binomial experiment if the question is, "What is the probability of getting four red chips?" A red chip can be classified as a "success" and nonred (meaning either blue or white) can be classified as a "failure." It is not a binomial experiment if the question is, "What is the probability of getting exactly five red, three blue, and two white chips?"

Always keep in mind that we do not have a binomial experiment unless the probability of a "success" is the same on each trial. Suppose, for example, that ten marbles are drawn at random from a bowl containing 20 red and 30 black marbles. This is a binomial experiment if each marble is replaced before another is selected. If the marbles are drawn without replacement, the probability of drawing a red marble changes each time a marble is taken from the bowl and we no longer have a binomial experiment.

Independence of trials is the most difficult property of a binomial experiment to identify or to assure. Strictly speaking, statistical independence requires that we are able to compute the probability of X "successes" by multiplying the probabilities of the possible outcomes of individual binomial trials. In practice, it is assumed that trials are independent if the outcome of one trial in no way affects the outcome of any other trial. This seems a reasonable assumption in the case of the inspector checking light bulbs, for example. Experiments can be carried out in such a way that the assumption of independence is not justified, however. Suppose a marketing research firm estimates that 10 percent of the people in a given city watch a certain television show each week. The firm wants to check this estimate and sets up the following binomial experiment to do so. A total of n persons are to be contacted during the time of the show. Each person is classified as "watching" or "not watching" the show. An interviewer is to call a family, ask the person who answers the phone if he is watching the show, write down the reply, and call another family. The interviewer does not read his instructions carefully, however. He calls a family, writes down the reply of the person who answers the phone, and then asks to speak with another member of the family. If each member of the family is considered a "trial," it is doubtful whether the trials can be considered independent.

6.1.2 *Binomial Probabilities*

Consider a *binomial trial* that consists of selecting a marble at random from a bowl containing 30 red (R) and 70 black (B) marbles. Let Y be a random variable that is equal to 0 if the marble is black and 1 if the marble is red. The probability of drawing a red marble is denoted by p and, therefore, the probability of drawing a black marble must be $(1 - p)$. The expected value of Y is

$$E(Y) = (1 - p)(0) + p(1) = p = 0.3$$

Now suppose a *binomial experiment* is performed that consists of $n = 4$ such binomial trials in which the marbles are drawn at random and with replacement. This experiment has 16 different possible outcomes. These are:

1. *RRRR*	9. *BRRR*
2. *RRRB*	10. *BRRB*
3. *RRBR*	11. *BRBR*
4. *RRBB*	12. *BRBB*
5. *RBRR*	13. *BBRR*
6. *RBRB*	14. *BBRB*
7. *RBBR*	15. *BBBR*
8. *RBBB*	16. *BBBB*

The probability of each of these outcomes can be obtained by using the multiplication rule of probability for independent events. Thus

$$P(RRBR) = pp(1 - p)p = (0.3)(0.3)(0.7)(0.3) = 0.0189$$

Both the multiplication rule for independent events and the addition rule for mutually exclusive events are used to find the probability of any given number of red marbles when the order of the outcomes is ignored. Thus, ignoring the order of outcomes, we have

$$P(3R \text{ and } 1B) = P(RRRB) + P(RRBR) + P(RBRR) + P(BRRR) = 0.0756$$

Notice that each of the individual probabilities in the sum is equal to $(1 - p)ppp$. The probability $P(3R \text{ and } 1B)$ could be obtained very easily if the number of individual items in the sum could be computed without enumerating all possible outcomes as was done here. Such enumeration is extremely tedious if the number of trials is very large and, as will be demonstrated later, is not necessary.

Let X be a number equal to the number of red marbles observed during a particular binomial experiment. Thus X is the sum of values taken on by the random variable Y and is therefore a random variable itself (i.e., $X = Y_1 + Y_2 + \cdots + Y_n$). The random variable X is equal to 0, 1, 2, 3, or 4 in the present example and takes on the values 0, 1, 2, ..., n in the general case of a binomial experiment consisting of n trials. The probability distribution of X is called the *binomial probability distribution*. In the case of four marbles selected at random with $p = 0.3$, the binomial distribution is:

Number of red marbles, x_i	0	1	2	3	4
Probability, $f(x_i)$	0.2401	0.4116	0.2646	0.0756	0.0081

A histogram of this probability distribution is shown in Figure 6.1.

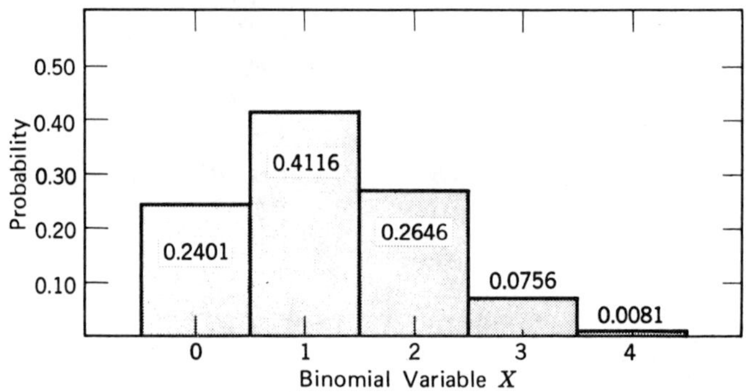

Figure 6.1 Binomial probability distribution for $n = 4$, $p = 0.3$.

In general, the probability of a particular order of x successes in n trials is

$$p^x(1-p)^{n-x}$$

where p is the probability of a success.

The general rule of permutations can now be used to obtain the number of different possible orders, each having x successes. There are x items with the characteristic "success" and $(n - x)$ items with the characteristic "failure." The general rule of permutations states that, given a set of n objects having x elements of one kind and $(n - x)$ elements of another, the number of permutations of n objects, taken all together, is

$$(6.1) \qquad {}_nP(x, n-x) = \frac{n!}{x!(n-x)!} = {}_nC_x$$

The numbers $n!/x!(n-x)!$ are called binomial coefficients. The individual probabilities of the binomial probability distribution can be written as

$$f(x) = {}_nC_x\, p^x(1-p)^{n-x} = \frac{n!}{x!\,(n-x)!}\, p^x(1-p)^{n-x} \tag{6.2}$$

In other words, the probability of x successes in n independent binomial trials is equal to the number of combinations of n things taken x at a time multiplied first by the probability of a "success" raised to the power x and then by the probability of a "failure" raised to the power $(n-x)$. When using the formula (6.2), it is helpful to remember that:

(a) $0! = 1$ (zero factorial is equal to 1)
(b) $p^0 = 1$ (any number raised to the zero power is equal to 1)

The probability $f(x)$ is one term in the binomial expansion $(p+q)^n$. Suppose, for the moment, that p and q are any quantities and n is a positive integer. It can be shown by actual multiplication that:

$$\begin{aligned}
(p+q)^0 &= 1 \\
(p+q)^1 &= p+q \\
(p+q)^2 &= p^2 + 2pq + q^2 \\
(p+q)^3 &= p^3 + 3p^2q + 3pq^2 + q^3 \\
(p+q)^4 &= p^4 + 4p^3q + 6p^2q^2 + 4pq^3 + q^4 \\
(p+q)^5 &= p^5 + 5p^4q + 10p^3q^2 + 10p^2q^3 + 5pq^4 + q^5
\end{aligned}$$

and so on.

The general pattern of the exponents is obvious. The terms in the expansions, apart from their coefficients, can be written by inspection. In general, the expansion of $(p+q)^n$ has $n+1$ terms which, without their coefficients, are equal to

$$p^n,\ p^{n-1}q,\ p^{n-2}q^2,\ p^{n-3}q^3,\ \ldots,\ p^xq^{n-x},\ \ldots,\ pq^{n-1},\ q^n$$

We now need to find the coefficients of these terms. Note that each term contains n factors that can be arranged in a number of different ways. For example, if $n = 5$, then the term p^4q consists of four p's and one q. This term can be obtained in five different ways:

$$pppp q \qquad pppqp \qquad ppqpp \qquad pqppp \qquad qpppp$$

As was mentioned previously, the number of different ways of arranging n factors, x of which have characteristic p and $(n-x)$ of which have characteristic q, can be obtained by using the permutation formula

$${}_nP(x, n-x) = \frac{n!}{(n-x)!\,x!} = {}_nC_x$$

Thus, the coefficient of the term p^4q is ${}_5C_4 = 5!/4!\,1! = 5$.

The *binomial formula* is therefore equal to

$$(p+q)^n = p^n + np^{n-1}q + \frac{n(n-1)}{1 \cdot 2} p^{n-2}q^2 + \cdots + \frac{n!}{(n-x)!\,x!} p^x q^{(n-x)} + \cdots + npq^{n-1} + q^n$$

If p and q are probabilities such that $q = 1 - p$, then the terms

$$f(x) = \frac{n!}{(n-x)!\,x!} p^x q^{n-x}$$

are themselves probabilities because $f(x) \geq 0$ for all x and $\sum f(x) = 1$. The fact that $f(x) \geq 0$ is obvious because ${}_nC_x \geq 0$ and $p^x q^{(n-x)} \geq 0$ for all values of x. To prove that $\sum_x f(x) = 1$, we use the fact that $q = 1 - p$ and, therefore,

$$\sum_x f(x) = (p+q)^n = (p + 1 - p)^n = 1^n = 1$$

In summary, the probabilities of the binomial probability distribution are terms in the binomial expansion of $(p+q)^n$, where p and q are probabilities such that $q = 1 - p$ and n is any positive integer.

To illustrate the use of formula (6.2), suppose that a telephone survey company is interviewing people for telephone survey work. The company knows that an average interviewer can obtain answers from 80 percent of the people contacted. In order to test the people applying for jobs, the firm has them phone eight different families and try to obtain answers to a short questionnaire. What is the probability of x successful interviews out of $n = 8$ tries if the probability of a successful interview is $p = 0.8$? Using formula (6.2), we obtain the probabilities given in Table 6.1.

Table 6.1 The Binomial Probability Distribution for $n = 8$, $p = 0.8$

(1) Number of "Failures"	(2) Number of "Successes"	(3) Probability $f(x)$	(4) Cumulative Probability $\sum f(x)$
8	0	$1(0.8)^0(0.2)^8 = 0.0000$	0.0000
7	1	$8(0.8)^1(0.2)^7 = 0.0001$	0.0001
6	2	$28(0.8)^2(0.2)^6 = 0.0011$	0.0012
5	3	$56(0.8)^3(0.2)^5 = 0.0092$	0.0104
4	4	$70(0.8)^4(0.2)^4 = 0.0459$	0.0563
3	5	$56(0.8)^5(0.2)^3 = 0.1468$	0.2031
2	6	$28(0.8)^6(0.2)^2 = 0.2936$	0.4967
1	7	$8(0.8)^7(0.2)^1 = 0.3355$	0.8322
0	8	$1(0.8)^8(0.2)^0 = 0.1678$	1.0000

Suppose the company's rule is to hire only those applicants who obtain six or more interviews out of eight. If $p = 0.8$ for everyone who applied for a job, then in the long run, 29.36 percent of the applicants would obtain six interviews, 33.55 percent seven interviews, and 16.78 percent eight interviews in eight calls. Thus, on the average, the company could expect to hire $(29.36 + 33.55 + 16.78) = 79.69$ percent of the people interviewed.

When n is large, the computation necessary to obtain binomial probabilities is very tedious, even with the help of formula (6.2). Fortunately, tables of the binomial probability distribution are available. Table V, in Appendix I, provides binomial probability distributions for n as large as 15 and $p = 0.05, 0.10, 0.15, \ldots, 0.45, 0.50$. When working with problems in which the probability of a "success," p, is greater than 0.5, the tabulated probabilities can be defined as $(1 - p)$ and the probability of $(n - x)$ "failures" can be obtained from the table. Once the probability distribution for "failures" is obtained, it can be converted to the probability distribution for the number of successes by replacing the number of failures, $(n - x)$, by the number of successes, x. (See Table 6.1, Columns (1) and (2), for an example.) It is shown later that for large n, say $n > 40$, the binomial probability distribution can be approximated by the normal probability distribution.

6.1.3 The Mean and Variance of a Binomial Distribution

The mean and standard deviation of a binomial distribution are especially easy to compute. By definition, the mathematical expectation of a binomial random variable X is

$$E(X) = \sum_{i=1}^{n} x_i({}_nC_{x_i}\, p^{x_i}(1 - p)^{n-x_i}) \tag{6.3}$$

The definition (6.3) can be evaluated by recalling that X is the sum of n independent random variables Y, each of which takes on the values 1 or 0 with probabilities p and $(1 - p)$, respectively, and has an expected value of p.[1] Therefore,

$$X = Y_1 + Y_2 + \cdots + Y_n \tag{6.4}$$

and

$$E(X) = E(Y_1) + E(Y_2) + \cdots + E(Y_n) = \underbrace{p + p + \cdots + p}_{n} = np \tag{6.5}$$

[1] Let Y be a binomial random variable that takes on the values 0 or 1 with probabilities $(1 - p)$ and p, respectively, Then $E(Y) = (1 - p)(0) + p(1) = p$.

In summary, the expected number of successes in a binomial experiment consisting of n independent trials each with a probability of success equal to p is $E(X) = np$.

To illustrate this, suppose a fair coin is thrown 100 times. We expect to observe $np = (100)(0.5) = 50$ heads. If the coin is actually thrown 100 times, we may observe 45, 60, 73, or any number of heads between 0 and 100; but, if the experiment is repeated a large number of times, we expect the average number of heads to be 50 out of 100.

By definition, the variance of a binomial random variable X is

$$\text{(6.6)} \qquad \text{Var}(X) = \sum_{i=1}^{n} (x_i - np)^2 ({}_nC_{x_i}\, p^{x_i}(1-p)^{n-x_i})$$

Again, recall that X is the sum of n independent random variables Y. The variance of each Y is equal to

$$\text{(6.7)} \qquad \begin{aligned} \text{Var}(Y) &= (1 - E(Y))^2 p + (0 - E(Y))^2(1-p) \\ &= (1-p)^2 p + (0-p)^2(1-p) \\ &= p(1-p) \end{aligned}$$

Letting $X = Y_1 + Y_2 + \cdots + Y_n$ and using the fact that the Y_i are independent random variables, we know that the variance of X is given by

$$\text{(6.8)} \qquad \text{Var}(X) = \text{Var}(Y_1) + \text{Var}(Y_2) + \cdots + \text{Var}(Y_n)$$

and therefore is equal to

$$\text{(6.9)} \qquad \text{Var}(X) = \underbrace{p(1-p) + p(1-p) + \cdots + p(1-p)}_{n} = np(1-p)$$

Table 6.2

(1) Number of Heads, x	(2) Squared Deviations $(x - np)^2$	(3) Probability $f(x)$	(4) Product of Columns (2) and (3)
0	4	0.0625	0.2500
1	1	0.2500	0.2500
2	0	0.3570	0
3	1	0.2500	0.2500
4	4	0.0625	0.2500
		Var(X) = Total =	1.0000

In other words, the variance of the number of successes in a binomial experiment is equal to the number of trials multiplied first by the probability of a "success" and then by the probability of a "failure."

As an example, suppose a fair coin is thrown four times. The variance of the number of heads is $np(1 - p) = 4(0.5)(0.5) = 1$. We can obtain this answer directly from the definition (6.6). The computations are presented in Table 6.2.

The standard deviation of the number of successes in a binomial experiment is obtained by taking the square root of the variance. Thus

$$\text{Standard deviation of } X = \sigma_x = \sqrt{np(1 - p)} \tag{6.10}$$

6.1.4 *Using the Binomial Distribution*

Example 1: A pharmaceutical company wants to test the effectiveness of a new vaccine in preventing a certain disease. It is estimated that 40 percent of the people exposed to the disease will contract it under normal circumstances. A group of 12 persons who have been exposed volunteer to be vaccinated. Two of the 12 people contract the disease after they are vaccinated.

An effective vaccine would reduce the probability that a person exposed to the disease will actually contract it. If the vaccine has no effect, the probability is $p = 0.4$ that any given person exposed to the disease will contract it. In this case, we would expect to find that $np = 12(0.4) = 4.8$ persons contract the disease on the average in a long series of experiments involving 12 persons. Is the fact that only 2 of the 12 people contracted the disease sufficient evidence to conclude that the vaccine is effective?

Using Table V, Appendix I, we find that the observed result (exactly 2 out of 12 ill) would occur about 6.39 percent of the time. Furthermore, the probability that 2 or less people will contract the disease, if the vaccine is not effective, is

$$P(x \leq 2) = P(0) + P(1) + P(2) = 0.0022 + 0.0174 + 0.0639 = 0.0835$$

The fact that only 2 out of 12 persons contracted the disease after being vaccinated is, therefore, not particularly convincing proof that the vaccine is effective. The company would probably want to continue its tests.

This situation is not a "pure" or "ideal" binomial situation. The probability $p = 0.40$ is an estimated probability that could be different for various age groups or for persons who have had a similar disease in the past. A sample of volunteers is not exactly a random sample. In short, the binomial distribution, in this case, provides a rough and quick approximation to the probability of the observed result under the rather restrictive assumptions made.

In a large-scale study, the company would no doubt want to consider other statistical tools that could be used to analyze the data and would test the vaccine in several different ways. For example, the vaccine may not be effective when administered after exposure to the disease but could possibly be effective if administered before such exposure.

Example 2: Two baseball fans, Jim and Larry, are discussing their favorite players. Jim believes that the Mudsville first baseman, Louie Slugger, must be one of the greatest hitters of all time. Louie has a batting average of 0.4 for the season. Larry does not share Jim's confidence in Louie so Jim offers to bet \$10 that Louie will get at least two hits during his next five times at bat. If you were Larry, would you take the bet?

This could be considered to be a binomial experiment. On the basis of his past record, Louie's expected number of hits is $np = 5(0.4) = 2$. The probability that he will get exactly two hits is

$$P(2) = \frac{5!}{2!3!}(0.4)^2(0.6)^3 = 0.3456$$

The probability that he will get *at least* two hits is

$$\begin{aligned} P(x \geq 2 \text{ hits}) &= P(2) + P(3) + P(4) + P(5) \\ &= 0.3456 + 0.2304 + 0.0768 + 0.0102 = 0.6630 \end{aligned}$$

The expected value of the bet for Jim is

$$\begin{aligned} \phi &= \text{Expected winnings} - \text{Expected loss} \\ &= 0.663(\$10) - 0.337(\$10) = \$3.26 \end{aligned}$$

Unless there is reason to believe that the probability of Louie getting a hit is less than 0.4, Larry would probably not take the bet. If Jim would agree to bet \$10 that Louie would get *over* two hits in his next five times at bat, the odds would be in Larry's favor.

Suppose that Jim and Larry agree that the probability of Louie getting a hit each time he goes to bat is $p = 0.4$. Jim further agrees to pay Larry J dollars if Louie gets *less than two* hits during his next five times at bat and Larry, in turn, agrees to pay Jim L dollars if Louie gets *two or more* hits. They want to select J and L so that the expected value of the bet is zero. The expected value of the bet for Jim is

$$\phi = 0.663(L) - 0.337(J)$$

Setting ϕ equal to zero and solving for J, we obtain

$$J = \left(\frac{0.663}{0.337}\right)(L) = 1.967(L)$$

Therefore, Larry pays Jim \$10 if Louie gets two or more hits and Jim pays Larry \$19.67, otherwise.

Example 3: A salesman calls on ten automobile parts stores every day, five days a week. During a six-week period (30 days), he keeps a record each day of the number of stores that buy something from him. The results are summarized in Table 6.3 and presented in Figure 6.2.

Table 6.3 Number of Stores that Purchase Automobile Parts

(1) Number of Stores that Buy Parts x_i	(2) Observed Frequency (Days) w_i	(3) Relative Frequency Column (2)/30	(4) Binomial Probability	(5) Expected Frequency 30[Column (4)] (Days)
0	0	0	0.0282	0.846
1	2	0.0667	0.1211	3.633
2	8	0.2667	0.2335	6.995
3	11	0.3666	0.2668	7.904
4	6	0.2000	0.2001	6.003
5	3	0.1000	0.1029	3.087
6	0	0	0.0368	1.104
7	0	0	0.0090	0.270
8	0	0	0.0014	0.042
9	0	0	0.0001	0.003
10	0	0	0.0000	0.000
Total	$\sum w_i = 30$			

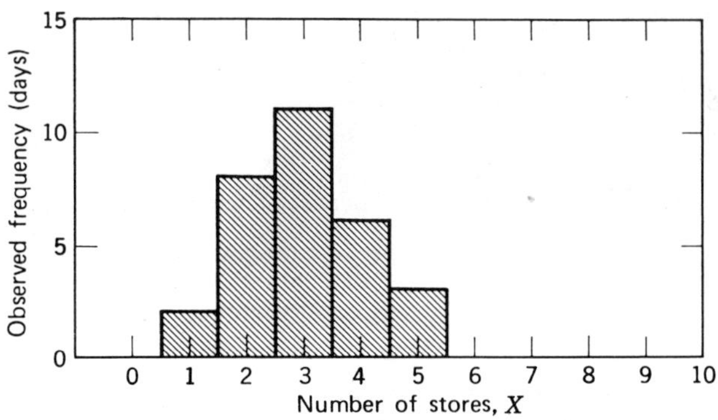

Figure 6.2 Observed distribution.

The average number of stores each day that buy parts is

$$\bar{X} = \frac{\sum x_i w_i}{\sum w_i} = \frac{90}{30} = 3$$

where x_i denotes the number of stores that buy parts on a given day and w_i is the number of days out of 30 that x_i stores bought parts. The estimated variance of the number of stores each day that buy parts is

$$s^2 = \frac{\sum_i (x_i - \bar{X})^2 w_i}{\sum w_i - 1} = \left(\frac{34}{29}\right) = 1.17$$

The salesman could consider his sales visits each day to be a binomial experiment with $n = 10$ (stores) and $p = 0.3$ (probability of a sale at any given store). The binomial probabilities for such an experiment are listed in Column (4) of Table 6.3. The probabilities are close to the observed relative frequencies for $x = 2$, 4, and 5. The largest difference between relative frequencies and binomial probabilities is for $x = 3$, in which case the difference is $(0.3666 - 0.2668) = 0.0998$.

The expected frequencies in Column (5) were obtained by multiplying the total number of days, 30, by the binomial probabilities in Column (4). The expected frequencies are the number of days out of 30 (on the average and in the long run) that the salesman could expect sales in x stores, *if* the number of stores buying parts each day is a binomial random variable with $n = 10$ and $p = 0.3$. The expected frequencies are presented in Figure 6.3.

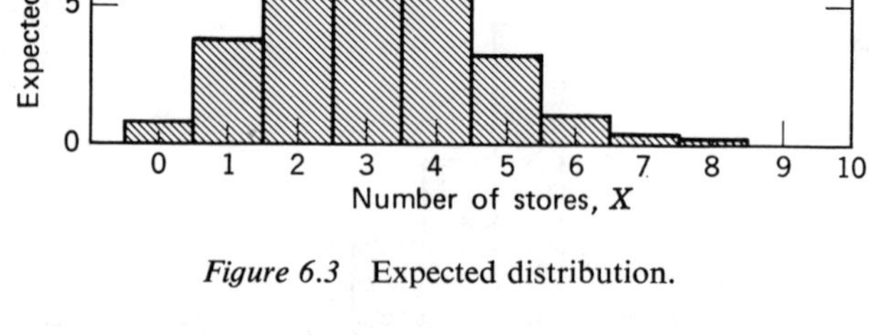

Figure 6.3 Expected distribution.

The expected value of the binomial random variable is $np = 10(0.3) = 3 =$ expected number of stores buying parts. The variance of the binomial distribution is $np(1 - p) = 10(0.3)(0.7) = 2.1$. The binomial distribution has the same mean as the observed distribution, in this case, but has a larger variance.

In future chapters, we shall discuss methods by which to determine if the observed frequencies and expected frequencies are similar enough to justify the assumption that x is a binomial random variable with $n = 10$ and $p = 0.3$.

6.1.5 Properties of the Binomial Distribution

It is obvious by now that the shape and location of binomial distributions change as p changes for a given n or as n changes for a given p. In this section, we shall study the changes that occur in the binomial distribution as p and n change and give special attention to the case in which n becomes large.

As p increases for a fixed n, the binomial distribution shifts to the right. This is demonstrated in Figure 6.4 where a series of line graphs have been

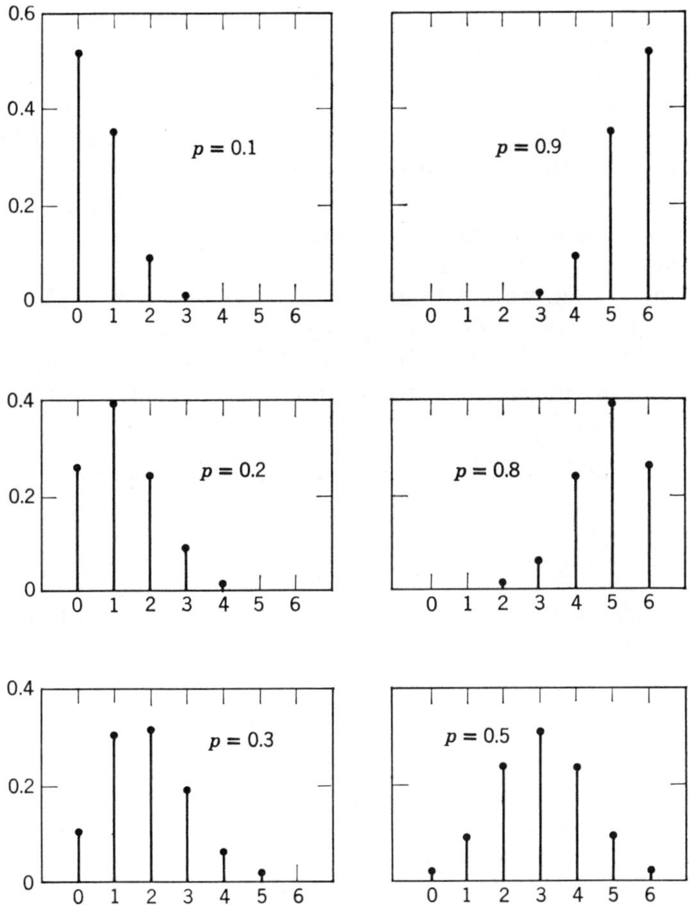

Figure 6.4 Binomial probability distributions for $n = 6$.

drawn for $n = 6$ and $p = 0.10, 0.20, 0.30, 0.50, 0.80$, and 0.90. The mode of the binomial distribution is equal to the value of X which has the largest probability. For example, if $n = 6$ and $p = 2$, the mode is equal to 1 while for $n = 6$ and $p = 0.8$ the mode is equal to 5. The mean and mode are equal if np is an integer. For example, when $n = 6$ and $p = 0.50$, the mean and mode are both equal to 3. For fixed n, both the mean and mode increase as p increases.

As n increases for a fixed p, the binomial distribution moves to the right, flattens, and spreads out. The mean of the binomial distribution, np, obviously increases as n increases with p held constant. For larger n, there are more possible outcomes of a binomial experiment and the probability associated with any particular outcome becomes smaller. As an extreme example, for $n = 1000$ and $p = 0.3$, the probability that the binomial variable X is *exactly* equal to 322 is effectively zero.

When n is large, we are not usually concerned with the probability associated with any particular value of X, but often we want to know the probability that X lies within a given range of values, say, $322 \leq X \leq 340$. Fortunately, such probabilities can be obtained easily by using the normal distribution to approximate the binomial distribution. This approximation will be illustrated later when we discuss the normal distribution. We shall need the following information at that time, however.

If the mean of a binomial random variable X is subtracted from every value that X can take on, we obtain a new random variable $W = (X - np)$. The expected value of W is equal to zero regardless of the value of n or p. The variance of W is equal to the variance of X. To verify these statements, we note that the mean, np, of a given binomial probability distribution is a constant and therefore

$$E(W) = E(X - np) = E(X) - np = np - np = 0$$

while

$$\text{Var}(W) = \text{Var}(X - np) = \text{Var}(X) = np(1 - p)$$

Next, divide W by the standard deviation of the binomial distribution, $\sqrt{np(1 - p)}$, and call the new random variable Z. That is,

$$Z = \frac{W}{\sqrt{np(1 - p)}} = \frac{X - np}{\sqrt{np(1 - p)}}$$

The expected value of Z is equal to zero and its variance is equal to one. To prove this statement, let $c = 1/\sqrt{np(1 - p)}$. The random variable Z can then be written as $Z = cW$. Applying the rules of mathematical expectations, we obtain

$$E(Z) = E(cW) = cE(W) = 0$$

and

$$\operatorname{Var}(Z) = \operatorname{Var}(cW) = c^2\operatorname{Var}(W) = \frac{1}{np(1-p)}\operatorname{Var}(W) = \frac{np(1-p)}{np(1-p)} = 1$$

The variable Z, called the standardized binomial random variable, plays an important role in the approximation procedure used later. The following theorem provides the justification for using the normal approximation to the binomial distribution.

DeMoivre-Laplace theorem. For any two numbers $z_1 < z_2$ and any fixed value of the probability of success p, $0 < p < 1$, if the number of completely independent trials in a binomial experiment is increased indefinitely, then the probability that the binomial variable X will satisfy the inequalities

$$z_1 < \frac{X - np}{\sqrt{np(1-p)}} < z_2$$

tends to the limit

$$\frac{1}{\sqrt{2\pi}} \int_{z_1}^{z_2} e^{-(z^2/2)}\, dz \tag{6.11}$$

The equation (6.11) is the equation of the normal distribution. The theorem simply asserts that

$$\lim_{n\to\infty} P\left(z_1 < \frac{X - np}{\sqrt{np(1-p)}} < z_2\right) = G(z_2) - G(z_1)$$

where the cumulative probability function $G(z_i)$ is defined by the equation

$$G(z_i) = \frac{1}{\sqrt{2\pi}} \int_{-\infty}^{z_i} e^{-(z^2/2)}\, dz \tag{6.12}$$

This equation, sprinkled with square roots, integrals, and exponents, appears unwieldy and unfriendly even though it describes a very beautiful distribution. Fortunately, it is not necessary to work with either equation (6.11) or (6.12) because tables of the normal distribution are commonly available and easy to use.

Although the DeMoivre-Laplace theorem applies to very large values of n, experience has shown that, if p is not too close to 0 or 1, the normal approximation to the binomial distribution is reasonably accurate for values of n as small as 40. If p is near 0.5, the normal approximation is often sufficiently accurate for values of n as small as 20.

The Poisson distribution can frequently be used to approximate the binomial distribution when n is large and p is very small. This distribution will be discussed next. Not only does it serve as an approximation to the binomial distribution, but it is also an important distribution in its own right.

6.2 THE POISSON DISTRIBUTION

A purchasing agent for a dress-making company is inspecting a shipment of cloth to determine if it meets the company's standards. He knows that there are normally about 16 defects per 400 square yards of cloth and considers this to be acceptable. The shipment of cloth consists of 200 bolts of 40 square yards each. What percentage of these 200 bolts can be expected to contain $x = 0, 1, 2, \ldots, 10$ defects if the average number of defects is 16 per 400 square yards?

The inspection process can be thought of as a special type of binomial experiment. The probability of finding more than one defect in any particular square foot of cloth is very small. Therefore, the inspection of one square foot of cloth can be thought of as a binomial trial which has two outcomes: either a defect is found (a success) or one is not found (a failure). If there are 16 defects per 400 square yards, then the probability of a defect in any randomly selected square *foot* is

$$p = \left(\frac{16}{9(400)}\right) = \frac{4}{900} = 0.004444$$

The number of trials performed in the inspection of a 40-yard bolt of cloth a square foot at a time is $9(40) = 360$. Thus, the expected number of defects per bolt of cloth is $np = 360(0.004444\ldots) = 1.6$ defects. The Poisson distribution can be used to compute the probability of any given number of defects per bolt of cloth. The probability of x defects per bolt can then be multiplied by 100 to obtain the expected percentage of bolts containing x defects.

The Poisson probability distribution is represented by the function

$$p(x) = \frac{e^{-m}m^x}{x!} \qquad \text{for } x = 0, 1, 2, 3, \ldots \tag{6.13}$$

where

e is equal to $2.71828\ldots$

m is a positive constant equal to the mean of the distribution

x is any positive integer

The Poisson distribution is a discrete distribution with a single parameter m. As m increases, the distribution shifts to the right. This is illustrated in Figure 6.5 for 4 values from $m = 0.3$ to $m = 4.0$

When the Poisson distribution is used to approximate the binomial distribution, x is the number of "successes" and m is equal to np. Strictly speaking, the Poisson distribution is the limiting distribution of the binomial distribution as n becomes infinitely large and p approaches zero in such a

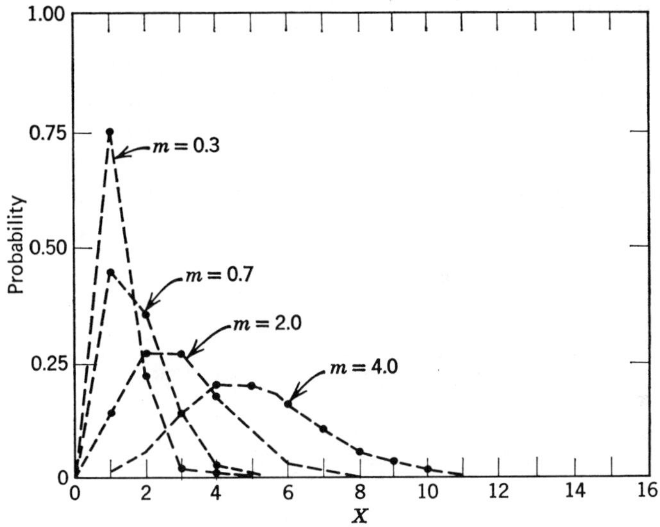

Figure 6.5 The Poisson distribution.

way that $np = m$ remains constant. Such situations are fairly common. The inspection of bolts of cloth is certainly such a situation. Suppose a trial is defined as the inspection of one square inch of cloth. The inspection of a 40-yard bolt of cloth then represents a binomial experiment consisting of $n =$ (144 square inches per square foot)(360 square feet per bolt) $= 51{,}840$ trials. The probability of finding a defect is $p = (0.004444/144) = 0.00003$. The mean number of defects per bolt, however, continues to be $np = 1.6$ defects. (Let the inspection area decrease to one-tenth of a square inch; n increases and p decreases but np remains equal to 1.6.)

In the inspection of cloth, we are interested in the number of "successes" per unit of *area*. The Poisson distribution is also used to compute the probability of x "successes" in a given *volume*, an interval of *time*, or a unit of *length*. For example, the Poisson distribution can serve as the expected distribution for the:

(a) Distribution of numbers of accidents per day by taxi drivers in a large city.
(b) Distribution of numbers of bacteria in a drop of water.
(c) Distribution of numbers of calls on a telephone line per minute.
(d) Distribution of numbers of typographical errors per line of typed material.
(e) Distribution of numbers of defective batteries per carton of 240 batteries.

This list could easily be expanded to letter (z). (Try to add at least three more examples.)

The Poisson distribution is not only a very useful distribution, but it is also very easy to use. Table VI, Appendix I, provides values of the Poisson probability distribution for various values of m from $m = 0.1$ to $m = 9.0$. The cloth inspection example will be used to illustrate the use of Table VI.

The expected number of defects per bolt of cloth is $m = 1.6$. Using (6.13), we have

$$p(0) = \frac{e^{-1.6}(1.6)^0}{0!} = \frac{1}{e^{1.6}} = 0.2019$$

$$p(1) = \frac{e^{-1.6}(1.6)^1}{1!} = \frac{1.6}{e^{1.6}} = 0.3230$$

$$p(2) = \frac{e^{-1.6}(1.6)^2}{2!} = \frac{2.56}{2e^{1.6}} = 0.2584$$

$$\cdots \qquad \cdots \qquad \cdots \qquad \cdots$$

$$p(6) = \frac{e^{-1.6}(1.6)^6}{6!} = \cdots = 0.0047$$

These probabilities can be obtained directly from Table VI, Appendix I. The probabilities along with the expected number of bolts of cloth out of 200 with x defects are listed in Table 6.4.

Suppose that the purchasing agent checks 10 percent of the bolts of cloth

Table 6.4 The Poisson Probability Distribution for $m = 1.6$

(1) Number of Defects, x	(2) Probability $p(x)$	(3) Expected Percent of Bolts with x Defects	(4) Expected Number of Bolts out of 200 with x Defects
0	0.2019	20.19	40.38
1	0.3230	32.30	64.60
2	0.2584	25.84	51.68
3	0.1378	13.78	27.56
4	0.0551	5.51	11.02
5	0.0176	1.76	3.52
6	0.0047	0.47	0.94
7	0.0011	0.11	0.22
8	0.0002	0.02	0.04
9	0.0000	0.00	0.00

to determine if they are acceptable and finds the information described in Table 6.5.

Table 6.5

Number of Defects x_i	Number of Bolts With x_i Defects w_i	Relative Frequency $w_i/20$
0	1	0.05
1	3	0.15
2	4	0.20
3	5	0.25
4	5	0.25
5	2	0.10
Totals	20	1.00

The average number of defects for the sample is

$$\overline{X} = \frac{\sum_i x_i w_i}{\sum_i w_i} = \left(\frac{56}{20}\right) = 2.80 \text{ defects per bolt of cloth}$$

Can he conclude on the basis of this sample that the true mean of the 200 bolts of cloth is significantly larger than 1.6 defects per bolt? We raise the question but postpone the answer until a later chapter. Instead, let us *assume* for the moment that the data do not "fit" a Poisson distribution with $m = 1.6$ and try to fit a new Poisson distribution to the data.

It can be shown that both the mean and variance of the Poisson distribution are equal to m. The easiest way to do this is with the help of moment generating functions and calculus. The interested reader can find the method in most books on mathematical statistics. For present purposes, we are only interested in the results. To use the function (6.13), it is only necessary to know the constant m in order to compute the Poisson probabilities for various values of X. Our best estimate of m in this case is $\hat{m} = 2.8$. Thus, the function

$$p(x) = \frac{e^{-2.8}(2.8)^x}{x!}$$

is used to "fit" a Poisson distribution to the observed data.

From the tables of the Poisson distribution for $m = 2.8$, we create Table 6.6. The probabilities for the "fitted" Poisson distribution with $m = 2.8$ do not match the observed relative frequencies exactly, but they certainly come much closer than the probabilities obtained from a Poisson distribution with $m = 1.6$.

Table 6.6

Number of Defects x_i	Probability $p(x_i)$	Observed Relative Frequency
0	0.0608	0.0500
1	0.1703	0.1500
2	0.2384	0.2000
3	0.2225	0.2500
4	0.1557	0.2500
5	0.0872	0.1000
6	0.0407	0
7	0.0163	0
8	0.0057	0
9	0.0018	0
10	0.0005	0
11	0.0001	0
12	0.0000	0

Several more examples of the use of the Poisson distribution will now be presented before turning to a discussion of the normal distribution.

Example 1: A cigarette company wants to promote the sale of its cigarettes (Brand X) with a special advertising campaign. Five out of every thousand cigarettes are rolled in gold foil and randomly mixed with the regular (special, king-sized, mentholated) cigarettes. The company offers to trade a new package of cigarettes for each gold cigarette a smoker finds in a package of Brand X. What is the probability that buyers of Brand X will find $X = 0, 1, 2, 3, \ldots$ gold cigarettes in a single package of 20?

This can be thought of as a binomial experiment in which there are $n = 20$ trials (in a pack) and the probability of finding a golden cigarette (a success) is $p = (5/1000) = 0.005$. The expected number of gold cigarettes per pack is therefore $m = np = 20(0.005) = 0.1$. Using the tables of the Poisson distribution (Table VI, Appendix I), we obtain

Number of Gold Cigarettes per Pack	*Probability*
0	0.9048
1	0.0905
2	0.0045
3	0.0002
4	0.0000

We find that slightly over 90 percent of the packages contain no golden cigarettes, about 9 percent contain one golden cigarette, and less than 0.5 percent contain two golden cigarettes. (From the company's viewpoint, the moral of the story is "buy them by the carton.")

Example 2: A radio parts company knows that, on the average, 4 percent of the 800 ohm, 5 watt resistors it manufactures are defective. The company sells the resistors to radio repair shops in boxes of 50. It begins to receive complaints from a few radio repair shops that they often find only 44 or 45 good resistors in a box. The company wants to consider the possibility of putting several extra resistors in each box in order to "assure" that the repair shops receive at least 50 good resistors per box. For simplicity, assume that the company puts 50 resistors in a box as usual and then adds Y extra resistors that have been tested and are known to be good. How many extra resistors should be added to each box if the company wants to be sure that at least 90 percent of the boxes have 50 good resistors?

We can consider the selection and testing of a resistor as a trial, the outcome of which can be classified as a "success" if the resistor is bad or a "failure" if the resistor is good. The probability of finding a bad resistor on any given trial is $p = 0.04$ and is assumed to remain constant on each trial. Since the outcome of one trial cannot affect the outcome of any other trial, it seems reasonable to assume that the trials are independent. Thus, testing the resistors in a box can be viewed as a binomial experiment with $n = 50$ and $p = 0.04$. Unfortunately, this particular binomial distribution is not tabled in Appendix I.

However, the Poisson distribution is frequently used to approximate the binomial distribution when n is large and p is small.

In this case, using the Poisson distribution with $np = m = 2.0$, we find the probability of three or less bad resistors to be 0.8571 and the probability of four or less bad resistors to be 0.9473. Therefore, the company would have to add four good resistors to be sure that *at least* 90 percent of the boxes would contain at least 50 good resistors. The Poisson distribution for $m = 2.0$ is given in Table 6.7.

An alternative would be to pack the boxes as usual and guarantee to replace any bad resistors that repairmen find in a box. If the average cost of replacing returned resistors is 20 cents each and the cost of adding extra resistors at the factory is 8 cents each, which system would cost the least?

Assuming that all bad resistors are returned, the "guarantee and replacement" plan would cost 4(20¢) = 80¢ per 100 resistors sold. (On the average, four out of 100 resistors are bad.) The "adding extra resistors at the factory" plan would cost 8(8¢) = 64¢ per 100 resistors sold. (The eight resistors added per 100 resistors are considered to be "given away.") If less than 80 percent of the bad resistors are returned under the "guarantee and replacement" plan, it would be cheaper not to pack extra resistors at the factory.

Table 6.7

Number of Bad Resistors x_i	Probability $p(x_i)$	Cumulative Probability $P(X \leq x_i)$
0	0.1353	0.1353
1	0.2707	0.4060
2	0.2707	0.6767
3	0.1804	0.8571
4	0.0902	0.9473
5	0.0361	0.9834
6	0.0120	0.9954
7	0.0034	0.9988
8	0.0009	0.9997
9	0.0002	0.9999

6.3 THE NORMAL DISTRIBUTION

If a statistician could select but one distribution to work with during his lifetime, he would almost surely select the *normal distribution*. We have seen that the binomial distribution approaches the normal distribution as the number of trials become large. We will show later that the **t**, **F**, and chi-square distributions are all related to the normal distribution. A large number of continuous random variables such as heights, weights, volumes, reaction times, and sizes are frequently found to be normally distributed. We venture to say that, although modern statisticians and applied economists could perhaps get along without electronic computers, it would be exceedingly difficult to do without the normal distribution.

In many respects, the normal distribution is the cornerstone of modern statistics. It is, therefore, very important that the material in this section be read with special care.

6.3.1 The Normal Curve

Normal distributions can be obtained from (or fitted to) many types of business and economic data. As an illustration, suppose a company has a machine that automatically fills ten-pound bags of sugar. The company checks this machine each week by selecting 100 bags of sugar at random and weighing these bags on a precision scale. The results of one such sample are summarized in Table 6.8 and presented as a relative frequency histogram in

Table 6.8 Sample Results of Sugar Bag Weights

Class Limits (Ounces)	Frequency (Bags)	Relative Frequency	Cumulative Relative Frequency
154–155	3	0.03	0.03
156–157	12	0.12	0.15
158–159	22	0.22	0.37
160–161	32	0.32	0.69
162–163	18	0.18	0.87
164–165	9	0.09	0.96
166–167	4	0.04	1.00
Totals	100	1.00	

Figure 6.6. A relative frequency polygon has been superimposed on the histogram to emphasize the similarity between the normal distribution in Figure 6.8 and the observed distribution in Figure 6.6. If the size of the class intervals were reduced, the resulting histogram would look even more like a smooth normal distribution. Such a histogram is presented in Figure 6.7.

In order to describe this distribution of weights fully, it was necessary to present a frequency table which, in turn, was used as a basis for the histogram in Figure 6.6. A summary description consisting of the mean and standard

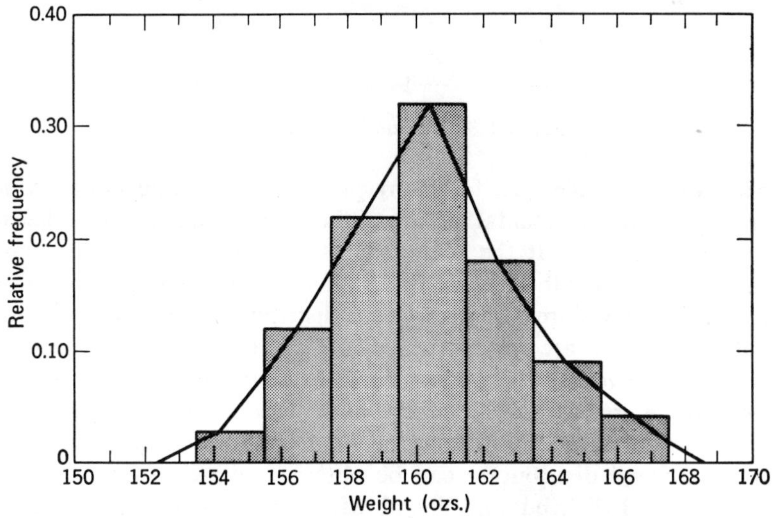

Figure 6.6 An observed distribution of sugar bag weights (2 oz class intervals).

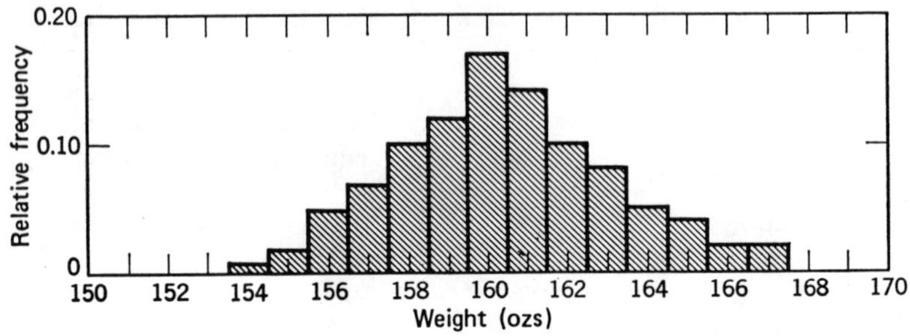

Figure 6.7 An observed distribution of sugar bag weights (1 oz class intervals).

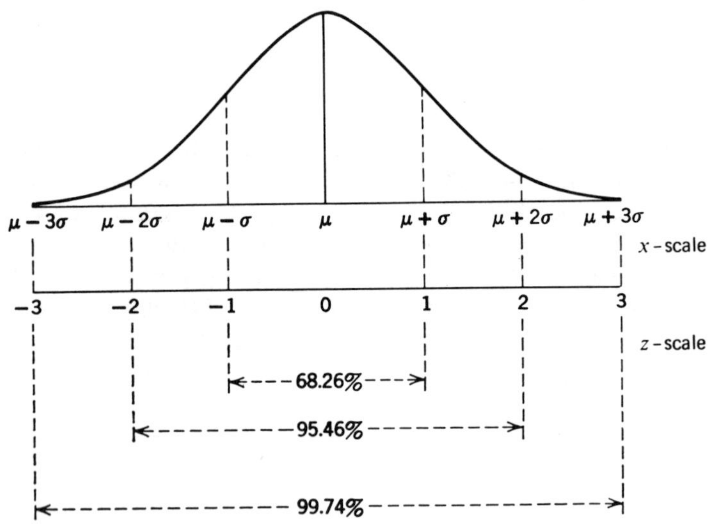

Figure 6.8 The normal distribution.

deviation would be adequate for many purposes but would not provide the reader with a complete mental picture of the distribution. On the other hand, if it could be stated that the observed distribution was "approximately a normal distribution with mean μ_0 and standard deviation σ_0," the informed reader would know a great deal about the distribution. He would know that normal distributions are symmetrical, bell-shaped distributions and that approximately 68 percent of the observations are within one standard deviation of the mean while approximately 95 percent of the observations are within two standard deviations of the mean. (See Figure 6.8.) Additional information about the normal distribution can be obtained quickly from tables of the standard normal distribution. The phrase "approximately a normal distribution" conveys a great deal of information to anyone familiar with modern statistics.

Sample distributions will be discussed in greater detail later, but first it is necessary to become familiar with the *theoretical* normal distribution.

The *distribution function* for a normal distribution with mean μ and standard deviation σ is

$$F(x) = \frac{1}{\sigma\sqrt{2\pi}} \int_{-\infty}^{x} \exp\left[-\frac{1}{2\sigma^2}(x-\mu)^2\right] dx \tag{6.14}$$

If a continuous random variable X is normally distributed with mean μ and standard deviation σ, then the probability that X lies between two numbers, say, x_0 and x_1 where $x_1 > x_0$, is given by

$$P(x_0 \le X \le x_1) = F(x_1) - F(x_0)$$

Fortunately, it is never necessary to work with the cumulative distribution function (6.14) directly because tables of the normal distribution are commonly available and easy to use.

The normal distribution is completely determined once the mean and standard deviation are specified. In other words, two parameters (μ and σ) determine the location and shape of the normal distribution. Two different normal distributions are drawn in Figure 6.9. Normal distributions with

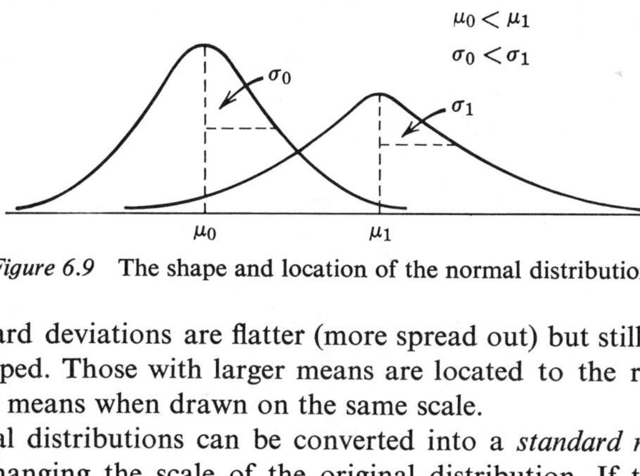

Figure 6.9 The shape and location of the normal distribution.

larger standard deviations are flatter (more spread out) but still symmetrical and bell-shaped. Those with larger means are located to the right of those with smaller means when drawn on the same scale.

All normal distributions can be converted into a *standard normal distribution* by changing the scale of the original distribution. If this were not possible, it would be necessary to construct a separate table of the normal distribution for each pair of values of μ and σ. The standard normal distribution has a mean of zero and a standard deviation equal to one. If X is a continuous random variable which is normally distributed with mean μ and variance σ^2, we indicate this by writing, "X is $N(\mu, \sigma^2)$." Applying this shorthand to the standard normal variable Z, we write, "Z is $N(0, 1)$." In other words, Z is normal, zero, one.[2]

[2] The shorthand notation "X is $n(x; \mu, \sigma^2)$" is used by some statisticians instead of "X is $N(\mu, \sigma^2)$."

If X is $N(\mu, \sigma^2)$, it can be converted to the standard normal variable Z by subtracting the mean μ from X and dividing the result by the standard deviation of X. Thus

$$Z = \frac{X - \mu}{\sigma} \tag{6.15}$$

The rules of expectations can be used to verify that Z is $N(0, 1)$. We find that

$$E(Z) = E\left[\frac{X - \mu}{\sigma}\right] = \frac{1}{\sigma}[E(X - \mu)] = \frac{1}{\sigma}[E(X) - \mu] = 0$$

and

$$\text{Var}(Z) = \text{Var}\left(\frac{X - \mu}{\sigma}\right) = \left(\frac{1}{\sigma^2}\right)\text{Var}(X - \mu) = \left(\frac{1}{\sigma^2}\right)\text{Var}(X) = 1$$

The procedure is the same as that discussed earlier in the case of the binomial distribution with mean $\mu = np$ and standard deviation $\sigma = \sqrt{np(1 - p)}$.

The change of scale converts μ on the x-scale to 0 on the z-scale, $\mu + \sigma$ on the x-scale to 1 on the z-scale, $\mu - 2\sigma$ to -2 on the z-scale, etc. This one-to-one relationship between values of X and values of Z is presented graphically in Figure 6.8.

The total area under the standard normal curve is equal to one. Therefore, probabilities can be represented as areas under the standard normal curve between two different values of Z. The shaded area in Figure 6.10, for

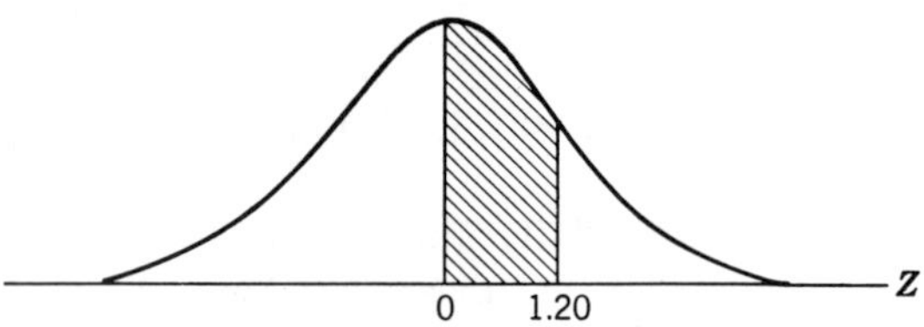

Figure 6.10

example, represents the probability that Z is greater than zero but less than 1.20.

The entries in Table I, Appendix I, are the areas under the normal curve between $z = 0$ and the values of Z given in the left-hand column of the table. Entries corresponding to negative values of Z are unnecessary because the normal curve is symmetrical. As a result, the probability that Z is between, say, -1.20 and zero is equal to the probability that Z lies between zero and 1.20. In other words, $P(0 \leq Z \leq 1.20) = P(-1.20 \leq Z \leq 0)$. Therefore, probabilities for any value of Z can be obtained from Table I, even though only positive values of Z are listed. Keep in mind that Z is a continuous variable and thus there is no probability associated with a particular value of Z. The probability that Z is exactly equal to, say, 1.20 is zero, but the probability that Z lies between zero and 1.20 is 0.3849. Furthermore, since the total area under the standard normal curve is equal to 1 and the curve is symmetrical, the area to the right (or left) of $z = 0$ must be equal to 0.5000. This fact is frequently useful when we want to find the probability that Z is greater than a given value. For example, suppose we want to determine the probability that Z is greater than 1.28. The probability that Z is between zero and 1.28 is obtained directly from Table I. This probability is then subtracted from 0.5000 to obtain the probability that Z is greater than 1.28. Thus

$$P(Z \geq 1.28) = 0.5000 - P(0 \leq Z \leq 1.28) = 0.5000 - 0.3997 = 0.1003$$

(What is the probability that Z is greater than 1.05?)

With a little practice, you will find Table I, Appendix I, to be very easy to use. It is a good practice at first to draw a picture of the normal distribution and sketch in the area you are interested in before looking up the probabilities in Table I. The following examples will illustrate how to use tables of the standard normal distribution.

Example 1: What is the probability that Z lies between 0 and 1.83? We are interested in the shaded area in Figure 6.11. This probability can be obtained from Table I directly and is equal to $P(0 \leq Z \leq 1.83) = 0.4664$.

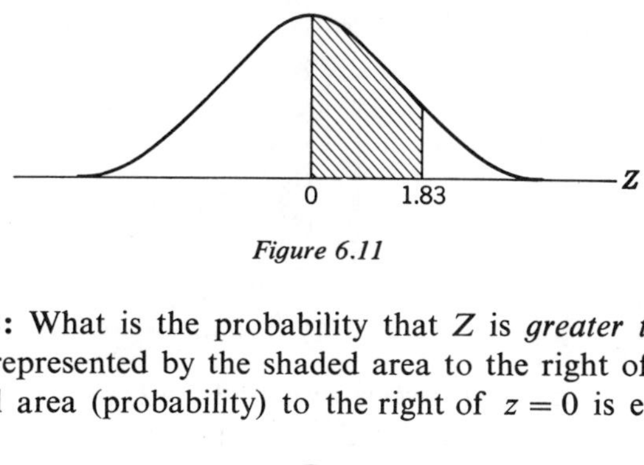

Figure 6.11

Example 2: What is the probability that Z is *greater than* 1.54? This probability is represented by the shaded area to the right of 1.54 in Figure 6.12. The total area (probability) to the right of $z = 0$ is equal to 0.5000.

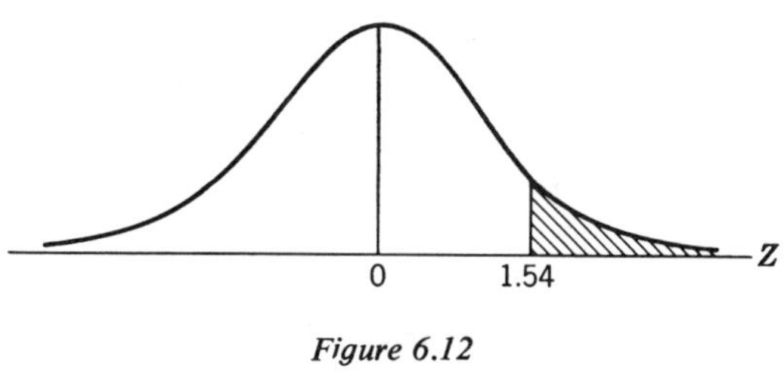

Figure 6.12

Therefore, we obtain the probability that Z is greater than 1.54 first by finding $P(0 \leq Z \leq 1.54)$ from Table I and then by subtracting this value from 0.5000. Thus

$$P(Z \geq 1.54) = P(Z \geq 0) - P(0 \leq Z \leq 1.54) = 0.5000 - 0.4382 = 0.0618$$

Example 3: What is the probability that Z is *greater than* -0.86? The area of interest is represented by the shaded area in Figure 6.13. Taking

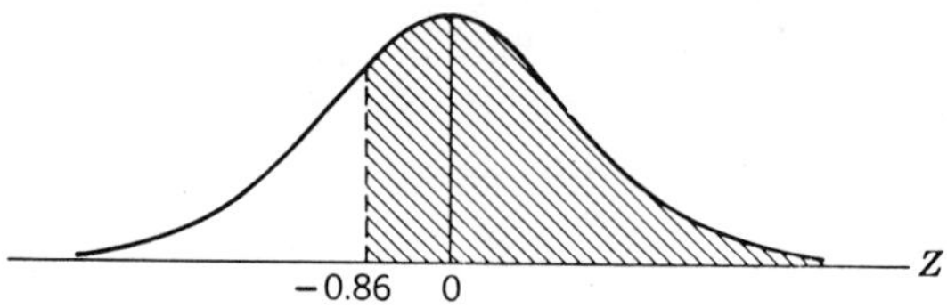

Figure 6.13

account of the symmetric of the normal distribution, we know that

$$P(Z \geq -0.86) = P(0 \leq Z \leq 0.86) + P(Z \geq 0) = 0.3051 + 0.5000 = 0.8051$$

In other words, first find the probability that Z lies between 0 and 0.86, which is equal to the probability that Z lies between 0 and -0.86. Then add the probability that Z is greater than zero.

Example 4: What is the probability that Z lies between 0.43 and 1.12? The shaded area in Figure 6.14 represents the required probability. First find

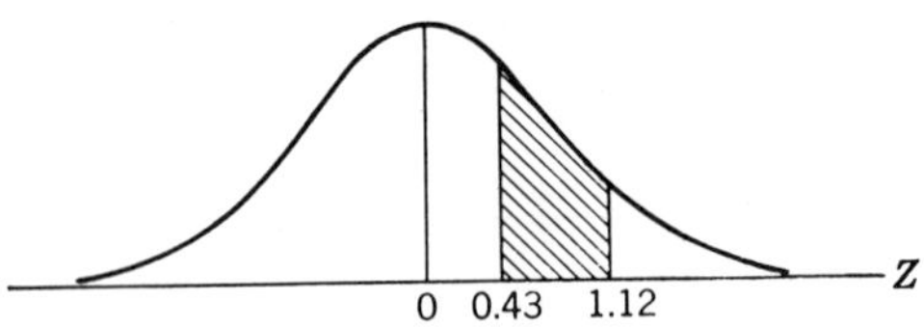

Figure 6.14

the probability that Z lies between zero and 1.12, and then subtract the probability that Z lies between 0 and 0.43. The result is

$$P(0.43 \leq Z \leq 1.12) = P(0 \leq Z \leq 1.12) - P(0 \leq Z \leq 0.43) = 0.3686 - 0.1664 = 0.2022$$

Example 5: What is the probability that Z is *less than* 0.77? This probability is presented as the shaded area in Figure 6.15. The answer is 0.7794. Explain how this answer was obtained.

In all of these examples, we have been dealing with the standard normal variable Z. Several examples for a normal variable X with mean μ and standard deviation σ will now be presented.

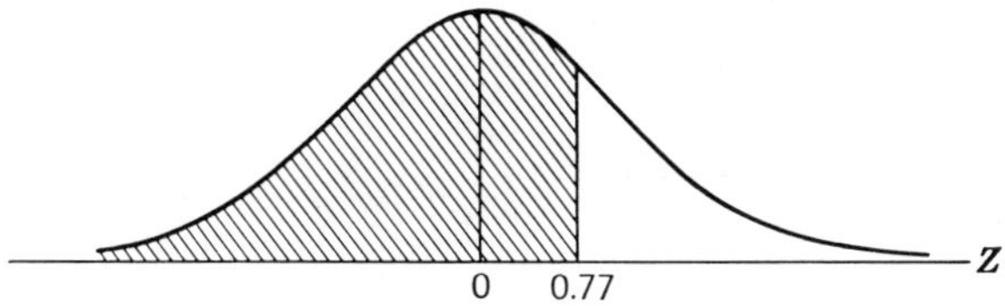

Figure 6.15

Example 6: Suppose that X is a continuous random variable with a normal distribution and has a mean of 12 and a standard deviation equal to 2. What is the probability that a value of X selected at random lies between 11 and 14? The area of interest is shown in Figure 6.16. The first step is to

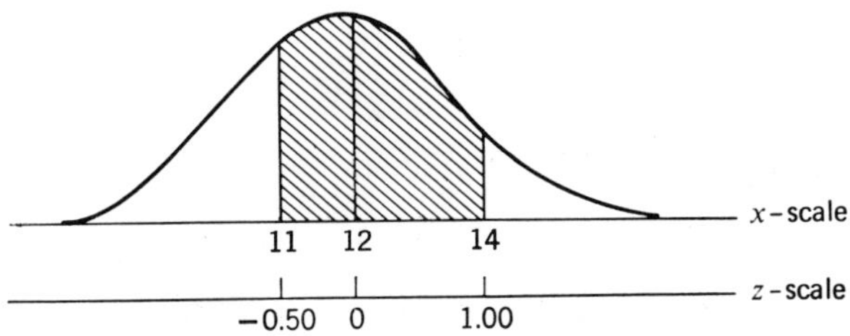

Figure 6.16

change from the x-scale to the z-scale. For $x = 11$, z is equal to

$$z = \frac{x - \mu}{\sigma} = \frac{11 - 12}{2} = \left(\frac{-1}{2}\right) = -0.50$$

and, for $x = 14$, we have

$$z = \frac{x - \mu}{\sigma} = \frac{14 - 12}{2} = \left(\frac{2}{2}\right) = 1.00$$

We are interested, therefore, in the probability that Z lies between -0.50 and 1.00. Thus

$$P(11 \leq X \leq 14) = P(-0.50 \leq Z \leq 1.00) = P(0 \leq Z \leq 0.50) + P(0 \leq Z \leq 1.00) = 0.1915 + 0.3413 = 0.5328$$

Example 7: Assume that X is $N(4, 9)$. What is the probability that a randomly selected X will lie between -2 and 5? This situation is presented in Figure 6.17. Keep in mind that the second number in the symbol $N(4, 9)$

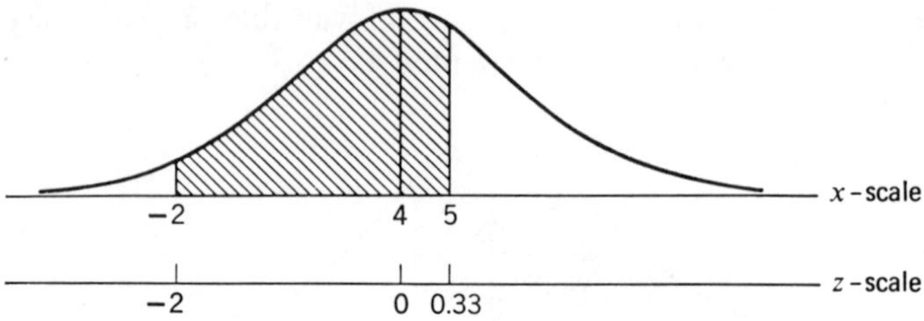

Figure 6.17

is the variance. The standard deviation is the square root of the variance. Performing a change of scale for $x = -2$, we obtain

$$z = \frac{-2-4}{3} = \left(\frac{-6}{3}\right) = -2$$

and, for $x = 5$, the Z value is

$$z = \frac{5-4}{3} = \left(\frac{1}{3}\right) = 0.33$$

The relevant probability is therefore

$$P(-2 \leq X \leq 5) = P(-2 \leq Z \leq 0.33) = 0.4772 + 0.1293 = 0.6065$$

Example 8: Let X be $N(20, 16)$. What is the probability that X is greater than 23? The area of interest is presented in Figure 6.18. The answer is

Figure 6.18

$P(X \geq 23) = 0.2266$. Change the X values to Z values, and show how this answer was obtained. Remember that the standard deviation in this case is equal to 4, not 16.

Before reading further, study Examples 1–8 until you understand how each of the probabilities were obtained. You will find it necessary to use the table of the standard normal distribution frequently in the following chapters.

Learning how to use it well now will save you a great deal of time later. We turn next to a discussion of several more realistic and complex examples of the use of the normal distribution.

6.3.2 *Using the Normal Distribution*

Example 1: A tropical fish specialist is interested in determining how long a certain type of tropical fish can survive under a particular toxic water condition. After a long series of experiments, the specialist estimates the average life of this type of tropical fish to be 90 days after it has been placed in toxic water. The standard deviation is estimated to be 20 days. The distribution of "days survived" appears to be approximately normal. A particular fish has survived for 110 days. What is the probability that this fish will live longer than 120 days?

Let X be the total number of days survived. If we assume that X is normally distributed with mean $\mu = 90$ and standard deviation $\sigma = 20$, the normal distribution can then be used to estimate the probability that X will be greater than 120 days, given that the fish has already survived 110 days.

Strictly speaking, X is not a continuous random variable since it takes on only discrete values such as 2, 30, or 80 days. The variable X could be defined as "time survived" and measured in fractions of days, hours, or minutes, however. In this case, it would be a continuous random variable.

At first glance, the random variable X, measured in days, is seen to have some of the characteristics of a binomial random variable. Each day constitutes a "trial," the outcome of which is either "survival" (a success) or "death" (a failure). The probability of a "success" is not constant on each trial, however. There is a high probability of survival during the first few days the fish is placed in the toxic water and a smaller probability of survival each day thereafter.

The probability that X is greater than 120, given that the fish has survived 110 days, is a conditional probability. It is therefore, necessary to use the multiplication rule for conditional probabilities discussed in Chapter 4. Consider two events B and C. The probability of B given C is

$$P(B \mid C) = \frac{P(B \cap C)}{P(C)}$$

Let B be the event "survives for more than 120 days" and C be "survives for more than 110 days." If the fish survives for 120 days, it must have survived for more than 110 days; therefore, $P(X \geq 120) = P(X \geq 120$ and $X \geq 110)$. In other words, $P(B)$ is equal to $P(B \cap C)$ in this case. The probabilities $P(X \geq 110)$ and $P(X \geq 120)$ can be obtained from tables of the standard normal distribution. Changing from the x-scale to the z-scale, we have, for $x = 110$,

$$z = \frac{110 - 90}{20} = 1.00$$

and, for $x = 120$,

$$z = \frac{120 - 90}{20} = 1.50$$

The areas of interest are shown in Figure 6.19. From Table I, Appendix I, we

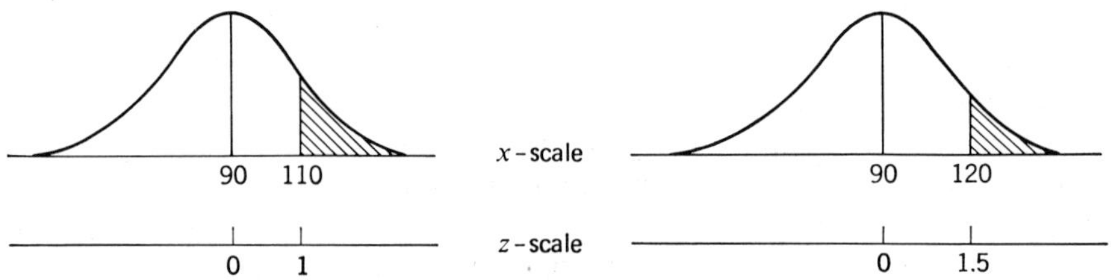

Figure 6.19

obtain

$$P(C) = P(X \geq 110) = P(Z \geq 1.00) = 0.5000 - 0.3413 = 0.1587$$

and

$$P(B) = P(X \geq 120) = P(Z \geq 1.50) = 0.5000 - 0.4332 = 0.0668$$

Therefore,

$$P(X \geq 120 \mid X \geq 110) = \frac{P(X \geq 120)}{P(X \geq 110)} \quad \frac{0.0668}{0.1587} = 0.4209$$

In other words, if a fish lives for 110 days, the probability is approximately 0.42 that it will live longer than 120 days.

Example 2: A fair coin is thrown four times. The binomial probabilities for 0, 1, 2, 3, and 4 heads are presented in Table 6.9. The graph of this distribution, with areas as probabilities, is shown in Figure 6.20.

The normal distribution can be used to approximate the binomial probabilities. Suppose, for example, that we want to estimate the probability that $x = 3$ heads are observed. We know that the expected value of X is equal to $\mu = np = 4(1/2) = 2$ while the standard deviation of X is equal to $\sigma = \sqrt{np(1-p)} = \sqrt{4(1/2)(1/2)} = 1$. Inspection of the histogram in Figure 6.20 shows that the probability of $x = 3$ is represented by the area under the

Table 6.9 Binomial Probabilities and Normal Approximations

Number of Heads, x	Binomial Probability $b(x)$	Normal Approximation to $b(x)$
0	0.0625	0.0668
1	0.2500	0.2417
2	0.3750	0.3830
3	0.2500	0.2417
4	0.0625	0.0668

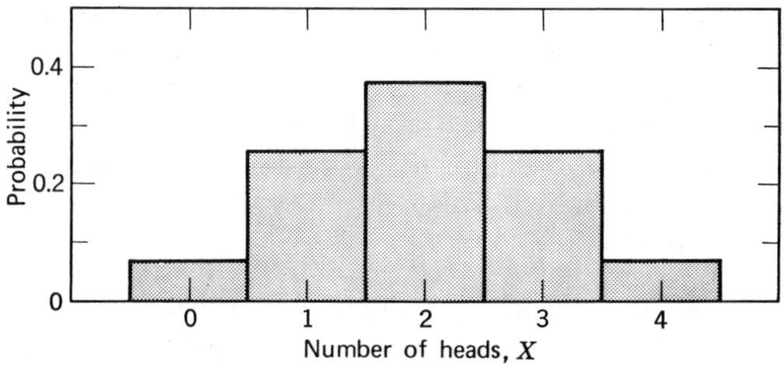

Figure 6.20 Binomial probabilities.

rectangle that has its left boundary at 2.5 and its right boundary at 3.5. Converting these values to values of the standard normal variable, we obtain

$$z = \frac{(x - 0.5) - \mu}{\sigma} = \frac{2.5 - 2}{1} = 0.5$$

and

$$z = \frac{(x + 0.5) - \mu}{\sigma} = \frac{3.5 - 2}{1} = 1.5$$

Therefore,

$$P(x = 3) = P(0.5 \leq Z \leq 1.5) = 0.4332 - 0.1915 = 0.2417$$

which is fairly close to the true binomial probability 0.2500.

As a second illustration, suppose the normal distribution is used to approximate the probability that X is greater than 2. This probability is represented in Figure 6.20 by the area in the rectangles to the right of 2.5. The Z value in this case is

$$z = \frac{2.5 - 2}{1} = 0.5$$

and

$$P(X > 2) = P(Z > 0.5) = 0.5000 - 0.1915 = 0.3085$$

The actual binomial probability is

$$P(X > 2) = P(x = 3) + P(x = 4) = 0.2500 + 0.0625 = 0.3125$$

The normal approximation to the binomial is within 1.3 percent of the true probability in this case. The normal approximation was improved slightly by including the tail area of the normal distribution that goes to infinity rather than stopping with the z value corresponding to 4.5, the right-hand boundary of the rectangle centered over $x = 4$ in Figure 6.20.

Generally speaking, the normal approximation to the binomial is very good unless p is very close to 0 or 1. The approximation tends to improve as the number of trials increases and tends to be more accurate for values of X near the mean np than for values in the tails of the binomial distribution. The addition and subtraction of 0.5 from the binomial variable X is necessary because the normal approximation treats the binomial distribution as if it were a continuous distribution. This refinement is usually not necessary if n is larger than about 40.

Example 3: A company estimates that approximately 30 percent of the special prize coupons it plans to mail for a sales promotion program will be returned. If the probability is $p = 0.3$ that a person receiving a coupon will return it, what is the probability that more than 165 coupons will be returned if 500 are mailed?

This is a binomial experiment with $n = 500$ and $p = 0.3$. The probability that more than $x = 165$ coupons are returned can be approximated by using the normal approximation to the binomial. Let X be the number of coupons returned. This distribution of X has mean $\mu = np = 150$ and standard deviation $\sigma = \sqrt{np(1-p)} = \sqrt{105} = 10.25$, since n is large, there is no need to subtract 0.5 from 165 in this case. Therefore, the standard normal variable is

$$z = \frac{165 - 150}{10.25} = 1.46$$

Using Table I, Appendix I, we obtain

$$P(X \geq 165) = P(Z \geq 1.46) = 0.5000 - 0.4279 = 0.0721$$

In other words, the probability is approximately 0.07 that more than 165 coupons of the 500 mailed will be returned to the company.

In a strict sense, the *assumption* that $p = 0.3$ for everyone who receives the coupons is not justified. Some people always return such coupons and others never do. By making such an assumption, however, the company can obtain a rough estimate of the probability that more than 165 coupons (or any other number) will be returned. Without the assumption that $p = 0.3$, the company would have to use a more costly method to estimate $P(X \geq 165)$. For example, it could mail 500 coupons to 100 different cities and determine the percentage of the cities from which more than 165 coupons were returned. The company may consider the rough estimate based on the assumption that $p = 0.3$ to be sufficiently accurate as a basis for such things as (1) deciding if it will have to hire extra help to process coupons returned or (2) judging the effectiveness of the sales promotion coupons.

Example 4: A tire manufacturer wants to estimate the expected road life of a new type of tire under normal driving conditions. Four hundred of the new tires are driven until they wear out. The results are presented in Table 6.10.

Table 6.10 Road Life of Tires

Class Limit (1000 Miles)	Class Midpoint	Frequency (Tires)	Relative Frequency	Cumulative Relative Frequency
13–15	14	20	0.050	0.050
16–18	17	40	0.100	0.150
19–21	20	50	0.125	0.275
22–24	23	70	0.175	0.450
25–27	26	80	0.200	0.650
28–30	29	60	0.150	0.800
31–33	32	40	0.100	0.900
34–36	35	30	0.075	0.975
37–39	38	10	0.025	1.000
Totals		400	1.000	

Analysis of the original data reveals that the average tire lasted 25,300 miles while the sample standard deviation was 6100 miles. The data are presented in the form of a cumulative relative frequency polygon in Figure 6.21.

A normal distribution can be fitted to the data by assuming that μ is equal to the sample mean and σ is equal to the sample standard deviation. The standard normal variable is then computed for different values of the road life of tires X by using the formula

$$z = \frac{x - 25.3}{6.1}$$

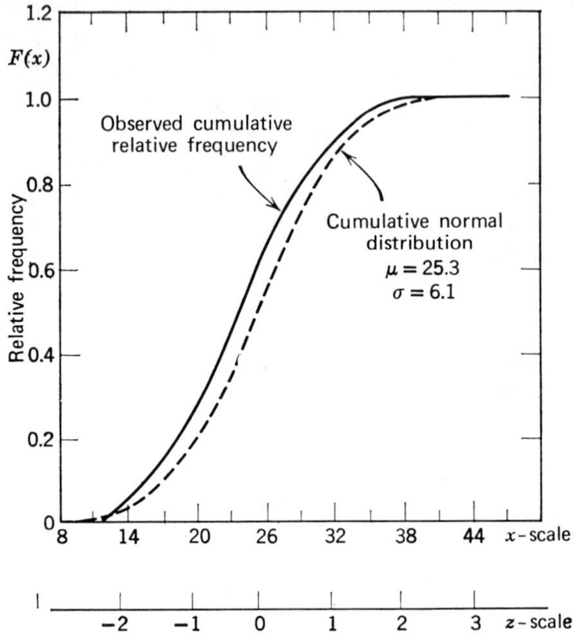

Figure 6.21 Cumulative relative frequency polygon of tire life.

The probabilities obtained are given in Table 6.11 and shown as a graph of the cumulative normal distribution in Figure 6.21.

We find that the normal distribution does not fit the observed data very

Table 6.11 Normal Distribution Fitted to Road Life of Tires

Class Midpoints x	Standard Normal Variable z	Cumulative Normal Distribution $F(x)$	Observed Cumulative Relative Frequency
14	−1.85	0.0322	0.050
17	−1.36	0.0869	0.150
20	−0.87	0.1922	0.275
23	−0.38	0.3520	0.450
26	0.11	0.5438	0.650
29	0.61	0.7291	0.800
32	1.10	0.8665	0.900
35	1.59	0.9441	0.975
38	2.08	0.9812	1.000

closely. The observed polygon is located to the left of the cumulative normal distribution for nearly all values of X. Inspection of the data indicates that more tires wear out before 25,300 miles than would be expected if the distribution of "tire lifes" was, in fact, normally distributed with $\mu = 25{,}300$ and $\sigma = 6100$ miles. The observed distribution is slightly skewed to the right.

6.4 THE CHI-SQUARE DISTRIBUTION

The chi-square distribution is useful when testing hypotheses about variances and when trying to determine if a sample distribution and a theoretical distribution are similar enough to justify the conclusion that the population from which the sample was drawn has a given theoretical distribution. For example, a die is thrown 300 times and the number of spots landing up each time is observed. In the long run, we expect to observe the six possible outcomes with equal frequency. In 300 throws, however, it is not very likely that each of the six possible outcomes occurs exactly 50 times. The chi-square distribution provides a method by which to compare the observed and expected outcomes and then determine if the observed results were sufficiently close to the expected results to justify the conclusion that the die was (or was not) a "fair" die. The purpose of this section is to provide a brief introduction to the chi-square distribution, which will be used in several different ways in later chapters.

If $y_1, y_2, \ldots, y_v$ are independently distributed random variables all of which are normally distributed with mean of 0 and standard deviation equal to 1, then the random variable

$$\chi_v^2 = \sum_{i=1}^{v} y_i^2 \tag{6.16}$$

is called the chi-square variable with v degrees of freedom. For example, if a set of n random variables $\{x_i\}$ are all $N(\mu, \sigma^2)$, then a set of n random variables $\{y_i\}$ can be computed such that all y_i are $N(0, 1)$ by using the formula

$$y_i = \frac{x_i - \mu}{\sigma}$$

The random variable

$$\chi_n^2 = \sum_{i=1}^{n} \left[\frac{x_i - \mu}{\sigma}\right]^2$$

is therefore distributed as chi-square with n degrees of freedom.

The term "degrees of freedom" is used frequently in statistics. The basic idea is relatively simple. The selection of a sample of n items from a population, for example, involves n degrees of freedom of choice. Suppose, however, that we consider only samples of values $x_1, x_2, \ldots, x_n$ which have a fixed sample mean, $\bar{x}$. If $(n-1)$ of the sample values are selected, the remaining value is determined. Thus, in this case, there are only $(n-1)$ degrees of freedom in the selection process. In general, if we require the sample values to satisfy k independent linear relations, only $(n-k)$ sample values can be selected freely and there are only $(n-k)$ degrees of freedom. (Note that the random variable χ_n^2 has only n degrees of freedom because the *population* mean μ and standard σ deviation do not constitute restrictions on the *sample* values x_i.)

Before going further, let us state a few characteristics of the χ^2 distribution. The variable χ^2 cannot be negative because it is the sum of positive numbers. Its mean increases as the number of degrees of freedom increase. The chi-square distribution is a continuous distribution whose shape and location depend on the number of degrees of freedom (df). When the number of degrees of freedom is small, the χ^2 distribution is skewed to the right. Curves of the χ^2 distribution for several different degrees of freedom are drawn in Figure 6.22.

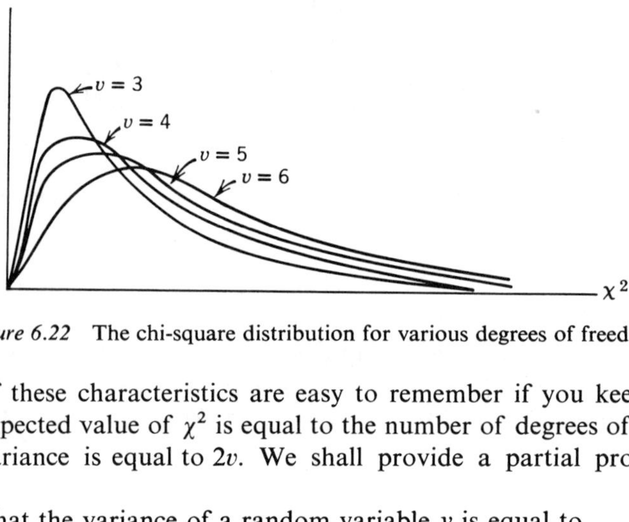

Figure 6.22 The chi-square distribution for various degrees of freedom.

Most of these characteristics are easy to remember if you keep in mind that the expected value of χ^2 is equal to the number of degrees of freedom v and its variance is equal to $2v$. We shall provide a partial proof of this statement.

Recall that the variance of a random variable y is equal to

$$\mathrm{Var}(y) = E(y^2) - [E(y)]^2$$

In particular, if y is normally distributed with mean zero and variance 1, then $E(y) = 0$ and $\mathrm{Var}(y) = E(y^2) = 1$. Furthermore, it can be shown that $E(y^4) = 3$, from which it follows that

$$\mathrm{Var}(y^2) = E(y^4) - [E(y^2)]^2 = 3 - 1 = 2$$

provided that y is $N(0, 1)$.

By definition, the chi-square variable is equal to the sum of v *independently distributed* random variables y_i which are $N(0, 1)$. That is

$$\chi_v^2 = y_1^2 + y_2^2 + y_3^2 + \cdots + y_v^2$$

Taking the expected value of χ_v^2, we obtain

$$\begin{aligned} E(\chi_v^2) &= E(y_1^2) + E(y_2^2) + E(y_3^2) + \cdots + E(y_v^2) \\ &= 1 \quad + 1 \quad + 1 \quad + \cdots + 1 \\ &= v \end{aligned}$$

Since the y_i are independent, the variance of the sum of the y_i^2 is equal to the sum of the variances. Thus

$$\begin{aligned} \mathrm{Var}(\chi_v^2) &= \mathrm{Var}(y_1^2) + \mathrm{Var}(y_2^2) + \mathrm{Var}(y_3^2) + \cdots + \mathrm{Var}(y_v^2) \\ &= 2 \quad + 2 \quad + 2 \quad + \cdots + 2 \\ &= 2v \end{aligned}$$

as was to be shown.[3]

To tabulate the chi-square distribution completely would require a different table for each degree of freedom. Complete tables are not necessary in most statistical applications, however. Tables of the chi-square distribution usually provide χ^2 values for various degrees of freedom between 1 and 100 and for a few of the most commonly used probabilities. Chi-square tables contain values of χ^2 in the body of the table rather than probabilities as in the case of tables of the standard normal distribution. The chi-square distribution with over 100 degrees of freedom is approximately a normal distribution and the quantity $z = \sqrt{2\chi^2} - \sqrt{2v - 1}$ can be treated as a standard normal variable, where v is the number of degrees of freedom.

Values of the χ^2 distribution for several different degrees of freedom and levels of probability are provided in Table IV, Appendix I. Part of a χ^2 table is presented in Table 6.12. Let α denote the probability that the chi-square variable takes a value larger than the tabled value of chi-square with v degrees of freedom. The tabled value of chi-square is denoted by $\chi^2_{\alpha,v}$. Therefore

$$P(\chi^2 \geq \chi^2_{\alpha,v}) = \alpha$$

For example, if $v = 10$, then 10 percent of the area under the chi-square distribution lies to the right of $\chi^2_{0.10,10} = 16.00$. This case is illustrated in Figure 6.23 where the probability $\alpha = 0.10$ is represented by the shaded area in the right-hand tail of the chi-square distribution.

[3] The derivation of the chi-square distribution can be found in most books on mathematical statistics. For example, see R. V. Hogg and A. T. Craig, *Introduction to Mathematical Statistics* (New York; The Macmillan Company, 1959), pp. 36–38.

Table 6.12 The Chi-Square Distribution

v	α 0.90	0.50	0.10	0.05
1	0.01	0.45	2.71	3.84
5	1.61	4.35	9.24	11.10
10	4.87	9.34	16.00	18.30
20	12.40	19.30	28.40	31.40
30	20.60	29.30	40.30	43.80
40	29.10	39.30	51.80	55.80

The chi-square table is in cumulative form. Thus, for ten degrees of freedom, 90 percent of the χ^2 distribution lies to the right of 4.87, 50 percent to the right of 9.34, and 5 percent to the right of 18.30. Alternatively, 10 percent of the χ^2 distribution for 10 degrees of freedom lies to the *left* of 4.87, 50 percent to the *left* of 9.34, and 95 percent to the *left* of 18.30.

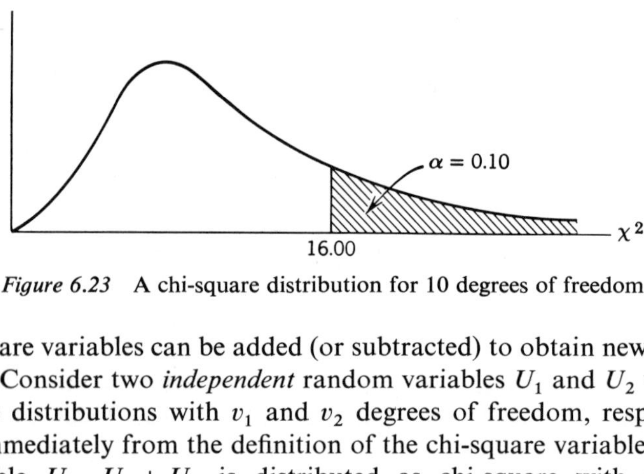

Figure 6.23 A chi-square distribution for 10 degrees of freedom.

Chi-square variables can be added (or subtracted) to obtain new chi-square variables. Consider two *independent* random variables U_1 and U_2 which have chi-square distributions with v_1 and v_2 degrees of freedom, respectively. It follows immediately from the definition of the chi-square variable (6.16) that the variable $U = U_1 + U_2$ is distributed as chi-square with $v = v_1 + v_2$ degrees of freedom. This additive property applies in general for k independent chi-square variables. We shall illustrate the use of this property later.

We shall now construct a random variable that has a chi-square distribution.

Let $x_1, x_2, \ldots, x_n$ represent a sample drawn from a normal population with mean μ and variance σ^2. The sample mean $\overline{X}$ and sample variance s^2 are defined as

$$\overline{X} = \left(\frac{1}{n}\right) \sum_{i=1}^{n} x_i \qquad \text{and} \qquad s^2 = \left(\frac{1}{n-1}\right) \sum_{i=1}^{n} (x_i - \overline{X})^2$$

The sum of squared deviations from the sample mean $\sum (x_i - \overline{X})^2$ plays an important role in the discussion that follows. We shall begin by showing that $\sum (x_i - \overline{X})^2$ can be partitioned into two parts: $\sum (x_i - \mu)^2$ and $n(\overline{X} - \mu)^2$.

If μ is added and subtracted from each term $(x_i - \overline{X})^2$, we have[4]

$$\begin{aligned} \sum (x_i - \overline{X})^2 &= \sum [((x_i - \mu) - (\overline{X} - \mu))^2] \\ &= \sum [(x_i - \mu)^2 - 2(\overline{X} - \mu)(x_i - \mu) + (\overline{X} - \mu)^2] \\ &= \sum (x_i - \mu)^2 - 2(\overline{X} - \mu) \sum (x_i - \mu) + n(\overline{X} - \mu)^2 \end{aligned}$$

The middle term in this expansion can be simplified by using the fact that $\sum x_i = n\overline{X}$. Thus

$$\begin{aligned} 2(\overline{X} - \mu) \sum (x_i - \mu) &= 2(\overline{X} - \mu)(\sum x_i - n\mu) \\ &= 2(\overline{X} - \mu)(n\overline{X} - n\mu) \\ &= 2n(\overline{X} - \mu)^2 \end{aligned}$$

The sum of squared deviations from the sample mean is equal, then, to

$$\sum (x_i - \overline{X})^2 = \sum (x_i - \mu)^2 - n(\overline{X} - \mu)^2$$

The sample variance can now be written as

$$s^2 = \frac{1}{n-1} [\sum (x_i - \mu)^2 - n(\overline{X} - \mu)^2]$$

On multiplying both sides by $(n - 1)/\sigma^2$, we have

$$\frac{(n-1)s^2}{\sigma^2} = \sum_{i=1}^{n} \left(\frac{x_i - \mu}{\sigma}\right)^2 - \left(\frac{\overline{X} - \mu}{\sigma/\sqrt{n}}\right)^2$$

The terms $y_i = (x_i - \mu)/\sigma$ are standard normal variables with mean 1 and variance equal to zero. We shall show later that the term

$$\frac{\overline{X} - \mu}{\sigma/\sqrt{n}}$$

is also $N(0, 1)$.

If $x_1, x_2, \ldots, x_n$ are independently distributed, then the quantity

$$\theta_n = \sum_{i=1}^{n} \left(\frac{x_i - \mu}{\sigma}\right)^2 = \sum_{i=1}^{n} y_i^2$$

[4] Note that $(\bar{X} - \mu)$ is a constant. Thus $\sum (\bar{X} - \mu)(X_i - \mu) = (\bar{X} - \mu)\sum (X_i - \mu)$ and $\sum (\bar{X} - \mu) = n(\bar{X} - \mu)$. A review of the rules of summations is presented in Section I of the Mathematical Appendix.

has a chi-square distribution with n degrees of freedom. Similarly, the quantity

$$\theta_1 = \left(\frac{\overline{X} - \mu}{\sigma/\sqrt{n}}\right)^2$$

has a chi-square distribution with 1 degree of freedom. Furthermore, it can be shown that θ_n and θ_1 are statistically independent.

It follows, therefore, from the definition of the chi-square variable (6.16) that

$$\frac{(n-1)s^2}{\sigma^2} = \theta_n - \theta_1$$

is distributed as chi-square with $v = n - 1$ degrees of freedom.

It has not been our intention to provide a rigorous proof that $(n-1)s^2/\sigma^2$ has a chi-square distribution with $n - 1$ degrees of freedom. Complete proofs can be found in most textbooks on mathematical statistics. We have outlined the form of the proof in some detail, however, because the variable $(n-1)s^2/\sigma^2$ will be used in several places later in this book.

To illustrate the use of the chi-square variable $(n-1)s^2/\sigma^2$, suppose a sample of six statistically independent observations is drawn from a population and the sample variance is found to be equal to 32. Is it very likely that the variance of the population from which the sample was drawn is equal to 16? To answer this question, first compute

$$\chi^2 = \frac{(n-1)s^2}{\sigma^2} = \frac{5(32)}{16} = 10$$

Then, using Table 6.12, it is seen that the probability of obtaining a value of χ^2 greater than 9.24 when the degrees of freedom are $v = (n - 1) = 5$ is equal to $\alpha = 0.10$. The observed value of χ^2 in this case was 10. The conclusion is that the sample was a rather unusual one if the population variance is actually equal to 16.

The chi-square distribution will be discussed in more detail in later chapters. We turn now to a discussion of a distribution involving the ratio of two chi-square variables.

6.5 THE *F* DISTRIBUTION

The *F* distribution is used to test hypotheses about (1) the equality of variances and (2) the equality of more than two means. It enables economists to judge how well production functions, demand functions, and consumption functions fit the observed data. It provides plant specialists with a method for determining which factors are important in explaining the variation in crop yields from different test plots. It helps quality control specialists decide if the variation in quality of parts produced by one type of machine is the same as (or significantly different than) the quality variation in parts produced by a second machine. The purpose of this section is to provide a brief introduction to the F distribution; examples of its uses will be presented in later chapters.

If U_1 and U_2 are independent random variables and U_1 has an χ^2 distribution with v_1 degrees of freedom while U_2 has an χ^2 distribution with v_2 degrees of freedom, then the distribution of the ratio

$$F = \frac{(U_1/v_1)}{(U_2/v_2)} = \left(\frac{U_1}{U_2}\right)\left(\frac{v_2}{v_1}\right) \tag{6.17}$$

is called an F distribution with v_1 and v_2 degrees of freedom. The ratio F is sometimes referred to as the *variance ratio.*

The F distribution was discovered by R. A. Fisher in the early 1920's. He tabulated the distribution in the form $Z = \log_e\sqrt{F}$. These tables were later revised by G. Snedecor using the transformation $F = e^{2Z}$. Snedecor called the ratio (6.17) the F ratio in honor of Fisher.[5]

Before considering the use of the F table, let us discuss a few characteristics of the F distribution. If (U_1/v_1) is very small and (U_2/v_2) very large, F will approach zero. Conversely, F may approach ∞. The degrees of freedom v_1 and v_2 are the only parameters of the F distribution. For small values of v_1 and v_2, the F distribution is skewed to the right. As the degrees of freedom v_1 and v_2 increase, however, the F distribution tends to normality.

Tables of the F distribution are slightly more complicated than those for the normal or the chi-square distribution because there are two sets of

Table 6.13 The F Distribution for $\alpha = 0.05$

	v_1				
v_2	4	8	10	16	20
4	6.39	6.04	5.96	5.84	5.80
8	3.84	3.44	3.35	3.20	3.15
10	3.48	3.07	2.98	2.82	2.77
15	3.06	2.64	2.54	2.39	2.33
20	2.87	2.45	2.35	2.18	2.12

[5] See G. W. Snedecor, *Statistical Methods*, Fifth Edition (Ames, Iowa: Iowa State College Press, 1957). p. 244.

degrees of freedom associated with the F ratio. A common procedure is to select a probability α located in the right-hand tail of the F distribution and then tabulate F_{α,v_1,v_2} for various values of v_1 and v_2. The value F_{α,v_1,v_2} is such that $P(F \geq F_{\alpha,v_1,v_2}) = \alpha$. Table III, Appendix I, provides the F values for α equal to 0.01 and 0.05. Part of Table III is presented as Table 6.13.

The F table in Appendix II locates the entire probability α in the right-hand tail of the F distribution. For example, 5 percent of the F distribution with $v_1 = 20$ and $v_2 = 15$ degrees of freedom lies to the right of 2.33. This is illustrated in Figure 6.24. Similarly, 5 percent of the F distribution with

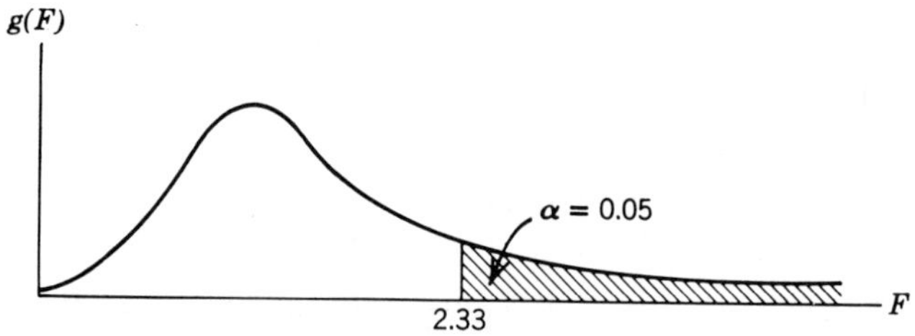

Figure 6.24 The F distribution for $v_1 = 20$ and $v_2 = 15$.

$v_1 = 10$ and $v_2 = 4$ degrees of freedom lies to the right of 5.96.

If values of F at the lower end of the F distribution are needed, they can be found by noting that the reciprocal of F has an F distribution with v_2 and v_1 degrees of freedom. This result follows directly from the definition of F in (6.17). In other words,

$$P(F \geq F_{\alpha,v_1,v_2}) = P\left(F \leq \frac{1}{F_{\alpha,v_2,v_1}}\right) = \alpha$$

To illustrate this, let F have an F distribution with $v_1 = 8$ and $v_2 = 10$ degrees of freedom. Then $P(F \geq 3.07) = 0.05$. To find the lower tail (or left tail) value of F corresponding to $\alpha = 0.05$, we first note that $F_{0.05,10,8} = 3.35$ and $(1/3.35) = 0.298$. Therefore, $P(F \leq 0.298) = 0.05$. The results are shown in Figure 6.25.

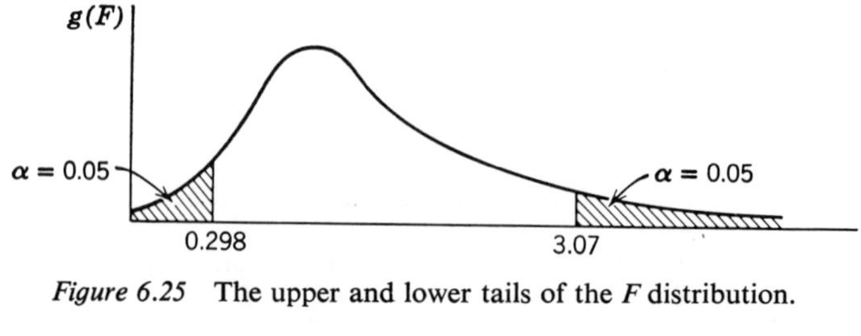

Figure 6.25 The upper and lower tails of the F distribution.

To illustrate one possible use of the F distribution, suppose that a company produces two types of glass rods, say Type A and Type B, which are identical except that they are made from different kinds of glass. Nine Type A glass rods are selected at random, carefully broken, and their breaking pressures recorded. The sample variance of breaking pressures for Type A glass rods is found to be equal to

$$s_a^2 = 66 \text{ pounds}$$

Nine Type B glass rods are also broken and the sample variance of breaking pressures is found to be

$$s_b^2 = 44 \text{ pounds}$$

Is it probable that the population variance of breaking pressures for Type A rods of σ_a^2 is equal to that of Type B rods σ_b^2?

You will recall that the ratios

$$U_a = \frac{(n-1)s_a^2}{\sigma_a^2} \quad \text{and} \quad U_b = \frac{(n-1)s_b^2}{\sigma_b^2}$$

are distributed as χ^2 with $(n-1)$ degrees of freedom. If $\sigma_a^2 = \sigma_b^2$, then the ratio

$$F = \frac{(U_a/(n-1))}{(U_b/(n-1))} = \frac{s_a^2}{s_b^2}$$

has an F distribution with $v_1 = (n-1)$ and $v_2 = (n-1)$ degrees of freedom. The observed value of F is $F = (66/44) = 1.50$. The tabled value of F for $\alpha = 0.05$ and $v_1 = v_2 = 8$ is 3.44. Since the observed value of 1.50 is considerably less than the tabled value of 3.44, we conclude that there is not enough evidence to reject the hypothesis that σ_a^2 is equal to σ_b^2.[6]

6.6 THE t DISTRIBUTION

The t distribution is important when working with samples of less than 100 observations. It is commonly used to determine whether sample averages are significantly different from hypothesized population means. An insurance company may use the t distribution to determine, on the basis of a small sample, if a certain type of driver travels more than 10,000 miles a year in the family car. An economist may use it to determine if the price elasticity of demand for shoes is significantly different from zero. A tax agency may

[6] A statistical hypothesis is an assumption about the frequency distribution of a random variable. Hypothesis testing will be discussed in greater detail in Chapter 8.

use it to obtain upper and lower boundaries for expected incomes of different classes of taxpayers. Examples of such uses of the t distribution will be discussed later. The main purpose of this section is to acquaint the reader with the t distribution and the use of t tables.

If y and U are independent random variables and y is normally distributed with mean zero and standard deviation equal to one while U is distributed as χ^2 with v degrees of freedom, then the distribution of the quotient

$$t = \frac{y}{(U/v)^{1/2}} \tag{6.18}$$

is called the t distribution with v degrees of freedom.

The random variable t can be either positive or negative because the normal variable y can take on both positive and negative values. The t distribution is a continuous distribution symmetric about $t = 0$, its expected value. The only parameter of the t distribution is the number of degrees of freedom v associated with the chi-square variable in the denominator of the t ratio (6.18). The variance of the t distribution is $v/(v - 2)$ provided that v is greater than 2. As the number of degrees of freedom increases, the variance $v/(v - 2)$ approaches 1 and the t distribution approaches the standard normal distribution.[7]

Since, as in the case of the chi-square and F distributions, the t distribution depends on the number of degrees of freedom, it is impractical to tabulate the distribution completely for each value of v. A common procedure is to select a probability to be distributed equally to the right- and left-hand tails of the t distribution and then table $t_{\alpha/2}$ for different values of v, where $t_{\alpha/2}$ is such that $P(t \geq t_{\alpha/2}) = \alpha/2$. This is illustrated in Figure 6.26. Table II,

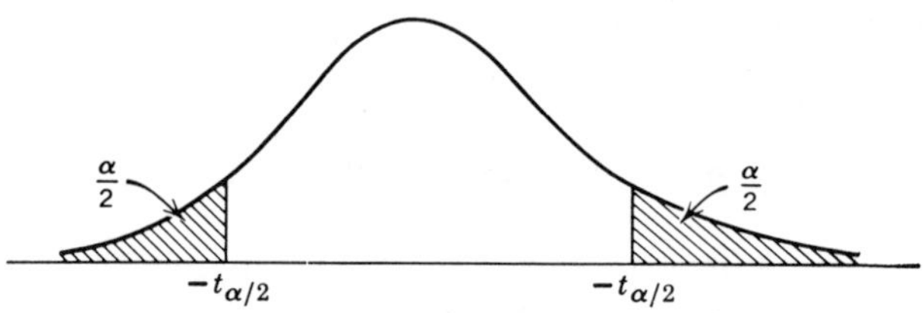

Figure 6.26 The t distribution.

Appendix I, lists t values for various degrees of freedom and for $\alpha = 0.50$, 0.25, 0.10, 0.05, 0.025, 0.01, and 0.005. Part of Table II is presented as

[7] A more rigorous discussion of the t distribution can be found in R. V. Hogg and A. T. Craig, *Introduction to Mathematical Statistics* (New York: The Macmillan Company, 1959), pp. 74–77.

Table 6.14 The t Distribution

	α		
v	0.10	0.05	0.01
1	6.3138	12.7060	63.6570
5	2.0150	2.5706	4.0321
10	1.8125	2.2281	3.1693
15	1.7530	2.1315	2.9467
25	1.7081	2.0595	2.7874
∞	1.6449	1.9600	2.5758

Table 6.14. For example, if $v = 15$, then the probability that t is greater than 1.753 is $\alpha/2 = 0.05$. Cumulative t tables and t tables in which the probability is located entirely in the right-hand tail of the distribution are also in common use. It is, therefore, a good idea, when using a new t table, to determine exactly what the probabilities listed at the top of the table refer to.

The t and F distributions are closely related. This is seen by squaring both sides of (6.18) to obtain

$$t^2 = \frac{y^2}{(U/v)}$$

Since y is normally distributed with mean zero and variance equal to one, the variable y^2 has a chi-square distribution with one degree of freedom. It is obvious, therefore, from the definition of F, that the ratio

$$t^2 = \frac{(y^2/1)}{(U/v)}$$

has an F distribution with 1 and v degrees of freedom.

This relationship can be verified with the help of the t and F tables in Appendix I. For example, if $\alpha = 0.05$ and $v = 6$, we find in Table II that $t_{0.05,6} = 2.4469$. The corresponding F value is $F_{0.05,1,6} = 5.99$. A check by multiplication will show that $(2.4469)^2 = 5.99$ as expected.

Let us now construct a random variable that has a t distribution. We begin with the sample mean

$$\bar{X} = \left(\frac{1}{n}\right) \sum_{i=1}^{n} x_i$$

where the x_i's are independent normally distributed random variables with mean μ and standard deviation σ. The expected value of $\bar{X}$ is

$$E(\bar{X}) = E\left[\frac{1}{n}\sum x_i\right] = \frac{1}{n}\sum E(x_i) = \frac{1}{n}\underbrace{(\mu + \mu + \cdots + \mu)}_{n} = \mu$$

and the variance of $\bar{X}$ is

$$\begin{aligned}\operatorname{Var}(\bar{X}) &= \operatorname{Var}\left(\frac{1}{n}\sum x_i\right) = \left(\frac{1}{n^2}\right)\operatorname{Var}\left(\sum x_i\right) = \frac{1}{n^2}\sum \operatorname{Var}(x_i) \\ &= \frac{1}{n^2}\underbrace{(\sigma^2 + \sigma^2 + \cdots + \sigma^2)}_{n} = \frac{n\sigma^2}{n^2} = \frac{\sigma^2}{n}\end{aligned}$$

The random variable $\bar{X}$ is the sum of normally distributed random variables and therefore has a normal distribution with mean μ and standard deviation $\sigma/\sqrt{n}$.[8] A new random variable y can be obtained by subtracting the mean μ from $\bar{X}$ and dividing by the standard deviation of $\bar{X}$. The random variable

$$y = \frac{\sqrt{n}(\bar{X} - \mu)}{\sigma}$$

is $N(0, 1)$.

It was shown earlier that the random variable $U = (n-1)s^2/\sigma^2$ is distributed as chi-square with $(n-1)$ degrees of freedom. If U is divided by $v = (n-1)$, we obtain $(U/v) = (s^2/\sigma^2)$. Furthermore, it can be shown that U is statistically independent of $\bar{X}$ and therefore independent of y.

Dividing y by the square root of (U/v), we obtain

$$t = \frac{y}{(U/v)^{1/2}} = \frac{\sqrt{n}(\bar{X} - \mu)/\sigma}{s/\sigma} = \frac{\sqrt{n}(\bar{X} - \mu)}{s} \tag{6.19}$$

which has a t distribution with $(n-1)$ degrees of freedom.

Notice that the t ratio (6.19) does not depend on the value of the population variance σ^2. As a result, the t distribution can be used to test hypotheses about the mean μ without knowledge of the variance σ^2. The t distribution is especially useful, then, when working with small samples which may not provide precise estimates of σ^2.

To illustrate the use of the t ratio (6.19), suppose that a consumer survey company randomly selects 16 families in a small midwestern town and asks

[8] For a proof of this statement, see H. D. Brunk, *An Introduction to Mathematical Statistics* (New York: Ginn and Company, 1960), p. 216.

them to carefully record the amount of money spent on fresh milk during the month of April. At the end of the month, the company finds that the average family in the *sample* spent \$12 for fresh milk during the month and that the standard deviation of expenditures on milk was $s =$ \$2. Is it very likely that the average family in the *town* spends only \$9 per month on fresh milk?

In summary, $\bar{X} = 12$, $s = 2$, $n = 16$, and $\mu = 9$. The t ratio is

$$t = \frac{4(12 - 9)}{2} = \frac{12}{2} = 6.00$$

For $v = (n - 1) = 15$ degrees of freedom, the probability that t is greater than 2.9467 is $\alpha/2 = 0.005$. The observed value of t, 6.00, is considerably larger than would be expected if the population mean is actually equal to $\mu =$ \$9. Therefore, we conclude that it is very unlikely that the average family in the town spends as little as \$9 a month on fresh milk.

The conclusions arrived at on the basis of the "t test" are justified only if the assumptions upon which the test is based are true. If the actual distribution of expenditures on fresh milk is not normally distributed, then, strictly speaking, the t test is not justified for small samples. If the sample is not a random sample, then the assumption that the observations are statistically independent is not justified and the conclusions based on the t test may not be correct. For example, if all families in the sample were located in the same neighborhood, the sample may not be a truly random sample. The neighborhood may have been one in which most families had children who drank a good deal of milk. The effect of violating the normality assumption is slight when making inferences about means *provided that* the sample is fairly large. When dealing with small samples, however, it is a good idea to check the normality assumption if possible. A review of similar samples or related research may provide evidence as to whether or not the population is normally distributed.

6.7 SUMMARY

We shall now compare and contrast the six distributions discussed in this chapter. It should be obvious that the normal distribution is by far the most important. The binomial, Poisson, chi-square, F, and t distributions are all related to the normal distribution in some way. This does not reduce their importance, however. Each distribution is important in its own right either because it can be used when the other cannot or because it can provide more accurate answers to probability problems than any of the others.

The distributions discussed in this chapter can be classified as either discrete or continuous and in terms of the range of the values they cover. The binomial and Poisson distributions are discrete distributions whereas the normal, chi-square, F and t distributions are continuous. A binomial variable X can take on any integer from 0 to the number of trials n associated with X. A Poisson variable can take on any integer from 0 to ∞. Chi-square and F variables can take on any number from 0 to ∞. The normal and t variables, on the other hand, have a range from $-\infty$ to $+\infty$.

None of the distributions discussed in this chapter has more than two parameters. The binomial distribution is completely determined by the number of trials n and the probability of a success p. The Poisson distribution has only one parameter: the constant m which is both the mean and variance of the Poisson distribution. The normal distribution, on the other hand, has two parameters: the mean μ and variance σ^2. The chi-square and t distributions have only one parameter v called the number of degrees of freedom. The F distribution has two parameters v_1 and v_2 corresponding to the number of degrees of freedom of the chi-square variables in the numerator and denominator of the F ratio.

All normal distributions can be transformed into the standard normal distribution using the z transformation

$$z = \frac{x - \mu}{\sigma}$$

where μ and σ are the mean and standard deviation, respectively, of the normal variable X. This transformation was used in several ways in this chapter. First, by using z, we were able to answer probability questions about many different normal distributions using only the table of the standard normal distribution. Second, the transformation played an important role in the normal approximation to the binomial distribution. And third, the transformation was used to show that the ratio $(n - 1)s^2/\sigma^2$ has a chi-square distribution. Several other uses could be added to this list, but these three should provide adequate illustration that the z transformation is an important one. Once it is understood, the discussion at several other points in the chapter will be much easier to follow.

The binomial and Poisson distributions are frequently used to obtain the probability of a given event and are used to "fit" observed data. In contrast; the chi-square, F, and t distributions are used to test hypotheses about the parameters of other distributions or to determine how well observed data fit a particular theoretical distribution. The standard normal distribution is used to obtain the probability of given events, for fitting observed data, and in hypothesis testing.

The Poisson distribution is sometimes used to approximate the binomial distribution when the number of trials n is large and the probability of a success p is small. In this case, the mean m of the Poisson distribution is set equal to the mean np of the binomial distribution. The Poisson distribution is an important distribution in its own right, however. If we know that a random variable X has a Poisson distribution with mean m, we can compute the probability of k successes without knowing n or p. When we say k successes in this case, we usually mean k successes in a given volume, a given area, or a given interval of time. This is quite a different thing than k successes in n binomial trials. If the mean m of the Poisson distribution is large, then the Poisson distribution will be approximately a normal distribution. As a result, for large values of m, it is possible to approximate the Poisson distribution by the normal distribution. The approximation procedure was not illustrated in this chapter and is mentioned here only to emphasize the close relationship that exists between the Poisson and normal distributions when m is large.

It was shown that, if a variable t has a t distribution with v degrees of freedom, then t^2 is distributed as F with 1 and v degrees of freedom. The t distribution approaches the standard normal distribution as v becomes large. The F distribution, on the other hand, approaches normality as its degrees of freedom v_1 and v_2 increase, but it cannot approach the standard normal distribution because it does not take on negative values.

The significance of many of the statements made in this chapter will not become evident until we have had the opportunity to consider more examples involving the use of these distributions. The chi-square, F, and t distributions have been discussed very briefly because we are not yet ready for a detailed analysis of hypothesis testing. These three distributions will be studied in greater detail later, however.

EXERCISES

Section 6.1

1. Define a binomial experiment and give two examples of such experiments.

2. A fair coin is thrown ten times. Let X denote the number of heads observed. Find:

(a) $P(X \geq 7)$.

(b) $P(X \leq 4)$.

(c) The probability that X is greater than 7, given that X is greater than 5.

3. Find the probability that in a family of six children there will be (a) at least two boys, (b) at least one boy and one girl. What assumption was necessary in order to answer (a) and (b)?

4. If 10 percent of the electric motors produced on an assembly line are defective, determine the probability that, out of five motors chosen at random:

(a) One will be defective.
(b) At least two will be defective.
(c) One or two will be defective.

5. What is the probability of rolling six twice on eight tosses of a die?
6. For a binomial experiment with $n = 4$ and $p = 0.6$, prove numerically that:
   (a) The binomial distribution is a probability distribution.
   (b) The mean of the binomial distribution is equal to np.
   (c) The variance of the binomial variable X is equal to $np(1-p)$.
7. Two hundred bolts are selected at random from a day's production of bolts at a large plant. If 20 percent of the bolts produced at the plant are defective, what is the probability that there are between 45 and 55 defective bolts in the random sample?
8. Four marbles are drawn at random from an urn containing six red and four black marbles. Let X be a random variable which denotes the total number of red marbles drawn. Construct the probability distribution of X if the marbles are drawn:
   (a) With replacement.
   (b) Without replacement.

Section 6.2

9. A Poisson distribution is given by $p(x) = (1.8)^x e^{-1.8}/x!$. Find:
   (a) $P(x = 1)$
   (b) $P(x \leq 2)$.
   (c) $P(x \geq 3)$.
10. If the probability of a screw being defective is $p = 0.008$, what is the probability that a box of 100 screws will contain one or more defective screws?
11. Suppose that the probability is 0.0002 that a person receiving a penicillin shot will have an unfavorable reaction to the shot. If 3000 people receive penicillin shots, what is the probability that 0, 1, 2, 3, 4, or 5 persons react unfavorably?
12. Let X be a discrete random variable which is equal to the number of "successes." Compute the probability distribution of X for $n = 12$ and $p = 0.10$:
    (a) Using the binomial distribution.
    (b) Using the Poisson distribution.
    (c) Graph the results of (a) and (b).
    (d) What is the mean and variance of X if it has a binomial distribution? If it has a Poisson distribution?
13. Plot the Poisson probabilities for $m = 0.5$, 1.6, 4.0, 8.0 on the same sheet of graph paper. If $m = 8.0$, find the probability that a Poisson variable X is greater that $X = 11$.

Section 6.3

14. If a continuous random variable X is normally distributed with a mean of 12 and a variance of 16, what is the probability that X is greater than 8 but less than 13?

15. An examination given to a large group of students shows a mean of 65 and a standard deviation of 10. If you want to give the top 15 percent a grade of *A*, the next 20 percent a grade of B, the middle 30 percent a grade of C, the next 25 percent a grade of D, and the lowest 10 percent a grade of F, what ranges of marks will be covered by each of these grades, assuming that the numerical marks follow the pattern of the normal curve?

16. An automobile battery manufacturer tests the company's best battery under normal operating conditions in Ohio. The results of the test are given below. Fit a normal distribution to the observed data.

Battery Life in Months

Class Limits (Months)	Frequency (Batteries)
13–15	10
16–18	20
19–21	40
22–24	60
25–27	40
28–30	20
31–33	10
Total	200 batteries

17. Suppose that the annual incomes of medical doctors in the United States are normally distributed with a mean income of $15,000 and a standard deviation of $3500.

(a) What is the probability that a randomly selected medical doctor will have an annual income of over $16,260?

(b) What is the probability that 16 randomly selected doctors will have an average income of over $16,260? (*Hint.* Remember that $\bar{X}$ is $N(\mu, \sigma^2/n)$.)

18. The scores made by candidates on an aptitude test are normally distributed with mean 500 and standard deviation 100.

(a) What percent of the candidates receive scores over 700? Between 400 and 600?

(b) What is the probability that, if we select a person at random from the group that did better than average (i.e., those with scores over 500), this person's score will be greater than 650?

19. The average height of male students at Big-U is 70 inches and the standard deviation is 3 inches.

(a) If the minimum height to try out for the basketball team is 72 inches, what proportion of the students would be eligible to try out for basketball?

(b) If 6 feet, 4 inches is the minimum height for the center position, what proportion of the students who are eligible for basketball would be eligible for the center position?

20. Suppose that 10 percent of the people in a large city have red hair. A public opinion survey group selects 200 people at random and asks them who they will vote for in the next election. What is the probability that less than 15 of these people will have red hair?

Section 6.4

21. What is the probability that χ^2 is greater than 40.30 if $v = 30$?

22. Is it very likely that the following randomly selected numbers are from a population with a variance of 2?

$$-4, 4, 3, 5, 5, -2, 12, 4, 5, 3$$

Section 6.5

23. If $v_1 = v_2 = 15$ and $\sigma_a^2 = \sigma_b^2$, then what is the probability that $F = (s_a^2/s_b^2)$ is greater than 2.40?

24. When working with the F ratio, $F = (s_a^2/s_b^2)$, explain why the largest sample variance is usually placed in the numerator?

Section 6.6

25. Change Table 6.14 to a t table in which the probability is located entirely in the right-hand tail of the t distribution.

26. Explain how tables of the F distribution could be used to construct a partial table of the t distribution. Derive a partial table of the t distribution from the F table in Appendix I.

Section 6.7

27. Find $E(s^2/\sigma^2)$ and $\text{Var}(s^2/\sigma^2)$.
Hint. Use the fact that $(n-1)s^2/\sigma^2$ is distributed as chi-square with $(n-1)$ degrees of freedom and therefore s^2/σ^2 is $\chi^2/(n-1)$.

28. If the variable Z is $N(0, 1)$, then $P(Z > 1.96) = 0.025$ and $P(Z^2 > 3.84) = 0.05$ because Z values less than -1.96 as well as those greater than 1.96 will result in values of Z^2 greater than 3.84.
(a) What type of distribution does Z^2 have?
(b) Sketch the probability density function of Z^2.
(c) If a variable F has an F distribution with 1 and $v = \infty$ degrees of freedom, explain why $P(Z^2 > 3.84) = P(F > 3.84)$.
(d) What is the probability that a variable distributed as chi-square with one degree of freedom is greater than 3.84?

29. Let U be a chi-square variable with ten degrees of freedom. Find:
(a) $P(3.25 < U < 20.48)$.
(b) $P(U > 18.31)$.

30. A random variable X has a Poisson distribution such that $P(x = 3) = P(x = 4)$; find $P(x = 2)$.

31. If U is a chi-square variable with $v = 128$ degrees of freedom, what is the probability that U is greater than 162? (*Hint.* Use $z = \sqrt{2\chi^2} - \sqrt{2v - 1}$.)

SELECTED REFERENCES

Feller, W., *An Introduction to Probability Theory and Its Applications*, John Wiley & Sons, Inc., New York, 1957. Provides greater detail on the binomial, Poisson, multinomial, and negative binomial distributions.

Spiegel, Murray R., *Theory and Problems of Statistics*, Schaum Publishing Co., New York, 1961. An excellent source of solved problems using theoretical distributions.

Wilks, S. S., *Elementary Statistical Analysis*, Princeton University Press, New Jersey, 1958. Provides an easy-to-read algebraic coverage of the binomial, Poisson, and normal distributions.

Extensive tables of the binomial distribution may be found in the following books.

Harvard Computer Laboratory, *Tables of the Cumulative Binomial Probability Distribution*, Harvard University Press, 1955. For $n = 1$ to 50 in steps of 1, $n = 50$ to 100 in steps of 2, $n = 100$ to 200 in steps of 10, $n = 200$ to 500 in steps of 20, and $n = 500$ to 1000 in steps of 50. For $p = 0.00$ to 0.50 in steps of 0.01.

National Bureau of Standards, *Tables of the Binomial Probability Distribution*, Applied Mathematics Series No. 6. Washington, D.C., U.S. Government Printing Office, 1950. For $n = 1$ to 50 in steps of 1 and $p = 0.00$ to 0.50 in steps of 0.01.

Romig, Harry G., *50–100 Binomial Tables*, John Wiley & Sons, Inc., New York, 1953. For $n = 50$ to 100 in steps of 5 and $p = 0.00$ to 0.50 in steps of 0.01.

CHAPTER 7

Sampling and Estimation

Samples are often used as a basis for making generalizations about populations. The accuracy of the generalizations depends, to a large extent, on how carefully a sample is designed and executed. A complete census of all items in a population is seldom necessary and is usually impractical. A good sample can often provide more precise information about a population than a complete census—and do so at a lower cost. A complete census is *impossible* when dealing with an infinite population and *infeasible* if the sampling (or inspection) process is destructive. Even if a complete census is possible and feasible, a sample may lead to more precise results if errors due to mistakes and carelessness are easier to control in the sample than in the census.

Sampling is both a science and an art. Statistical theory, sound judgment, practical experience, and imagination are all important ingredients in the design of samples. All aspects of this highly technical and complex subject cannot be covered in a single chapter. We can only hope to provide an awareness of (1) the sources of difficulties involved in sampling, (2) the commonly used sampling techniques, (3) the value of professional help in designing and carrying out samples and surveys, and (4) the importance of random sampling. With this introduction, you should be able to undertake simple samples and to tell when samples can (or cannot) be used to obtain the information that is needed.

There are usually several ways to use sample information to estimate population parameters. An explanation of how the commonly used estimators are derived and criteria by which to choose between alternative estimators are discussed in this chapter. We shall see that, although there are no perfect estimators, the more popular ones are often very good. Even the best estimators cannot provide good estimates, however, without data obtained from well-designed and carefully conducted samples. Thus, sampling and estimation are closely related topics.

A point estimate such as a sample mean can be accompanied by an interval estimate and a confidence statement that provide information about the precision of the point estimate. A sample mean of 50, for example, may be accompanied by the statement that we are "95 percent confident that the interval from 40 to 60 covers the true population mean." Such information helps the reader decide if the point estimate is accurate enough for his use. The concepts on which interval estimators are based can also be used to determine the size of sample necessary to achieve a desired degree of accuracy in the estimation of a population parameter. Several of the most commonly used interval estimators will be discussed at the end of this chapter.

7.1 SIMPLE RANDOM SAMPLING

The term "random sample" has been used several times in previous chapters but has not been carefully defined. The word *random* refers to the way in which the sample is selected rather than to the particular type of sample. Although the word random may imply that observations are selected in a haphazard or casual way, there is, in fact, nothing haphazard about truly random samples. They require careful design and execution to assure independence of observations. If the sample is not a random sample, there is usually no way to measure the degree of confidence that can be placed in inferences drawn from the sample. Generalizations about populations based on nonrandom samples are difficult to justify, especially if they are based on estimators that were derived under the assumption of randomness. Thus, the quality of randomness is a very important one in sampling. Fortunately, there are many ways to select random samples.

The sampling procedure used depends partly on the type of population from which the sample is drawn. In broad terms, populations can be classified as (1) finite and (2) infinite. The houses located in a given geographical area at any given time, name cards in a file, students in junior colleges in Ohio, and bags of coffee in a warehouse are all examples of finite populations. Finite populations can, in theory, be enumerated even though the enumeration in practice may be very expensive or time consuming. Fortunately, complete enumeration is seldom necessary when dealing with large finite populations.

The selection of random samples from a finite population can be a difficult and time-consuming job. For example, suppose you wanted to estimate the average age of students registered at a university. You might set up a booth outside the student union and ask the age of every tenth student that went by. This would not be a random sample, however, because some students are more likely to pass the student union than others. You might telephone 10 percent of the students living in the residence halls some evening. This still would not be a random sample because some students are more likely than others to study in the evening. One way to select a random sample would be to obtain a complete list of all the students registered at the university, assign a number to each student on the list, write the numbers on small slips of paper, place the numbers in a big box, and then select the numbers from the box with replacement and careful mixing after each selection. You could then determine the age of each student so selected. Even at a small university, this process would take several days.

Infinite populations are often associated with some type of process. The process of tossing a coin, for example, generates a set of outcomes called "heads" or "tails" that is infinite because the process, in theory, could be continued indefinitely. The process result consisting of a small part produced by a machine could be considered the outcome of a random selection from an infinite population consisting of all the results that would be generated if the identical process was continued indefinitely. A finite set of such process results would constitute a random sample from an infinite population.

Economists frequently use data which are time series of such economic magnitudes as national income, prices, and consumption expenditures *as if* the individual numbers of such series were the result of a random process carried out by "nature." It is assumed that "nature's" random process could have resulted in an infinite number of different outcomes.

When sampling from an infinite population, it is impossible to list the items in the population and there is less opportunity to use random selection methods such as drawing numbers from a hat. When throwing a coin, for example, randomness in a sample of n tosses is achieved if it is possible to throw the coin in such a way that the probability of heads remains the same on each toss and the outcome of one toss does not affect the outcome of another toss. When using time series data on national income and personal consumption to estimate a consumption function, the investigator can only assume that "nature" generated the data in part by a random process. When inspecting radios on an assembly line, there is no possibility of inspecting those radios which will be produced by the same production process tomorrow or next week. An inspector may decide to select every fifth or every tenth radio, but he cannot assure that every item in the population has an equal probability of being selected. The basic rule when sampling from infinite populations is to try to select items in such a way that the outcome of one selection (or trial) does not affect the outcome of any other selection.

A frequent practice in statistics is to assume that samples selected in certain ways will be random. This assumption is not always justified but is usually difficult to check. Analysis of the sequence of observations can occasionally be used to test the assumption of randomness. For example, suppose a coin is thrown 20 times and the sequence of outcomes is found to be

HHHHHHHHHHTTTTTTTTTT

This particular sample has only two "runs" and would be very rare if the sample was, in fact, generated by a random process. (A "run" is a sequence of one or more identical outcomes.) The sequence

HTHTHTHTHTHTHTHTHTHT

contains 20 runs, but it is equally likely to cause the investigator to question the assumption of randomness. Although there is no way to prove conclusively that a sample is random, statistical tests can occasionally help the investigator to decide whether or not to accept the randomness assumption.

The estimates of a population parameter obtained from several random samples drawn from the same population will seldom be identical. Differences between estimates can be accounted for in several ways. To begin with, there is always the possibility that errors, which are sometimes referred to as *nonsampling errors*, have been made in recording or analyzing the sample data. These can be controlled to some extent by carefully checking such things as the coding of data, calculations, inspection procedures, interviewing techniques, and card punching. It is up to the investigator to guard against this type of error. Statistical methods for measuring and controlling errors resulting from mistakes, carelessness, and nonresponse are being developed, but such methods cannot replace careful control in sampling.[1]

The estimates obtained from different random samples will differ even though all sample data are recorded and analyzed without errors. Suppose, for example, that the population consists of the ten integers: 1, 2, 3, 4, 5, 6, 7, 8, 9, 10. Furthermore, suppose that we select five random samples of four observations each from this population and compute the sample means. The results are as follows:

Sample Number	*Sample Observations*	*Sample Mean* $\overline{X}$
1	1, 9, 6, 2	4.50
2	10, 8, 7, 4	7.25
3	4, 8, 3, 9	6.00
4	3, 1, 4, 6	3.50
5	4, 1, 6, 9	5.00

[1] One "nonsampling error" that is extremely easy to overlook is that of not sampling from the population the investigator wants to study. If an estimate of the average income of residents of New York state is needed, for example, the random sample of residents should be selected from people living throughout the state—not just those living in Manhattan. One of the first principles of sampling is to define carefully the population in which you are interested.

The population mean is equal to $\mu = 5.5$. The sample-to-sample variation in the sample means is not surprising because the estimator $\overline{X}$ is a random variable. The expected variation in the sample means could be reduced by taking larger samples, but there would still be some variation in the results obtained from different random samples.

The precision with which a parameter is estimated is measured in terms of the standard deviation of the sampling distribution of the estimator. The smaller the standard deviation of an estimator, the greater the precision with which repeated samples from the same population will reproduce an estimate of a population parameter. The expected standard deviation of a sample mean based on 16 observations is twice as large as the expected standard deviation of a sample mean based on 64 observations. Thus, we expect the sample mean based on 64 observations to be a more precise estimate of the population mean. Increasing the sample size is, therefore, one way in which to reduce sample-to-sample variation.

There are costs associated with controls designed to reduce errors in coding, calculations, or card punching. Extra observations, which will increase the precision with which a parameter is estimated, also increase the total cost of drawing a sample. More precise estimation usually increases the value of the information obtained from a sample, but such increases in value may be difficult to determine in terms of dollars. There is no general rule as to the best way to allocate limited funds in an attempt to reduce errors or increase the precision of an estimator; experience and judgment usually play an important role.

Later in this chapter, we shall need a mathematical model appropriate for a discussion of random sampling. Our starting point for the construction of such a model is the concept of a population. The word "population" has been used in a rather general sense so far to refer to all of the items in a well-defined set such as cards in a given file, automobiles registered in Ohio, or the possible outcomes obtained from flipping a coin. In reality, the "population" we are actually interested in is usually a set of numbers. The numbers, of course, may be associated in some way with things such as cards in a file or automobiles in Ohio.

It seems reasonable, then, to define a *population* as the set of all possible numbers that a random variable X can take on. According to this definition, the distribution of the population is the distribution of X and the population is discrete or continuous according as X is discrete or continuous.

Next, we introduce a random variable X_1, defined as "the number determined by the first sample observation." If the first sample observation is drawn from the population associated with the random variable X, then X_1 could take on the same values as X. The values X_1 actually takes on depend partly on the way in which the first sample observation is drawn.

A second random variable X_2 can be defined as "the number determined by the second sample observation." Proceeding in this way, we could define n random variables $X_1, X_2, \ldots, X_n$ for a sample of size n. A particular sample would consist of n numbers $(x_1, x_2, \ldots, x_n)$.

If $X_1, X_2, \ldots, X_n$ are discrete random variables, then

$$P(X_1 = x_1, X_2 = x_2, \ldots, X_n = x_n) = f(x_1, x_2, \ldots, x_n)$$

where $f(x_1, x_2, \ldots, x_n)$ is the joint probability density function of the random variables $X_1, X_2, \ldots, X_n$. The form of the joint probability density function $f(x_1, x_2, \ldots, x_n)$ depends partly on the way in which the sample is drawn.

Suppose that X is a discrete random variable with probability density function $f(x) = P(X = x)$. If the sampling process is random, the random variables $X_1, X_2, \ldots, X_n$ are independent and each has the same distribution as X. In this case, for a "random sample" of size n, we have

$$P(X_1 = x_1, X_2 = x_2, \ldots, X_n = x_n) = f(x_1)f(x_2)\cdots f(x_n)$$

In other words, if a sample is random, the joint probability density function of the random variables $X_1, X_2, \ldots, X_n$ is equal to the product of their marginal probability density functions.

To illustrate these ideas, suppose that X is a discrete random variable that takes on the values 1 and 0 with probability p and $(1 - p)$, respectively. The probability density function of X is

$$f(x) = p^x(1 - p)^{1-x}$$

Thus, if $x = 1$, then $f(x) = p$, and if $x = 0$, $f(x) = (1 - p)$.

Now suppose that a sample of two observations is to be drawn. Let X_1 denote the number determined by the first sample observation and X_2 denote the number determined by the second. If the random variables X_1 and X_2 are independent and each has the same probability density function as X, then

$$P(X_1 = 1, X_2 = 0) = f(1, 0) = f(1)f(0) = p(1 - p)$$
$$P(X_1 = 1, X_2 = 1) = f(1, 1) = f(1)f(1) = p^2$$
$$P(X_1 = 0, X_2 = 0) = f(0, 0) = f(0)f(0) = (1 - p)^2$$
$$P(X_1 = 0, X_2 = 1) = f(0, 1) = f(0)f(1) = (1 - p)p$$

The joint probability density function of X_1 and X_2 can, therefore, be written as

$$f(x_1, x_2) = f(x_1)f(x_2) = (p^{x_1+x_2})(1 - p)^{2-x_1-x_2}$$

when a random sampling process is used. [Show that if X_1 has the same probability density function as X, that is, $f(x_1) = f(x)$, but $P(X_2 = 1) = 1$, then $f(x_1, x_2) \neq f(x_1)f(x_2)$.]

We are now ready to discuss several specific sample designs.

7.2 SAMPLE DESIGNS

A sample design is a plan that specifies the way in which the sample is to be selected from a given population. Several of the most widely used sample designs will be discussed briefly in this section. Which sample design to use in any given survey depends on things such as the type of population, the type of information needed, and the funds available to carry out the survey.

7.2.1 Probability Sampling

Probability samples are samples in which items are selected from the population according to known probabilities. The probability mechanism by which the items are selected is specified prior to sampling and leaves no opportunity for the investigator to decide which population items should be included in the sample. In a well-designed sample, the estimators of population parameters would also be chosen prior to the collection of data.

When designing a probability sample, the investigator can determine the sample size necessary to obtain a given degree of precision in the estimation of a parameter. Furthermore, there is less opportunity for inspectors or interviewers to introduce biases when collecting sample data because their judgments are not involved in the selection of sample items.

To illustrate one way in which a probability sample can be selected, suppose that a card file contains a set of numbered cards which give the name and IQ of 800 students in a high school. The principal of the high school wants to estimate the average IQ of the students on the basis of a random sample. He decides to use the arithmetic mean to compute the average IQ and to select a random sample of 40 cards with the help of a table of random numbers. One page of such a table is presented as Table VIII in Appendix I. A table of random numbers (or random digits) consists of groups of digits that are generated by some random device. He randomly selects a page and starting point in the table of random numbers and then uses the random numbers to decide which cards in the file are to be selected. For example, suppose that the first column of Table VIII, Appendix I, was selected as the starting place. The first number is 039; therefore, card number 39 is included in the sample. The next card included in the sample is numbered 385; the third card is 175. This process continues until 40 cards have been selected.

This was a very simple sampling situation. The population was relatively small and well defined, the arithmetic mean is a good estimator of the population mean and easy to work with, there was little opportunity for mistakes, and it was easy to assure that every item in the population has an equal (and known) probability of being included in the sample at each stage of the sampling process. At no point did the principal have to decide which cards would be included in the sample. He selected the random mechanism (in this case a table of random numbers), and the random mechanism selected the cards.

It is not always possible to obtain a list of all items in a finite population. For example, an up-to-date listing of all residents of the United States does not exist. Conceptually, such a list could be constructed, but in practice the expense would be prohibitive. If a complete list of the population is not available, then simple random sampling (such as selecting cards from a file) is not possible and other probability sampling procedures must be used.

Stratified, systematic, area, and multistage sampling are a few of the most widely used types of probability sampling. They are designed to solve certain types of sampling problems or to improve the precision of an estimator. Keep in mind that each of these schemes can be used in combination with the others. The following sections will serve as an introduction; more detailed information can be obtained from the selected references listed at the end of this chapter.

7.2.2 Stratified Sampling

In stratified sampling, the population is divided into subpopulations, each of which is sampled independently. For example, suppose a company wants to estimate the amount of life assurance carried by its employees. It believes that employees working in the office may carry more life insurance than those working in the plant or elsewhere and decides to classify employees according to three subpopulations (or strata): (1) plant, (2) office, and (3) other. The number of employees in these three subpopulations is given in Table 7.1.

Table 7.1 Stratification of Employees

(1) Stratum j	(2) Classification of Employees	(3) Number of Employees	(4) Proportion of Employees in Stratum j, p_j
1	Plant	800	0.80
2	Office	150	0.15
3	Other	50	0.05
	Total	1000	1.00

Let n_j represent the number of employees in a sample from stratum j. We shall consider two ways by which to choose the sample sizes n_j. The first is *proportional sampling* and the second, *optimal sampling*.

In proportional sampling, the number of sample items selected from stratum j is proportional to the number of items in stratum j. In other words,

$$n_j = np_j \tag{7.1}$$

where

$n = \sum_j n_j$ is the total number of items in the sample

p_j is the proportion of the items in the population which are in stratum j

The p_j's are listed in Column (4) of Table 7.1.

If a sample of 80 employees were selected, proportional sampling would result in 64 employees selected from the plant, 12 from the office, and 4 from "other" (sales personnel, night watchmen, etc.).

Let μ_j denote the average amount of life insurance carried by employees in stratum j and μ represent the weighted average of the μ_j. That is,

$$\mu = \sum_j p_j \mu_j \tag{7.2}$$

The sample mean for stratum j is equal to

$$\bar{X}_j = \frac{1}{n_j} \sum_{i=1}^{n_j} x_{ij} \tag{7.3}$$

In other words, $\bar{X}_j$ represents the average amount of life insurance carried by employees in the *sample* drawn from stratum j.

The expected value of $\bar{X}_j$ is equal to $E(\bar{X}_j) = \mu_j$. Assuming that the sample (or subsample) drawn from stratum j is random, the variance of $\bar{X}_j$ is equal to

$$\mathrm{Var}(\bar{X}_j) = \frac{\sigma_j^2}{n_j} \tag{7.4}$$

where σ_j^2 is the variance in the amount of life insurance carried by the employees in stratum j.

The overall (proportional) sample mean $\bar{X}_p$ is the weighted average of the stratum sample means. That is,

$$\bar{X}_p = \sum_j p_j \bar{X}_j = \frac{1}{n} \sum_j \sum_{i=1}^{n_j} x_{ij} \tag{7.5}$$

The expected value of $\overline{X}_p$ is

$$E(\overline{X}_p) = \sum_j E(p_j \overline{X}_j) = \sum_j p_j E(\overline{X}_j) = \sum_j p_j \mu_j = \mu \tag{7.6}$$

where μ is the weighted average of the stratum means.

If the sampling process is random, the $\overline{X}_j$'s are independent random variables. In this case, the variance of the overall (proportional) sample mean is equal to

$$\begin{aligned} \operatorname{Var}(\overline{X}_p) &= \sum_j \operatorname{Var}(p_j \overline{X}_j) = \sum_j p_j^2 \operatorname{Var}(\overline{X}_j) = \sum_j \left(\frac{n_j}{n}\right)^2 \frac{\sigma_j^2}{n_j} \\ &= \frac{1}{n} \sum_j p_j \sigma_j^2 \end{aligned} \tag{7.7}$$

If a simple random sample of size n were drawn from all employees without regard to the strata which have been established, the (simple random) sample mean would be

$$\overline{X}_r = \frac{1}{n} \sum_{i=1}^{n} x_i \tag{7.8}$$

It can be shown that the variance of $\overline{X}_r$ is equal to

$$\operatorname{Var}(\overline{X}_r) = \frac{1}{n} \sum_j p_j \sigma_j^2 + \frac{1}{n} \sum_j p_j(\mu_j - \mu)^2 \tag{7.9}$$

The variance of the (simple random) sample mean $\overline{X}_r$ is larger than the variance of the overall (proportional) sample mean $\overline{X}_p$ by the amount

$$\frac{1}{n} \sum_j p_j(\mu_j - \mu)^2$$

Proportional sampling from a stratified population, therefore, provides a more precise estimate of the population mean *if* there is a difference in the means of the subpopulations. If the subpopulations have identical distributions, there is nothing to be gained from proportional sampling.

It is obvious, therefore, that strata should be selected in such a way that the items in each stratum differ as much as possible with respect to the characteristic of interest. The items within each stratum, on the other hand, should be as homogeneous as possible. Preliminary surveys and information from other surveys can be useful in deciding which strata to use.

If optimal sampling is used, the number of sample items selected from stratum j is equal to

$$n_j = \frac{n p_j \sigma_j}{\sum_j p_j \sigma_j} \tag{7.10}$$

for all j.

The sample mean for stratum j is still defined by (7.3) except that n_j is now defined by (7.10) instead of (7.1). The overall (optimal) sample mean $\bar{X}_o$ is the weighted average of these $\bar{X}_j$. The term "optimal sampling" is used because the variance of the overall (optimal) sample mean is minimized when the n_j are selected according to (7.10).

It can be shown that the variance of $\bar{X}_o$ is equal to

$$\text{(7.11)} \qquad \operatorname{Var}(\bar{X}_o) = \frac{1}{n}\sum_j p_j \sigma_j^2 - \frac{1}{n}\sum_j p_j(\sigma_j - \bar{\sigma})^2$$

where $\bar{\sigma} = \sum_j p_j \sigma_j$ is the weighted standard deviation. Selecting the sample sizes according to (7.10) rather than according to proportional sampling reduces the variance of the overall sample mean by the amount

$$\frac{1}{n}\sum_j p_j(\sigma_j - \bar{\sigma})^2$$

If there are no differences in the variances of the subpopulations, the variance of the overall sample mean will be the same for optimal sampling and proportional sampling.

In practice, the standard deviations of the subpopulations, σ_j, are usually not known. It is difficult, therefore, to justify the use of optimal sampling unless the σ_j can be estimated fairly accurately. Proportional sampling may be less precise than optimal sampling, but it requires much less information and can be used when optimal sampling cannot.

Although numerous problems arise in designing a stratified random sample, stratification of populations remains a very useful and commonly used sampling technique.

7.2.3 Systematic Sampling

In a *systematic sample*, every kth element in an ordered population is included in the sample. The starting point is randomly chosen among the first k elements and the sampling continues until n items have been selected. For example, a telephone survey could be conducted by calling every 50th name in the telephone directory.

Systematic sampling is a relatively simple technique and, in some cases, may be more efficient than simple random sampling. The main difficulty is that systematic samples are not always random samples. If the population is ordered in a systematic way with respect to the characteristic the investigator is interested in, then it is possible that only certain types of items will be included in the population, or at least more of certain types than others. For instance, if a systematic sample were selected from a list of a company's employees in which office employees were listed first and plant employees second, then the proportion of office employees included in the sample would be greater than the proportion of office employees in the population unless observations were selected up to the end of the list.

In many cases, systematic samples can be treated *as if* they were random samples. Slight modifications, such as occasionally selecting a new starting point or randomly changing the sampling interval, can be used to avoid problems such as the sampling interval coinciding with a periodic arrangement in the population.

7.2.4 Area Sampling

Area sampling is a type of cluster sampling in which the clusters are geographical subdivisions such as city blocks, townships, or counties. The areas (sampling units) to be included in the sample are selected at random, and the elementary units from which the desired information is to be obtained are then randomly selected from lists of all units within each area. The selection of sample units by stages is called *multistage sampling*. An area sample is not necessarily a multistage sample, however. For example, once the areas are selected, all of the elementary units in the areas may be included in the sample, in which case there is no second stage in the sampling procedure.

Area sampling eliminates the problem of listing every item in a large, finite population. It is, therefore, a very useful way to obtain probability samples for nationwide surveys. Government agencies, for example, make frequent use of area sampling.

Area sampling may require larger samples than other sample designs in order to obtain the same level of precision in estimation, but this disadvantage is often more than offset by the cost reduction that results from being able to concentrate interviewers in a limited number of geographical locations. This is especially true for national surveys.

7.2.5 Judgment Sampling

In a judgment sample, judgment plays an important role in the selection of items to be included in the sample. Such samples are very useful at times, especially if the investigator is well acquainted with the population and is consistently able to select "representative" items for the sample. The precision of estimates obtained from judgment samples cannot be evaluated on the basis of the sample itself. The investigator must use his own judgment in deciding if an estimate is "good" or "bad."

Probability samples from one population can easily be converted to judgment samples for another population if the investigator decides that the results of the probability sample can be applied to both the original and new population. For example, suppose that a random sample of a company's *office* employees is selected and the average age of the employees in the sample is computed. If the investigator decides to use this sample estimate as the average age of both office and plant employees, the original sample becomes a judgment sample. In other words, in the investigator's judgment the office employees who were included in the probability sample can be considered to be "representative" of both office and plant employees.

Quota sampling is a type of judgment sampling. In a quota sample, quotas are set up according to given criteria but, within the quotas, the selection of sample items depends on personal judgments. For example, in a radio listening survey, the interviewers may be told to interview 500 people living in a certain area and that out of every 100 persons interviewed 50 are to be housewives, 30 farmers, and 20 children under the age of 15. Within these quotas, the interviewer is free to select the people interviewed. The cost per person interviewed may be relatively small for a quota sample but there are numerous opportunities for biases which may invalidate the results. Interviewers may miss farmers working in the fields or only talk with those housewives who are at home. If a person refuses to respond, the interviewer simply selects someone else. Interviewers may prefer to work in certain sections of a city or may find it convenient to interview people only during the day. A few of these potential sources of bias can be eliminated by establishing quotas according to several criteria. For instance, in a radio listening survey the investigator could require that one-half of the housewives, farmers, and children be interviewed after 6:00 P.M.

7.3 THE CENTRAL LIMIT THEOREM

We have mentioned several times that the sample mean $\bar{X}$ is a random variable and, as such, has its own distribution. The distribution of $\bar{X}$ is usually referred to as the sampling distribution of means. We have also pointed out that, if n sample observations are drawn from a *normal* distribution with mean μ and variance σ^2, the distribution of the sample mean $\bar{X}$ is normal with mean μ and variance σ^2/n. The quantity $\sigma/\sqrt{n}$ is frequently called the standard error of means. If the variance of the normally distributed population is known, it is easy to determine the probability that the sample mean will differ from the population mean by any given amount. For example, suppose that a sample of 64 observations is to be drawn from a normal population with $\sigma^2 = 144$ and unknown mean μ. Furthermore, suppose that we want to determine the probability that the sample observations will result in a sample mean $\bar{X}$ that will differ from the population mean μ by less than one unit. In other words, before the sample is drawn, we want to determine

$$P(-1 < \overline{X} - \mu < 1)$$

In order to use the table of the standard normal variable Z, we must divide each member of the inequalities by the standard deviation of $\overline{X}$ which is equal to

$$\frac{\sigma}{\sqrt{n}} = \sqrt{\frac{144}{64}} = 1.5$$

Making this division and using Table I, Appendix I, we find

$$P\left(\frac{-1}{1.5} < \frac{\overline{X} - \mu}{1.5} < \frac{1}{1.5}\right) = P(-0.66 < Z < 0.66) = 0.4908$$

In other words, if we were to select a large number of samples consisting of 64 observations, approximately 49 percent of the sample means would be within one unit of the population mean. A more precise estimate of the population mean could be obtained by drawing a larger sample.

In many practical problems, however, the populations from which we are sampling are not normally distributed. In such cases, is it possible to make any general statement about the distribution of the sample mean? Fortunately, the answer to this question is "Yes!"

To begin with, we know that for random samples the mean of the sampling distribution of means is equal to the mean of the population from which the samples are drawn. Furthermore, the variance of the sampling distribution of means is equal to the variance of the population σ^2 divided by the sample size n. To illustrate these statements, suppose that a population consists of the four numbers 1, 2, 4, 5. The 16 possible samples of size 2 which can be drawn with replacement from this population are:

(1, 1)	(1, 2)	(1, 4)	(1, 5)
(2, 1)	(2, 2)	(2, 4)	(2, 5)
(4, 1)	(4, 2)	(4, 4)	(4, 5)
(5, 1)	(5, 2)	(5, 4)	(5, 5)

The corresponding sample means are:

1.0	1.5	2.5	3.0
1.5	2.0	3.0	3.5
2.5	3.0	4.0	4.5
3.0	3.5	4.5	5.0

The mean and variance of the population are

$$\mu = \frac{1 + 2 + 4 + 5}{4} = 3.0$$

and

$$\sigma^2 = \frac{(1-3)^2 + (2-3)^2 + (4-3)^2 + (5-3)^2}{4} = 2.5$$

The mean of the sampling distribution of means is

$$\mu_{\bar{X}} = \frac{1}{16} \sum_{i=1}^{16} \bar{X}_i = \frac{48}{16} = 3.0$$

The variance of the sampling distribution of means is

$$\sigma_{\bar{X}}^2 = \frac{1}{16} \sum_{i=1}^{16} (\bar{X}_i - \mu_{\bar{X}})^2 = \frac{20}{16} = 1.25$$

which is the same as

$$\sigma_{\bar{X}}^2 = \frac{\sigma^2}{n} = \frac{2.5}{2} = 1.25$$

In this case, it is easy to make probability statements about $\bar{X}$ because the entire sampling distribution of means is known. For example, the probability that $\bar{X}$ is greater than, say, 4.0 is

$$P(\bar{X} > 4.0) = P(\bar{X} = 4.5) + P(\bar{X} = 5.0) = \frac{2}{16} + \frac{1}{16} = \frac{3}{16}$$

When dealing with large populations, however, it may be impractical or impossible to enumerate all possible sample means. In such cases, it is impossible to make probability statements about $\bar{X}$ unless we can specify the *form* of the distribution of $\bar{X}$. Fortunately, this is often possible.

In general, if $\bar{X}$ is the sample mean of a sample of size n from a population with mean μ and standard deviation σ, then for large samples the distribution of the quantity

$$\frac{\bar{X} - \mu}{\sigma/\sqrt{n}}$$

is approximately a normal distribution with mean zero and variance equal to one. This result is of basic importance in statistics. It increases the usefulness of the normal distribution and contributes to the popularity of the arithmetic mean. Presented more precisely, it is known as the *Central Limit Theorem* which states that:

If $\overline{X}$ is the sample mean of a sample of size n from a population with mean μ and standard deviation σ, and Z is the standard normal variable, then for any real numbers a and b with $a < b$, the probability

$$P\left(a < \frac{\overline{X} - \mu}{\sigma/\sqrt{n}} < b\right)$$

approaches the probability $P(a < Z < b)$ as the sample size n approaches infinity.

The amazing feature of the Central Limit Theorem is that nothing is said about the form of the distribution of the population. The theorem applies to both discrete and continuous distributions. The distribution from which the sample is drawn may be skewed to the right or left or be bimodal, but the distribution

$$\frac{\overline{X} - \mu}{\sigma/\sqrt{n}}$$

still approaches a standard normal distribution as the sample size increases. Furthermore, experience suggests that, unless the distribution of the population is very unusual, the probability

$$P\left(a < \frac{\overline{X} - \mu}{\sigma/\sqrt{n}} < b\right)$$

can frequently be treated as if it were equal to $P(a < z < b)$ even when the sample size n is as small as 30 or 40 observations.

The Central Limit Theorem presented here is concerned only with the sample mean $\overline{X}$. It is possible to generalize the theorem to other kinds of functions of random variables. A special case was presented in Chapter 6, for example, when the normal approximation to the binomial distribution was discussed.

A rigorous proof of the Central Limit Theorem is beyond the scope of this book. The following numerical example will serve to illustrate the basis of the theorem, however.

Suppose a population consists of 100,000 integers, 1, 2, 3, 4, 5, 6, 7, 8, 9, and 10, each occurring with equal frequency. The distribution of this population is shown in Figure 7.1. This discrete distribution bears little resemblance to a normal distribution. The mean of the distribution is equal to $\mu = 5.5$ while the variance is equal to $\sigma^2 = 8.25$.

Let us now draw 70 random samples of size 20 from this population and compute the sample mean $\overline{X}$ for each sample. The results are presented as a frequency distribution in Table 7.2 and are shown in Figure 7.2.

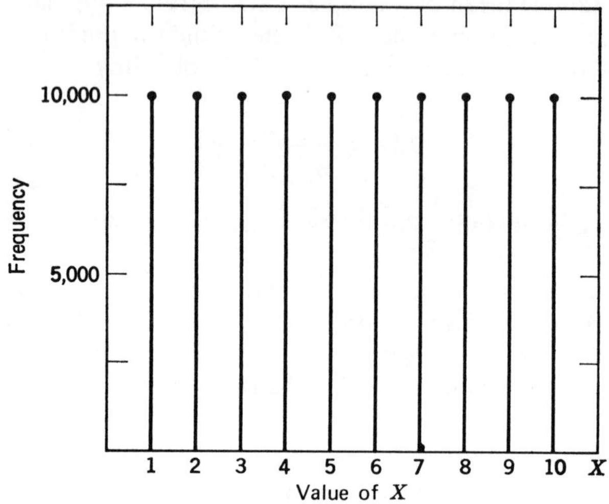

Figure 7.1 A distribution of integers.

Table 7.2 Frequency Distribution of 70 Sample Means

Class Limits	Frequency
4.60–4.79	4
4.80–4.99	5
5.00–5.19	4
5.20–5.39	9
5.40–5.59	22
5.60–5.79	12
5.80–5.99	6
6.00–6.19	5
6.20–6.39	2
6.40–6.59	1
	Total = 70 sample means

The mean of the 70 sample means is

$$\overline{\overline{X}} = \frac{1}{70} \sum_{i=1}^{70} \overline{X}_i = 5.55$$

and the sample variance of the sample means is equal to

$$s_{\overline{X}}^2 = \frac{1}{69} \sum_{i=1}^{70} (\overline{X}_i - \overline{\overline{X}})^2 = 0.34$$

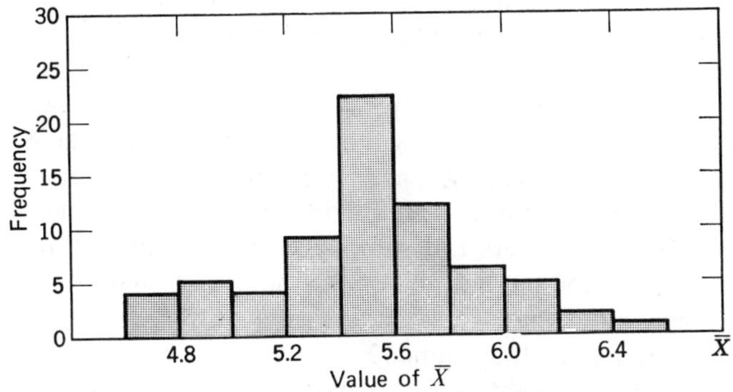

Figure 7.2 A frequency distribution of 70 sample means.

The reason for dividing $\sum (\overline{X}_i - \overline{\overline{X}})^2$ by $(n - 1) = 69$ instead of $n = 70$ is to obtain an unbiased estimate of $\sigma_{\overline{X}}{}^2$. This will be illustrated later in the discussion related to equation (7.22).

The frequency distribution of the 70 sample means certainly bears a greater resemblance to a normal distribution than the distribution from which the observations were drawn. If more samples of 20 observations were obtained and added to the 70 already available, we would expect the new frequency distribution to bear an even-closer resemblance to a normal distribution than the frequency distribution shown in Figure 7.2.

When the sample size is equal to 20, the distribution of $\overline{X}$ is a discrete distribution. The sample means can take on values such as 4.80, 4.85, or 5.10, but they cannot take on values such as 4.82, 4.825, 4.88, or 5.14. The reason for this is that the minimum possible change in the sum of the observations $\sum_{i=1}^{20} X_i$ is one unit when we are sampling from the distribution shown in Figure 7.1. Therefore, the minimum possible change between sample means is $1/20 = 0.05$. As the sample size is increased, however, the distribution of $\overline{X}$ would tend to approach a continuous distribution because the minimum possible change between means, $1/n$, becomes smaller.

If all possible samples of size $n = 20$ were drawn from the original population, the resulting distribution of sample means would have a mean equal to

$$E(\overline{X}) = \mu = 5.5$$

and a variance equal to

$$E[(\overline{X} - \mu)^2] = \frac{\sigma^2}{n} = \frac{8.25}{20} = 0.4125$$

The mean of the sample means $\overline{\overline{X}} = 5.55$, of the frequency distribution of the 70 sample means, is somewhat larger than the mean of the distribution of $\overline{X}$.

The variance of the 70 sample means $s_{\bar{X}}^2 = 0.34$, on the other hand, is somewhat smaller than the variance of the distribution of $\bar{X}$ for $n = 20$.

As an illustration of the use of the Central Limit Theorem, suppose that an airline wishes to compute the probability that the average weight of passengers flying on one of its jets will exceed 155 pounds if all 81 seats are filled. A random sample of 5000 passengers selected from persons on such jets results in a sample mean of 150 pounds and a standard deviation of 21 pounds. The airline considers this sample to be large enough to justify the assumption that $\mu = 150$ pounds and $\sigma = 21$ pounds. Using this information, what is the probability that the average weight of 81 persons flying on one of its jets is greater than 155 pounds?

The distribution of weights of male passengers of, say, age 23 may be a normal distribution, but the distribution of "passenger weights" need not be normal. Thus, it would be difficult to provide a satisfactory answer to the probability question if it were not for the Central Limit Theorem. Using the standard normal transformation with $\bar{X} = 155$, we find

$$z = \frac{155 - 150}{(21/\sqrt{81})} = 2.14$$

Therefore,

$$P(\bar{X} > 155) = P(Z > 2.14) = 0.5000 - 0.4838 = 0.0162$$

In other words, the probability is 0.0162 that the average weight per person will exceed 155 pounds for a random sample of 81 passengers. This probability statement is approximately correct even if the distribution of passenger weights is not a normal distribution, but it would not apply if the sample of 81 passengers is nonrandom. For example, a charter flight of 81 passengers consisting of a professional football team and their "biggest" fans could not be considered a random sample.

7.4 POINT ESTIMATION

The estimation of population parameters is one of the primary purposes of scientific experimentation. Point estimation, therefore, is a very important part of statistics. Many different point estimators have been discussed in previous chapters. The mean, median, and mode of a sample can all be used as point estimators of the population mean. The sample standard deviation is a point estimator of the standard deviation of a population. A relative frequency can serve as a point estimator for a probability. In this section we shall discuss the characteristics of "good" estimators and present a technique that is frequently used to derive point estimators.

7.4.1 Properties of Estimators

A point estimator is a function of a set of sample observations. The estimates obtained from such a function will vary from sample to sample. Thus, each estimator has its own distribution. Intuitively, it seems obvious that the distribution of a "good" estimator should be concentrated as closely as possible to the true value of the population parameter. Suppose, for example, that θ is the true value of a population parameter and that $\hat{\theta}_1$, $\hat{\theta}_2$, $\hat{\theta}_3$ are different estimators of θ with probability density functions as illustrated in Figure 7.3.[2] The estimator $\hat{\theta}_1$ would usually be considered to be a better

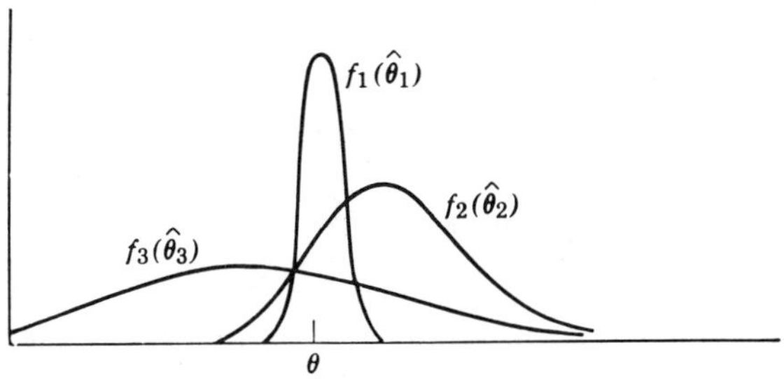

Figure 7.3 Distribution of estimators.

estimator of θ than either $\hat{\theta}_2$ or $\hat{\theta}_3$ because its distribution is concentrated near the true value of θ.

Intuition alone is not always an adequate guide in the selection of estimators. Therefore, we shall present several properties that are frequently used when comparing estimators.

An estimator $\hat{\theta}$ for a parameter θ is said to be *unbiased* if its expected value is equal to the true value of θ. That is, if

$$E(\hat{\theta}) = \theta \tag{7.12}$$

Other things being equal, unbiased estimators are preferred to biased ones but, if the choice is between a biased estimator whose distribution is concentrated near θ and an unbiased estimator with a highly dispersed distribution, the biased estimator may be preferred, especially if the direction and magnitude of the bias can be determined.

An estimator $\hat{\theta}$ for a parameter θ is said to be *consistent* if the value of the estimator approaches the value of the parameter as the sample size becomes

[2] A "hat" ^ over the symbol for the parameter indicates that we are dealing with an estimator of the parameter. A parameter may have several estimators in which case subscripts can be used to distinguish between them.

infinitely large. More precisely, if for any given small number $\varepsilon > 0$, the probability $P(|\hat{\theta} - \theta| < \varepsilon)$ that the estimator $\hat{\theta}$ will differ by less than ε from the true value θ tends to unity as n approaches infinity, we say that $\hat{\theta}$ is a consistent estimator of θ.

Consistent estimators are not necessarily unbiased for finite sample sizes. Nor are unbiased estimators necessarily consistent. The sample variance

$$s^2 = \frac{1}{n}\sum (X_i - \bar{X})^2$$

is a consistent estimator of the population variance σ^2 even though the expected value of s^2 is smaller than σ^2. The first observation in a sample of size n drawn from a population with mean μ and variance σ^2 may be used as an estimator of μ. This estimator is unbiased, since its expected value is equal to μ. It is not consistent, however, because the variance of its distribution is always equal to σ^2 regardless of the size of the sample.

The sample mean $\bar{X} = \frac{1}{n}\sum X_i$ is a consistent and unbiased estimator. It is unbiased because $E(\bar{X}) = \mu$. It is consistent because its standard deviation, $\sigma/\sqrt{n}$, approaches zero as the sample size n approaches infinity and, as a result, the distribution of $\bar{X}$ becomes concentrated at the population mean μ.

It can be shown under certain conditions that the variance of an unbiased estimator can never be less than a certain number, say, c, that depends on the population distribution and the sample size. If the variance of an unbiased estimator $\hat{\theta}$ is equal to this number c, than $\hat{\theta}$ is said to be *efficient* in the sense that it has the smallest possible variance. Alternatively, $\hat{\theta}$ is sometimes called the *minimum variance estimator* or the *best estimator* of θ. It can be shown, for example, that if $\hat{\theta}$ is an estimator of the mean μ of a normally distributed population, the variance of $\hat{\theta}$ cannot be smaller than $c = \sigma^2/n$. Since the variance of the sample mean $\bar{X}$ is equal to σ^2/n, we may conclude that $\bar{X}$ is a minimum variance estimator of μ.

Estimators can also be compared on the basis of their *relative efficiency*. Suppose that $\hat{\theta}_1$ and $\hat{\theta}_2$ are different estimators of θ. The relative efficiency of $\hat{\theta}_2$, as compared with θ_1, is defined by the ratio

$$R = \frac{E(\hat{\theta}_1 - \theta)^2}{E(\hat{\theta}_2 - \theta)^2}$$

If the ratio R is greater than 1, $\hat{\theta}_2$ is *relatively* more "efficient" than $\hat{\theta}_1$. If R is less than 1, then $\hat{\theta}_1$ is relatively more "efficient" than $\hat{\theta}_2$. The word "efficient" is used here in a somewhat different sense than previously. First, the estimators $\hat{\theta}_1$ and $\hat{\theta}_2$ need not be unbiased. Second, relative efficiency comparisons involve only two estimators rather than an entire set. Finally, the ratio R is the ratio of the variances of $\hat{\theta}_1$ and $\hat{\theta}_2$ only in the event that $\hat{\theta}_1$ and $\hat{\theta}_2$ are unbiased.

It is often possible to construct an estimator $\hat{\theta}$ of θ, such that $\sqrt{n}(\hat{\theta} - \theta)$ is asymptotically normal with mean zero. If the asymptotic variance of $\hat{\theta}$ is at least as small as that of any other asymptotically normal and (asymptotically) unbiased estimator of θ, than $\hat{\theta}$ is said to be *asymptotically efficient*. Such estimators are often referred to as "efficient" estimators. There may be several estimators of θ that are asymptotically efficient.[3]

It can be shown, under certain conditions, that both the sample mean and the sample median are asymptotically normal and unbiased estimators of the population mean. The asymptotic variance of the sample median is greater than the asymptotic variance of the mean, however. Thus, the sample median is not an asymptotically efficient estimator of the population mean. It can be shown, however, that the variance of the sample mean is at least as small as that of any other asymptotically normal and unbiased estimator of the population mean and, therefore, $\bar{X}$ is an asymptotically efficient estimator of μ.

Asymptotically efficient estimators need not be unbiased for finite samples. They are necessarily consistent estimators, however, because they are unbiased in the limit.

An estimator $\hat{\theta}$ is said to be *sufficient* if it summarizes all of the information contained in the sample about the parameter θ. More precisely, $\hat{\theta}$ is sufficient if the conditional distribution of the sample variables for a given value of $\hat{\theta}$ is independent of the parameter θ. This means that the joint density function of the sample $x_1, x_2, \ldots, x_n$ can be written as

$$f(x_1, x_2, \ldots, x_n) = g(x_1, x_2, \ldots, x_n \mid \hat{\theta})h(\hat{\theta}; \theta) \tag{7.13}$$

where the function g does not involve θ. This will be illustrated (in the following section) using the binomial distribution.

Sufficient estimators are the best kind but, in many cases, they do not exist. The sample mean is a sufficient estimator of the population mean of some distributions.

These concepts are easily extended to cases of more than one parameter. For example, a set of estimators $\hat{\theta}_1, \hat{\theta}_2, \ldots, \hat{\theta}_k$ for the parameters $\theta_1, \theta_2, \ldots, \theta_k$ is said to be unbiased if

$$E(\hat{\theta}_i) = \theta_i \qquad \text{for } i = 1, 2, \ldots, k$$

[3] A random variable such as $\sqrt{n}(\hat{\theta} - \theta)$ is said to be asymptotically normal if the distribution of the random variable approaches a normal distribution as the sample size n approaches infinity. The asymptotic variance of $\hat{\theta}$ is the value that $\text{Var}(\hat{\theta})$ approaches as the sample size n approaches infinity. An estimator $\hat{\theta}$ is said to be asymptotically unbiased if $E(\hat{\theta})$ approaches θ as n approaches infinity.

How would you define a consistent *set* of estimators?

The statistical properties of the commonly used estimators are well known. Therefore, it is seldom necessary in applied economics to prove that estimators such as the sample mean or sample variance are (or are not) unbiased, efficient, sufficient, or consistent. You should understand the meaning of these terms, however, because they are used frequently to describe estimators.

7.4.2 *The Maximum Likelihood Method*

The principle of maximum likelihood provides a general method of deriving estimators which, under conditions often satisfied in random sampling, have several desirable properties. Roughly speaking, the principle of maximum likelihood, as it applies to estimation, is to select that estimator which will maximize the probability of obtaining the sample actually observed. More precisely, if $f(x_1, x_2, \ldots, x_n; \theta)$ is the joint density function for a random sample of size n drawn from a population with an unknown parameter θ, then the maximum likelihood estimate of θ is the number $\hat{\theta}$ that maximizes the joint density function when it is regarded as a function of θ for fixed $x_1, x_2, \ldots, x_n$.

This is much simpler than it sounds. Suppose that we want to estimate the proportion of defective radio tubes in a carton containing 300 tubes using a random sample of size 3. The number of bad tubes x in the sample has the binomial distribution

$$b(x) = \frac{3!}{x!(3-x)!} p^x(1-p)^{3-x}$$

where p is the probability of selecting a bad tube on any given trial as well as the proportion of defective tubes. The estimator $\hat{p} = x/n$ is the maximum likelihood estimate of the probability (and proportion) p. In other words, $\hat{p}$ maximizes the probability $b(x)$ for the value of x observed. The binomial probabilities for $n = 3$ and various values of p are listed in Table 7.3. The estimator $\hat{p}$ for $n = 3$ is

$$\begin{array}{llll} \hat{p} = 0 & \text{if } x = 0 & \hat{p} = 0.66 & \text{if } x = 2 \\ \hat{p} = 0.33 & \text{if } x = 1 & \hat{p} = 1.00 & \text{if } x = 3 \end{array}$$

Inspection of Table 7.3 suggests that $\hat{p}$ does, in fact, maximize $b(x)$ for the value of x observed, since for a given x, say, x^*, the probabilities $b(x^*)$, for values of p other than $\hat{p}$, are always less than the probability $b(x^*)$ obtained by using $\hat{p} = x^*/n$. An explanation of the way in which $\hat{p}$ was obtained will be given shortly.

Table 7.3 Binomial Probabilities for $n = 3$

$b(x)$ for	X			
	0	1	2	3
$\hat{p} = 0$	1.0000	0	0	0
$p = 0.25$	0.4219	0.4210	0.1406	0.0156
$p = 0.30$	0.3430	0.4410	0.1890	0.0270
$\hat{p} = 0.33$	0.2964	0.4444	0.2222	0.0371
$p = 0.40$	0.2160	0.4320	0.2880	0.0640
$p = 0.50$	0.1250	0.3750	0.3750	0.1250
$p = 0.60$	0.0640	0.2880	0.4320	0.2160
$\hat{p} = 0.66$	0.0371	0.2222	0.4444	0.2964
$p = 0.75$	0.0156	0.1406	0.4219	0.4219
$\hat{p} = 1.0$	0	0	0	1.0000

If the observations in a sample are independent, then the joint density function of the sample values $x_1, x_2, \ldots, x_n$ can be written as

$$L(\theta) = f(x_1, x_2, \ldots, x_n; \theta) = \prod_{i=1}^{n} f(x_i; \theta) \tag{7.14}$$

where the symbol $\prod$ tells us to multiply $f(x_1; \theta)$ by $f(x_2; \theta)$, the product by $f(x_3; \theta)$, etc., up to $f(x_n; \theta)$. In other words,

$$\prod_{i=1}^{n} f(x_i; \theta) = f(x_1; \theta) f(x_2; \theta) f(x_3; \theta) \ldots f(x_n; \theta)$$

The expression (7.14), regarded as a function of θ for fixed x_i, is called the likelihood function of θ. The maximum likelihood estimate of θ is the number $\hat{\theta}$ that maximizes the likelihood function (7.14).

This will be illustrated to begin with by obtaining the maximum likelihood estimate of the probability of a success for a binomial variable. Let y be a binomial variable which takes on the values 1 and 0 with probabilities p and $(1 - p)$, respectively. The distribution of y can be written as

$$f(y; p) = p^y (1 - p)^{1-y}$$

A random sample of binomial trials consists of a sequence of zeros and ones denoted by $y_1, y_2, \ldots, y_n$. The likelihood function of the sequence is

$$L(\theta) = \prod_{i=1}^{n} p^{y_i}(1 - p)^{1-y_i} = p^x (1 - p)^{n-x} \tag{7.15}$$

where $x = \sum_{i=1}^{n} y_i$.

The value of p that maximizes (7.15) is the maximum-likelihood estimator $\hat{p}$. The customary procedure is to write (7.15) as a logarithm to the base e so that

$$\ln L(\theta) = x \ln(p) + (n - x)\ln(1 - p) \tag{7.16}$$

The value of p that maximizes the logarithm of the likelihood function also maximizes the likelihood function. The derivative of $\ln L(\theta)$ with respect to p is equal to

$$\frac{d \ln L(\theta)}{dp} = \frac{x}{p} - \frac{(n - x)}{(1 - p)} \tag{7.17}$$

The value of p which maximizes the likelihood function is found by setting the expression (7.17) equal to zero and solving for p. Doing this, we obtain the estimator

$$\hat{p} = \frac{x}{n} = \frac{1}{n}\sum y_i \tag{7.18}$$

which is simply the average number of "successes" observed in a series of n trials.

The expected value of the sum of the binomial trials x is equal to np. The expected value of $\hat{p}$ is equal, then, to

$$E(\hat{p}) = E\left(\frac{x}{n}\right) = \frac{1}{n}E(x) = p$$

The estimator $\hat{p}$ is therefore an unbiased estimator of p.

The variance of x is equal to $np(1 - p)$. The variance of $\hat{p}$ is equal to

$$\mathrm{Var}(\hat{p}) = \mathrm{Var}\left(\frac{x}{n}\right) = \frac{1}{n^2}\mathrm{Var}(x) = \frac{p(1 - p)}{n}$$

As n becomes infinitely large, the variance of $\hat{p}$ approaches zero and the distribution of $\hat{p}$ becomes concentrated at p. Thus, the estimator $\hat{p}$ is a consistent estimator of p.

A random sample of size n drawn from a normal distribution has a likelihood function

$$L(\theta) = \prod_{i=1}^{n}\left[\frac{1}{2\pi\sigma^2}\right]^{n/2}\exp\left[-\frac{\sum(x_i - \mu)^2}{2\sigma^2}\right]$$

The logarithm of the likelihood function is

$$\ln L(\theta) = -\frac{n}{2}\ln 2\pi - \frac{n}{2}\ln \sigma^2 - \frac{1}{2\sigma^2}\sum(x_i - \mu)^2$$

Differentiating $\ln L(\theta)$ with respect to μ and σ^2, we obtain

$$\frac{\partial \ln L(\theta)}{\partial \mu} = \frac{1}{\sigma^2} \sum (x_i - \mu)$$

and

$$\frac{\partial \ln L(\theta)}{\partial \sigma^2} = -\frac{n}{2}\frac{1}{\sigma^2} + \frac{1}{2\sigma^4} \sum (x_i - \mu)^2$$

Setting these derivatives equal to zero and solving the resulting equations for the maximum likelihood estimators $\hat{\mu}$ and $\hat{\sigma}^2$, we find

(7.19) $$\hat{\mu} = \frac{1}{n} \sum x_i = \bar{X}$$

and

(7.20) $$\hat{\sigma}^2 = \frac{1}{n} \sum (x_i - \bar{X})^2$$

We have already shown that the sample mean is an unbiased estimator of μ.

It can be shown that the expected value of $\hat{\sigma}^2$ is equal to

(7.21) $$E(\hat{\sigma}^2) = \frac{n-1}{n}\sigma^2 = \sigma^2 - \frac{\sigma^2}{n}$$

The maximum-likelihood estimator of σ^2 therefore tends to underestimate the true value of σ^2. As the sample size n increases, the bias $(-\sigma^2/n)$ becomes smaller, however. Thus, $\hat{\sigma}^2$ is asymptotically unbiased but is biased for finite samples.

If $\hat{\sigma}^2$ is multiplied by $n/(n-1)$, we obtain[4]

(7.22) $$s^2 = \frac{1}{n-1} \sum (x_i - \bar{X})^2$$

which is an unbiased estimator of σ^2.

[4] The expected value of $\hat{\sigma}^2$ given by (7.21) is based on the assumption that the sampling is done *with* replacement. If a sample of size n is drawn *without* replacement from a finite population consisting of N items, the expected value of $\hat{\sigma}^2$ is equal to

$$E(\hat{\sigma}^2) = \sigma^2 - \frac{\sigma^2}{n}\left[1 - \frac{n-1}{N-1}\right] = \frac{n-1}{n}\frac{N}{N-1}\sigma^2$$

Thus,

$$\frac{n}{n-1}\frac{N-1}{N}\hat{\sigma}^2 = \frac{N-1}{N}\left[\frac{1}{n-1}\sum_i (x_i - \bar{X})^2\right]$$

is an unbiased estimator of σ^2 when sampling without replacement from a finite population of size N.

Even though maximum likelihood estimators are not necessarily unbiased, their distributions are always centered near the true value of the parameter. Thus, if they are biased, they can usually be modified slightly so as to obtain an unbiased estimator.

Under fairly general conditions, maximum-likelihood estimators are consistent and asymptotically efficient estimators. The maximum-likelihood estimators are always sufficient estimators *if* sufficient estimators exist. The proof of these statements requires a rather advanced mathematical discussion and will not be attempted here. Instead, we shall illustrate that the maximum-likelihood estimator $\hat{p} = x/n$ is a sufficient estimator.

If the binomial variable y_i is 1 or 0 depending on whether the ith trial is a success or failure, the sample $y_1, y_2, \ldots, y_n$ shows how many successes occurred and in what order. The joint density function of the sample is

$$f(y_1, y_2, \ldots, y_n; p) = p^x(1-p)^{n-x} \tag{7.23}$$

where x is equal to $\sum y_i$. The order of the successes does not provide any evidence concerning the value of p. Once the total number of successes $\sum y_i$ is known to be equal to x, each of the $n!/x!\,(n-x)!$ possible orders of the x successes is equally likely regardless of p. Therefore, given x (which is the same as saying "given $\hat{p}$" because $\hat{p} = x/n$), the probability of the observed sample is

$$g(y_1, y_2, \ldots, y_n \mid \hat{p}) = \frac{x!\,(n-x)!}{n!} \tag{7.24}$$

The probability of observing a particular value of x (which is the same as the probability of obtaining a particular value of $\hat{p}$) is given by

$$h(\hat{p}; p) = \frac{n!}{x!\,(n-x)!}\, p^x(1-p)^{n-x} \tag{7.25}$$

Similarly, when sampling without replacement, the variance of the sample mean is equal to

$$\text{Var}(\bar{X}) = \frac{\sigma^2}{n}\left[\frac{N-n}{N-1}\right]$$

In this case, an unbiased estimator of $\text{Var}(\bar{X})$ is given by

$$s_{\bar{X}}^2 = \left[\frac{N-n}{N}\right]\frac{\hat{\sigma}^2}{n-1}$$

where $\hat{\sigma}^2$ is given by (7.20). The larger the sample size n relative to the population size N, the more important the "correction factor" $(N-n)/N$.

The joint density function (7.23) can then be written as the product of (7.24) and (7.25). That is,

$$f(y_1, y_2, \ldots, y_n; p) = g(y_1, y_2, \ldots, y_n \mid \hat{p})h(\hat{p}; p)$$

Therefore, $\hat{p} = x/n$ must be a sufficient estimator of p.

The following examples will serve to illustrate several of the main ideas in this section.

7.4.3 Some Examples of Point Estimation

Example 1: A telephone office selects a random sample of 100 telephone subscribers and counts the number of telephone connections to a wrong number made by each subscriber during the month of June. The sample results are found in Columns (1) and (2) of Table 7.4. Similar samples drawn by other offices indicate that the distribution of the number of connections to the wrong number has a Poisson distribution. In other words, if X is a

Table 7.4 Connections to a Wrong Number

(1) Number of Wrong Connections k	(2) Number of Subscribers with k Wrong Connections n_k	(3) (1) · (2) $(k)(n_k)$	(4) $E(n_k) =$ $100 \cdot p(k \mid m = 5.2)$
0	0	0	0.55
1	3	3	2.87
2	8	16	7.46
3	15	45	12.93
4	16	64	16.81
5	19	95	17.48
6	13	78	15.15
7	9	63	11.25
8	7	56	7.31
9	4	36	4.23
10	3	30	2.20
11	2	22	1.04
12	1	12	0.45
Over 13	0	0	0.28
Total	100	520	100.00

random variable representing the number of wrong connections made by a subscriber, then the probability that X is equal to k is

$$p(k) = \frac{e^{-m}m^k}{k!}$$

where the mean m can be thought of as the number of wrong connections a subscriber would "normally" make during the month.

In order to determine how well the data in Table 7.4 fit a Poisson distribution, it is necessary first to estimate the mean m of the distribution. We shall begin by deriving the maximum likelihood estimator of m.

Suppose that a random sample of n observed values $x_1, x_2, \ldots, x_n$ is to be used to estimate m. If the x_i's are all drawn from a Poisson distribution having mean m, then

$$p(x_i) = \frac{e^{-m}m^{x_i}}{x_i!}$$

and the likelihood function is

$$L(\theta) = \prod_{i=1}^{n} p(x_i) = \prod_{i=1}^{n} \frac{e^{-m}m^{x_i}}{x_i!} = \frac{e^{-nm}m^{\sum x_i}}{\prod(x_i!)}$$

Taking the logarithm of $L(\theta)$, we obtain

$$\ln L(\theta) = -nm + \sum x_i \ln(m) - \log \prod (x_i!)$$

The derivative of $\ln L(\theta)$ with respect to m is

$$\frac{d \ln L(\theta)}{dm} = -n + \frac{\sum x_i}{m} \tag{7.26}$$

Setting (7.26) equal to zero and solving for the estimator $\hat{m}$, we find

$$\hat{m} = \frac{1}{n}\sum x_i = \bar{X}$$

In words, the sample mean is the maximum likelihood estimator of the Poisson parameter m.

Using the data in Table 7.4, we compute

$$\sum x_i = \sum_k (k)(n_k) = 520$$

$$n = \sum_k n_k = 100$$

and

$$\hat{m} = \left(\frac{520}{100}\right) = 5.2$$

If we assume that the "number of wrong connections" has a Poisson distribution with $m = 5.2$, the probabilities $p(k)$ can be obtained from Table VI, Appendix I. In this example, the expected number of subscribers making k wrong connections during the month is equal to

$$E(n_k) = n \cdot p(k) = 100 \cdot p(k \mid m = 5.2)$$

The values $E(n_k)$ are found in Column (4) of Table 7.4. There is a strikingly close agreement between the observed values and the expected values in Table 7.4.

Example 2: A manufacturer of projection lamps for movie projectors develops a new type of lamp which, under normal operating conditions, he believes will always last for at least c hours, never last for more than d hours, and is equally likely to fail at any time between c and d hours. If X represents the time that a projection lamp lasts, then the probability density function of X is

$$f(x) = \begin{cases} \dfrac{1}{d - c} & \text{for } c < x < d \\ 0 & \text{otherwise} \end{cases} \tag{7.27}$$

The manufacturer tests ten of the lamps and records the time each lamp lasts. The results of the test are as follows:

Lamp Number	*Hours of Useful Life* x_i
1	9.5
2	9.7
3	8.2
4	6.5
5	6.0
6	7.8
7	10.0
8	9.3
9	8.6
10	9.7
Total	85.3

Using these data, compute the maximum likelihood estimates of c and d.

The likelihood function for a sample of size n, in this case, is

$$L(\theta) = \prod_{i=1}^{n} \frac{1}{(d-c)} = \frac{1}{(d-c)^n}$$

The maximum of this function cannot be obtained by differentiation because $L(\theta)$ is an ever-decreasing function of $(d - c)$. It is evident, however, that in order to make $L(\theta)$ as large as possible we need only make $(d - c)$ as small as possible. But, clearly, c can be no larger than the smallest sample observation nor can d be smaller than the largest observation in the sample. Thus, the smallest possible value of $(d - c)$ obtained from the sample is equal to the largest of the sample values, $\max_i \{x_i\}$, minus the smallest of the sample values, $\min_i \{x_i\}$. Thus, the maximum likelihood estimates of c and d for the sample of ten lamps are

$$\hat{d} = \max_i\{x_i\} = 10.0$$

and

$$\hat{c} = \min_i\{x_i\} = 6.0$$

If X has the uniform distribution (7.27), then the expected value of X is equal to

$$E(X) = \int_c^d xf(x)\,dx = \int_c^d \frac{x}{(d-c)}\,dx = \frac{d+c}{2}$$

Two possible estimators of $E(X)$ are

$$\bar{X} = \frac{1}{n}\sum x_i \quad \text{and} \quad M = \left(\frac{\max_i\{x_i\} + \min_i\{x_i\}}{2}\right)$$

These estimators can provide very different estimates of $E(X)$. For the sample of ten lamps, we have

$$\bar{X} = \frac{85.3}{10} = 8.53 \quad \text{and} \quad M = \frac{10+6}{2} = 8.0 \text{ hours}$$

Which estimator would you use to estimate $E(X)$?

7.5 INTERVAL ESTIMATION

In many applications, a point estimate of a parameter is not very meaningful without some measure of the possible error in the estimate. In such cases, it is helpful to provide the reader with an interval about the point estimate together with a measure of assurance that the interval covers the true parameter. Such intervals are called *confidence intervals* or *interval estimates*. The estimation of confidence intervals is called *interval estimation*. The concept of a confidence interval is also useful when estimating the sample size required to obtain a given degree of precision in the estimation of population parameters.

7.5.1 Interval Estimators for Means

A confidence interval consists of two numbers L and U, with U greater than L, computed in such a way that

$$P(L < \theta < U) = 1 - \alpha \tag{7.28}$$

where θ is a population parameter and $1 - \alpha$ is a probability representing the desired level of confidence. The goal of interval estimation is to provide intervals $(U < \theta < L)$ which are such that $1 - \alpha$ percent of the intervals will cover the population parameter θ for which they are estimated. The selection of the *confidence coefficient* $1 - \alpha$ depends on the investigator. The most commonly used values of $1 - \alpha$ are 0.90, 0.95, and 0.99. The interval estimator of the population mean μ will be derived to illustrate these ideas.

From a table of the normal distribution, we find that the probability is $1 - \alpha = 0.95$ that a standard normal variable Z will satisfy the inequalities

$$-1.96 < Z < 1.96$$

The sample mean $\overline{X}$ of a random sample drawn from a normal population with mean μ and variance σ^2 is a normal variable with mean μ and standard deviation $\sigma/\sqrt{n}$. The quantity

$$\frac{\overline{X} - \mu}{\sigma/\sqrt{n}}$$

is therefore a standard normal variable, and the probability is $1 - \alpha = 0.95$ that it will satisfy the inequalities

$$-1.96 < \frac{\overline{X} - \mu}{\sigma/\sqrt{n}} < 1.96$$

Writing this as an inequality for μ, we obtain

$$\overline{X} - 1.96 \frac{\sigma}{\sqrt{n}} < \mu < \overline{X} + 1.96 \frac{\sigma}{\sqrt{n}} \tag{7.29}$$

which is the 95 percent interval estimator for the population mean μ. If confidence intervals for μ are computed using (7.29), then in the long run, 95 percent of these confidence intervals will contain the true value of the parameter μ.

To obtain the interval estimator (7.29), it was assumed that the random sample, from which $\bar{X}$ was computed, was drawn from a normal distribution. According to the central limit theorem, $\bar{X}$ is approximately normal with mean μ and standard deviation $\sigma/\sqrt{n}$ even if the random sample is not obtained from a normal distribution *provided that* the sample size is large. Thus, (7.29) can serve as an interval estimator for the mean of a wide variety of distributions provided that $\bar{X}$ is computed from large samples.

Once a sample is taken and the numbers $L = \bar{X} - 1.96\ \sigma/\sqrt{n}$ and $U = \bar{X} + 1.96\ \sigma/\sqrt{n}$ are computed, the probability statement $P(L < \mu < U) = 0.95$ is meaningless because μ either is or is not in the interval defined by L and U. Suppose, for example, that ten samples of 36 observations are drawn from a normal population with $\mu = 20$ and standard deviation $\sigma = 12$. The sample means for the ten samples as well as the 95 percent confidence intervals are

Table 7.5 Confidence Intervals for Ten Samples

Sample Number	Sample Mean, $\bar{X}$	95 Percent Confidence Interval L	U
1	17.00	14.08	20.92
2	21.00	17.08	24.92
3	23.50	19.58	27.42
4	19.40	15.48	23.32
5	20.20	16.28	24.12
6	18.00	14.08	21.92
7**	15.50	11.58	19.42
8	16.90	12.98	20.82
9	22.30	18.38	26.22
10	21.00	17.08	24.92

given in Table 7.5 and presented in Figure 7.4. The confidence intervals for these samples are given by

$$L = \bar{X} - 1.96\left(\frac{12}{\sqrt{36}}\right) = \bar{X} - 3.92$$

and

$$U = \bar{X} + 1.96\left(\frac{12}{\sqrt{36}}\right) = \bar{X} + 3.92$$

The confidence interval for sample number 7 is the only one that does not cover the true value of μ. In addition to the constants σ and n, confidence

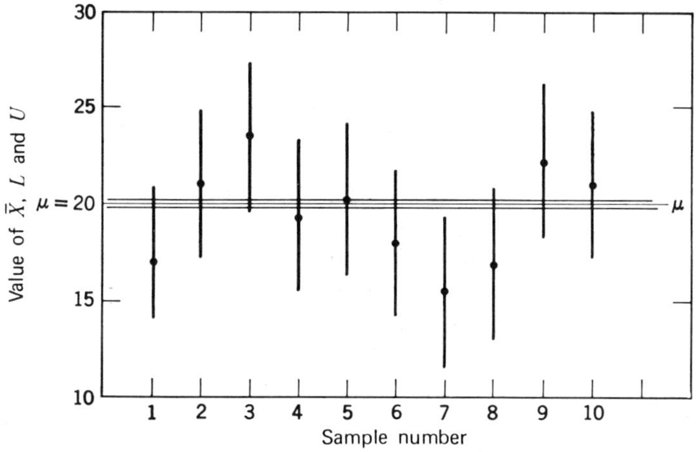

Figure 7.4 Confidence intervals for 10 samples.

intervals depend on the random variable $\overline{X}$ and are therefore random intervals. The variation in the location of the confidence intervals is easily seen in Figure 7.4.

The general interval estimator for μ can be written as

$$\overline{X} - z_c \frac{\sigma}{\sqrt{n}} < \mu < \overline{X} + z_c \frac{\sigma}{\sqrt{n}} \tag{7.30}$$

where z_c is a standard normal variable which depends on the level of confidence desired. The interval estimator (7.30) is used for large samples from an infinite population or when sampling with replacement from a finite population. For samples of size n drawn without replacement from a finite population of size N, the interval estimator

$$\overline{X} - z_c \frac{\sigma}{\sqrt{n}} \sqrt{\frac{N-n}{N-1}} < \mu < \overline{X} + z_c \frac{\sigma}{\sqrt{n}} \sqrt{\frac{N-n}{N-1}} \tag{7.31}$$

can be used provided that the standard deviation of the population is known.

When dealing with small samples drawn with replacement from a population with unknown variance, the interval estimator for the population mean is obtained from (7.30) by replacing z_c with the t value t_c for $(n-1)$ degrees of freedom and σ with the sample standard deviation s. The new interval estimator for μ is

$$\overline{X} - t_c \frac{s}{\sqrt{n}} < \mu < \overline{X} + t_c \frac{s}{\sqrt{n}} \tag{7.32}$$

A better understanding of interval estimation can be obtained by carefully studying the following examples.

Example 1: On the basis of a sample of 144 observations, safety experts estimated the average reaction time of truck drivers responding to a stop signal to be 2.00 seconds and the sample standard deviation to be 0.60 seconds. Compute a 99 percent confidence interval for the mean reaction time μ.

The sample is a relatively large one drawn from an infinite population. Therefore, (7.30) can be used to obtain the confidence interval. The standard normal variable z_c for $1 - \alpha = 0.99$ is 2.58 while $s/\sqrt{n} = \sigma/\sqrt{n} = (0.60/\sqrt{144}) = 0.05$. The 99 percent confidence interval is therefore equal to

$$1.8710 < \mu < 2.1290$$

The 90 percent confidence interval would be

$$2.00 - 1.65(0.05) < \mu < 2.00 + 1.65(0.05)$$

or

$$1.9175 < \mu < 2.0825$$

which is smaller than the 99 percent confidence interval.

In general, as the confidence coefficient $1 - \alpha$ is increased, the confidence interval becomes wider.

Example 2: A food inspector selects a random sample of four packages of Brand X breakfast food from the shelves of a supermarket. The contents of each package are weighed and the average weight found to be $\bar{X} = 6.00$ ounces while the sample standard deviation is $s = 0.20$ ounces. The net weight marked on every box of Brand X is 6.5 ounces. Does the 99 percent confidence interval for the true mean weight μ cover the advertised net weight of 6.5 ounces?

The sample is very small relative to the population and the population variance is unknown. Therefore, the t distribution is used to obtain the 99 percent confidence interval. The value of t_c for $(n - 1) = 3$ degrees of freedom is 5.841 and $s/\sqrt{n} = 0.10$. Using (7.32), we obtain the confidence interval

$$5.4159 < \mu < 6.5841$$

which covers the advertised net weight of 6.5 ounces.

The t value $t_c = 5.841$ is relatively large and results in a wide confidence interval for μ. This is a reflection of the fact that the food inspector has selected a high confidence level and has very little information to work with because the sample size is very small and the standard deviation of the population is unknown. If the sample size had been $n = 16$ and it were found that $\overline{X} = 6.00$ with $s = 0.20$, then the 99 percent confidence interval would have been

$$6.00 - 2.947(0.05) < \mu < 6.00 + 2.947(0.05)$$

or

$$5.8526 < \mu < 6.1474$$

which would not have covered the advertised weight of 6.5 ounces.

Using $n = 16$, $\overline{X} = 6.00$, and $s = 0.20$, compute the 90 percent confidence interval for μ.

Example 3: Suppose that a manufacturer of radio antenna towers delivers 100 towers to a radio supply company which had ordered towers capable of withstanding winds up to 80 miles per hour. The radio supply company wants to determine if the towers meet this specification. A random sample of three towers is selected; the towers are set up and tested. The company estimates that, on the average, the towers were able to withstand winds up to 76 m.p.h. with a sample standard deviation $s = 2$ m.p.h. If the population mean is 80 m.p.h., will it be covered by a 95 percent confidence interval computed for this sample?

The random sample was drawn from a finite population without replacement. The sample standard deviation of $\overline{X}$ is therefore equal to

$$\frac{s}{\sqrt{n}}\sqrt{\frac{N-n}{N-1}} = \frac{2}{\sqrt{3}}\sqrt{\frac{97}{99}} = 1.137$$

Since the standard deviation of the population is unknown and the sample is very small, the t distribution is used to obtain the confidence interval. The t value with $(n - 1) = 2$ degrees of freedom for a 95 percent confidence interval is $t_c = 4.303$. The 95 percent confidence interval is therefore equal to

$$71.11 < \mu < 80.89$$

which covers $\mu = 80$. The 90 percent confidence interval, on the other hand, would not cover $\mu = 80$ m.p.h.

Notice that the correction factor $\sqrt{(N-n)/(N-1)}$ is slightly smaller than 1.00 and therefore narrows the confidence interval by a small amount. When sampling without replacement from a finite population, the confidence interval narrows as the sample size increases because $\sqrt{(N-n)/(N-1)}$ decreases as well as $\sigma/\sqrt{n}$. Thus, sampling without replacement from a finite population is more "efficient" than sampling with replacement.

7.5.2 Interval Estimators for Proportions

The estimation of proportions is an important part of many statistical studies. A public opinion poll may wish to determine the proportion of voters who will vote for candidate Jones, a marketing research firm may wish to know the proportion of housewives who prefer Brand G to all other brands, or a builders' supply firm may want to estimate the proportion of defective ceiling tile in a carload shipment. In each case, the population contains a large number of items that can be classified as having or not having the characteristic being studied. The desired proportion would be estimated by the relative frequency x/n, where x is the number of items having the characteristic and n is the sample size. If random sampling is used, it seems reasonable to assume that x is a binomial variable. This assumption, which is used in many situations where the goal is to estimate a proportion, will be made in the following discussion.

It has been shown that the statistic $\hat{p} = x/n$ is a good estimator of the population proportion (or probability) p. Even so, there are many situations in which the investigator may want to compute a confidence interval for p. The procedure used to obtain an interval estimator for p is similar to that used to obtain the interval estimator (7.29) for the population mean.

The random variable $\hat{p}$ is unbiased with variance equal to

$$\text{Var}(\hat{p}) = \frac{p(1-p)}{n} \tag{7.33}$$

The normal approximation to the binomial distribution is easily modified to apply to $\hat{p}$. The standard normal variable for $\hat{p}$ is

$$z = \frac{\hat{p} - p}{\sqrt{p(1-p)/n}}$$

and, for large samples, the probability is approximately 0.95 that z satisfies the inequalities

$$-1.96 < \frac{\hat{p} - p}{\sqrt{p(1-p)/n}} < 1.96$$

Using $\hat{p} = x/n$ and rearranging terms, we obtain

$$\frac{x}{n} - 1.96\sqrt{\frac{p(1-p)}{n}} < p < \frac{x}{n} + 1.96\sqrt{\frac{p(1-p)}{n}} \tag{7.34}$$

Further manipulation is necessary in order to obtain an inequality which does not contain p in both the left- and right-hand terms. Such manipulation is not undertaken here, however, because special tables are available which quickly provide the confidence intervals for p once x/n is known.

Table VII, Appendix I, can be used to obtain 95 percent confidence intervals for the proportion p. Suppose that $x = 30$ in a random sample of size 100. To use Table VII, first find $x/n = 30/100 = 0.3$ on the horizontal scale; next, go up vertically from this point to the *two curves* labeled 100; then move

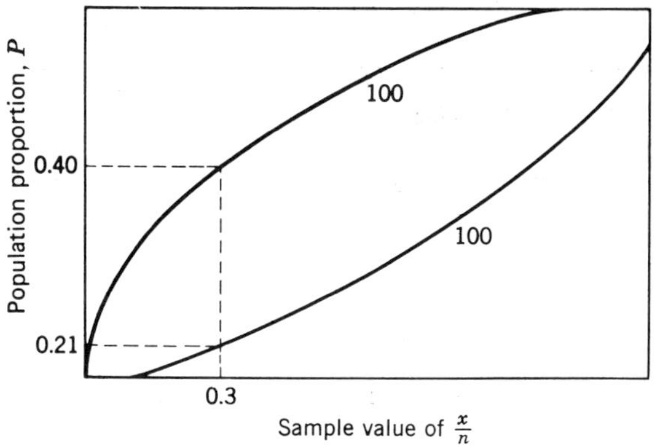

Figure 7.5 95 percent confidence intervals for porportions.

horizontally to the left from both lines to find the upper and lower values of the confidence interval for p. This is illustrated in Figure 7.5. The confidence interval in this case is

$$0.21 < p < 0.40$$

Interpolation can be used to estimate confidence intervals for values of n not given in Table VII. Suppose, for example, that a purchasing agent wants to obtain a 95 percent confidence interval for the proportion of defective electric motors in a large shipment. Eighty of the motors are selected at random and tested. Eight are found to be defective. From Table VII for $x/n = 0.10$, we find the confidence interval for p for $n = 50$ to be

$$0.030 < p < 0.225$$

and, for $n = 100$, we have

$$0.050 < p < 0.175$$

Using $(80 - 50)/(100 - 50) = 3/5$ and interpolating, the confidence interval for $n = 80$ is found to be

$$0.030 + \frac{3}{5}(0.05 - 0.03) < p < 0.225 - \frac{3}{5}(0.225 - 0.175)$$

or

$$0.042 < p < 0.195$$

If a table such as Table VII is not available, the standard deviation to p can be approximated by substituting x/n for p in (7.33). Thus, the interval estimator for p is approximated by

$$\frac{x}{n} - z_c\sqrt{\frac{(x/n)(1 - x/n)}{n}} < p < \frac{x}{n} + z_c\sqrt{\frac{(x/n)(1 - x/n)}{n}} \tag{7.35}$$

where z_c is a standard normal variable that depends on the confidence level desired.

To illustrate the use of (7.35), suppose that $x = 30$ in a random sample of size 100. The 95 percent confidence interval obtained from (7.35) is

$$0.30 - 1.96\sqrt{\frac{(0.30)(0.70)}{100}} < p < 0.30 + 1.96\sqrt{\frac{(0.30)(0.70)}{100}}$$

or

$$0.21 < p < 0.39$$

which is very close to the values obtained previously from Table VII.

7.5.3 *Interval Estimators for Differences of Means and Proportions*

The procedure used in the previous sections to obtain interval estimators for proportions and means can be used to obtain interval estimators for many other parameters. In general, if $\hat{\theta}$ is an estimator of a parameter θ and has a sampling distribution which is approximately normal, then the interval estimator for θ is given by

$$\hat{\theta} - z_c \sigma_{\hat{\theta}} < \theta < \hat{\theta} + z_c \sigma_{\hat{\theta}} \tag{7.36}$$

where z_c is a standard normal variable the value of which depends on the confidence level desired, $\sigma_{\hat{\theta}}$ is the standard deviation of $\hat{\theta}$, and it is assumed that $\hat{\theta}$ is obtained from a random sample.

To illustrate the use of (7.36), suppose that an interval estimator is needed for the difference between the means of two infinite populations. In this case, the parameter is $\theta = \mu_1 - \mu_2$ and the estimator is the difference of the sample means $\hat{\theta} = \overline{X}_1 - \overline{X}_2$. If $\overline{X}_1$ and $\overline{X}_2$ are based on random samples, the variance of $(\overline{X}_1 - \overline{X}_2)$ is

$$\mathrm{Var}(\hat{\theta}) = \mathrm{Var}(\overline{X}_1) + \mathrm{Var}(\overline{X}_2) = \frac{\sigma_1{}^2}{n_1} + \frac{\sigma_2{}^2}{n_2}$$

where $\sigma_1{}^2$ and $\sigma_2{}^2$ are the variances of the two populations while n_1 and n_2 are the sizes of the samples used to compute $\overline{X}_1$ and $\overline{X}_2$, respectively. The interval estimator for the difference of two population means, for large populations, is given by

$$\overline{X}_1 - \overline{X}_2 - z_c\sqrt{\frac{\sigma_1{}^2}{n_1} + \frac{\sigma_2{}^2}{n_2}} < \mu_1 - \mu_2 < \overline{X}_1 - \overline{X}_2 + z_c\sqrt{\frac{\sigma_1{}^2}{n_1} + \frac{\sigma_2{}^2}{n_2}} \tag{7.37}$$

When working with small samples and unknown population variances, the t distribution would be used to compute the interval estimator for the difference $\mu_1 - \mu_2$. The interval estimator, in this case, is

(7.38)

$$\overline{X}_1 - \overline{X}_2 - t_c s_p \sqrt{\left(\frac{1}{n_1}\right) + \left(\frac{1}{n_2}\right)} < \mu_1 - \mu_2 < \overline{X}_1 - \overline{X}_2 + t_c s_p \sqrt{\left(\frac{1}{n_1}\right) + \left(\frac{1}{n_2}\right)}$$

where t_c is a t value with $(n_1 + n_2 - 2)$ degrees of freedom which depends on the desired level of confidence and $s_p{}^2$ is the pooled variance of the samples:

$$s_p{}^2 = \frac{(n_1 - 1)s_1{}^2 + (n_2 - 1)s_2{}^2}{n_1 + n_2 - 2}$$

As an example, suppose a marketing specialist wants to determine if the "average" driver in Seattle purchases more gasoline during July than the "average" driver in Reno. A random sample of nine drivers is selected in each city, and their gasoline purchases are recorded. The results are presented in Table 7.6.

Using the data in Table 7.6, we obtain

$$\overline{X}_1 = \frac{1}{n_1} \sum_{i=1}^{n_1} x_i = \frac{360}{9} = 40 \text{ gallons}$$

$$\overline{X}_2 = \frac{1}{n_2} \sum_{i=1}^{n_2} x_i = \frac{288}{9} = 32 \text{ gallons}$$

$$s_1{}^2 = \frac{1}{n_1 - 1} \sum_{i=1}^{n_1} (x_i - \overline{X}_1)^2 = \frac{312}{8} = 39.00$$

$$s_2{}^2 = \frac{1}{n_2 - 1} \sum_{i=1}^{n_2} (x_i - \overline{X}_2)^2 = \frac{682}{8} = 85.25$$

$$s_p{}^2 = \frac{312 + 682}{16} = 62.125$$

and

$$s_p \sqrt{\frac{1}{n_1} + \frac{1}{n_2}} = 3.72 \text{ gallons}$$

Since the sample sizes are small and the population variances are unknown, the t distribution is used to obtain the confidence interval for $\mu_1 - \mu_2$. For

Table 7.6 Gasoline Purchases

Driver	Gasoline Purchased (in Gallons) Seattle	Reno
1	40	30
2	46	24
3	50	16
4	36	25
5	38	35
6	34	40
7	42	46
8	44	38
9	30	34
Totals	360	288

$\alpha = 0.05$ and 16 degrees of freedom, the t value is 2.12. The 95 percent confidence interval for $\mu_1 - \mu_2$ is therefore equal to

$$8 - 2.12(3.72) < \mu_1 - \mu_2 < 8 + 2.12(3.72)$$

or

$$0.27 < \mu_1 - \mu_2 < 15.73$$

Since this interval does not cover $(\mu_1 - \mu_2) = 0$, we conclude, on the basis of these samples, that the average driver in Seattle may have purchased more gasoline in July than the average driver in Reno. The evidence is not particularly convincing, however, due to the small size of the samples used.

If $\overline{X}_1 - \overline{X}_2$ and $\mu_1 - \mu_2$ are replaced by $\overline{X}_1 + \overline{X}_2$ and $\mu_1 + \mu_2$, the formulas (7.37) and (7.38) provide interval estimators for the sum of two population means.

An approximate interval estimator for the difference between two population proportions, where the populations are infinite, can be obtained from (7.36) by letting $\theta = p_1 - p_2$ and $\hat{\theta} = \hat{p}_1 - \hat{p}_2$.

If random samples of size n_1 and n_2 are used to estimate p_1 and p_2, then the variance of the estimator $\hat{\theta}$ is

$$\text{Var}(\hat{\theta}) = \text{Var}(\hat{p}_1) + \text{Var}(\hat{p}_2) \doteq \sqrt{\frac{p_1(1 - p_1)}{n_1}} + \sqrt{\frac{p_2(1 - p_2)}{n_2}}$$

An estimator of $\text{Var}(\hat{\theta})$ is obtained by substituting $\hat{p}_1 = x_1/n_1$ and $\hat{p}_2 = x_2/n_2$ for p_1 and p_2, respectively.

The approximate interval estimator is equal to

$$\hat{p}_1 - \hat{p}_2 - z_c\sqrt{\frac{\hat{p}_1(1-\hat{p}_1)}{n_1} + \frac{\hat{p}_2(1-\hat{p}_2)}{n_2}} < p_1 - p_2 < \hat{p}_1 - \hat{p}_2 + z_c\sqrt{\frac{\hat{p}_1(1-\hat{p}_1)}{n_1} + \frac{\hat{p}_2(1-\hat{p}_2)}{n_2}} \tag{7.39}$$

where the value of z_c depends on the desired confidence level.

As an illustration of the use of (7.39), let p_1 and p_2 represent the proportions of voters in Boston and Dallas who plan to vote for a particular candidate for national office. Furthermore, suppose a poll taker wants to compute a 90 percent confidence interval for $p_1 - p_2$. Samples of 120 voters are drawn in Boston and Dallas. It is found that $x_1 = 90$ voters in Boston and $x_2 = 78$ voters in Dallas plan to vote for the candidate. Using these figures, we compute

$$\hat{p}_1 = \frac{x_1}{n_1} = \frac{90}{120} = 0.75$$

$$\hat{p}_2 = \frac{x_2}{n_2} = \frac{78}{120} = 0.65$$

$$\hat{p}_1 - \hat{p}_2 = 0.75 - 0.65 = 0.10$$

$$\sqrt{\frac{\hat{p}_1(1-\hat{p}_1)}{n_1} + \frac{\hat{p}_2(1-\hat{p}_2)}{n_2}} = \sqrt{\frac{0.1875}{120} + \frac{0.2275}{120}} = 0.059$$

and

$$z_c = 1.65$$

The approximate 90 percent confidence interval for $p_1 - p_2$ is equal to

$$0.10 - 1.65(0.059) < p_1 - p_2 < 0.10 + 1.65(0.059)$$

or

$$0.003 < p_1 - p_2 < 0.197$$

Although $\hat{p}_1 - \hat{p}_2 = 0.10$, it appears that we cannot be especially confident that the difference in the true population proportions is greater than zero.

7.5.4 *An Interval Estimator for σ^2*

It was shown in Chapter 6 that the ratio $(n-1)s^2/\sigma^2$ is distributed as χ^2 with $(n-1)$ degrees of freedom. Thus, two values of χ^2 can be chosen such that the probability is $(1-\alpha)$ that $(n-1)s^2/\sigma^2$ will satisfy the inequalities

$$\chi_c^2 < \frac{(n-1)s^2}{\sigma^2} < \chi_\ell^2$$

The equivalent inequalities for σ^2 are

$$\frac{(n-1)s^2}{\chi_{\ell}^{\,2}} < \sigma^2 < \frac{(n-1)_{\ell}^{\,2}}{\chi_c^{\,2}} \tag{7.40}$$

and can be used to compute confidence intervals for the population variance σ^2. To illustrate the use of (7.40), suppose that a sample of 14 observations yields a sample variance $s^2 = 9$. The appropriate values of χ^2 for a 95 percent confidence interval are obtained from Table IV, Appendix I, using $(n-1) = 13$ degrees of freedom. We find that $\chi_c^{\,2} = 5.00$ and $\chi_{\ell}^{\,2} = 24.73$. The resulting confidence interval for σ^2 is

$$\frac{(13)(9)}{24.73} < \sigma^2 < \frac{(13)(9)}{5.00}$$

or

$$4.73 < \sigma^2 < 23.40$$

This confidence interval appears to be rather wide. Remember, however, that the sample size is small and the confidence interval is for the variance rather than the standard deviation. The confidence interval for the standard deviation is obtained by taking the square roots of the confidence limits for σ^2. The 95 percent confidence interval for σ, in this case, is

$$2.17 < \sigma < 4.84$$

The confidence interval for σ^2 is not symmetrical around $s^2 = 9$ because the chi-square distribution is skewed to the right for small samples.

The confidence interval obtained for σ^2 is a "two-sided confidence interval" which places upper and lower values on σ^2. The values of χ^2 used for two-sided confidence intervals are selected to allocate one-half of the probability α to each tail of the χ^2 distribution. For example, if $(1-\alpha) = 0.95$, then $\alpha/2 = 0.025$ is placed in the lower and upper tails of the χ^2 distribution to

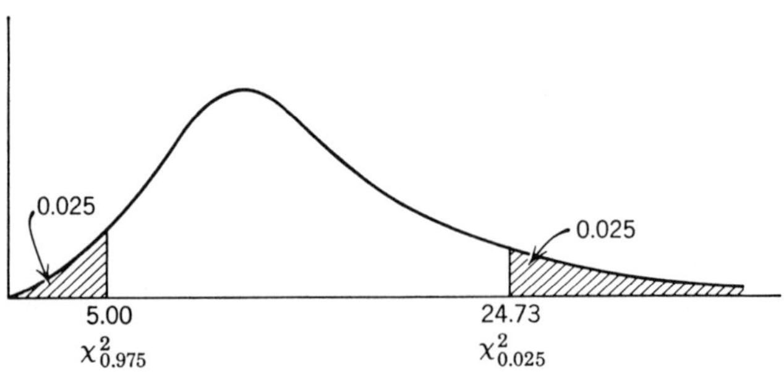

Figure 7.6 The chi-square distribution.

compute a two-sided 95 percent confidence interval for σ^2. This is shown in Figure 7.6 for $(n-1) = 31$ degrees of freedom.

In some applications, the investigator is concerned only with an upper confidence limit for the population variance. In this case, a "one-sided confidence interval" which places the entire probability α in the right-hand tail of the χ^2 distribution is preferred. The one-tailed 97.5 percent confidence interval, for the example in which $n = 14$ and $s^2 = 9$, is

$$0 < \sigma^2 < 23.40$$

and the 95 percent confidence interval is

$$0 < \sigma^2 < \frac{(13)(9)}{5.89}$$

or

$$0 < \sigma^2 < 19.87$$

The lower limit of a confidence interval for σ^2 cannot be less than 0 because the population variance is always positive.

The general formula (7.36) was not used to derive the interval estimator for σ^2 because, for small samples, the estimator s^2 does not have a normal distribution. If σ is the standard deviation of a normally distributed population and the sample size is large, say, $n \geq 100$, the approximate interval estimator for σ is given by

$$s - z_c \frac{s}{\sqrt{2n}} < \sigma < s + z_c \frac{s}{\sqrt{2n}} \tag{7.41}$$

where s is the standard deviation of the sample. That is,

$$s = \sqrt{\frac{1}{n-1} \sum_{i=1}^{n} (x_i - \overline{X})^2}$$

To illustrate the use of (7.41), let us assume that $n = 128$ and $s^2 = 9$. The 95 percent confidence interval for σ is equal to

$$3 - \frac{(1.96)(3)}{16} < \sigma < 3 + \frac{(1.96)(3)}{16}$$

or

$$2.633 < \sigma < 3.367$$

This confidence interval is symmetrical around $s = 3$ and reflects the fact that, for large samples, the χ^2 distribution is approximately a normal distribution.

7.5.5 Determination of Sample Size

In previous sections, it was assumed that a sample of size n was available and that, on the basis of the sample information, the investigator wanted to estimate the precision with which a population parameter had been estimated. Frequently, in the design of samples, the investigator knows the degree of precision he would like to achieve and wants to estimate the sample size required to obtain the given level of precision. This problem is approached in the same way as that of finding interval estimators for parameters.

Suppose that a market analyst wants to estimate the average price of grade A eggs sold in the Chicago area. On the basis of previous studies, he believes the standard deviation of egg prices to be $\sigma = 16$ cents. The analyst would like to be 95 percent "confident" that the error E in his estimate of the true average price of eggs μ is not more than plus or minus 2 cents.

If the sample mean $\bar{X}$ is used as a point estimator of μ, the probability is 0.95 that the error E is less than $1.96(\sigma/\sqrt{n})$. Therefore, the analyst should select that sample size n for which $1.96(\sigma/\sqrt{n}) = E$. Solving this equation for n, we obtain

$$n = \left[\frac{1.96\sigma}{E}\right]^2$$

In this case, $\sigma = 16$, $E = 2$, and the required sample size is

$$n = \left[\frac{(1.96)(16)}{2}\right]^2$$

or approximately 246 observations.

The general formula for the sample size required to obtain a given level of precision in the estimation of a population mean is

$$n = \left[\frac{z_c\sigma}{E}\right]^2 \tag{7.42}$$

where the value of the standard normal variable z_c depends on how "confident" the researcher wants to be that the sample mean $\bar{X}$ is within plus or minus E of the true mean μ. For example, if the researcher wants to be 99 percent "confident" that the error is less than E, he sets z_c equal to 2.58. The equation (7.42) applies only if the sample to be drawn is to be a random sample large enough to justify the assumption that the sampling distribution of $\bar{X}$ is approximately normal.

From (7.42) it is easily seen that, other things being equal, the required sample size increases as:

1. The desired level of confidence (and thus z_c) increases.
2. The acceptable error E decreases.

The researcher usually has to estimate the standard deviation of the population when determining the required sample size. If his estimate of σ is too low, the estimated sample size will be too small to give him the precision required. If there is no information available about σ, a two-stage sampling procedure may be used in which σ is estimated on the basis of a small preliminary sample; this information may be used to determine how many additional observations are needed to achieve the desired degree of precision.

As a second illustration, suppose that a purchasing agent plans to estimate the proportion of defective transistors in a large shipment. Furthermore, suppose that he wants to be 95 percent confident that the sample proportion $p = x/n$ will be within plus or minus $E = 0.04$ of the true proportion p. How large a sample should be drawn?

For large random samples, we can be 95 percent confident that an estimate of p will be within plus or minus

$$E = 1.96\sqrt{\frac{p(1-p)}{n}}$$

of the true value of p. Solving for n, we obtain

$$n = p(1-p)\left[\frac{1.96}{E}\right]^2$$

This formula cannot be used directly to determine the sample size because it involves the unknown quantity p that we want to estimate. If absolutely no information is available about p, we could proceed to find the value of p that maximizes $p(1-p)$ and thereby maximizes the sample size n needed to attain the desired degree of precision in the estimation of p.

To find the value of p that maximizes $p(1-p)$, we let $y = p(1-p) = p - p^2$ and differentiate y with respect to p to obtain

$$\frac{dy}{dp} = 1 - 2p$$

Setting dy/dp equal to zero and solving for p, we obtain $p = 1/2$. The maximum possible value of $p(1-p)$ is therefore equal to 1/4, and we can write

$$n \leq (0.25)\left[\frac{1.96}{E}\right]^2$$

When $E = 0.04$, we have

$$n \leq (0.25)\left[\frac{1.96}{0.04}\right]^2 = (0.25)(2401) = 600$$

A sample of 600 transistors is the largest sample required to attain the desired degree of precision in the estimation of p.

If, on the basis of past experience, the purchasing agent is sure that p is not greater than 0.2, then the sample size need not be larger than

$$n = (0.2)(0.8)\left[\frac{1.96}{0.04}\right]^2 = 384 \text{ transistors}$$

Use of such additional information about p would save both time and expense.

7.5.6 *Using Interval Estimators*

Several examples of the use of various types of interval estimators will be presented in this section. After reading each example, change the basic numbers and calculate new confidence intervals using your data.

Example 1: A manufacturer of television sets is able to use either of two types of picture tubes in a new television model being developed. The manufacturer would like to be 90 percent "confident" that he has selected the tube type that has the longest expected life. A random sample of 100 Type I tubes and 100 Type II tubes is obtained, and the tubes are tested. The sample statistics are:

	$\overline{X}$	s	
Type I	3600	200	(hours)
Type II	3500	200	(hours)

A confidence interval for the difference in the population means $\mu_1 - \mu_2$ can be obtained from either (7.37) or (7.38). Using (7.37), and assuming that $s_i^2 = \sigma_i^2$, we find that

$$\overline{X}_1 - \overline{X}_2 = 3600 - 3500 = 100 \text{ hours}$$

$$z_c = 1.64 \text{ for a 90 percent confidence interval}$$

$$\sqrt{\frac{\sigma_1{}^2}{n_1} + \frac{\sigma_2{}^2}{n_2}} = \sqrt{\frac{80{,}000}{100}} \doteq 28.28 \text{ hours}$$

and the 90 percent confidence interval is

$$100 - 1.64(28.28) < \mu_1 - \mu_2 < 100 + 1.64(28.28)$$

or

$$53.62 < \mu_1 - \mu_2 < 146.38$$

This confidence interval does not cover zero. The manufacturer could, therefore, be 90 percent "confident" that the Type I tube has a longer expected life. (Could he be 99 percent "confident?")

Suppose that the manufacturer had known that the standard deviation of picture tube lives was 200 hours and had wanted to be 95 percent "confident" that the error E in his estimate of the difference $\mu_1 - \mu_2$ was not more than 100 hours. How large a sample of each type of tube would have been necessary? This question is easily answered if we assume that an equal number of each type of tube is to be selected.

Since

$$E = z_c \sqrt{\frac{\sigma_1{}^2}{n_1} + \frac{\sigma_2{}^2}{n_2}}$$

and $n_1 = n_2 = n$, the required sample size is

$$n = \frac{[z_c]^2[\sigma_1{}^2 + \sigma_2{}^2]}{E^2} = \frac{(1.96)^2(80{,}000)}{10{,}000} = 30.73$$

Thus, a random sample of 31 Type I tubes and 31 Type II tubes would be adequate to provide the manufacturer with the desired precision in estimation.

Example 2: A marketing research firm wants to estimate the proportions of men and women who are familiar with a nationally advertised product. In a random sample of 100 men and 200 women, it is found that 20 men and 60 women are familiar with the product. The numbers needed to compute the 95 percent confidence interval for the difference in the proportions of men and women familiar with the product are:

$$\hat{p}_1 = \frac{x_1}{n_1} = \left(\frac{60}{200}\right) = 0.3 = \text{the proportion of women familiar with the product}$$

$$\hat{p}_2 = \frac{x_2}{n_2} = \left(\frac{20}{100}\right) = 0.2 = \text{the proportion of men familiar with the product}$$

$$\hat{p}_1 - \hat{p}_2 = 0.10$$

and

$$\sqrt{\frac{\hat{p}_1(1-\hat{p}_1)}{n_1} + \frac{\hat{p}_2(1-\hat{p}_2)}{n_2}} = 0.0538 = \text{the estimated standard deviation of } p_1 - p_2$$

Using (7.39), the 95 percent confidence interval for the difference $p_1 - p_2$ is

$$0.10 - 1.96(0.0538) < p_1 - p_2 < 0.10 + 1.96(0.0538)$$

or

$$-0.0054 < p_1 - p_2 < 0.2054$$

Since this confidence interval covers zero, the firm *cannot* be 95 percent "confident" that there is a difference in the proportions of men and women familiar with the product. (It could be 90 percent "confident" that a difference exists, however.)

Using Table VII, Appendix I, we find that the 95 percent confidence interval for p_1 is $0.22 < p_1 < 0.38$ and for p_2 is $0.12 < p_2 < 0.29$. These individual confidence intervals *cannot* be used to obtain a confidence interval for the difference $p_1 - p_2$, however. The confidence intervals for p_1 and p_2 are based on the individual (marginal) probability distributions of p_1 and p_2, respectively, while the confidence interval for $p_1 - p_2$ is based on the *joint* probability distribution of p_1 and p_2.

Example 3: A manufacturer of flashlight batteries wants to estimate the standard deviation of the life of a certain type of battery when the battery is used continuously. When designing the sample, the manufacturer decides that he would like to be 95 percent "confident" that the estimate of the standard deviation is within 10 percent of the true standard deviation. How many flashlight batteries should be tested?

For large samples, the 95 percent confidence interval for σ is given by $s \pm 1.96\,(s/\sqrt{2n})$. The percentage error is approximately

$$\frac{1.96(s/\sqrt{2n})}{s}(100) = \frac{196}{\sqrt{2n}}$$

The manufacturer wants to select n so that $196/\sqrt{2n} = 10$. Squaring both sides and solving for n, we obtain

$$n = \frac{(196)^2}{(2)(100)} = 194.08$$

Thus, a random sample of 195 batteries would be sufficient to achieve the desired level of precision.

Notice that no information about the magnitude of σ was used to estimate the sample size because the manufacturer was interested only in the percentage error rather than the absolute error of estimation. If the manufacturer wants to be 95 percent confident that the estimate of the standard deviation is within $E = 0.5$ hours of the true standard deviation, the required sample size is approximately

$$n = \frac{1}{2}\left[\frac{1.96\sigma}{0.5}\right]^2$$

and cannot be estimated without knowing σ.

Suppose that a random sample of only 25 batteries is tested and the sample variance found to be $s^2 = 9.00$ hours. Using (7.40), the 95 percent confidence interval for σ^2 is

$$\frac{(24)(9)}{39.36} < \sigma^2 < \frac{(24)(9)}{12.40}$$

or

$$5.48 < \sigma^2 < 17.42$$

The 95 percent confidence interval for the standard deviation is therefore equal to

$$2.34 < \sigma < 4.18$$

What is the 95 percent confidence interval for σ, if a random sample of 196 batteries yields a sample variance $s^2 = 9$?

EXERCISES

Section 7.1

1. Define the term "random sample" for a sample selected from:
   (a) An infinite population.
   (b) A finite population.
2. A market analyst wants to estimate the average price of eggs in Salt Lake City.
   (a) How could he select a random sample of egg prices?
   (b) What types of errors is he likely to encounter?
   (c) How could he reduce these errors?

Section 7.2

3. Explain the difference between probability sampling and judgment sampling.
4. What are advantages and disadvantages of:
   (a) Stratified sampling.
   (b) Area sampling.
   (c) Quota sampling.
5. A radio listening survey is being designed in order to determine the average number of hours that people listen to the radio each day. A random sample of 500 observations is to be selected from a community of 20,000 people. Similar surveys in other communities suggest that there are considerable differences in the listening habits of men, women, and children. Census data show that there are approximately 6000 men, 6500 women, and 7500 children (under age 15) in the community.
   (a) How should the 500 observations be allocated between these groups? Explain your answer in detail.

(b) Is there any advantage to the allocation that you suggest if there is no difference in the average number of hours that men, women, and children listen to the radio each day?

(c) What type of sample design would you use to select the sample? What are the advantages and disadvantages of the sample design that you suggest?

6. What type of sample design would you use to estimate the prices of food items included in the consumers' price index for San Francisco?

7. What type of sample survey would you use to estimate the percentage of unemployed adults in Chicago? How could you "check" your results?

Section 7.3

8. Suppose that a population is normally distributed with mean $\mu = 10$ and variance $\sigma^2 = 16$.

   (a) What is the probability that a randomly selected item will have a value larger than 12?

   (b) What is the probability that the sample mean of nine observations will be larger than 12?

9. A sample of 25 observations is drawn at random from a population with mean 25 and variance 10. What is the probability that the sample mean will be between 24 and 27? What assumption is implicit in the method used to calculate this probability?

Section 7.4

10. Define a consistent *set* of estimators.

11. The first observation in a random sample could be used as an estimator of the population mean.

    (a) Is this estimator unbiased? Explain.

    (b) Is the estimator consistent? Explain why or why not.

    (c) Is the estimator efficient? Explain why or why not.

12. It is *known* that a deck of 50 cards contains either 20 *or* 30 red cards, the remainder being black. A random sample of three cards is selected from the deck with replacement. What is the maximum likelihood estimator of the proportion of red cards in the deck?

Section 7.5

13. A random sample of 144 observations yields a sample mean $\bar{X} = 160$ and a sample variance $s^2 = 100$.

    (a) Compute a 95 percent confidence interval for the population mean.

    (b) Compute a 90 percent confidence interval for the population mean.

    (c) If the researcher wants to be 95 percent "confident" that his estimate is within plus or minus 1.2 of the true population mean, how many *additional* observations are required?

14. The standard deviation of the median of a sample of size n from a population with mean μ and variance σ^2 is $1.2533\ (\sigma/\sqrt{n})$. A sample of size 144 yields a *median* of 66 and a standard deviation equal to 18. Compute a 90 percent confidence interval for the population mean.
15. If an investigator knows that a population has a variance equal to 144, how large a sample is required if he wants to be 95 percent "confident" that:
   (a) The sample mean $\bar{X}$ will be within three units of the population mean?
   (b) The sample *median* will be within three units of the population mean?
16. Two independent random samples are taken to determine the proportion of voters that favor an increase in the gasoline tax. The first sample consists of 100 Republicans and the second of 100 Democrats. It is found that 50 of the Republicans and 60 of the Democrats favor the tax increase.
   (a) Compute a 95 percent confidence interval for the difference in the proportions of Republicans and Democrats who favor the tax increase.
   (b) On the basis of this confidence interval, could you conclude that there is no difference in the proportion of Democrats and Republicans favoring the tax increase?
17. A coin is thrown 500 times and yields 220 heads and 280 tails. Compute a 90 percent confidence interval for the probability of a head. Does this appear to be a "fair" coin?
18. The breaking strengths of 30-pound test fishing line for a sample of six spools of line were found to be 34, 33, 26, 32, 28, and 27 pounds. Compute a 95 percent confidence interval for the population variance σ^2.
19. A hybrid seed corn company plants two new lines of hybrid corn on five different farms. The yields in bushels per acre were:

Hybrid I	90	85	95	76	80
Hybrid II	84	87	90	92	90

   (a) Estimate a 90 percent confidence interval for the difference between the two mean yields.
   (b) What type of "population" is the company working with in this test?
   (c) What assumptions were made in order to estimate the confidence interval in part (a)?
20. What happens to the length of the confidence intervals for μ, for p, and for σ as:
   (a) The sample size increases?
   (b) The confidence level increases?
21. The 95 percent confidence limits for the arithmetic mean of a population are \$20 and \$30. Which of the following is correct?
   (a) Approximately 95 out of 100 sample means drawn at random from this population would fall between \$20 and \$30.
   (b) Approximately 95 out of 100 population means drawn at random from this population would fall between \$20 and \$30.
   (c) In the long run, the interval from \$20 to \$30 will include the true population mean 95 percent of the time.
   (d) Approximately 95 out of 100 "95 percent confidence intervals" will include the population mean in the long run.

22. An automobile dealer claims that, of all automobiles purchased by couples, the color is chosen by the wife in more than 80 percent of the cases. To test his claim, you take a random sample of 400 couples who have bought new automobiles within the last year and find that in 310 cases the wife chose the color. If the true proportion is 80 percent, would it be included in a 95 percent confidence interval based on your sample?

23. A statistician calculates the difference between the third quarter 1967 earnings after taxes and those of the third quarter 1966 for 100 large corporations chosen randomly and finds that the average increase in net profits for this particular sample is 18 percent and the sample standard deviation is 20 percent. You are told that one could be "reasonably certain" the true average increase in net profits for the period for all large corporations was greater than 13 percent but less than 23 percent.

(a) What confidence coefficient did this statistician employ?

(b) Rewrite the last sentence of the above paragraph in such a way that the term "reasonably certain" is replaced by a more precise expression. Explain the meaning of your revised statement in nontechnical terms as fully as possible.

SELECTED REFERENCES

For additional information on sample designs see:

Cochran, W. G., *Sampling Techniques*, John Wiley and Sons, Inc., New York, 1953.

Deming, W. G., *Some Theory of Sampling*, John Wiley and Sons, Inc., New York, 1950.

Neter, J. and W. Wasserman, *Fundamental Statistics for Business and Economics*, Allyn & Bacon, Inc., Boston, 1961.

For a more general discussion of point and interval estimation see:

Brunk, H. D., *An Introduction to Mathematical Statistics*, Ginn and Company, Boston, 1960.

Mood, A. L., *Introduction to the Theory of Statistics*, McGraw-Hill Book Company, Inc., New York, 1950.

CHAPTER 8

Testing Statistical Hypotheses

The tools of statistical inference discussed so far have been developed over several decades as scientists in many fields of endeavor have continually searched for better methods with which to analyze and evaluate statistical data. Since the main purpose of many scientific experiments is to draw conclusions about the behavior of the world around us, it is not surprising that scientists have developed ways to use these statistical tools for testing hypotheses. The techniques of hypothesis testing are highly developed and widely accepted at the present time. There is the important advantage that they allow us to evaluate the risks associated with different decision rules.

The businessman may find the techniques of testing hypotheses useful when setting up quality control programs or in evaluating the effectiveness of advertising campaigns. The economist can use the techniques when deciding whether or not a price elasticity is significantly different from zero or if the interest rate has a significant effect on credit purchases. The engineer may use them to determine if a change in a production process results in a significant improvement in a product. For example, does a certain change in a sheet steel rolling process make the steel easier to bend?

In recent years, efforts have been made to develop the techniques of hypothesis testing into a more general theory of decision making under uncertainty. From the businessman's viewpoint, such a theory should take into account the cash values associated with different decisions and the fact that most decision-making situations involve a *set* of decisions. Such a large order is not easily filled, but progress is being made. Most of the discussion that follows is concerned with the simpler task of how to test hypotheses about parameters of frequency distributions.

8.1 INTRODUCTION TO TESTING HYPOTHESES

In the simplest case, a *statistical hypothesis* is a statement about the frequency distribution of a random variable. A *test* of a statistical hypothesis is a procedure by which a decision is made to "accept" or "reject" the hypothesis under consideration. Obviously, a hypothesis has to be specified carefully before it can be tested. For example, the hypothesis that "the price of bread in Chicago is high" cannot be tested unless the meaning of the word "high" is made clear. The hypothesis that "the average price of bread in Chicago is $0.35 for a one pound loaf" is testable. A random sample of bread prices could be obtained and rules formulated for deciding whether to accept or reject the hypothesis on the basis of the sample outcome.

8.1.1 Types of Errors

There are two types of mistakes possible in testing statistical hypotheses. We may reject a hypothesis when it is, in fact, true or we may accept a hypothesis when it is, in fact, false. The rejection of a true hypothesis is called a *Type I error*. The acceptance of a false hypothesis is referred to as a *Type II error*. Let H denote the hypothesis being tested. The testing situation is described schematically in the following table.

Decision	H is True	H is False
Accept H	Correct Decision	Type II Error
Reject H	Type I Error	Correct Decision

The decision maker usually wants to keep the probabilities of making both types of errors as small as possible. Unfortunately, this is a difficult task because, for a given sample size, the probability of making a Type II error increases as the probability of making a Type I error is reduced. Given the decision rule, however, it is generally possible to reduce both types of errors simultaneously by increasing the size of the sample. There is usually a cost associated with additional sample observations. Thus, the degree to which it is possible to reduce errors depends partly upon how large a sample the decision maker can afford to take.

In addition to the "costs" associated with reducing the probabilities of errors, there are "costs" associated with making errors. Suppose, for example, that a manufacturer of television picture tubes accepts the hypothesis that the average life of a certain type of picture tube is three years when, in fact, the average life is only 18 months. If, on the basis of accepting the hypothesis, the manufacturer decides to guarantee the picture tubes for two years, it will cost him more money to replace defective tubes than he anticipates.

In some cases, a cash values can be assigned to each of the four possible "outcomes" of a test. These cash values depend, among other things, on the action to be taken as a result of accepting or rejecting the hypothesis, H_0, being tested and on the alternative hypothesis accepted if H_0 is rejected. In the case of the television tube manufacturer, if the alternative hypothesis is that tubes last four years and the action to be taken, if H_0 is rejected, is to guarantee the tubes for three years, the cash values associated with the four possible "outcomes" will be quite different from the set that would be used if the alternative hypothesis is that picture tubes last only one year and the action to be taken, if H_0 is rejected, is to guarantee the tubes for three months.

In many cases, it is possible to conclude that one type of error is more serious than the other even though cash value cannot be assigned to the different "outcomes." For example, a pharmaceutical company trying to decide whether or not to release a new drug is likely to consider a decision to release a toxic drug to be more serious than a decision *not* to release a helpful drug. The company may have no way to place a cash value on such decisions, but the ability to rank its actions in terms of "seriousness" can be helpful when deciding on the hypothesis to be tested and the testing procedure to be used.

To illustrate these ideas, suppose that a wholesaler wants to decide whether or not to accept a large shipment of power saws. He is willing to accept the saws if at least 80 percent of them meet certain specifications. The wholesaler decides to take a random sample of ten saws, test each saw in the sample, and accept the shipment if seven or more saws meet his specifications. The testing procedure can be thought of as a binomial experiment with $n = 10$ independent trials in which a saw either meets or fails to meet the wholesaler's specifications. The probability, p, that a randomly selected saw meets the wholesaler's specifications is equal to the proportion of acceptable saws in the shipment. This probability is nearly constant on each trial provided the sample is small relative to the total number of saws in the shipment. A reasonable hypothesis to test in this case is that the proportion of acceptable saws is 0.8. The hypothesis to be tested is commonly denoted by H_0. Thus, we have

$$H_0: p = 0.8$$

If the wholesaler's decision rule is to accept H_0 when the number of acceptable saws X is seven or more, what is the probability that he will reject H_0 when, in fact, H_0 is true? In other words, what is the probability of a Type I error associated with this decision rule? The answer is very simple. The probability of a Type I error in this case is equal to the probability that X is less than seven *given that* $p = 0.8$. Using the table of the binomial distribution (Table V, Appendix I), we find that the probability of a Type I error is

$$p(X < 7 \mid p = 0.8) = 0.1209 = p(X > 3 \mid p = 0.2)$$

In other words, if 80 percent of the saws are acceptable, the probability is 0.1209 that less than seven of the saws in a random sample of ten will meet the specifications.

The wholesaler could reduce the probability of a Type I error by using the decision rule:

$$\begin{array}{ll} \text{Accept } H_0 & \text{if} \quad X \geq 6 \\ \text{Reject } H_0 & \text{if} \quad X < 6 \end{array}$$

The probability of a Type I error for this decision rule is

$$P(X < 6 \mid p = 0.8) = 0.0328 = p(X > 4 \mid p = 0.2)$$

Thus, if the wholesaler wants to keep the probability of a Type I error below a certain level, say, below 0.05, he would prefer the decision rule "reject H_0 if $X < 6$" to the rule "reject H_0 if $X < 7$."

In order to calculate the Type II error, the wholesaler has to specify exactly the way in which H_0 is false. The true proportion could be $p = 0.9$, $p = 0.5$, or $p = 0.3$. The Type II error is very high if $p = 0.9$ but is not very serious because the wholesaler would be delighted if 90 percent of the saws were acceptable. If $p = 0.3$, the probability of a Type II error is very low, but such an error would be considered rather serious. In this example, a Type II error is "serious" only if the true value of p is below 0.8. The further the true value of p is below 0.8, the more serious a Type II error becomes. Fortunately, this is partially offset by the fact that, for values of p below 0.8, the probability of a Type II error for a given decision rule decreases as the difference between the true value of p and 0.8 increases.

Suppose that the wholesaler uses the decision rule "reject H_0 if $X < 7$" and decides that the alternative hypothesis to be accepted if H_0 is rejected is

$$H_a\colon p = 0.6$$

What is the probability of accepting H_0 when, in fact, the alternative H_a is true? In other words, what is the probability of a Type II error if $p = 0.6$? Using the table of the binomial distribution, we find the probability of a Type II error to be

$$P(X \geq 7 \mid p = 0.6) = 0.3822 = p(X \leq 3 \mid p = 0.4)$$

The Type II error for the decision rule "reject H_0 if $X < 6$," on the other hand, is

$$P(X \geq 6 \mid p = 0.6) = 0.6330 = p(X \leq 4 \mid p = 0.4)$$

These results are summarized in the following table:

	Probability of	
Decision Rule: Reject H_0 if	Type I Error	Type II Error
$X < 7$	0.1209	0.3822
$X < 6$	0.0328	0.6330

Given the sample size, the wholesaler is able to reduce the probability of a Type I error by changing his decision rule, but the probability of a Type II error is increased in the process. The only way in which he can decrease both types of errors simultaneously is to take a larger sample and use a new decision rule. For example, if the sample size is increased to 17 saws and the decision rule changed to "reject H_0 if $X < 12$," the probability of a Type I error is

$$P(X < 12 \mid p = 0.8) = 0.1057 = p(X > 5 \mid p = 0.2)$$

and the probability of a Type II error, given $p = 0.6$, is

$$P(X \geq 12 \mid p = 0.6) = 0.2639 = p(X \leq 5 \mid p = 0.4)$$

This is an improvement over the decision rule "reject H_0 if $X < 7$" based on a sample of ten saws because the probabilities of both types of errors have been reduced. Whether or not it is worthwhile to use a larger sample would depend on the cost of testing the extra five saws and the costs associated with making a wrong decision.

8.1.2 The Choice of a Test

The standard procedure when testing a statistical hypothesis is to begin by setting up a hypothesis to be tested H_0 and an alternative hypothesis H_a which is to be accepted if H_0 is rejected. The way in which H_0 and H_a are stated depends on the parameters of interest and the type of information which the investigator has or is planning to obtain. Suppose, for example, than an economist plans to estimate a linear demand function

$$Q = c + dP$$

where Q is quantity demanded, P is the price of the items, and c and d are constants that determine the location of the function. The economist may be concerned primarily about the value of d. On the basis of economic theory, he would expect the quantity demanded to decrease as the price of the product increases and, therefore, that d is less than zero. The hypothesis tested in this case would involve only d, and the alternative hypothesis used would incorporate the information that d is probably less than zero. Thus, the economist may decide to test $H_0\colon d = 0$ against the alternative hypothesis $H_a\colon d < 0$.

In most cases (but not all), a person testing a statistical hypothesis considers one of the possible errors discussed in the previous section to be more "serious" than the other. The hypothesis tested is then set up in such a way that the more serious error is to reject the hypothesis when it is true. The standard procedure is to select arbitrarily a small probability α called the level of significance and to require that the probability of committing a Type I error does not exceed α. The most commonly used values of α are 0.10, 0.05, and 0.01.

The *critical region* of a test (or the *region of rejection*) is the set of sample points which, if observed, would lead to the rejection of the hypothesis. For example, if the hypothesis

$$H_0\colon X \quad \text{is a binomial variable with} \quad p = 0.8 \quad \text{and} \quad n = 10$$

is rejected when X is less than 6, the critical region of the test consists of the set of sample points: 0, 1, 2, 3, 4, and 5.

Generally, a large number of critical regions satisfy the condition that the probability of committing a Type I error does not exceed α. The investigator usually wants to select that critical region, corresponding to the chosen level of significance, which has the smallest probability of a Type II error. In order to determine the probability of accepting the hypothesis tested when it is false, it is necessary to specify the way in which the hypothesis is false. It is therefore impossible to select the "best" critical region unless the form of the alternative hypothesis is known.

To illustrate these ideas, suppose that a firm which sells grass seed wants the percentage P of noxious weed seed in a certain brand of grass seed to be less than 1 percent. It considers the rejection of the hypothesis $H_0 : P = 1$ percent to be rather serious if H_0 is, in fact, true. Suppose that the firm's decision to accept or reject H_0 is to be based on the percentage $\hat{P}$ of noxious weed seed in a randomly selected sample of grass seed. If the alternative hypothesis is $H_a\colon P < 1$ percent, the firm would want to reject H_0 and accept H_a if the sample percentage is significantly less than 1 percent. The appropriate critical region in this case would consist of values of $\hat{P}$ which are small relative to 1 percent. The dividing line between "large" and "small" values of P depends on the level of significance of the test.

If, on the other hand, the alternative hypothesis is $H_a: P > 1$ percent, then values of $\hat{P}$ less than 1 percent would lead the firm to believe that the probability that H_0 is true is greater than the probability that H_a is true. Therefore, only values of $\hat{P}$ greater than 1 would lead to the rejection of H_0 in favor of the alternative that P is greater than 1 percent. The appropriate critical region in this case would consist of "large" values of $\hat{P}$.

If the alternative hypothesis is $H_a: P \neq 1$ percent, then the critical region would consist of some values of $\hat{P}$ less than 1 percent and other values of $\hat{P}$ which were greater than 1 percent.

The alternative hypothesis can take three different forms:

1. $H_a: P < 1$ percent.
2. $H_a: P > 1$ percent.
3. $H_a: P \neq 1$ percent.

Which alternative hypothesis should the firm use?

The answer is that it should use the hypothesis which it wants to accept if H_0 is rejected. In this case, the firm would probably use the alternative $H_a: P > 1$ percent and reject H_0 if the sample percentage $\hat{P}$ was significantly greater than 1 percent. The following numerical example should help to clarify these concepts.

Let us suppose that we have a friend named Tom who believes he can read minds—at least 90 percent of the time. We decide to develop a test of his ability. A number is to be selected at random from a table of random numbers, we are to concentrate on the number selected, and Tom is to tell us if the number is odd or even. This test is to be carried out eight times before we decide if Tom actually has the ability to read minds or not. Let p be the proportion of times that Tom is able to correctly identify the number as odd or even.

If Tom's claim is true, then $p = 0.9$. If he is merely guessing, then it seems reasonable to assume that he should be correct about 50 percent of the time, in which case p is equal to 0.5, so we decide to test the hypothesis $H_0: p = 0.5$ against the alternative $H_a: p > 0.5$.

The testing procedure can be viewed as a binomial experiment with $n = 8$ and $p = 0.5$ if H_0 is true. Let X be the number of times Tom correctly identifies a number. The probability distribution of X is presented in Table 8.1. Our decision to accept or reject H_0 will be based on the observed value of X. Suppose that we decide that we would like the probability of a Type I error to be less than 0.15. One possible critical region satisfying this condition consists of the values of X greater than 5. In other words, the probability that X is equal to 6 or 7 or 8, if H_0 is true, is

$$P(X > 5 \mid p = 0.5) = p(6) + p(7) + p(8) = 0.1445$$

The probability of accepting H_0, if the true value of p is 0.9, is equal to

$$P(X \leq 5 \mid p = 0.9) = 0.0381 = p(X \geq 3 \mid p = 0.1)$$

Therefore, if Tom can actually read minds "90 percent of the time," then it is unlikely that H_0 will be accepted when the critical region consists of values of X greater than 5.

If Tom is only able to read minds "70 percent of the time," the probability of accepting H_0 when it is false is

$$P(X \leq 5 \mid p = 0.7) = 0.4482$$

Table 8.1 The Binomial Distribution for $n = 8$

X	$p(x)$ if $p = 0.5$	$p(x)$ if $p = 0.9$	$p(x)$ if $p = 0.7$	$p(x)$ if $p = 0.2$
0	0.0039	0.0000	0.0001	0.1678
1	0.0312	0.0000	0.0012	0.3355
2	0.1094	0.0000	0.0100	0.2936
3	0.2188	0.0004	0.0467	0.1468
4	0.2734	0.0046	0.1361	0.0459
5	0.2188	0.0331	0.2541	0.0092
6	0.1094	0.1488	0.2965	0.0011
7	0.0312	0.3826	0.1977	0.0001
8	0.0039	0.4305	0.0576	0.0000

Another possible critical region consists of the values 0, 1, and 2. The probability of rejecting H_0 when it is true in this case is

$$P(X < 3 \mid p = 0.5) = p(0) + p(1) + p(2) = 0.1445$$

Thus, the probability of a Type I error for the new test is the same as the probability of a Type I error for the old test. The new test is not a useful one, however, because the probability is high of accepting H_0 when H_0 is false if p is actually greater than 0.5. For example, if $p = 0.7$, then

$$P(X \geq 3 \mid p = 0.7) = 0.887$$

In other words, using the new test, we are likely to conclude that Tom has no ability to read minds when, in fact, he does have some ability. The original test based on a critical region consisting of the values 6, 7, and 8 is said to be a more "powerful" test.

The *power of the test* is equal to the probability β of rejecting the hypothesis tested when it is false. Thus

$$\beta = \text{power of the test} = 1 - (\text{probability of a Type II error})$$

Obviously, the smaller the probability of a Type II error, the greater the power of the test. The investigator, therefore, wants to select the critical region which will give the most powerful test for a given level of significance.

Suppose that the hypothesis being tested is $H_0: \theta = \theta_0$ where θ represents a parameter of a distribution of a random variable. The *power function* of a test of H_0 is a function of θ, say, $\beta(\theta)$, that gives the power of the test for all values that θ could possibly take on. A typical power function for a test of H_0 against the alternative $H_a: \theta \neq \theta_0$ is shown in Figure 8.1. At the point $\theta = \theta_0$,

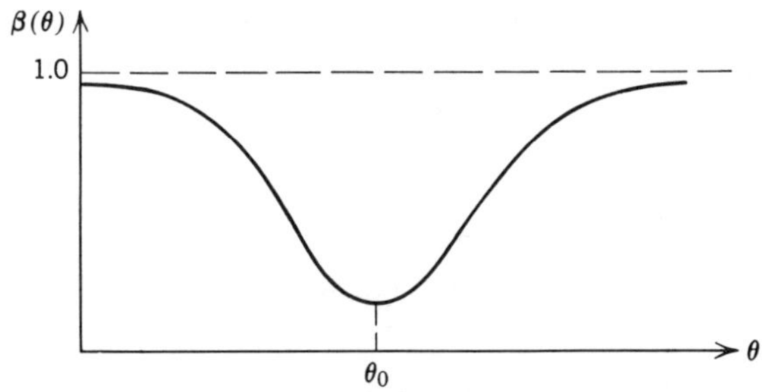

Figure 8.1 The power function of a test.

the power of the test is equal to the probability of a Type I error. If the true value of θ is near θ_0, it may not be too serious if the test procedure fails to reject H_0. For values of θ far from θ_0, we would like the probability of rejecting H_0 to be relatively large.

Let us compute the power function for the test of Tom's ability to read minds. Our decision rule is to reject $H_0: p = 0.5$ and accept $H_a: p > 0.5$ if Tom correctly identifies a number as odd or even more than five times in eight tries. The power function is therefore equal to

$$\beta(p) = 1 - P(X \leq 5 \mid p)$$

The probabilities $P(X \leq 5 \mid p)$, for various values of p, can be obtained from the tables of the binomial distribution, Table V, Appendix I. The power of the test for several values of p are presented below:

$$\begin{aligned}
\beta(0.2) &= 1 - P(X \leq 5 \mid p = 0.2) = 1 - 0.9988 = 0.0012 \\
\beta(0.4) &= 1 - P(X \leq 5 \mid p = 0.4) = 1 - 0.9501 = 0.0499 \\
\beta(0.6) &= 1 - P(X \leq 5 \mid p = 0.6) = 1 - 0.6846 = 0.3154 \\
\beta(0.7) &= 1 - P(X \leq 5 \mid p = 0.7) = 1 - 0.4482 = 0.5518 \\
\beta(0.8) &= 1 - P(X \leq 5 \mid p = 0.8) = 1 - 0.2031 = 0.7969 \\
\beta(0.9) &= 1 - P(X \leq 5 \mid p = 0.9) = 1 - 0.0381 = 0.9619
\end{aligned}$$

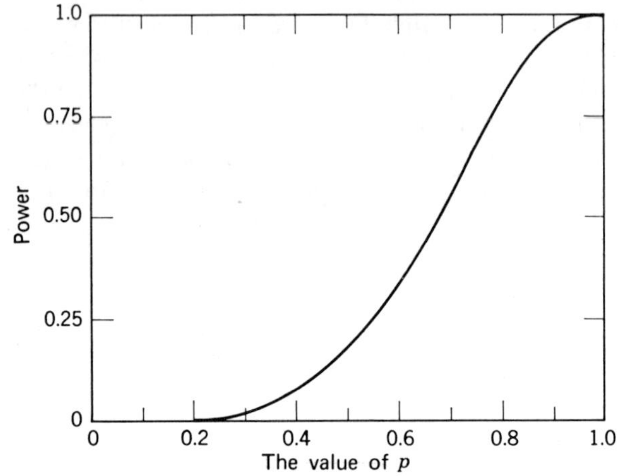

Figure 8.2 The power function of an upper-tail test.

The power function of the test is shown in Figure 8.2.

Given the number of trials n and the fact that we want $\alpha \leq 0.15$, the upper-tail critical region consisting of the values 6, 7, and 8 provides the most powerful test of $H_0: p = 0.5$ for values of p greater than 0.5. The power of the test for a given value of α could be increased by taking a larger sample. Similarly, the power of the test for a given value of n could be increased if we were to increase α.

Different tests have different power functions. Suppose that $\beta(\theta)$ represents the power function of one test and $\beta_1(\theta)$ represents the power function of any other test. If the power function $\beta(\theta)$ is always greater than or equal to the power function $\beta_1(\theta)$ of any other test, we say that the test corresponding to $\beta(\theta)$ is a *uniformly most powerful test*. For example, power function A in Figure 8.3 is the power function of a uniformly most powerful test provided there are only three possible tests of the hypothesis $H_0: \theta = \theta_0$. Unfortunately, uniformly most powerful tests for all values of θ are rare. We shall discuss the power of a test in greater detail later in this chapter.

* *8.1.3 Sequential Testing*

The sequential sampling approach to hypothesis testing is somewhat different from the standard approach described above. The main difference between the two approaches is that the sample size is fixed in advance in the

* This section may be omitted without loss of continuity.

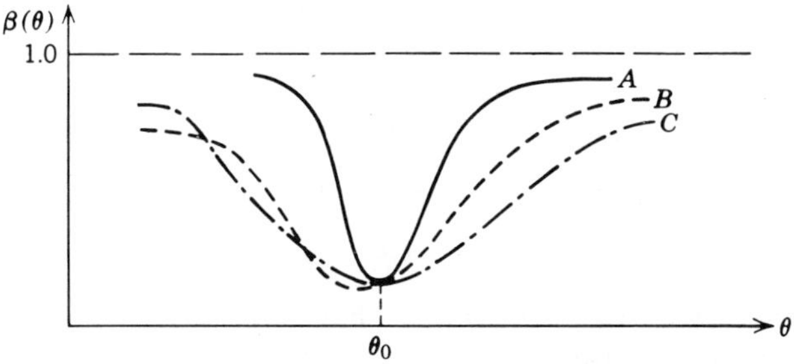

Figure 8.3 The power functions of three tests.

standard approach but depends on the particular sequence of observations in the sequential approach. The advantage of the sequential approach to hypothesis testing is that the expected sample size required to obtain given levels of Type I and Type II errors is often smaller than the required sample size when using the standard approach. For example, suppose that the Navy wants to determine if various lots of artillery shells in an ammunition dump meet Navy specifications after several years of storage. Furthermore, suppose that the Navy sets up a standard test which consists of firing 25 shells and rejecting the hypothesis that the lot is satisfactory if two of the shells do not fire. Using a sequential test, the procedure would be to fire shells until two shells did not fire or until 25 shells had been fired, whichever came first. If less than two shells had not fired by the time 25 shells had been fired, the lot would be considered satisfactory; otherwise it would be rejected as soon as two shells had not fired. Obviously, in this case the sequential test could never require a larger sample of shells than the standard test procedure.

The main features of sequential sampling can be illustrated with the help of a special deck of cards. Suppose that a deck of 50 cards is known to contain either 20 or 30 red cards, the remainder being black. You are invited to play a game that consists of selecting a card from the deck. If the card is red, you win \$5; otherwise, you lose \$5. Thus, the odds are in your favor if there are 30 red cards in the deck. You are given an opportunity to select a few cards randomly with replacement before you decide whether or not to play the game.

Sequential sampling could be used to help you make your decision. Let $y = 1$ if the card is red and $y = 0$ if the card is black. The probability of drawing a red card is known to be either

$$p = 0.4 \text{ when there are 20 red cards in the deck or}$$

$$p = 0.6 \text{ when there are 30 red cards in the deck}$$

The probability density function of y is therefore equal to

$$f_1(y) = 0.4^y(0.6)^{1-y} \quad \text{if} \quad p = 0.4$$

or

$$f_2(y) = 0.6^y(0.4)^{1-y} \quad \text{if} \quad p = 0.6$$

The joint density function of y for a sample of size n is

$$\phi_i = f_i(y_1)f_i(y_2) \cdots f_i(y_n) \qquad \text{for } i = 1, 2$$

Thus, ϕ_1 represents the probability that the sequence of sample values $y_1, y_2, \ldots, y_n$ will be observed if the probability of a red card is 0.4 and ϕ_2 represents the corresponding probability for $p = 0.6$.

The *likelihood ratio* obtained by dividing ϕ_2 by ϕ_1 is equal to

$$\lambda_n = \left(\frac{\phi_2}{\phi_1}\right)$$

If $p = 0.6$, then the probability is high that the sequence of sample values will be such that ϕ_2 is large relative to ϕ_1. If, on the other hand, $p = 0.4$, we expect to find ϕ_1 to be large relative to ϕ_2. Thus, the hypothesis that $p = 0.4$ would be rejected if λ_n is too large and accepted if λ_n is sufficiently small.

In sequential sampling, two positive numbers A and B, with A greater than B, are selected prior to sampling. After each sample item is drawn, the likelihood ratio is computed. If λ_n is greater than A, the hypothesis $p = 0.4$ is rejected; if λ_n is smaller than B, the hypothesis $p = 0.4$ is accepted; if neither, another sample item is drawn.

The numbers A and B depend on two probabilities specified by the investigator. Suppose that you want to be sure that the probability of rejecting the hypothesis $p = 0.4$ is not greater than $\alpha = 0.10$ if, in fact, $p = 0.4$ and, at the same time, you would like to assure that the probability is not more than $\beta = 0.40$ of failing to reject the hypothesis $p = 0.4$ if, in fact, $p = 0.6$. The appropriate values for A and B are approximately equal to

$$A = \frac{1-\beta}{\alpha} = 6 \qquad \text{and} \qquad B = \frac{\beta}{1-\alpha} = 0.44$$

You are now ready to start drawing cards. Suppose that the first card is red. The likelihood ratio is

$$\lambda_1 = \frac{(0.6)^1(0.4)^0}{(0.4)^1(0.6)^0} = \frac{0.6}{0.4} = 1.50$$

Since $B < \lambda_1 < A$, another card is drawn. If the second card is red, then

$$\lambda_2 = \frac{(0.6)^1(0.4)^0(0.6)^1(0.4)^0}{(0.4)^1(0.6)^0(0.4)^1(0.6)^0} = \frac{0.36}{0.16} = 2.25$$

Again, we have $B < \lambda_2 < A$ and another card is selected. If the third card is red, we have

$$\lambda_3 = \frac{(0.6)^3}{(0.4)^3} = \frac{0.216}{0.064} = 3.38$$

The sampling continues until λ_n is either greater than A or smaller than B.

Sequential sampling is useful when the investigator wants to test a hypothesis about a population using as few sample observations as possible. This would often be the case when sample observations are very expensive or when sample items are destroyed in a testing process. A random mechanism by which to select sample items is still needed to assure that the observations in the sample are independent.

8.1.4 The Value of a Test

In most hypothesis testing situations, there is one action associated with the acceptance of H_0 and another action to be taken if H_0 is rejected. In many cases, the investigator is concerned about the profits or losses associated with these actions and interested in the cost of reducing the possible errors by taking a larger sample. The businessman, for example, is likely to ask, "Which sample size and decision rule will result in the largest expected profit?" There are several ways in which this question could be answered. We will discuss one possible approach.

Consider a wholsesaler who has to decide whether or not to accept a shipment of saws. We will assume that:

1. The shipment consists of 1000 saws.
2. The profit on each good saw is \$10.
3. The loss on each bad saw is \$20.
4. The cost of testing a saw is \$50, but the saw can be sold after it is tested.

The profit on a shipment of saws is therefore equal to

$$\pi = 10G - 20B - 50n$$

where

π is profit
G is the number of good saws
B is the number of bad saws
n is the sample size

The wholesaler wants to test the hypothesis $H_0: p = 0.8$ against the alternative $H_a: p = 0.6$ where p is the proportion of good saws. The profits and loss under H_0 and H_a for a sample of size 10 are presented in Table 8.2.

Table 8.2

	$H_0: p = 0.8$ is True	$H_a: p = 0.6$ is True
Profit on good saws	\$8000	\$6000
Loss on bad saws	−4000	−8000
Cost of testing ten saws	−500	−500
Profit or loss (−) if shipment is accepted	\$3500	−\$2500
Loss (sampling costs) if shipment is rejected	−500	−500

The probabilities associated with the different outcomes of the hypothesis testing situation for the decision rule "reject H_0 if the number of good saws X is less than 7" when $n = 10$ are:

Decision	$H_0: p = 0.8$ is True	$H_a: p = 0.6$ is True
Accept H_0	0.8791	0.3822
Reject H_0	0.1209	0.6178

The "conditional" expected profit, given that H_0 is true, is therefore equal to

$$E(\pi \mid H_0) = 0.8791(3{,}500) - 0.1209(500) = \$3016.40$$

and given that H_a is true, we have

$$E(\pi \mid H_a) = -0.3822(2500) - 0.6178(500) = -\$1264.40$$

In order to compute "unconditional" expected profits, the wholesaler has to be able to assign personal probabilities to the possibilities that H_0 and H_a are true. These probabilities may be based on experience with past purchases of saws from the same manufacturer, on visual inspection of the conditions of the saws, or any other relevant information the wholesaler may have. We will assume that the wholesaler has very little information about the shipment of saws and considers each possibility to be equally likely. In this case, we have

$$E(\pi) = 0.5(3016.40) - 0.5(1264.40) = \$876.00$$

In other words, the "unconditional" expected profit of the decision to select a sample of size 10 and reject H_0 if X is less than 7 is \$876.00.

An alternative decision rule for a sample of size 10 is to reject H_0 if $X < 6$. The probabilities associated with this decision rule are:

Decision	$H_0: p = 0.8$ is True	$H_a: p = 0.6$ is True
Accept H_0	0.9672	0.6330
Reject H_0	0.0328	0.3670

Thus, we obtain

$$E(\pi \mid H_0) = 0.9672(3500) - 0.0328(500) = \$3368.80$$
$$E(\pi \mid H_a) = -0.6330(2500) - 0.3670(500) = -\$1766.00$$
$$E(\pi) = 0.5(3368.80) - 0.5(1766.00) = \$801.40$$

The decision rule "reject H_0 if $X < 6$" reduces the probability of a Type I error, but the increase in the probability of a Type II error results in a reduction of expected profits.

The profitability of alternative testing procedures can be evaluated in the same way. Suppose that a sample of 15 saws is selected and the decision rule "reject H_0 if $X < 10$" is used. The probabilities in this case are:

Decision	$H_0: p = 0.8$ is True	$H_a: p = 0.6$ is True
Accept H_0	0.9389	0.4032
Reject H_0	0.0611	0.5968

The profits and losses used previously are now reduced by \$250, the cost of testing five additional saws. They are therefore equal to:

	$H_0: p = 0.8$	$H_a: p = 0.6$
Profit or loss (−) if shipment is accepted	\$3250	− \$2750
Loss (sampling costs) if shipment is rejected	−750	−750

Using this new set of numbers, we obtain

$$E(\pi \mid H_0) = \$3005.60$$
$$E(\pi \mid H_a) = -\$1556.40$$
$$E(\pi) = 0.5(3005.60) - 0.5(1556.40) = \$724.60$$

The expected profit under the new decision rule based on a sample of size 15 is less than the expected profit of either decision rule based on a sample of size 10. In this particular case, the expected value of the additional information is less than the cost of taking the additional observations. This suggests that the best procedure may be to accept the shipment *without* taking a sample. If the shipment is always accepted without sampling, the probability of a Type I error is zero, the probability of a Type II error is 1.00, and the expected profit when the wholesaler considers H_0 and H_a equally likely is

$$E(\pi) = 0.5(4000) - 0.5(2000) = \$1000$$

The decision to accept the shipment without taking a sample has the highest expected profit of all decision rules considered so far.

Suppose, however, that the personal probabilities of the wholesaler were

$$\text{probability that } H_0 \text{ is true} = p_0 = 0.3$$

and

$$\text{probability that } H_a \text{ is true} = p_a = 0.7$$

The expected profits for several different sample sizes and decision rules, using these new personal probabilities, are listed in Table 8.3.

Table 8.3

Test Number	Sample Size n	Reject H_0 if X is Less Than	Expected Profit $E(\pi)$
1	10	6	−\$225.56
2	10	7	+ 19.84
3	10	8	+ 79.28 "best" test
4	10	9	− 113.86
5	15	10	− 187.80
6	0	Always reject H_0	0.0
7	0	Always accept H_0	− 200.00

The most "profitable" test of the seven considered is to take a sample of size 10 and reject the lot if less than eight saws are acceptable. The expected profit in this case is \$78.56. There are many other tests that the wholesaler could use and some of them could have an even larger expected value.

The procedure discussed in this subsection would not be acceptable to everyone because the use of personal probabilities is required. Even if the procedure is acceptable, it may be infeasible for more realistic problems because of the tremendous amount of computation required. Such procedures can often be simplified or computerized and are likely to become more acceptable in the future.

8.2 TESTS CONCERNING MEANS

There are many situations in which the decision to take (or not to take) a certain action depends primarily on the average value of a random variable. For example, a company may not be willing to guarantee its vacuum cleaners for 90 days unless it is confident that on the average, its vacuum cleaners work for at least six months. Similarly, a purchasing agent may be unwilling to accept a shipment of cardboard boxes unless he is confident that the average bursting strength is at least 175 pounds per square inch. Or, a manufacturer may be interested in knowing if a new type of machine for cutting drills will reduce his costs by at least five cents per drill before he decides to purchase the machine. In each of these cases, the decision maker could base his decision on the results of a test of a hypothesis about a mean. In this section, we shall consider tests of hypotheses involving a single mean and the difference between two means.

8.2.1 Tests Involving a Single Mean

Let us begin with the case in which an investigator wants to test the hypothesis

$$H_0: \mu = \mu_0$$

against the alternative

$$H_a: \mu > \mu_0$$

where μ_0 is a given value of the population mean μ. It seems reasonable to reject H_0 in favor of H_a if the sample mean $\overline{X}$ is "too large" relative to μ_0. A critical region would be obtained, therefore, by selecting a value of $\overline{X}$, say, $\overline{X}_\alpha$, such that the probability is equal to α that the sample mean is greater than $\overline{X}_\alpha$ if, in fact, H_0 is true. In other words, select $\overline{X}_\alpha$ such that

$$P(\overline{X} > \overline{X}_\alpha \mid H_0) = \alpha$$

where α is the level of significance of the test. The location of $\overline{X}_\alpha$ is shown in Figure 8.4. This type of test is called an *upper-tail test* of H_0 because the

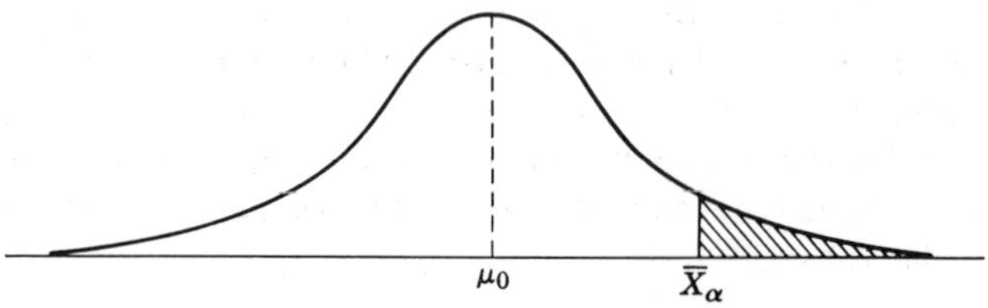

Figure 8.4 The region of rejection for H_0.

critical region consists only of the right-hand (or upper) tail of the distribution of $\bar{X}$.

If $\bar{X}$ has a normal distribution, the table of the standard normal distribution can be used to find the critical value of $\bar{X}$. Suppose, for example, that $\alpha = 0.05$. We know that the probability that the standard normal variable z is greater than 1.64 is 0.05 and can, therefore, use the standard normal transformation

$$1.64 = \frac{\bar{X}_\alpha - \mu_0}{\sigma/\sqrt{n}}$$

to find the critical value $\bar{X}_\alpha$. Doing this, we obtain

$$\bar{X}_\alpha = \mu_0 + 1.64\left(\frac{\sigma}{\sqrt{n}}\right)$$

The hypothesis H_0 is rejected in favour of H_a: $\mu > \mu_0$ if the sample mean $\bar{X}$ is greater than $\mu_0 + 1.64(\sigma/\sqrt{n})$.

The alternative is to compute

$$z = \frac{\bar{X} - \mu_0}{\sigma/\sqrt{n}}$$

after $\bar{X}$ is known and to compare this number with 1.64. The hypothesis would be rejected at the 5 percent level of significance if z was greater than 1.64 and accepted otherwise.

To illustrate the procedure for testing a hypothesis involving a single mean, suppose a truck drivers' union believes that the average annual income of its members last year was over \$7000 and wants to check this assumption. It selects a random sample of 256 truck drivers and asks them how much they earned last year. An analysis of the data shows that the average income was \$7100 while the sample standard deviation was \$1600. The hypothesis tested is

$$H_0: \mu = \$7000$$

If H_0 is rejected, the alternative

$$H_a: \mu > \$7000$$

is to be accepted. Suppose that the union decides to test H_0 at the 5 percent level of significance and to assume that $\sigma = s = \$1600$. The critical value of $\overline{X}$ is equal, therefore, to

$$\overline{X}_\alpha = \mu_0 + 1.64\left(\frac{\sigma}{\sqrt{n}}\right) = \$7000 + 1.64\left(\frac{\$1600}{\sqrt{256}}\right) = \$7164$$

Since the observed value of \$7100 is less than the critical value \$7164, the hypothesis H_0 is accepted. In other words, even though the sample mean $\overline{X}$ is larger than $\mu = \$7000$, the difference between $\overline{X}$ and μ is not large enough to allow the union to conclude that the average annual income was "significantly" larger than \$7000.

The alternative approach is to compute

$$z = \frac{\overline{X} - \mu_0}{\sigma/\sqrt{n}} = \frac{7100 - 7000}{1600/\sqrt{256}} = 1.00$$

Since the observed z of 1.00 is less than the critical value 1.64, the hypothesis H_0 is accepted.

The power function for the test is obtained by computing the probability that $\overline{X}$ is greater than \$7164 using several different values of μ. Doing this, we obtain

$$P(\overline{X} > 7164 \mid \mu = 6800) = P(Z > 3.64) = 0.0000$$
$$P(\overline{X} > 7164 \mid \mu = 7100) = P(Z > 0.64) = 0.2611$$
$$P(\overline{X} > 7164 \mid \mu = 7200) = P(Z > -0.36) = 0.6406$$
$$P(\overline{X} > 7164 \mid \mu = 7300) = P(Z > -1.36) = 0.9131$$
$$P(\overline{X} > 7164 \mid \mu = 7400) = P(Z > -2.36) = 0.9909$$

The power function for the test of H_0 is presented in Figure 8.5. A clearer idea of the procedure being used to compute the power of the test can be obtained by inspection of Figure 8.6. Under H_0, it is assumed that the distribution of $\overline{X}$ has a mean $\mu_0 = \$7000$. The dark shaded area to the right of $\overline{X}_\alpha$ represents the probability of a Type I error. The critical region of the test was selected so that the probability of a Type I error is equal to $\alpha = 0.05$. One possible distribution under H_a is presented in Figure 8.6. This distribution has mean μ_a. The power of the test for this alternative is equal to the probability that $\overline{X}$ is greater than $\overline{X}_\alpha$ given μ_a is the true mean. This probability is represented by the area to the right of $\overline{X}_\alpha$ under the distribution with mean μ_a. To find these probabilities, the $\overline{X}$ values are converted to z values

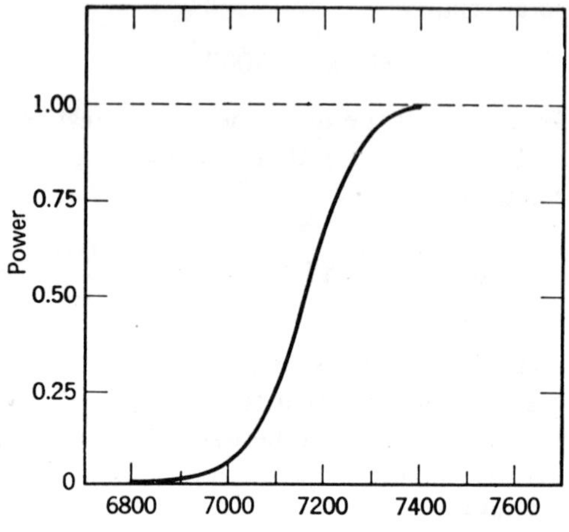

Figure 8.5 The power function for the test of H_0.

and the table of the standard normal variable is used to obtain the corresponding probabilities. (A review of the section on the normal distribution in Chapter 6 will be helpful at this point.)

The test procedure just described is the most powerful test of H_0 when the alternative hypothesis is H_a: $\mu > \mu_0$. Suppose, however, that the union wanted to test H_0 against the alternative H_a: $\mu \neq \mu_0$. In this case, it seems reasonable to reject H_0 in favor of H_a if the sample mean $\bar{X}$ is either "too large" or "too small" relative to μ_0. Therefore, the region of rejection has to be divided into two parts. The smallest possible region of acceptance of H_0 against the alternative H_a: $\mu \neq \mu_0$ is obtained when the probability α is divided equally

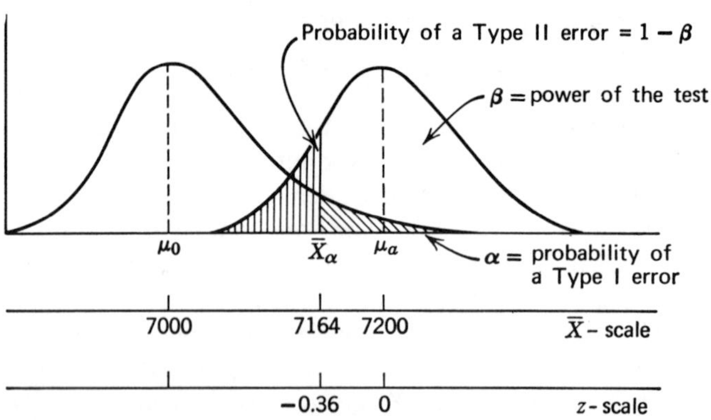

Figure 8.6 The distribution of $\bar{X}$ under H_0 and H_a.

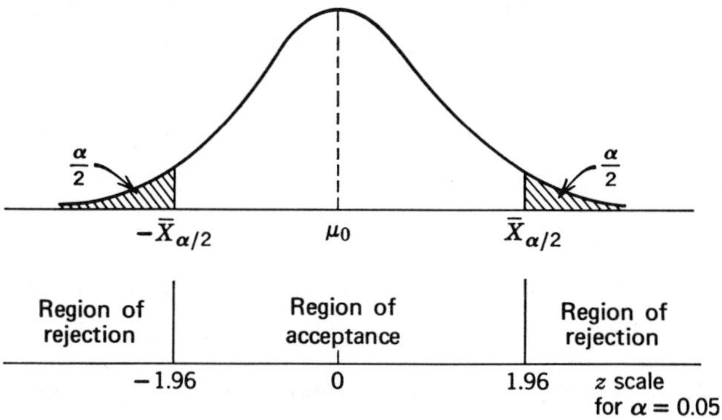

Figure 8.7 A two-tailed test of H_0.

between the tails of the distribution of $\bar{X}$. This division is shown in Figure 8.7. The critical values of $\bar{X}$ for $\alpha = 0.05$ can be obtained from the equations

$$-1.96 = \frac{\bar{X}_{\alpha/2} - \mu_0}{\sigma/\sqrt{n}} \qquad \text{(for the lower region of rejection)}$$

and

$$1.96 = \frac{\bar{X}_{\alpha/2} - \mu_0}{\sigma/\sqrt{n}} \qquad \text{(for the upper region of rejection)}$$

This type of test is called a *two-tailed test* of H_0 because the critical region is divided between the upper and lower tails of the distribution of $\bar{X}$.

The alternative approach is to reject H_0 if the observed

$$z = \frac{\bar{X} - \mu_0}{\sigma/\sqrt{n}}$$

is less than -1.96 or greater than 1.96 when the test is being made at the 5 percent level of significance.

Using the sample data on the annual incomes of truck drivers, we find the critical values of $\bar{X}$ for the two-tailed test to be

$$\text{lower value } \bar{X}_{\alpha/2} = \$7000 - 1.96\left(\frac{\$1600}{\sqrt{256}}\right) = \$6804$$

and

$$\text{upper value } \bar{X}_{\alpha/2} = \$7000 + 1.96\left(\frac{\$1600}{\sqrt{256}}\right) = \$7196$$

Since the sample mean \$7100 is greater than \$6804 but less than \$7196, the hypothesis H_0 is accepted.

The procedure for computing the power of the test is the same as before except that now there are two regions of rejection to take into account. Thus, for $\mu = \$6800$, we have

$$\begin{aligned}\beta &= P(\bar{X} < 6804 \mid \mu = 6800) + P(\bar{X} > 7196 \mid \mu = 6800)\\ &= P(Z < 0.04) + P(Z > 3.96) = 0.5160\end{aligned}$$

and for $\mu = 7100$, we obtain

$$\begin{aligned}\beta &= P(\bar{X} < 6804 \mid \mu = 7100) + P(\bar{X} > 7196 \mid \mu = 7100)\\ &= P(Z < -2.96) + P(Z > 0.96) = 0.1700\end{aligned}$$

The power function in Figure 8.8 is obtained by following the same procedure

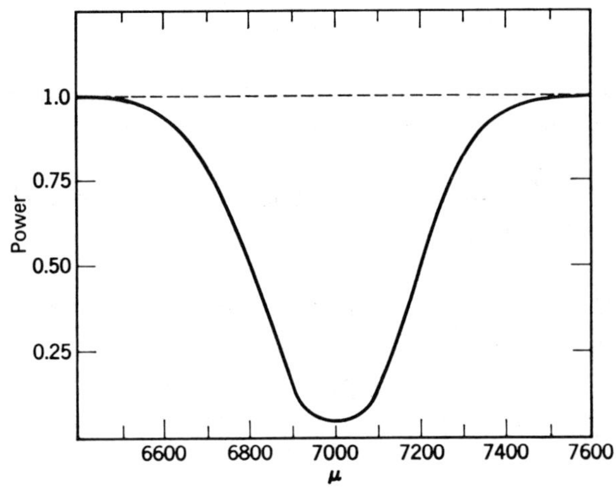

Figure 8.8 The power function of a two-tailed test of H_0.

for several other values of μ and plotting the results. The power function, in this case, is symmetrical around $\mu = \$7000$. Comparing the power function of the two-tailed test of H_0 in Figure 8.8 with the power function of the one-tailed test of H_0 in Figure 8.5, we find that the one-tailed test is more powerful for values of μ above \$7000 while the two-tailed test is considerably more powerful for values of μ below \$7000. In other words,

$$\begin{aligned}P(\bar{X} < 6804 \mid \mu > 7000) &+ P(\bar{X} > 7196 \mid \mu > 7000)\\ &\le P(\bar{X} > 7164 \mid \mu > 7000)\end{aligned}$$

As the sample size increases, $\sigma/\sqrt{n}$ decreases and the distribution of $\bar{X}$ becomes concentrated closer to the true value of μ. The level of significance of a test is usually controlled and equal to α regardless of the sample size. As a result, the size of the region of acceptance decreases as the sample size increases, and the power of a test for any given value of μ tends to increase. When designing a sample for testing a hypothesis, it is often helpful to compute the power function of the test to be used. This can be done without any sample information if σ is known. You can then determine the sample size necessary to obtain a given level of power for different values of μ.

The test procedures presented so far in this section are usually acceptable when dealing with large samples because the distribution of $\bar{X}$ is approximately normal when n is large even if the sample observations themselves are not normally distributed. When dealing with small samples from populations which are approximately normally distributed, the t distribution is used to test hypotheses about means. The test statistic for the hypothesis $H_0\colon \mu = \mu_0$ is

$$t = \frac{\bar{X} - \mu_0}{s/\sqrt{n}}$$

when the sample size n is small. For a test of H_0 against the alternative $H_a\colon \mu \neq \mu_0$, the hypothesis tested is rejected if the observed value of t is less than $-t_\alpha$ or greater than $+t_\alpha$ where t_α is obtained from Table II, Appendix I, using the α level of significance and n—1 degrees of freedom.

To illustrate the t test, suppose that a manufacturer wants to test the accuracy of a new machine which, when it is adjusted properly, automatically rolls 1760 feet of wire on to spools.

If the manufactuer has no information as to whether the machine tends to put too much or too little wire on a spool, he is likely to test the hypothesis

$$H_0\colon \mu = 1760 \text{ feet}$$

against the alternative

$$H_a\colon \mu \neq 1760 \text{ feet}$$

To carry out such a test, the manufacturer selects a random sample of 25 spools of wire wound by the machine, measures the amount of wire on each spool, and finds that the average amount is $\bar{X} = 1700$ feet while the sample standard deviation is 100 feet. The observed value of t is therefore, equal to

$$t = \frac{1700 - 1760}{100/\sqrt{25}} = \left(\frac{-60}{20}\right) = -3.00$$

The critical values of t for $\alpha = 0.05$ and 24 degrees of freedom are -2.0639 and $+2.0639$. Since the observed value of t is less than -2.0639, the hypothesis H_0 is rejected at the 5 percent level of significance in favor of the alternative $H_a\colon \mu \neq 1760$. The manufacturer would conclude that the machine was not putting enough wire on the spool and adjust it accordingly.

Suppose that the manufacturer had wanted to test the hypothesis H_0

against the alternative $H_a\colon \mu < 1760$. In this case, a one-tail test is used and H_0 is rejected if the observed value of t is less than the critical value $-t_{2\alpha}$ where $t_{2\alpha}$ is obtained from Table II using the α level of significance and $n-1$ degrees of freedom. For example, if $\alpha = 0.05$, then $-t_{2\alpha}$ with 24 degrees of freedom is equal to -1.7109.

If the sample mean had been equal to 1722 feet, rather than 1700, the observed value of t would have been

$$t = \frac{1722 - 1760}{100/\sqrt{25}} = \left(\frac{-38}{20}\right) = -1.90$$

This value of t would result in the acceptance of H_0 when tested against the alternative hypothesis $H_a\colon \mu \neq 1760$ and the rejection of H_0 when tested against $H_a\colon \mu < 1760$ provided that $\alpha = 0.05$ in both cases. This type of situation is not unusual. Therefore, to avoid the temptation to select the test to fit the results, select the hypothesis to be tested, the alternative hypothesis, and the level of significance before drawing the sample.

8.2.2 Differences Between Means

There are many practical problems in which we want to determine if there is a significant difference between the means of two distributions. To mention a few examples, a hybrid seed company may want to determine if there is a difference in yields between two varieties of seed corn, a truck line may wish to know if there is a significant difference in the expected useful life of two types of truck batteries, or a government agency may want to determine if there is an actual difference in the percentage of income spent on food by white and blue collar workers.

In their simplest form, tests of hypotheses involving two means are basically the same as tests involving only one mean. The hypothesis tested is usually written as

$$H_0\colon \mu_1 = \mu_2 \qquad \text{or} \qquad H_0\colon \mu_1 - \mu_2 = 0$$

Mathematically, the two forms are identical. The alternative hypothesis usually takes one of three possible forms:

$$(1)\quad H_a\colon \mu_1 \neq \mu_2, \qquad (2)\quad H_a\colon \mu_1 < \mu_2, \qquad (3)\quad H_a\colon \mu_1 > \mu_2$$

A two-tail test is used if H_a takes form (1) and one-tail tests are used if H_a takes form (2) or (3).

To illustrate the testing procedure, let us assume that a company wants to determine if there is a significant difference in the puncture voltage between a type of 1000 volt capacitor it now produces and one that it is planning to produce with a new process. It wants to test the hypothesis

$$H_0\colon \mu_1 - \mu_2 = 0$$

where μ_1 is the average puncture voltage of the new capacitors it may produce and μ_2 is the average puncture voltage of its standard capacitors. The company believes that μ_1 is greater than μ_2 and would be delighted if sample evidence supported this belief. It is concerned, however, about the possibility that μ_1 may be less than μ_2, in which case the company does not want to use the new process for making capacitors. The appropriate alternative hypothesis is

$$H_a: \mu_1 - \mu_2 > 0$$

A test of H_0 would be based on the difference between the mean puncture voltage of a sample of the new capacitors $\bar{X}_1$ and the mean puncture voltage of a sample of the standard capacitors $\bar{X}_2$. The most powerful test of H_0 against H_a is obtained by using a one-tail test which places the entire region of rejection in the upper tail of the distribution of $\hat{\theta} = (\bar{X}_1 - \bar{X}_2)$. The problem is to find a value of $\hat{\theta}$, say, $\hat{\theta}_\alpha$, such that the probability is equal to α that $\hat{\theta}$ will be larger than $\hat{\theta}_\alpha$ if H_0 is true. The number $\hat{\theta}_\alpha$ depends on the distribution of $(\bar{X}_1 - \bar{X}_2)$. If the sample observations on the new and standard capacitors possess independent normal distributions with means μ_1 and μ_2 and standard deviations σ_1 and σ_2, then the variable $\hat{\theta} = (\bar{X}_1 - \bar{X}_2)$ is normally distributed with mean $(\mu_1 - \mu_2)$ and standard deviation

$$\sigma_{\hat{\theta}} = \sqrt{\frac{\sigma_1^2}{n_1} + \frac{\sigma_2^2}{n_2}} \tag{8.1}$$

where n_1 and n_2 are the sample size on which $\bar{X}_1$ and $\bar{X}_2$ are based.

For large samples, the distribution of $(\bar{X}_1 - \bar{X}_2)$ is approximately normal even if the sample observations are not normally distributed. In this case, the standard normal distribution can be used to find $\hat{\theta}_\alpha$. For a one-tail test with $\alpha = 0.05$, $\hat{\theta}_\alpha$ is equal to

$$\hat{\theta}_\alpha = 1.64\, \sigma_{\hat{\theta}}$$

because, if H_0 is true, $\mu_1 - \mu_2 = 0$.

Suppose that a random sample of 100 standard capacitors and 100 new capacitors are tested and the following results are obtained.

New Capacitors	*Standard Capacitors*
$\bar{X}_1 = 1072$ volts	$\bar{X}_2 = 1060$ volts
$s_1^2 = 3200$ volts	$s_2^2 = 3200$ volts

The sample variances can be used as estimates of σ_1^2 and σ_2^2. The estimated value of $\sigma_{\hat{\theta}}$ is therefore equal to

$$s_{\hat{\theta}} = \sqrt{\frac{3200}{100} + \frac{3200}{100}} = 8.0$$

For large samples, $s_{\hat{\theta}}$ is used *as if* it were equal to the true value of $\sigma_{\hat{\theta}}$ to obtain

$$\hat{\theta}_\alpha = 1.64(8.0) = 13.12$$

The difference in the sample means is

$$\bar{X}_1 - \bar{X}_2 = 1072 - 1060 = 12$$

which is less than 13.12 and results in the acceptance of H_0 at the 5 percent level of significance. In other words, when $\alpha = 0.05$, the average puncture voltage of the sample of new capacitors is not enough greater than the average puncture voltage of the sample of standard capacitors to allow the company to conclude that μ_1 is actually greater than μ_2.

If the company had tested H_0 using $\alpha = 0.10$, the region of rejection would have consisted of values of $(\bar{X}_1 - \bar{X}_2)$ greater than

$$\hat{\theta}_\alpha = 1.28(8.0) = 10.24$$

and H_0 would have been rejected.

The testing situation is shown in Figure 8.9. Notice that, as the level of significance α becomes smaller, the critical value $\hat{\theta}_\alpha$ moves to the right and the probability of a Type II error increases while the power of the test decreases.

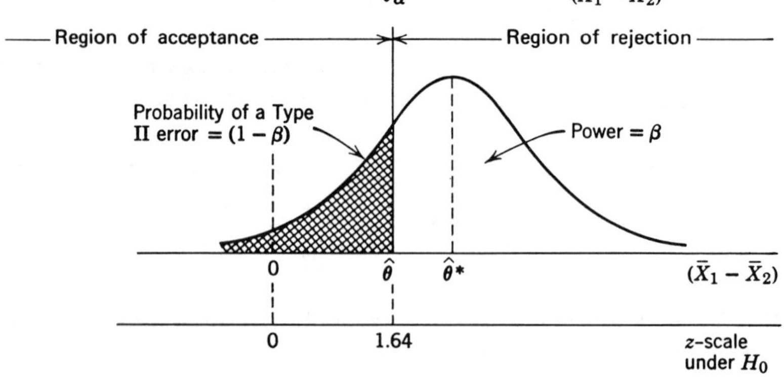

Figure 8.9 A test of $H_0 : \mu_1 - \mu_2 = 0$.

If additional or prior information is available about one of the distributions, the test of a hypothesis involving two means can sometimes be reduced to a test of a hypothesis involving only one mean. Suppose, for example, that on the basis of several thousand previous tests the capacitor company "knew" that the average rupture voltage of its standard capacitors was 1060 volts. It could then test the hypothesis

$$H_0: \mu_1 = 1060$$

against the alternative

$$H_a: \mu_1 > 1060$$

The test, in this case, is based on

$$z = \frac{\bar{X}_1 - 1060}{s_1/\sqrt{n}} = \frac{1072 - 1060}{(56.6/10)} = 2.12$$

which is greater than 1.64 and, therefore, leads to the rejection of H_0 at the 5 percent level of significance. The use of the additional information results in a more powerful test in this case because s_1 can be used instead of s_θ.

The t is used for testing the difference of two means when sample sizes are small. The t test is more restrictive because it requires that the observations used to compute $\bar{X}_1$ and $\bar{X}_2$ be drawn from independent normal distributions with *equal standard deviations*. For small samples, the hypothesis $H_0: \mu_1 = \mu_2$ can be tested by using

$$t = \frac{\bar{X}_1 - \bar{X}_2}{\sqrt{(n_1 - 1)s_1{}^2 + (n_2 - 1)s_2{}^2}} \sqrt{\frac{n_1 n_2(n_1 + n_2 - 2)}{n_1 + n_2}} \tag{8.2}$$

which has a t distribution with $v = n_1 + n_2 - 2$ degrees of freedom.

Suppose, for example, that the capacitor company had taken samples of size 11 of the standard and new capacitors in order to test $H_0: \mu_1 - \mu_2 = 0$ against the alternative $H_a: \mu_1 - \mu_2 > 0$ and had obtained the sample results presented earlier. The t value, in this case, is

$$t = \frac{1072 - 1060}{\sqrt{10(3200) + 10(3200)}} \sqrt{\frac{121(20)}{22}} = 0.497$$

The tabled value of t with 20 degrees of freedom for a one-tail test at the 5 percent level of significance is 1.7247. Since the observed value of t is less than 1.7247, the hypothesis H_0 is accepted at the 5 percent level of significance.

The fact that a hypothesis is accepted does not mean that it is true but simply that, on the basis of the available evidence, we cannot conclude that the hypothesis is false. On the other hand, the rejection of a hypothesis does not prove that it is false because there is always the possibility that we have made a Type I error. Thus, we neither prove nor disprove anything by testing hypotheses. The procedure used, however, does allow us to judge how strong the evidence is, for or against. When testing hypotheses about means, for example, large positive or negative values of t or z indicate that either the hypothesis tested is false or that an extremely unusual sample was selected. Most people accept the first alternative but, by the very way the test was constructed, the alternative hypothesis will be accepted α percent of the time in repeated tests. Keep in mind, also, that the correctness of this conclusion depends to a large extent on the quality of the sample and the correctness of the many assumptions usually made when testing hypotheses.

8.3 TESTS CONCERNING PROPORTIONS

Tests of hypotheses concerning proportions are basically the same as tests concerning means. The procedure to use when dealing with small samples is based on the binomial distribution and was discussed in Section 8.1. The normal approximation to the binomial distribution is used when sample sizes are large.

To illustrate the steps in the test procedure, suppose that a consumer opinion company wants to test the claim of a senator that 70 percent of the people in his state favor his stand on proposal A. The company decides to test the hypothesis

$$H_0 : p = 0.7$$

against the alternative

$$H_a : p < 0.7$$

with $\alpha = 0.10$. A random sample of 225 voters is chosen from the senator's home state, and it is found that 150 of them favor the senator's action.

The normal variable

$$Z = \frac{X - np}{\sqrt{np(1 - p)}}$$

is used to test hypotheses about proportions when the sample size is large. For a one-tail test with the region of rejection in the left-hand tail of the binomial distribution, the hypothesis H_0 is rejected at the 10 percent level of significance if Z is less than -1.28. If H_0 is true, then the expected value of X is $np = (225)(0.7) = 157.5$, the standard deviation is $\sqrt{np(1 - p)} = 6.87$, and we obtain

$$z = \frac{150 - 157.5}{6.87} = -1.09$$

which does not lead to the rejection of H_0.

There are many problems in which we want to determine if there is a significant difference between two sample proportions. For large samples, the normal approximation to the binomial distribution is used as a basis for such tests.

Let X_1 and X_2 represent the number of "successes" observed in two large independent samples of size n_1 and n_2, respectively. The sampling distribution of the difference between the sample proportions

$$\left(\frac{X_1}{n_1}\right) - \left(\frac{X_2}{n_2}\right)$$

is approximately a normal distribution with mean $p_1 - p_2$ and standard deviation

$$\sqrt{\frac{p_1(1 - p_1)}{n_1} + \frac{p_2(1 - p_2)}{n_2}}$$

The usual hypothesis tested is $H_0: p_1 = p_2$. Since p_1 and p_2 are not usually known, the sample data are pooled to provide an estimate of their common value p. The estimator of p is

$$\hat{p} = \frac{X_1 + X_2}{n_1 + n_2} \tag{8.3}$$

The estimator of the standard deviation of the sampling distribution is equal to

$$\hat{\sigma}_{p_1 - p_2} = \sqrt{\hat{p}(1 - \hat{p})\left(\frac{1}{n_1} + \frac{1}{n_2}\right)} \tag{8.4}$$

where $\hat{p}$ is given by (8.3). The procedure used to obtain estimates of p and the standard deviation is justified provided that H_0 is true.

The test statistic used to determine if H_0 is accepted or rejected is

$$Z = \frac{(X_1/n_1) - (X_2/n_2)}{\sqrt{\hat{p}(1 - \hat{p})(1/n_1 + 1/n_2)}} \tag{8.5}$$

The values of Z which result in the rejection of H_0 depend on the form of the alternative hypothesis and the level of significance used.

To illustrate the test procedure, suppose that an appliance manufacturer believes that men are just as likely as women to recall its colorful advertisement in a recent issue of a well-known magazine. The manufacturer decides to test the hypothesis

$$H_0: p_1 = p_2$$

against the alternative

$$H_a: p_1 \neq p_2$$

at the 10 percent level of significance, where p_1 is the proportion of men who read the magazine and recall the advertisement and p_2 is the corresponding proportion of women. The results of independent random samples of men and women are as follows.

Men	*Women*
$X_1 = 50$	$X_2 = 70$
$n_1 = 200$	$n_2 = 200$

where X_1 is the number of men who recall the advertisement and X_2 is the corresponding number of women.

Substituting the numerical values into the appropriate formulas, we find

$$\hat{p} = \frac{50 + 70}{200 + 200} = 0.30$$

$$\hat{\sigma}_{p_1 - p_2} = \sqrt{(0.3)(0.7)\left(\frac{2}{200}\right)} = 0.0458$$

and

$$z = \frac{0.25 - 0.35}{0.0458} = -2.18$$

When the hypothesis H_0 is tested against the two-tailed alternative, $H_a: p_1 \neq p_2$, H_0 is accepted at the 10 percent level of significance if $-1.64 \leq Z \leq 1.64$. Since the observed value -2.18 is not in the region of acceptance, we reject H_0 and accept H_a.

When testing H_0 against one-side alternatives, it is a good idea to draw a picture of the sampling distribution and sketch in the regions of acceptance and rejection. If this is done, you are less likely to mislabel X_1 and X_2 or to use the wrong alternative hypothesis.

Two of the most important words in the previous discussion are "independent samples." If the samples are not independent, the theory used to obtain the test for equality of two proportions does not apply. In the advertisement recall test, for example, the investigator would not want to include both husband and wife from the same family in his samples. A family considering buying an appliance may be more conscious of such advertisements in which case the husband's recall of an advertisement is not likely to be completely "independent" from that of his wife. The requirement that samples be independent is usually not difficult to satisfy when designing a sample but can be easily overlooked.

8.4 TESTS CONCERNING VARIANCES

The variation in a population sometimes is of as much concern as the average. To mention a few examples, a steel manufacturer may want the variation in the thickness of a certain type of sheet steel to be within narrow tolerance limits, the United States Treasury may need to know the extent of year-to-year variation in personal incomes in order to estimate the effect of an income-averaging proposal on tax revenues, or the manager of a supermarket may be concerned about the variation in the number of customers per hour in his store. In many such cases, an estimate of the variance or a confidence interval on the standard deviation may provide all the information needed. At other times, theory or "rules of thumb" may suggest the need for tests of hypotheses about variances. In principle, these tests are the same as those involving means or proportions.

You will recall that, if a random sample is drawn from a normal distribution, the ratio

$$\frac{(n-1)s^2}{\sigma^2}$$

is distributed as chi-square with $n - 1$ degrees of freedom. This ratio is used as the basis for testing the hypotheses

$$H_0: \sigma^2 = \sigma_0{}^2$$

where $\sigma_0{}^2$ is a given value of the population variance.

As an illustration, suppose that a feed company wants to determine if a new feeding program will reduce the variation in the weight gained by feeder cattle. From past experience, the company "knows" that the variance of weight gains of cattle fed 90 days on its standard feeding program is 1600 pounds. It decides, therefore, to test the hypothesis

$$H_0: \sigma^2 = 1600$$

against the alternative

$$H_a: \sigma^2 < 1600$$

and to make the test at the 0.025 level of significance. A random sample of 30 feeder cattle of nearly identical weights is selected, and the cattle are fed on the new feeding program for 90 days. At the end of the feeding period, the cattle are carefully weighed and the variance of the weights is computed. The company finds that the sample variance is $s^2 = 1000$ pounds. The observed value of chi-square is therefore equal to

$$\chi^2 = \frac{(29)(1000)}{1600} = 18.12$$

and is not small enough to allow the rejection of H_0 at the 2.5 percent level of significance.

The chi-square distribution drawn in Figure 8.10 will help you visualize the test situation. The region of rejection is located in the lower tail of the chi-square distribution because the company wants to be able to determine if the new feeding program reduces the variance in gains. The chi-square value $\chi^2_{0.975,29} = 16.05$ divides the regions of acceptance and rejection. If H_0 is true, the probability is 0.025 that the ratio $((n-1)s^2/\sigma^2)$ will be less than 16.05 and thereby lead to the rejection of H_0.

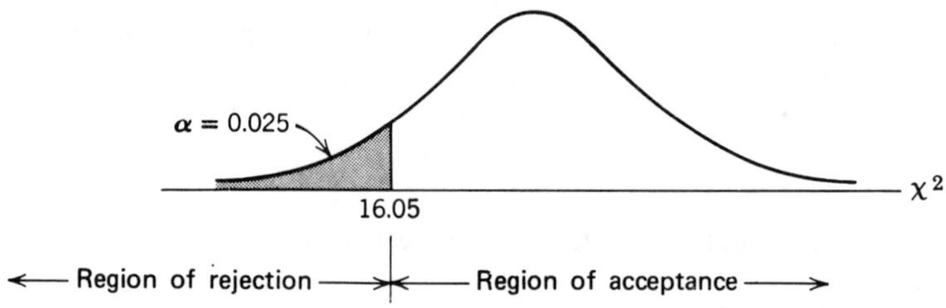

Figure 8.10 The chi-square distribution.

The feed company is likely to be pleased with the results and even more confident that the new feeding program does reduce the variation in weight gains but, according to the "rules of the game," the company does not yet have sufficient evidence to reject H_0 at the level of significance which it chose. (H_0 would be rejected at the 10 percent level, however.)

The F distribution is used when testing hypotheses about the difference of two variances. The hypothesis tested is usually of the form

$$H_0: \sigma_1^2 = \sigma_2^2$$

where σ_1^2 and σ_2^2 are the variances of two populations.

The statistic

$$F = \frac{s_l^2}{s_s^2}$$

with $(n_l - 1)$ and $(n_s - 1)$ degrees of freedom is used to test H_0. The numbers s_l^2 and s_s^2 are the sample variances obtained from *independent* samples from the two populations. If the samples are large, the sample variances should be close to the true values of σ_1^2 and σ_2^2, in which case the value of F should be near one if H_0 is true. Therefore, observed values of F near zero or substantially larger than one would lead to the rejection of H_0. Tables of the F distribution provide only the large values of F in the right-hand tail of the distribution. Thus, it is always necessary to place the larger sample variance s_l^2 in the numerator. (The subscript l on s_l^2 is changed to one or two depending on whether s_l^2 is an estimate of σ_l^2 or σ_2^2.) Given the level of significance of the test, the smaller the sample sizes the larger the observed value of F required to reject H_0 because small samples provide less precise estimates of $\sigma_1{}^2$ and $\sigma_2{}^2$.

To illustrate the F test, suppose the feed company mentioned earlier has two new feed programs that it wants to test. The company decides to test the hypothesis

$$H_0: \sigma_1{}^2 = \sigma_2{}^2$$

at the 5 percent level of significance against the alternative

$$H_a: \sigma_1{}^2 \neq \sigma_2{}^2$$

where $\sigma_1{}^2$ and $\sigma_2{}^2$ are the variances in weight gains of feeder cattle during a 90-day feeding period. A random sample of 50 feeder cattle is selected and then randomly divided into two groups with 25 cattle in each group. The cattle in group 1 are placed on feeding program 1 and those in group 2 on program 2. At the end of the 90 days, the cattle are weighted and the sample variances are computed. It is found that

$$s_1{}^2 = 900 \text{ pounds} = s_s{}^2$$

and

$$s_2{}^2 = 1400 \text{ pounds} = s_l{}^2$$

The observed value of F is therefore equal to

$$F = \left(\frac{1400}{900}\right) = 1.555$$

The tabled value of F for $\alpha = 0.05$ with 24 and 24 degrees of freedom is $F_{0.05,24,24} = 1.98$. The region of rejection is therefore equal to $F \geq 1.98$. The observed value of F is less than 1.98, and H_0 is not rejected.

There are two reasons the company may be interested in testing the hypothesis of equal variances. In some cases, it may be interested primarily in the variances themselves. In other cases, it may be interested primarily in differences in the average weight gains and simply proceed to check the assumption of equal variances before applying the usual t test. If the assumption of equal variances is rejected, a t test of the hypothesis $H_0: \mu_1 = \mu_2$ is not justified, and the company would want to consider other methods for comparing the means.

8.5 TESTS OF INDEPENDENCE AND GOODNESS OF FIT

There are times when we want to compare observed frequencies with expected frequencies and determine if the differences can be attributed to chance or whether they indicate that the expected frequencies do not "fit" the observed results. For example, suppose a die is thrown 60 times. If the die is "fair," each of the six possible outcomes is expected to occur with equal frequency. The observed frequencies are seldom exactly equal to the expected frequencies, however. Such differences may be due to chance or because the die is not "fair." A test based on the chi-square distribution is commonly used to determine if the discrepancies are small enough to be classified as "due to chance."

Some general notation needs to be introduced before the chi-square test is presented. Suppose that a sample observation can be associated with one of s mutually exclusive events $E_1, E_2, \ldots, E_s$. In the case of a die, there are six events corresponding to the number of spots on each face of the die. Furthermore, suppose that in a sample of size n, these events are *observed to occur* with frequencies $n_1, n_2, \ldots, n_s$ and *expected to occur* with frequencies $e_1, e_2, \ldots, e_s$ where $\sum_j n_j = \sum_j e_j = n$. This information is presented in the following table:

Event	E_1	E_2	...	E_s	Row Total
Observed frequency	n_1	n_2	...	n_s	n
Expected frequency	e_1	e_2	...	e_s	n

The chi-square statistic used to measure the discrepancy between the observed and expected frequencies is equal to

$$\chi^2 = \frac{(n_1 - e_1)^2}{e_1} + \frac{(n_2 - e_2)^2}{e_2} + \cdots + \frac{(n_s - e_s)^2}{e_s} = \sum_j \frac{(n_j - e_j)^2}{e_j} \tag{8.6}$$

and has $(s - 1)$ degrees of freedom provided it is not necessary to estimate any population parameters in order to obtain the expected frequencies.

If the observed frequencies are exactly equal to the expected frequencies, the observed value of chi-square will be equal to zero. The value of chi-square increases as the discrepancy between the observed and expected frequencies increases. Large values of chi-square would lead, therefore, to the rejection of the assumptions used to obtain the expected frequencies.

The die example will be used to illustrate the chi-square test. Suppose that the results of throwing the die 60 times are as shown in Table 8.4.
If the die is "fair," the probability is 0.05 that a value of chi-square larger than

$$\chi^2_{0.05,5} = 11.07$$

Table 8.4

Number of Spots	1	2	3	4	5	6	Row Total
Observed Frequency	8	6	12	14	6	14	60
Expected Frequency	10	10	10	10	10	10	60
$\dfrac{(n_j - e_j)^2}{e_j}$	0.4	1.6	0.4	1.6	1.6	1.6	7.2

will be observed. The acutal observed value is 7.2 and, therefore, the hypothesis that the die is "fair" cannot be rejected at the 5 percent level of significance.

The equation (8.6) can be simplified and written as

$$\chi^2 = \sum_j \left[\frac{n_j^2}{e_j}\right] - n \tag{8.7}$$

The use of equation (8.7) will usually speed the computation of the chi-square value.

The chi-square test presented above can be generalized easily. Suppose for example, that we want to decide whether the students in four different schools have significant differences in opinion concerning a certain action taken by the United Nations. A random sample of students at each school is selected, and the students are asked if they favor or oppose the action. The sample data are recorded in Table 8.5.

Table 8.5 Student Opinion on an Action by the United Nations

	School Number				
Opinion	1	2	3	4	Row Total
Favor the action	80	70	60	90	300
Oppose the action	20	30	40	10	100
Column total	100	100	100	100	400

If the proportion of students favoring the action is the same in each school, we could say that there are no differences in opinions about the action. The hypothesis to be tested could be written

$$H_0: p_1 = p_2 = p_3 = p_4 = p$$

where p_i is the proportion of students in favor of the action in school i and p is the common value of the p_i.

The alternative hypothesis is simply that H_0 is not true. No attempt is made to specify the alternative hypothesis exactly because there are a great number of ways in which H_0 can be false and the chi-square test does not provide any information as to how H_0 is false if H_0 is rejected. The value of the common proportion p is usually unknown. The best estimate of p, in this case, is the total number of students in favor of the action divided by the total sample size. Thus, we have $\hat{p} = (300/400) = 0.75$.

Some general notation will again be introduced before the chi-square test is presented. Suppose that a sample observation can be classified in one of s mutually exclusive ways according to criterion A and in one of r mutally exclusive ways according to criterion B. This is called a two-way classification. Furthermore, suppose that n_{ij} of the sample observations in a sample of size n are classified as being in the ith class according to criterion A and the jth class according to criterion B while e_{ij} are the expected frequencies in these classes. Tables containing the classified observed frequencies are called *r by s contingency tables*. Table 8.5, for example, is a 2 by 4 contingency table in which students are classified according to whether they favor or oppose the United Nations' action and according to their school. Table 8.6 is an r by s contingency table written in general notation.

Table 8.6 An r by s Contingency Table

	Classification A				
	A_1	A_2	...	A_s	Row Totals
B_1	n_{11} (e_{11})	n_{12} (e_{12})	...	n_{1s} (e_{1s})	$n_{1\cdot}$
B_2	n_{21} (e_{21})	n_{22} (e_{22})	...	n_{2s} (e_{2s})	$n_{2\cdot}$
...	...	...	...	...	...
B_r	n_{r1} (e_{r1})	n_{r2} (e_{r2})	...	n_{rs} (e_{rs})	$n_{r\cdot}$
	$n_{\cdot 1}$	$n_{\cdot 2}$	...	$n_{\cdot s}$	n

The first subscript (i) on the observed frequencies n_{ij} is the row number and the second subscript (j) is the column number. For example, n_{21} is in

Row 2, Column 1. A dot in place of the row subscript is a shorthand notation indicating that the number is the sum of all the numbers in the column while a dot in place of the column subscript indicates that the number is the sum of all the numbers in the row. In other words,

$$n_{.j} = \sum_i n_{ij} \qquad \text{and} \qquad n_{i.} = \sum_j n_{ij}$$

The total number of observations is equal to

$$n = n_{..} = \sum_j n_{.j} = \sum_i n_{i.} = \sum_i \sum_j n_{ij}$$

This notation is used frequently in the following discussion. If you are not familiar with the notation, pause a moment and practice using it with the numbers in Table 8.5. For example, the row total for Row 2 in Table 8.5 is

$$\sum_j n_{2j} = 20 + 30 + 40 + 10 = 100 = n_{2.}$$

and the column total for Column 3 is

$$\sum_i n_{i3} = 60 + 40 = 100 = n_{.3}$$

What are $n_{1.}$, $n_{.4}$, n_{21}, and n_{13} equal to?

The expected frequencies e_{ij} are found by multiplying the Row i total $n_{i.}$ by the Column j total $n_{.j}$ and dividing the product by the grand total n. This gives us

$$e_{ij} = \frac{n_{i.}n_{.j}}{n} \tag{8.8}$$

The expected frequencies for Table 8.5 are therefore equal to

$$e_{11} = \frac{(100)(300)}{400} = 75 = e_{12} = e_{13} = e_{14}$$

and

$$e_{21} = \frac{(100)(100)}{400} = 25 = e_{22} = e_{23} = e_{24}$$

The expected frequencies in each row are equal because an equal number of students was selected from each school. Notice that $e_{1j} = n_{.j}\hat{p}$ and $e_{2j} = n_{.j}(1 - \hat{p})$ where $\hat{p} = 300/400$.

If the hypothesis tested is true, the test statistic

$$\chi^2 = \sum_{i,j} \frac{(n_{ij} - e_{ij})^2}{e_{ij}} \tag{8.9}$$

is distributed as chi-square with $v = (r - 1)(s - 1)$ degrees of freedom. The hypothesis tested is rejected if the observed value of chi-square is greater than the tabled value of chi-square $\chi^2_{\alpha,v}$ where α is the level of significance of the test.

The data in Table 8.5 are used to illustrate the chi-square test. The observed value of chi-square is

$$\chi^2 = \frac{1}{75}(5^2 + 5^2 + 15^2 + 15^2) + \frac{1}{25}(5^2 + 5^2 + 15^2 + 15^2) = 26.66$$

The tabled value of chi-square for $\alpha = 0.05$ and $v = 3$ degrees of freedom is 7.815. The hypothesis that the proportions of students in favor of the action are the same for all schools is rejected, therefore, at the 5 percent level of significance.

When an r by s contingency table has more than two rows, the hypothesis tested usually takes the form

$$H_0\colon p_{ij} = p_{i\cdot}p_{\cdot j}$$

and the alternative hypothesis is

$$H_a\colon p_{ij} \neq p_{i\cdot}p_{\cdot j}$$

where the p's represent probabilities. The hypothesis H_0 states that the probability that a sample observation is located in the i,jth cell is equal to the probability that an observation is located in Row i multiplied by the probability that an observation is located in Column j. In other words, the classification of a sample observation according to criterion A is *independent* of its classification according to criterion B, if H_0 is true.

To test H_0, the chi-square value (8.9) is computed and then compared with the tabled value of chi-square with $v = (r - 1)(s - 1)$ degrees of freedom. If the observed value of chi-square is larger than the tabled value of chi-square, H_0 is rejected.

As an illustration, suppose an automobile company wants to determine if the type of car that a buyer purchases is independent of where he lives. The company decides to use three classes of cars and to classify buyers according to whether they live in the country, a small town, or a large city. The results of a random sample of 400 people are presented in Table 8.7.

The hypothesis tested is $H_0\colon p_{ij} = p_{i\cdot}p_{\cdot j}$. The same hypothesis, written in terms of the number of observations expected in the i,jth cell, is $H_0\colon np_{ij} = np_{i\cdot}p_{\cdot j}$ where n is the sample size. The probabilities $p_{i\cdot}$ and $p_{\cdot j}$ are usually unknown but can be estimated from the sample data. The estimates are

$$\hat{p}_{i\cdot} = \left(\frac{n_{i\cdot}}{n}\right) \quad \text{and} \quad \hat{p}_{\cdot j} = \left(\frac{n_{\cdot j}}{n}\right)$$

Table 8.7 Type of Car Purchased

Location of Buyer's Home	Type of Car			
	Compact Model	Standard Model	Deluxe Model	Row Total
Country	40 (40)	50 (50)	10 (10)	100
Small town	50 (60)	90 (75)	10 (15)	150
Large city	70 (60)	60 (75)	20 (15)	150
Column total	160	200	40	400

If H_0 is true, the expected frequencies are equal to

$$e_{ij} = n\hat{p}_{i\cdot}\hat{p}_{\cdot j} = \frac{(n_{i\cdot})(n_{\cdot j})}{n}$$

The expected frequencies are given in parentheses in Table 8.7. For example,

$$e_{22} = \frac{(n_{2\cdot})(n_{\cdot 2})}{n} = \frac{(150)(200)}{400} = 75$$

Notice that the sum of expected frequencies in any row is equal to the row total and the sum of expected frequencies in any column is equal to the column total. Therefore, it is only necessary to compute $(r - 1)(s - 1)$ expected frequencies and the remaining e_{ij} can be obtained by subtraction from the totals of the appropriate rows of columns.

Using the data in Table 8.7 and formula (8.9), we obtain $\chi^2 = 12.66$. The tabled value of chi-square for $\alpha = 0.05$ and $v = 4$ degrees of freedom is 9.4877. Since the observed value is greater than the tabled value, we reject H_0 at the 5 percent level of significance. The chi-square test does not tell us exactly how H_0 is false, but inspection of Table 8.7 and the individual terms of formula (8.9) presented in Table 8.8 provides information as to which differences were the most important in leading to the rejection of H_0. We find that buyers living in small towns buy substantially more standard model cars than expected while buyers in large cities buy fewer.

The chi-square test procedure described here is an approximate test which works best when the sample size is fairly large. When dealing with small samples or large contingency tables, there are several "rules of thumb" that should be observed. The first rule is to pool the data under different classifications if the expected frequency in any cell is less than five. The data should

Table 8.8 The $(n_{ij} - e_{ij})^2/e_{ij}$ Values

Location of Buyer's Home	Type of Car		
	Compact Model	Standard Model	Deluxe Model
Country	0	0	0
Small town	1.66	3.00	1.66
Large city	1.66	3.00	1.66

be pooled in such a way that the evidence against the hypothesis shows up the best unless the investigator believes that an alternative pooling arrangement is justified.

To illustrate a pooling operation, suppose the automobile company had initially used four classifications for cars and obtained the results presented in Table 8.9. The expected values in the station wagon column are all less than five. Therefore, the station wagon category should be combined with one of the other classes of cars. In this case, it seems reasonable to combine the deluxe model and station wagon categories, since both types of cars would be in the "high price" class. The result would be a table identical to Table 8.7 except that the third category would be labeled "Deluxe Model and Station Wagons." Combining the original two categories into one reduces the degrees of freedom from six to four.

Table 8.9 Type of Car Purchases

Location of Buyer's Home	Compact Model	Standard Model	Deluxe Model	Station Wagons	Row Total
Country	40	50	6	4	100
Small town	50	90	7	3	150
Large city	70	60	17	3	150
Column total	160	200	30	10	400

The second "rule of thumb" applies primarily to 2 by 2 contingency tables. It states that, when the sample size n is less than 40, a more accurate chi-square test is obtained by subtracting 0.5 from the absolute differences between the observed and expected frequencies before squaring the result and dividing by e_{ij}. The corrected test statistic is equal to

$$\chi_c^2 = \sum_{i,j} \frac{(|n_{ij} - e_{ij}| - 0.5)^2}{e_{ij}}$$

where $|n_{ij} - e_{ij}|$ indicates that the differences are considered to be positive numbers (absolute values) from which the correction factor 0.5 is subtracted. The corrected χ^2 will always be smaller than the uncorrected χ^2.

To illustrate the use of the corrected χ^2, suppose a consumer research company wants to determine if there is a difference between two classes of consumers insofar as their preference for two brands of the same product is concerned. A sample of 30 consumers yields the results shown in Table 8.10.

Table 8.10

	Preference		
Consumer Class	Brand A	Brand B	Row Total
Class I	9 (6.5)	6 (8.5)	15
Class II	4 (6.5)	11 (8.5)	15
Column total	13	17	30

The corrected χ^2 value is

$$\chi_c^2 = \frac{2^2}{6.5} + \frac{2^2}{6.5} + \frac{2^2}{8.5} + \frac{2^2}{8.5} = 2.17$$

The tabled value of χ^2 for $\alpha = 0.05$ and $v = 1$ degree of freedom is 3.8415. Thus, the consumer research company could not reject the hypothesis that brand preference is independent of consumer class at the 5 percent level of significance.

The uncorrected χ^2 value for a 2 by 2 table is quickly obtained from the formula

$$\chi^2 = \frac{n(n_{11}n_{22} - n_{12}n_{21})^2}{n_{.1}n_{.2}n_{1.}n_{2.}} \tag{8.10}$$

Using (8.10), we obtain an uncorrected χ^2 value equal to 3.39 which is also less than the critical value 3.8415.

So far, the chi-square criterion has been used to test hypotheses involving several proportions. It can also be used to test hypotheses concerning a single proportion. Suppose a coin is thrown 300 times and we observe 170 heads and 130 tails. The chi-square test could be used to test the hypothesis that the coin is fair. The results are summarized in the following table:

Event	*Heads*	*Tails*
Observed frequency	170	130
Expected frequency	150	150

Using the test statistic (8.6), we obtain

$$\chi^2 = \frac{(170 - 150)^2}{150} + \frac{(130 - 150)^2}{150} = 5.33$$

The tabled value of chi-square for $\alpha = 0.05$ and $v = 1$ is 3.84. Thus, we reject the hypothesis that the coin is fair. An alternative approach would be to use the standard normal approximation to the binomial distribution. The test statistic would then be equal to

$$z = \frac{X - np}{\sqrt{np(1 - p)}} = \frac{170 - 150}{\sqrt{300(0.5)(0.5)}} = 2.309$$

which again leads to the rejection of the hypothesis that the coin is fair at the 5 percent level of significance.

Notice that

$$z^2 = (2.309)^2 = 5.33 = \chi^2$$

which is exactly what we would expect because, by definition, the chi-square value is a summation of squared standard normal variables. In this case, there is only one standard normal variable in the sum.

The chi-square values of repeated experiments can be added and the sum used to test the hypothesis under consideration. For example, suppose the automobile company that wanted to determine if the type of car purchased was independent of where the buyer lives had selected three different random samples and computed the χ^2 value for each sample. Furthermore, suppose the following three values of chi-square were obtained:

$$\chi_1^2 = 12.66 \qquad v_1 = 4$$
$$\chi_2^2 = 7.34 \qquad v_2 = 4$$
$$\chi_3^2 = 8.00 \qquad v_3 = 4$$

Adding these results, we obtain

$$\chi^2 = 12.66 + 7.34 + 8.00 = 28.00$$

which has $v = v_1 + v_2 + v_3 = 12$ degrees of freedom. The tabled value of chi-square for $\alpha = 0.05$ and $v = 12$ is equal to 21.026. Therefore, the company could reject the hypothesis of independence at the 5 percent level of significance. There is no advantage to computing three values of χ^2 if all of the data from each sample are at hand but, when only the χ^2 values are available, the additive property of the chi-square distribution allows the results of the separate experiments to be quickly pooled to provide a test based on all available data.

The chi-square criterion (8.6) may be used to determine if it is justifiable to approximate an observed frequency distribution by means of a theoretical frequency distribution. The parameters of the theoretical distribution are estimated using the observed data. Expected frequencies are then obtained from the theoretical distribution on the assumption that the estimated parameters are the true values of the population parameters. The observed and expected frequencies are used to compute the χ^2 criterion (8.6). Large values of χ^2 lead to the rejection of the hypothesis that the observed data "fit" the theoretical distribution used to obtain the expected frequencies.

Suppose, for example, that a radio announcer makes six phone calls during his music program each day and asks people to identify the song that was just played on the program. This could be thought of as a binomial situation in which there are $n = 6$ phone calls (independent trials) each day and the probability that a person identifies the song (a "success") is constant on each phone call. A record of the number of times that people successfully identified songs each day during a 384 day period is presented in Table 8.11.

During the 384 day period, there was a total of $N = (6)(384) = 2304$ phone calls, and $X = \sum X_i n_i = 1152$ persons correctly identified the song. The

Table 8.11 Frequency Distribution of the Number of Correctly Identified Songs

Number of Songs Correctly Identified X_i	Observed Frequencies n_i	Theoretical Probabilities $b(X_i)$	Expected Frequencies $e_i = (384)b(X_i)$
0	6	1/64	6
1	60	6/64	36
2	66	15/64	90
3	132	20/64	120
4	60	15/64	90
5	36	6/64	36
6	24	1/64	6
Totals	384	1.00	384

probability that a song will be correctly identified is estimated to be

$$\hat{p} = \frac{X}{N} = \frac{(1152)}{(2304)} = 0.5$$

Assuming that the distribution of the number of correctly identified phone calls is a binomial distribution with $n = 6$ and $p = 0.5$, we can compute the binomial probabilities $b(X_i)$ for $X_i = 0, 1, \ldots, 6$ and obtain the expected number of correctly identified phone calls (e_i) by multiplying these probabilities times the number of days for which data are available (384). The results are shown in Table 8.11.

Using the chi-square formula (8.6), we obtain

$$\chi^2 = \frac{(6-6)^2}{6} + \frac{(60-36)^2}{36} + \frac{(66-90)^2}{90} + \frac{(132-120)^2}{120}$$

$$+ \frac{(60-90)^2}{90} + \frac{(36-36)^2}{36} + \frac{(24-6)^2}{6} = 87.6$$

The tabled value of chi-square for $\alpha = 0.05$ and $v = 5$ degrees of freedom is equal to 11.07. We conclude that, on the basis of the observed data, it would not be justified to assume that the distribution of the number of correctly identified songs is a binomial distribution with $p = 0.5$.

The number of degrees of freedom for a chi-square goodness of fit test is equal to $v = k - 1 - m$ where k is the number of terms $(n_i - e_i)^2/e_i$ computed and m is the number of population parameters estimated. In the example just considered, there were $k = 7$ terms (one for each cell in Table 8.11) and it was necessary to estimate the probability that a person would correctly identify a song, p. Had we assumed that p was equal to 0.7, say, and computed the expected frequencies using this assumed value, then the number of degrees of freedom would have been $v = (k - 1) = 6$ because the sample data were not used to estimate the parameter p. When fitting a normal distribution to observed data, it is necessary to estimate the mean and standard deviation from the sample data. In this case, the chi-square test for goodness of fit would have $v = k - 3$ degrees of freedom.

EXERCISES

Section 8.1

1. A research worker is considering the following plans for a two-tail test of $H_0: \mu = 70$:

(1) $n = 40$, $\alpha = 0.01$ (n is sample size, α is the probability of a Type I error)

(2) $n = 100, \quad \alpha = 0.01$
(3) $n = 200, \quad \alpha = 0.02$
(4) $n = 200, \quad \alpha = 0.05$
(a) Which plan will give him the highest Type II error?
(b) Which plan will give him the lowest Type II error?
(c) Which will incur the least risk of rejecting the hypothesis if it is true?
(d) If he accepts the hypothesis, which will allow him to make the strongest statement?

2. Explain why you would agree or disagree with the following statement. "If a 5 percent level of significance is consistently used in testing hypotheses, then 5 percent of those hypotheses which are false will be accepted."

3. In order to determine if a coin is "fair," a gambler decides to throw the coin 10 times and reject the hypothesis that $p = 0.5$ if he observes 0, 1, 2, 8, 9, or 10 heads.
(a) What is the probability of a Type I error?
(b) What is the probability of a Type II error if the probability of a head is actually equal to $p = 0.8$?

4. Compare the nonsequential and sequential approach to the problem of making inferences about populations on the basis of sample data.

5. Suppose a sample of 100 observations provides a 95 percent confidence interval for the population mean equal to $50 \leq \mu \leq 60$.
(a) What was the sample mean equal to?
(b) Would you accept or reject the hypothesis $H_0 : \mu = 59$ against the alternative $H_a : \mu \neq 59$ at the 5 percent level of significance?

6. The hypothesis $H_0 : \mu = 50$ is to be tested against the alternative hypothesis $H_a : \mu < 50$ at the 5 percent level of significance. It is known that $\sigma = 20$. A sample of 100 observations is to be drawn.
(a) An *operating characteristic curve* (or OC curve) is a graph showing the probabilities of Type II errors under various alternative hypotheses. Graph the OC curve for a test of H_0.
(b) Graph the power function for a test of H_0.
(c) If the sample mean is $\bar{X} = 47$, is H_0 accepted or rejected?

7. Two statements, A and B, are made about the world. If statement A is true, B is false and, if B is true, then A is false. Two random experiments can be used to decide which statement is true. Experiment I has five possible outcomes: x_1, x_2, x_3, x_4, x_5; and experiment II has outcomes: y_1, y_2, y_3, y_4, y_5. The probability distribution of the outcomes of experiments I and II under the condition that statement A is true and under the condition that statement B is true are as follows.

Outcomes of Experiment I

		x_1	x_2	x_3	x_4	x_5
True	A	0.10	0.15	0.50	0.15	0.10
Statement	B	0.40	0.25	0.15	0.10	0.10

Outcomes of Experiment II

		y_1	y_2	y_3	y_4	y_5
True	A	0.10	0.15	0.50	0.15	0.10
Statement	B	0.10	0.15	0.50	0.15	0.10

(a) Construct two tests of the hypothesis, H_0 : statement A is true, against the alternative, H_a : statement B is true. One test should use experiment I while the second test should use experiment II. Both tests should be constructed in such a way that $P(\text{reject } H_0 \mid H_0 \text{ is true}) = 0.20$.

(b) Which of the tests constructed in part (a) is better? Explain how you arrived at this conclusion.

(c) Suppose that you are "90 percent certain" that statement A is true. Experiment I is performed and the outcome is x_2. Using Bayes theorem, what is the posterior probability that statement A is true?

(d) Will either of the tests constructed in part (a) prove beyond all doubt that either A or B is true?

8. Explain why you would agree or disagree with each of the following statements.

(a) If the hypothesis being tested is true, the probability of a Type II error is zero. If the hypothesis tested in false, it is possible that the probability of a Type II error is zero.

(b) Once a level of significance is determined, the critical region of a test does not change if the size of sample used in the test changes.

Section 8.2

9. Students are randomly assigned to two sections of an economics course. Different books and different teaching techniques are used in the two sections. The final examination given in each section is identical and the grades were as follows:

	Number of Students	*Average Grade*	*Variance of Grades*
Section A	36	62	420
Section B	25	58	350

Test the hypothesis that there is no significant difference in the test achievement. Use the 1 percent level of significance.

10. A sample of 25 observations yields a sample mean $\bar{X} = 100$ and a sample standard deviation $s = 25$.

(a) Construct a 95 percent confidence interval for the population mean.

(b) Test the hypothesis that the sample was drawn from a population with $\mu = 110$. Use $\alpha = 0.05$.

11. Two workers were randomly assigned to different machines on successive days. Their output each day was as follows.

Day	*Worker A*	*Worker B*
Monday	6	8
Tuesday	10	7
Wednesday	8	5
Thursday	14	6
Friday	7	4
$\bar{X}_i$	$=9$	6
$\sum_i (X_i - \bar{X}_i)^2$	$=40$	10

(a) What assumptions would be necessary in order to test the hypothesis that the workers' average outputs are the same? Do all of these assumptions appear to be justified? Explain why or why not.

(b) If only one worker were to be kept on the job, which would you choose on the basis of productivity alone? Explain why you would make this decision.

12. Using the data of the following random sample, test the hypothesis that the sample was drawn from a population with mean equal to eight:

$$7, 8, 4, 10, 7, 6$$

13. A feed company wants to test the hypothesis that the mean weight of six-month-old pigs fed on a new ration is equal to 160 pounds. The company decides to reject the hypothesis if the pigs weigh less than 155 pounds. On the basis of previous tests, the company knows that the standard deviation of the population of weights for six-month-old pigs is 25 pounds.

(a) What is the probability that the company will make a Type I error?

(b) If the true population mean is 156 pounds, what is the probability of making a Type II error?

14. A tire dealer is interested in purchasing tires with an average life of 15,000 miles. A tire manufacturer informs the dealer that he has tires that have a life of 15,000 miles. The dealer selects a random sample of 25 tires and finds

$$s^2 = 625{,}000 \text{ miles}$$

$$\bar{X} = 25{,}000 \text{ miles}$$

(a) Test the hypothesis $H_0 : \mu = 15$ thousand miles, against the alternative hypothesis $H_a : \mu \neq 15$ thousand miles, at the 10 percent level of significance.

(b) What would have been a better alternative hypothesis in this case?

15. A sample of 100 items is used to test the hypothesis

$$H_0 : \mu = W$$

against the alternative hypothesis

$$H_a : \mu > W$$

at the 5 percent level of significance. The sample standard deviation is equal to 40. If the true mean is 100, the probability of accepting H_0 when it is false is equal to 0.33. Find the value of W.

Section 8.3

16. A coin is flipped 200 times and yields 90 heads.
  (a) Construct a 90 percent confidence interval for the true probability of throwing a head, p.
  (b) Test the hypothesis $H_0: p = 0.5$ against the alternative $H_a: p \neq 0.5$ at the 5 percent level of significance. Test H_0 using the chi-square distribution and also by using the standard normal distribution. Explain why both tests give the same results.

17. A hardware wholesaler is considering buying a large quantity of flashlight bulbs. He does not want to accept a shipment of bulbs which has over 1 percent defective bulbs, however. In order to test the hypothesis that the proportion of defective bulbs is 0.01, he decides to select a random sample of 100 bulbs to be tested from a shipment of 10,000 bulbs.
  (a) If he uses the decision rule, "reject the shipment if there are more than two defective bulbs in the sample of 100," what is the probability that he will reject a true hypothesis? (Since p is small, use the Poisson distribution to answer this question.)
  (b) Using the decision rule in part (a), what is the probability that the wholesaler will buy a shipment of bulbs when 2 percent of the bulbs are actually defective?

18. If 70 percent or more of the old bricks in a brick yard meet certain standards, it will be profitable to hire workmen to sort the bricks. A random sample of 100 bricks is examined and 60 are found to conform to the standards. Would you hire workmen to sort the bricks?

19. A public opinion company selects a random sample of 200 voters from New York and 400 voters from California and asks each person if he or she will vote for Candidate A in the next national election. It finds that 50 of the New York voters and 120 of the California voters plan to vote for Candidate A. Is there a significant difference between the proportion of voters who support Candidate A in these two states? Use the 10 percent level of significance.

Section 8.4

20. Two samples yield the following information:

		Sample A	*Sample B*
Sample size	n	11	9
	$\sum X_i$	110	72
	$(\sum X_i)^2$	12,100	5,184
	$\sum X_i^2$	1,400	1,056
Sample mean	$\bar{X}$	10	8

Test the hypothesis

$$H_0 : \sigma_A{}^2 = \sigma_B{}^2$$

against the alternative

$$H_a : \sigma_A{}^2 \neq \sigma_B{}^2$$

at the 5 percent level of significance.

21. The sample standard deviation obtained from a sample of 21 observations is 10. Is it likely that this sample was drawn from a population with variance equal to 40?

Section 8.5

22. Four coins are flipped 32 times, and the number of coins landing heads is counted. The results are:

Number of heads:	0	1	2	3	4
Number of times:	6	10	10	5	1

Does the binomial distribution with probability of heads equal to one-half provide a good fit to the observed distribution at the 5 percent level of significance?

23. A company's records show that during the past year 12 workers had accidents in plant A, 18 in plant B, and 24 in plant C. One-hundred people were employed in each plant under similar work conditions. Is there any association between plants and the number of accidents? Use the 5 percent level of significance.

24. The progeny of a cross between two species of angelfish were classified into four groups, the numbers being 20, 40, 50, and 70. According to a genetic model, these numbers should be in the ratio 2:3:4:5. Are the data consistent with the model at the 5 percent level of significance?

25. In a toothpaste test, children were randomly divided into three groups. Each group used a different type of toothpaste for one year. The results of the test are presented in the following table:

	Number of New Cavities		
Brand of toothpaste	0–1	2–3	4 or Over
A	50	20	10
B	60	10	10
C	70	10	10

Are the data consistent with the statement that there are "significant differences between the brands of toothpaste with respect to their ability to reduce the number of cavities?"

SELECTED REFERENCES

Chernoff, H. and L. E. Moses, *Elementary Decision Theory*, John Wiley and Sons, Inc., New York, 1959.

Lehmann, E. L., *Testing Statistical Hypotheses*, John Wiley and Sons, Inc., New York, 1959. (An advanced treatment of the subject.)

Neyman, J., *First Course in Probability and Statistics*, Henry Holt and Company, New York, 1950. (An intermediate text with emphasis on testing hypotheses.)

Raiffa, H. and R. Schlaifer, *Applied Statistical Decision Theory*, Harvard Business School, Division of Research, Boston, 1961.

Schlaifer, R., *Probability and Statistics for Business Decisions*, McGraw-Hill Book Company, New York, 1959.

CHAPTER 9

Simple Regression

Regression analysis is a branch of statistical theory that is widely used in almost all of the scientific disciplines. In economics, it is the basic technique for measuring or estimating the relationships among economic variables which constitute the essence of economic theory and economic life. In this chapter, we shall discuss simple regression techniques involving two variables. The basic ideas presented here will be extended in a later chapter to regression analysis involving more than two variables. Keep in mind that there is no magic in the techniques of regression analysis and nothing to save us from the consequences of inaccurate data or illogical selection of variables.

9.1 THE REGRESSION RELATIONSHIP

The fundamental purpose of simple regression analysis is to estimate the relationship between two variables X and Y. Economists are usually interested in the estimated relationship for prediction purposes. Suppose, for example, that we wanted to predict the quantity of wheat that would be purchased at different farm prices. Such information would be useful to the Department of Agriculture when setting a support price for wheat. We may decide to estimate a simple demand curve for wheat using price as one variable and quantity of wheat purchased as the second. Regression techniques could be used to estimate the demand curve from a given set of observations on these two variables.

Once the relationship between the variables X and Y has been estimated, we may wish to know the closeness of the relationship or to estimate the accuracy of predictions based on the relationship. Is the relationship consistent with what we would expect on the basis of theory? How confident can we be that a predicted value will be close to the actual value? More precise versions of such questions can be answered using the basic ideas already developed in the discussions on hypotheses testing and confidence intervals.

It is impossible to estimate the relationship between two variables without first making some assumptions about the form of the relationship. Some forms can be ruled out immediately. In other cases, we can check our assumptions about the form of the relationship after the relationship has been estimated.

Let us begin with linear functions. Such functions are of considerable importance in economics because they are relatively easy to work with and can often be used to approximate nonlinear functions. Algebraically, linear functions have the form

$$Y = \alpha + \beta X \tag{9.1}$$

where α and β are constants. The constant α is called the *intercept* coefficient. It tells us the level of Y when X is equal to zero. The constant β is the *slope* coefficient and is equal to the change in Y divided by the corresponding change in X. Geometrically, a linear function may be presented as a straight line such as is shown in Figure 9.1.

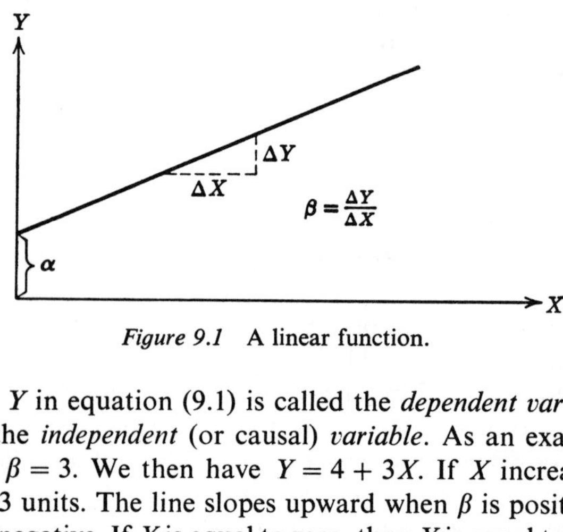

Figure 9.1 A linear function.

The variable Y in equation (9.1) is called the *dependent variable*, and X is referred to as the *independent* (or causal) *variable*. As an example, suppose that $\alpha = 4$ and $\beta = 3$. We then have $Y = 4 + 3X$. If X increases by 1 unit, Y increases by 3 units. The line slopes upward when β is positive and downward when β is negative. If X is equal to zero, then Y is equal to 4, the value of the intercept coefficient.

In economics, the designation of one variable as dependent and the other as independent is somewhat artificial at times. If, for example, (9.1) represents a Keynesian consumption function, then it is merely one equation in a mathematical model describing the relationships between important national aggregates such as income, investment, and the supply of money. In this case, consumption (Y) plays an important role in determining income (X). The terminology is often more applicable in the case of microeconomic relationships. In the theory of consumer behavior, for example, the individual consumer considers the price (X) as fixed and adjusts his consumption (Y) so as to maximize his utility.

The applied economist is likely to find that a linear functional relationship such as (9.1) does not describe the data with which he is working. Suppose he has a set of observations on the number of radios assembled daily (Y) by ten men (X) working on an assembly line. He may find that, although the number of men was the same each day, there is day-to-day variation in the number of radios assembled. The linear function (9.1) does not allow for such variations and, therefore, cannot provide an adequate description of the relationship between the variables X and Y.

Suppose, however, that we assume that observed data were generated by a *linear stochastic equation* such as

$$Y = \alpha + \beta X + u \tag{9.2}$$

where u is an error term. This equation allows Y to be larger or smaller than $\alpha + \beta X$ depending on whether u is positive or negative. Ideally, one wants the error term u to be small and uncorrelated with X so that one can change X without a change in u and determine what will happen on the average to Y. We will use (9.2) as the basic equation for the regression models developed in this chapter.

In general, the input data for simple regression analysis can be viewed as a paired array of numbers such as

$$Y_1, Y_2, \ldots, Y_n$$
$$X_1, X_2, \ldots, X_n$$

where the pair (Y_i, X_i) is the ith observation on variables Y and X. There is an error term u_i associated with each observation (Y_i, X_i). Suppose the ob-

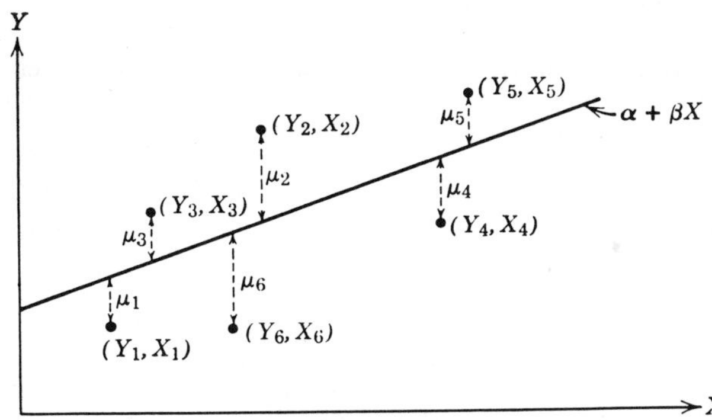

Figure 9.2 A scatter diagram.

servation (Y_i, X_i) is presented in a *scatter diagram* such as Figure 9.2. If the values of the parameters α and β were known, we could draw in the *regression function* $\alpha + \beta X$. The error terms are equal to the vertical distance between the observed points (Y_i, X_i) and the corresponding points on the regression line $(\alpha + \beta X_i, X_i)$. This is illustrated in Figure 9.2.

We are now ready to develop our basic regression model. In general, a *regression model* consists of a set of assumptions about the distribution of the error terms and the relationships between the variables X and Y. To begin with, we shall assume that the errors u are independent random variables with expected values equal to zero and variance equal to σ_u^2 for all values of X. Stated algebraically,[1] the assumptions are

(9.3) $$E(u_i) = 0 \qquad \text{for all } i$$

(9.4) $$\text{Var}(u_i) = \sigma_u^2 \qquad \text{for all } i$$

(9.5) $$\text{Cov}(u_i, u_j) = 0 \qquad \text{for } i \neq j$$

It is further assumed that the independent variable X is fixed on successive observations and that the dependent variable Y is a linear function of X.

Since the expected value of u is zero, the expected value of the dependent variable Y for a given value of X is given by the regression function $\alpha + \beta X$. This can be verified as follows:

(9.6) $$E(Y \mid X) = E(\alpha + \beta X + u) = \alpha + \beta X + E(u) = \alpha + \beta X$$

The symbol $E(Y \mid X)$ is read "the expected value of Y given X."

The variance of Y for a given value of X is equal to

(9.7) $$\text{Var}(Y \mid X) = \text{Var}(\alpha + \beta X + u) = \text{Var}(u) = \sigma_u^2$$

The assumption that $\text{Var}(Y \mid X)$ is the same for all values of X is a very useful one because it allows us to use all of the observations on X and Y to estimate σ_u^2. When working with economic data, we often have too few values

[1] A short review of the concept of covariance may be helpful here. Let W and Z be two random variables with a joint probability density function and expected values μ_w and μ_z, respectively. The mathematical expectation

$$E[(W - \mu_w)(Z - \mu_z)] = E(WZ) - \mu_w \mu_z$$

is called the covariance of W and Z and denoted by $\text{Cov}(W, Z)$. The covariance provides a measure of the closeness of the relationship between W and Z. If $\mu_w = \mu_z = 0$, the covariance of W and Z is equal to the expected value of the product of W and Z.

Since $E(u_i) = E(u_j) = 0$, we have $\text{Cov}(u_i, u_j) = E(u_i u_j)$. Assumption (9.5) could therefore be written as $E(u_i u_j) = 0$ and is satisfied if a particular error on one observation does not affect the value of the error on any other observation. Zero covariance does not imply independence of u_i and u_j, but independence of μ_i and μ_j would imply a zero covariance.

of Y for any given value of X to determine if the assumption of constant variance is justified. This assumption will be discussed in greater detail later.

From (9.6) and (9.7) it follows that, for a given value of X, the dependent variable Y has a conditional probability density function with mean $\alpha + \beta X$ and variance $\sigma_u{}^2$. Such a density function is shown in Figure 9.3. The location

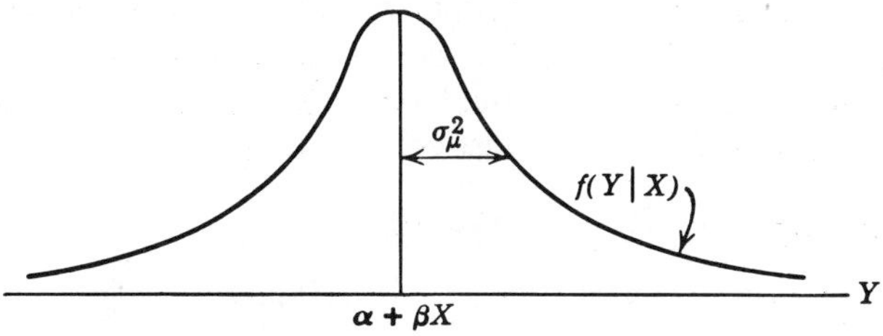

Figure 9.3 A conditional probability density function.

of the conditional probability density function $f(Y|X)$ changes as the value of X changes. This is shown geometrically in the three-dimension diagram in Figure 9.4. Notice that the mean of the conditional probability density function moves out along the regression function $\alpha + \beta X$.

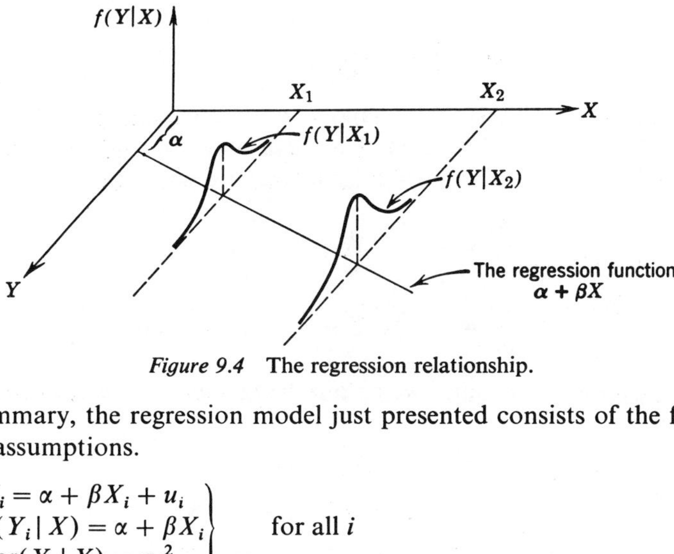

Figure 9.4 The regression relationship.

In summary, the regression model just presented consists of the following sets of assumptions.

(a) $Y_i = \alpha + \beta X_i + u_i$ } for all i
(b) $E(Y_i | X) = \alpha + \beta X_i$ } for all i
(c) $\text{Var}(Y_i | X) = \sigma_u{}^2$ } for all i
(d) $\text{Cov}(u_i, u_j) = 0$ for $i \neq j$
(e) The variable X is fixed on successive observations.

The exact form of the distribution of the errors is not specified in this model. The assumption that the errors are normally distributed is not needed at this point but will be added later.

9.2 ESTIMATION OF THE REGRESSION PARAMETERS

If the entire population of values (Y_i, X_i) is known, it is possible to compute the exact values of the regression parameters α, β, and σ_u^2. We are usually working with samples, in which case, the statistical problem is how best to estimate the parameters α, β, and σ_u^2. The method most widely used for fitting a straight line is the *method of least squares*. The least squares regression line is not necessarily the "best" line in some ultimate sense, but it does have several desirable statistical properties.

9.2.1 *Intercept and Slope Coefficients*

Suppose we let (a, b) be estimators of (α, β) and

$$e_i = Y_i - a - bX_i$$

be a deviation from the fitted regression lines for an observed (Y_i, X_i). In other words, e_i is an estimator of the error term u_i. The method of least squares requires that the estimators (a, b) be selected in such a way that the sum of squared deviations from the fitted regression line be as small as possible. We select a and b so as to minimize the sum of squared deviations:

(9.8)
$$\phi = \sum_{i=1}^{n} e_i^2 = \sum_{i=1}^{n} (Y_i - a - bX_i)^2$$

To minimize ϕ with respect to a and b, we differentiate ϕ partially with respect to a and b, equate the partial derivatives to zero, and solve the resulting equations for a and b. The partial derivatives are given by

$$\frac{\partial \phi}{\partial a} = -2 \sum (Y_i - a - bX_i) = 0$$

$$\frac{\partial \phi}{\partial b} = -2 \sum (Y_i - a - bX_i)X_i = 0$$

Simplifying these expressions, we obtain the *normal equations* of least squares regression

(9.9)
$$\sum Y_i = na + b \sum X_i$$

(9.10)
$$\sum Y_i X_i = a \sum X_i + b \sum X_i^2$$

where n is the number of paired observations. The normal equations are linear equations and can be solved simultaneously for a and b. From (9.9), we obtain

(9.11)
$$a = \frac{\sum Y_i}{n} - b\left(\frac{\sum X_i}{n}\right) = \bar{Y} - b\bar{X}$$

Substituting this result into equation (9.10), we find

$$b = \frac{n \sum Y_i X_i - \sum Y_i \sum X_i}{n \sum X_i^2 - (\sum X_i)^2} \tag{9.12}$$

This value of b is then used to compute the value of a using (9.11). The equations (9.11) and (9.12) not only define the estimators a and b but also serve as computing formulas.[2]

The estimated regression line is written as $\hat{Y} = a + bX$. The symbol $\hat{Y}$ is used because the value of Y obtained from the estimated regression line for a given value of X is an estimate of $E(Y \mid X)$.

As an illustration of the use of formulas (9.11) and (9.12), suppose we have $n = 14$ observations on income (X) and consumption (Y) and want to estimate a linear consumption function. We first compute the following quantities:

$$\begin{aligned}
&\text{(a)} \quad \textstyle\sum X_i &&= 3{,}500\\
&\text{(b)} \quad \textstyle\sum Y_i &&= 2{,}931.44\\
&\text{(c)} \quad \textstyle\sum X_i^2 &&= 885{,}714.28\\
&\text{(d)} \quad \textstyle\sum Y_i^2 &&= 636{,}944\\
&\text{(e)} \quad (\textstyle\sum X_i)^2 &&= 12{,}250{,}000\\
&\text{(f)} \quad \textstyle\sum X_i Y_i &&= 741{,}430
\end{aligned}$$

Next, the estimate of β is obtained using formula (9.12). We find

$$b = \frac{(14)(741{,}430) - (3500)(2931.44)}{(14)(885{,}714.28) - 12{,}250{,}000} = \frac{119{,}980}{150{,}000} \doteq 0.80$$

[2] If Y, X, and u are all random variables and X and u are stochastically independent, the relationship between the parameter β and the estimator b can be investigated as follows. Take the expected value of equation (9.2) and solve for α to obtain

$$\alpha = E(Y) - \beta E(X)$$

Multiplying equation (9.2) by X and taking the expected value gives

$$E(XY) = \alpha E(X) + \beta E(X^2)$$

since $E(uX) = E(u)E(X) = 0$ under the assumption of independence of X and u. Substitute for α to obtain

$$E(XY) = [E(Y) - \beta E(X)]E(X) + \beta E(X^2)$$

and solve for β to find

$$\beta = \frac{E(XY) - E(X)E(Y)}{E(X^2) - (E(X))^2} = \frac{\text{Cov}(X, Y)}{\text{Var}(X)}$$

The sign of β depends on the sign of Cov(X, Y) since Var (X) is always positive. The least squares estimator b given by (9.12) can therefore be thought of as the ratio of the estimator of Cov(X, Y) to the estimator of Var(X) in this case.

Once the value of b is available, the estimate of α is computed from (9.11). We find

$$a = \frac{2931.44}{14} - \frac{(0.80)(3500)}{14} = 209.39 - (0.80)(250) = 9.39$$

The consumption function obtained by the method of least squares is therefore equal to $\hat{Y} = 9.39 + 0.80X$ where X and $\hat{Y}$ are measured in millions of dollars. The slope coefficient, 0.80, provides an estimate of the marginal propensity to consume. According to our estimates, consumption will increase by \$800,000 if income increases by \$1,000,000. The intercept, 9.39, is an estimate of the expected level of consumption if income (X) was equal to zero.

The statistical characteristics of estimators can be used as criteria for evaluating the methods by which estimators are obtained. In Chapter 7, for example, it was pointed out that estimators obtained by the method of maximum-likelihood are not always unbiased but they are, among other things, efficient in the sense of having the smallest possible variance. The method of maximum-likelihood is therefore considered to be a reasonably good method for obtaining estimators. If we had specified the form of the distribution of the error terms in our regression model, the method of maximum-likelihood could have been used to obtain estimators of α and β. The method of least squares, on the other hand, can be applied without such specification. Furthermore, it can be shown that, for the regression model summarized at the end of the previous section, the least squares estimators a and b are the *b*est, *l*inear, *u*nbiased *e*stimators of α and β. By "unbiased," we mean that the expected value of a is equal to α and the expected value of b is equal to β. By "linear," we mean that a and b are linear functions of the dependent variable Y. Finally, the least squares estimators are "best" in the sense that the variances of a and b are at least as small as, or smaller than, the variances of any other linear, unbiased estimators of α and β. The proof of these statements is rather lengthy without the help of matrix algebra and will not be undertaken here. We conclude, however, that the method of least squares is recommended on two counts. First, it is easy to apply. Second, in the case of the regression model being discussed, the method of least squares provides estimators with several desirable statistical properties.

If the assumption that the error terms are normally distributed is added to the regression model presented at the end of the previous section, then the method of maximum-likelihood can be used to obtain estimators for α and β. Since $E(u_i) = 0$ and $\text{Var}(u_i) = \sigma_u^2$, the probability density function of u is given by

$$f(u_i) = \frac{1}{\sigma_u\sqrt{2\pi}} \exp\left[-\frac{u_i^2}{2\sigma_u^2}\right] \tag{9.13}$$

which is the usual normal density function. Substituting $Y_i - \alpha - \beta X_i$ for u_i, we find the likelihood function for a sample of size n to be equal to

(9.14) $$L(\theta) = \frac{1}{(\sigma_u\sqrt{2\pi})^n} \exp\left[-\frac{\sum (Y_i - \alpha - \beta X_i)^2}{2\sigma_u^2}\right]$$

The maximum-likelihood estimators of α and β are found by maximizing $L(\theta)$ with respect to α and β. First, taking the log of $L(\theta)$ to the base e, we obtain

$$\ln L(\theta) = -n \ln (\sigma_u\sqrt{2\pi}) - \frac{1}{2\sigma_u^2} \sum (Y_i - \alpha - \beta X_i)^2$$

Differentiating partially with respect to α and β gives

$$\frac{\partial(\ln L(\theta))}{\partial\alpha} = \frac{1}{\sigma_u^2} \sum (Y_i - \alpha - \beta X_i)$$

$$\frac{\partial(\ln L(\theta))}{\partial\beta} = \frac{1}{\sigma_u^2} \sum X_i(Y_i - \alpha - \beta X_i)$$

Equating these expressions to zero and simplifying, we obtain the normal equations

$$\sum Y_i = n\hat{\alpha} + \hat{\beta} \sum X_i$$

$$\sum X_i Y_i = \hat{\alpha} \sum X_i + \hat{\beta} \sum X_i^2$$

where the hats, ^, are placed over the parameters to indicate that $\hat{\alpha}$ and $\hat{\beta}$ are estimators. The maximum likelihood estimators of α and β are therefore identical to the least squares estimators.

9.2.2 The Standard Error of Estimate

Intuitively, it seems plausible to use the residuals from the least squares regression line to estimate the variance of the error terms σ_u^2. The sample variance of these residuals is equal to

(9.15) $$\hat{\sigma}_u^2 = \frac{\sum (e_i - \bar{e})^2}{n}$$

where $\bar{e} = \sum e_i/n$. Notice, however, that

$$\sum e_i = \sum (Y_i - a - bX_i) = \sum Y_i - na - b \sum X_i = 0$$

by the first normal equation (9.9). Therefore, $\hat{\sigma}_u^2$ can be written as

(9.16) $$\hat{\sigma}_u^2 = \frac{\sum e_i^2}{n}$$

Under the assumption that the errors are normally distributed, this estimator is identical to the maximum-likelihood estimator of σ_u^2. It should not be surprising, therefore, to learn that $\hat{\sigma}_u^2$ is a biased estimator.

An unbiased estimator is obtained by multiplying $\hat{\sigma}_u^2$ by $n/(n-2)$. The new estimator is commonly denoted by $s_{y.x}^2$ and its square root

$$s_{y.x} = \sqrt{\frac{\sum e_i^2}{n-2}} = \sqrt{\frac{\sum (Y_i - a - bX_i)^2}{n-2}} \tag{9.17}$$

is called the *standard error of estimate* or the *sample standard deviation of regression.*

The standard error of estimate is a measure of the variability of the conditional distribution of Y for fixed values of X. All of the residuals from the fitted regression line are used when computing $s_{y.x}$ because all of the conditional distributions are assumed to have the same variance. The use of $n-2$ in the denominator of (9.17) is a reflection of the fact that two parameters, α and β, have to be estimated before the residuals e_i can be computed. As a result, there are only $n-2$ degrees of freedom associated with the quantity $\sum e_i^2$

The computational formula for the standard error of estimate is given by

$$s_{y.x} = \sqrt{\frac{\sum Y_i^2 - a\sum Y_i - b\sum X_i Y_i}{n-2}} \tag{9.18}$$

Using formula (9.18) and the consumption function data presented in Section 9.2.1, we find the standard error of estimate is equal to

$$s_{y.x} = \sqrt{\frac{636{,}944 - 9.39(2{,}931.44) - (0.8)(741{,}430)}{12}} = 36.82$$

Several uses of the standard error of estimate will be discussed in the following section.

9.3 STATISTICAL INFERENCE AND PREDICTION

9.3.1 *Statistical Tests Involving* α, β, *and* σ_u^2

Suppose that the values of the independent variable X are held fixed and repeated samples of n observations (Y_i, X_i) are taken. The values of the dependent variable Y associated with the fixed values of X will vary from sample to sample. As a result, the estimates of α and β obtained from the least squares estimators a and b will also vary. The distribution of the resulting values of a and b is the joint sampling distribution of the least squares estimators. We need to know the means, variances, and forms of the two marginal distributions of a and b before we can develop tests of hypotheses about α and β.

Since the slope coefficient is usually of more interest to economists than the intercept coefficient, we shall give somewhat more attention to the distribution of b than to that of a.

In the previous section, we mentioned that the least squares estimator b is a linear function of the observed values of the dependent variable Y. Applying the rules of summation to the definition (9.12) in Section 9.2.1, we obtain

$$b = \sum w_i Y_i \tag{9.19}$$

where the constants w_i have the form

$$w_i = \frac{nX_i - \sum X_i}{n \sum X_i^2 - (\sum X_i)^2} = \frac{X_i - \overline{X}}{\sum (X_i - \overline{X})^2} \tag{9.20}$$

Since $Y_i = \alpha + \beta X_i + u_i$, (9.19) can be written as

$$b = \sum w_i(\alpha + \beta X_i + u_i) = \alpha \sum w_i + \beta \sum w_i X_i + \sum w_i u_i$$

Recalling that the sum of deviations from the mean $\sum (X_i - \overline{X})$ is equal to zero, we find from (9.20) that $\sum w_i = 0$. Multiplying (9.20) by X_i and summing we find that $\sum w_i X_i = 1$. Hence

$$b = \beta + \sum w_i u_i$$

The expected value of b is therefore equal to

$$E(b) = \beta + \sum w_i E(u_i) = \beta$$

because $E(u_i)$ is equal to zero for all i. We have thus shown that b is an unbiased estimator of β.

To calculate the variance of b, we note that, under the assumption that the covariances of the variables Y_i for given X are zero, we have

$$\text{Var}(b) = \text{Var}(\sum w_i Y_i) = \sum w_i^2 \text{Var}(Y_i \mid X) = \sigma_u^2 \sum w_i^2$$

Substituting for w_i^2, we obtain

$$\sum w_i^2 = \sum \left[\frac{(X_i - \overline{X})}{\sum (X_i - \overline{X})^2} \right]^2 = \frac{1}{\sum (X_i - \overline{X})^2} = \frac{n}{n \sum X_i^2 - (\sum X_i)^2}$$

Thus the variance of b is equal to

$$\text{Var}(b) = \frac{n\sigma_u^2}{n \sum X_i^2 - (\sum X_i)^2} \tag{9.21}$$

In a similar manner, it can be shown that the mean of the marginal distribution of a is

$$E(a) = \alpha \tag{9.22}$$

and the variance is equal to

$$\text{Var}(a) = \frac{\sigma_u^2 \sum X_i^2}{n \sum X_i^2 - (\sum X_i)^2} \tag{9.23}$$

Estimators for the variances are obtained from (9.21) and (9.23) by replacing σ_u^2 with $s_{y \cdot x}^2$. Making these substitutions and taking the square roots, we obtain the standard error of estimate of b

$$s_b = \sqrt{\frac{n s_{y \cdot x}^2}{n \sum X_i^2 - (\sum X_i)^2}}$$

and the standard error of estimate of a

$$s_a = \sqrt{\frac{s_{y \cdot x}^2 (\sum X_i^2)}{n \sum X_i^2 - (\sum X_i)^2}}$$

If the error terms are normally distributed, then the distributions of the estimators a and b are normal distributions with the means and variances given earlier. In this case, the quantity

$$t = \frac{a - \alpha}{s_a} \tag{9.24}$$

has a t distribution with $n - 2$ degrees of freedom and can be used to perform tests of hypotheses about α. Similarly, the quantity

$$t = \frac{b - \beta}{s_b} \tag{9.25}$$

has a t distribution with $n - 2$ degrees of freedom and serves as a test function for hypotheses about β.

The $100(1 - \varepsilon)$ percent confidence interval estimator for α is given by

$$a \pm t_\varepsilon s_a \tag{9.26}$$

and the $100(1 - \varepsilon)$ percent confidence interval estimator for β by

$$b \pm t_\varepsilon s_b \tag{9.27}$$

Tests of hypotheses about σ_u^2 are based on the chi-square distribution. The quantity

$$\chi^2 = \frac{(n - 2)s_{y \cdot x}^2}{\sigma_u^2} \tag{9.28}$$

has a chi-square distribution with $n - 2$ degrees of freedom. The discussion in Section 8.4 on tests of hypotheses about variances applies to the use of (9.28).

The $100(1 - \varepsilon)$ percent confidence interval estimator for σ_u^2 is given by

$$\frac{(n-2)s_{y.x}^2}{\chi_{\varepsilon/2}^2} \le \sigma_u^2 \le \frac{(n-2)s_{y.x}^2}{\chi_{(1-\varepsilon/2)}^2} \tag{9.29}$$

Additional information on confidence intervals for variances can be found in Section 7.5.

To illustrate the use of s_b, let us assume that $Y = \alpha + \beta X$ represents a production function in which X is the quantity of fertilizer used and Y is the quantity of hay produced. The coefficient β is the extra output (hay) that results from a one-unit increase in input (fertilizer). In other words, β is the marginal physical product of the variable input fertilizer.

Suppose that the data from 12 test plots are used to compute the production function

$$\begin{aligned} Y = {} & 2.38 + 0.32X \\ & (0.30) \quad (0.08) \end{aligned}$$

where Y is measured in tons of hay per acre and X in hundreds of pounds of fertilizer used per acre. The standard errors of estimate s_a and s_b are given in the parentheses below the least squares estimates. If the slope of the true regression line is zero, then a change in the amount of fertilizer used has no effect on the yield of hay. Graphically, the true regression line would be a horizontal line at $Y = \alpha$. The estimates $b = 0.32$ and $s_b = 0.08$ can be used to test the hypothesis that the slope of the true regression line is equal to zero. To test $H_0: \beta = 0$ against the alternative hypothesis $H_a: \beta \neq 0$ at the 5 percent level of significance, we first compute

$$t = \frac{b - \beta}{s_b} = \frac{0.32 - 0}{0.08} = 4.0$$

The tabled value of t with $n - 2 = 10$ degrees of freedom is $t_{0.05} = 2.2281$. Since the observed value of t, (4.0), exceeds the tabled value of t, the hypothesis that fertilizer has no affect on hay yields is rejected at the 5 percent level of significance.

When deciding whether or not to use fertilizer, the farmer may want to compute a confidence interval for β. If the increase in yield of hay due to the use of fertilizer is small, it may not pay to use fertilizer. For example, suppose that hay is worth \$15 per ton and fertilizer costs \$3 per hundred pounds. If 100 pounds of fertilizer increases the hay yield by 0.32 tons, the total return to fertilizer is (\$15)(0.32) = \$4.80 and the net return is \$4.80 − \$3.00 = \$1.80. If the true increase in hay yields per acre is only 0.16 tons per hundred pounds of fertilizer used, the expected net return to fertilizer is (\$15)(0.16) − \$3 = −\$0.60 in which case the farmer would not want to use fertilizer even though it does increase yields. Using the production function data given above, we find the 90 percent confidence interval for the marginal physical product of fertilizer to be

$$0.32 - (1.8125)(0.08) \leq \beta \leq 0.32 + (1.8125)(0.08)$$
$$0.175 \leq \beta \leq 0.465$$

A farmer deciding whether or not to use fertilizer could be "90 percent confident" that the interval 0.175 to 0.465 tons of hay per acre covers the true marginal physical product of fertilizer. More precisely, if a large number of production functions for hay were computed from random samples of 12 observations and formula (9.27) is used to compute confidence intervals for β with $t = 1.8125$, we expect 90 percent of these confidence intervals to cover the true value of β.

9.3.2 Predicting $E(Y \mid X)$ and Y

We are often interested in predicting the mean value of Y corresponding to some given value of X, say, X_0. Such predictions may be presented as point or interval predictions. Earlier in this chapter, it was pointed out that $\hat{Y} = a + bX$ is an estimator of $E(Y \mid X)$. We shall now find the mean and variance of the distribution of $\hat{Y}$ and show how $\hat{Y}$ can be used in making predictions.

Let Y_0 denote the value of Y corresponding to the given X_0. The mean of the distribution of $\hat{Y}_0$ is equal to

$$E(\hat{Y}_0) = E(a + bX_0) = E(a) + E(b)X_0 = \alpha + \beta X_0 = E(Y_0 \mid X_0)$$

where a and b are the least squares estimators. This result follows directly from the fact that a and b are unbiased estimators of α and β. In other words, $\hat{Y}_0 = a + bX_0$ is an unbiased point estimator of the mean value of Y corresponding to the given value X_0.

To find the variance of $\hat{Y}_0$, we note that

$$\hat{Y}_0 = a + bX_0 = \bar{Y} - b\bar{X} + bX_0 = \bar{Y} + b(X_0 - \bar{X})$$

Thus, the variance of $\hat{Y}_0$ is

$$\begin{aligned} \text{Var}(\hat{Y}_0) &= \text{Var}(\bar{Y} + b(X_0 - \bar{X})) \\ &= \text{Var}(\bar{Y}) + \text{Var}(b(X_0 - \bar{X})) \\ &= \text{Var}(\bar{Y}) + (X_0 - \bar{X})^2 \,\text{Var}(b) \end{aligned} \tag{9.30}$$

Since the observations Y_i are assumed to be independent, we have

$$\text{Var}(\bar{Y}) = \text{Var}\left(\frac{\sum Y_i}{n}\right) = \frac{\sigma_u^{\ 2}}{n}$$

Thus (9.30) can be written as

$$\text{Var}(\hat{Y}_0) = \frac{\sigma_u^{\ 2}}{n} + \frac{(X_0 - \bar{X})^2 \sigma_u^{\ 2}}{\sum (X_i - \bar{X})^2} \tag{9.31}$$

The second term on the right-hand side of (9.31) is equal to zero if X_0 is equal to $\bar{X}$. The greater the difference between X_0 and $\bar{X}$, the larger $(X_0 - \bar{X})^2$ and the variance of $\hat{Y}_0$.

An estimator of $\text{Var}(\hat{Y}_0)$ can be obtained by substituting $s_{y.x}^2$ for $\sigma_u^{\ 2}$ in (9.31). Making the substitution, expanding $\sum (X_i - \bar{X})^2$, and taking the square root, we obtain

$$s_{\hat{y}} = \sqrt{\frac{s_{y.x}^2}{n} + \frac{n(X_0 - \bar{X})^2 s_{y.x}^2}{n \sum X_i^2 - (\sum X_i)^2}} \tag{9.32}$$

If the error terms are normally distributed, then $\hat{Y}_0$ has a normal distribution with mean $\alpha + \beta X_0$ and variance given by (9.31). In this case, a $(100)(1 - \varepsilon)$ percent confidence interval estimator for $E(Y_0 \mid X_0)$ is given by

$$\hat{Y}_0 - t_\varepsilon s_{\hat{y}} \leq E(Y_0 \mid X_0) \leq \hat{Y}_0 + t_\varepsilon s_{\hat{y}} \tag{9.33}$$

In some cases, we may be interested in obtaining a confidence interval for the individual Y_0 value associated with X_0. If, for example, a new pair of observations (Y_0, X_0) is obtained, we may wish to determine whether they could have been generated by the assumed linear model. This information may be of special interest if X_0 does not lie within the range of the sample observations. The desired confidence interval estimator is based on the distribution of the difference $\hat{Y}_0 - Y_0$. If Y_0 is generated by the assumed linear model, then $Y_0 = \alpha + \beta X_0 + u_0$. Therefore,

$$E(\hat{Y}_0 - Y_0) = E(\hat{Y}_0) - E(Y_0) = \alpha + \beta X_0 - \alpha - \beta X_0 = 0$$

and we see that $\hat{Y}_0$ is an unbiased estimator of the individual Y_0 value as well as the mean $E(Y_0 \mid X_0)$.

The variance of $(\hat{Y}_0 - Y_0)$ is equal to

$$\begin{aligned} \text{Var}(\hat{Y}_0 - Y_0) &= \text{Var}(\hat{Y}_0) + \text{Var}(Y_0) \\ &= \frac{\sigma_u^{\ 2}}{n} + \frac{(X_0 - \bar{X})^2 \sigma_u^{\ 2}}{\sum (X_i - \bar{X})^2} + \sigma_u^{\ 2} \end{aligned} \tag{9.34}$$

An estimator of $\text{Var}(\hat{Y}_0 - Y_0)$ is obtained by substituting $s_{y.x}^2$ for $\sigma_u^{\ 2}$ in (9.34).

If the error terms are normally distributed, then the difference $(\hat{Y}_0 - Y_0)$ has a normal distribution and the quantity

$$t = \frac{\hat{Y}_0 - Y_0}{s_{y.x}\sqrt{1 + 1/n + (X_0 - \bar{X})^2/\sum (X_i - \bar{X})^2}}$$

has the t distribution with $n - 2$ degrees of freedom. Thus, a $100(1 - \varepsilon)$ confidence interval estimator for Y_0 is given by

$$\hat{Y}_0 \pm t_\varepsilon s_{y.x}\sqrt{1 + \frac{1}{n} + \frac{(X_0 - \bar{X})^2}{\sum (X_i - \bar{X})^2}} \tag{9.35}$$

Note that the variance of $\hat{Y}_0 - Y_0$ increases the further X_0 lies from the mean of the sample values used to compute a and b. Thus, a confidence interval for Y_0 will be wider the larger the difference $(X_0 - \bar{X})$.

As an illustration, suppose that we estimate from family budget studies that $\hat{Y} = 500 + 0.30X$ where X is family income and Y is food expenditures measured in dollars. Given that the Jones family has an annual income of \$10,000, determine the 95 percent confidence interval for its food expenditures. We are not interested here in a confidence interval for food expenditures by by the average family with a \$10,000 income; we are concerned with a particular family—the Jones. Thus we would use equation (9.35) rather than (9.33) to find the required confidence interval.

9.3.3 *Some Regression Problems*

Example 1: To illustrate the mechanics of regression analysis, suppose that the following set of numbers consistutes a *population* of paired observations.

X	Y	X	Y
4	11	9	24
3	8	6	17
11	27	3	14
2	11	8	18
6	14	5	14
9	22	13	28
3	10	1	5
7	16		

Furthermore, suppose that the six observations listed below represent a *random sample* from the above population.

X	6	4	8	3	1	5
Y	17	11	18	8	5	14

It is assumed that the relationship between X and Y has the form

$$Y_i = \alpha + \beta X_i + u_i$$

and that the regression model described in Section 9.1 holds.

In order to obtain the least squares estimates of α and β from the sample data, we first compute

$$\begin{array}{llll} \sum X_i & = 27 & \sum Y_i^2 & = 1019 \\ \sum Y_i & = 73 & \sum X_i Y_i & = 389 \\ \sum X_i^2 & = 151 & \overline{X} & = 4.5 \\ (\sum X_i)^2 & = 729 & \overline{Y} & = 12.17 \end{array}$$

$$n \sum X_i^2 - (\sum X_i)^2 = 177$$

Using formula (9.12), we obtain

$$b = \frac{(6)(389) - (27)(73)}{(6)(151) - 729} = 2.05$$

From formula (9.11), we compute

$$a = 12.17 - (2.05)(4.5) = 2.945$$

The estimated regresssion line is therefore equal to

$$\hat{Y} = 2.945 + 2.05X$$

and is shown in Figure 9.5.

Suppose that another pair of observations (X, Y) is to be selected at random from the population and we want to predict the value of Y. If no information about X is available, the predicted value of Y would be $\overline{Y} = 12.17$.

In some cases, we may want to assume that X will be equal to some particular value, say, X_0, and then predict Y given that $X = X_0$. The estimated regression line would be used in this case. If, for example, we assume that $X = 3$, then the predicted value of Y is

$$\hat{Y} = 2.945 + (2.05)(3) = 9.095$$

The actual value of Y may be 8, 10, or 14 but, given the limited information available about the population, our best estimate of Y, given that $X = 3$, is 9.095.

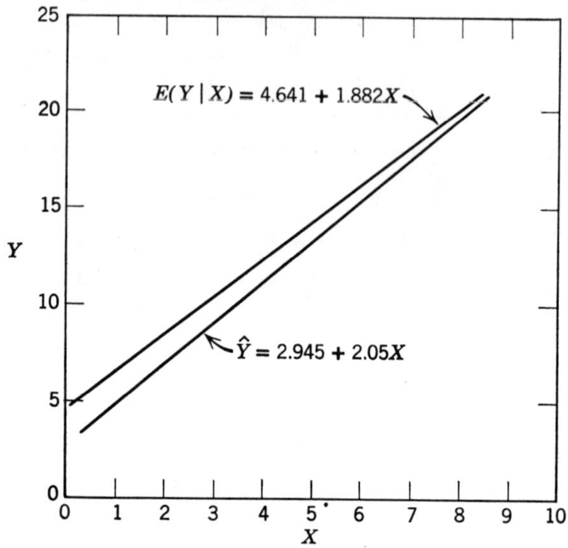

Figure 9.5 Actual and estimated regression lines.

Information about X is of little value when attempting to predict Y if there is no relationship between X and Y. Even if X and Y are related, we may be unable to obtain a sufficiently accurate estimate of the relationship from the available data. If our estimated relationship differs substantially from the true relationship, then our ability to predict Y given X may, in fact, be reduced. Suppose, for example, that the true relationship is $Y = 10 - 2X + u$ while the estimated relationship is $\hat{Y} = 12 + 2X$. Predictions of Y based on the estimated relationship obviously would be very misleading. An investigator can frequently determine if an estimated relationship between X and Y will be useful for prediction purposes by testing the hypothesis that β is equal to zero. We shall use the estimated regression line $\hat{Y} = 2.945 + 2.05X$ to illustrate such a test, but first we must estimate $s_{y.x}$ and s_b.

The standard error of estimate $s_{y.x}$ is obtained from formula (9.18) and is found to be equal to

$$s_{y.x} = \sqrt{\frac{1019 - (2.945)(73) - (2.05)(389)}{4}} = \sqrt{1.64} = 1.28$$

The standard error of estimate of b is therefore equal to

$$s_b = \sqrt{\frac{(6)(1.64)}{177}} = 0.236$$

The 95 percent confidence interval for β is

$$2.05 - 2.77(0.236) \leq \beta \leq 2.05 + 2.77(0.236)$$
$$1.40 \leq \beta \leq 2.70$$

Since this confidence interval does not cover zero, we conclude that it is rather unlikely that $\beta = 0$. Thus information about the value of X is expected to be useful when predicting Y.

Using all of the items in the population, we find

$$\alpha = 4.641 \qquad \text{and} \qquad \beta = 1.882$$

The expected value of Y for a given value of X is therefore equal to

$$E(Y \mid X) = 4.641 + 1.882X$$

Even if the true values of α and β are known, we cannot predict the individual values of Y precisely unless the values of the error terms are known. Given $X = 3$, for example, we find

$$E(Y \mid X) = 4.641 + (1.882)(3.0) = 10.287$$

More precise prediction of Y may be possible if, by modifying the basic model, we are able to "explain" part of the variation in Y which is left "unexplained" by our simple regression model.

Example 2: The first two columns of Table 9.1 contain time series observations on per capita consumption of beef and the retail price of beef deflated by a consumer price index. According to the theory of consumption, an equal percentage change in the prices of all commodities and in money income should not disturb the pattern of quantities demanded or consumed. Hence, during a period such as 1949–1960 in which the Consumer Price Index rose more than 20 percent, we would expect that the actual price of beef would be less closely associated with changes in its per capita consumption than would the deflated price, which shows changes in the price of beef *relative to* the average price level for all consumer goods and services.

The close inverse relationship between changes in per capita consumption of beef and changes in the deflated retail price of beef over time is seen in Section A of Figure 9.6. Under the assumption that the demand relationship connecting beef consumption and price is a straight line, we find the least squares simple regression line to be

$$\hat{Q} = 158.52 - 1.27P$$

where $\hat{Q}$ is the predicted per capita consumption of beef for any given value of the deflated retail price of beef, P. The estimated regression line is shown in the scatter diagram in Section B of Figure 9.6. The slope coefficient,

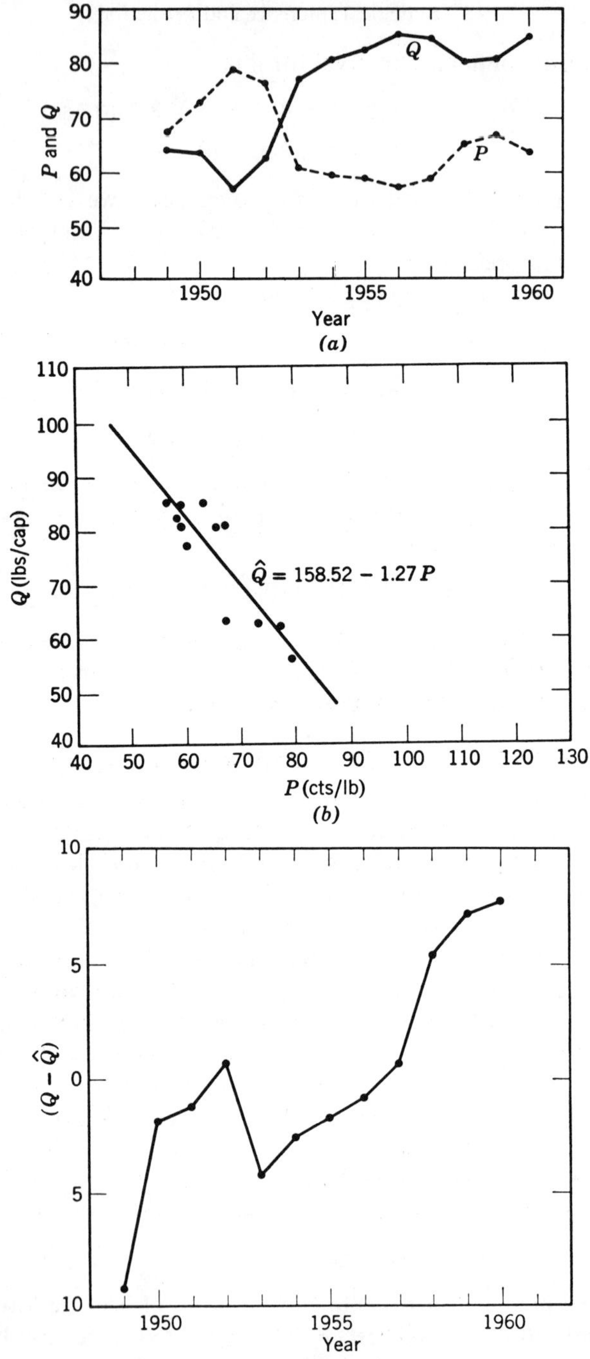

Figure 9.6 Beef consumption and prices: 1949–1960. (*a*) Original data (time series). (*b*) Original date (scatter diagram). (*c*) Residuals, $(Q - \hat{Q})$.

-1.27, implies that a 1.27 pound decrease in predicted per capita consumption, $\hat{Q}$, is associated with every one-cent increase in the deflated retail price of beef.

Table 9.1 Beef Consumption and Price: 1949–1960

Year	(1) Consumption Per Capita Q pounds	(2) Retail Price[a] P Cents/pound	(3) Estimated Consumption $\hat{Q}$ pounds/Capita	(4) Residuals $(Q - \hat{Q})$
1949	63.9	67.2	73.2	−9.3
1950	63.4	73.3	65.4	−2.0
1951	56.1	79.5	57.5	−1.4
1952	62.2	76.3	61.6	0.6
1953	77.6	60.4	81.8	−4.2
1954	80.1	59.7	82.7	−2.6
1955	82.0	59.0	83.6	−1.6
1956	85.4	56.8	86.4	−1.0
1957	84.6	58.7	84.0	0.6
1958	80.5	65.6	75.1	5.4
1959	81.4	66.4	74.2	7.2
1960	85.2	63.8	77.5	7.7

[a] Deflated by the Consumer Price Index, 1957–1959 = 100.

$\sum Q_i = 902.4$ $\qquad$ $\sum Q_i \sum P_i = 709{,}918$

$\sum P_i = 786.7$ $\qquad$ $(\sum P_i)^2 = 618{,}897$

$\sum Q_i^2 = 69{,}096$ $\qquad$ $\bar{Q} = 75.20$

$\sum P_i^2 = 52{,}178$ $\qquad$ $\bar{P} = 65.56$

$\sum Q_i P_i = 58{,}392$

The differences between the actual and predicted quantities $(Q - \hat{Q})$ are shown in Column (4) of Table 9.1 and in Section C of Figure 9.6. There appears to be some upward trend in the residuals, but we shall make no further comment on this for the present.

Example 3: The data in Table 9.2 are based on an auto trip. The gas tank was filled at the beginning of the trip and was refilled six times along the way. At each refueling stop, the number or miles traveled and the number of gallons of gasoline needed to refill the tank were recorded. The circumstances of this example are very similar to an experiment in which the values of the independent variable are controlled. The dependent variable Y was the quantity of gas required to refill the tank. The independent variable X was the miles traveled between refills.

Table 9.2 Gasoline Mileage

Observation Number i	Gallons of Gasoline Used Y	Miles Traveled X
1	11.3	184
2	12.9	204
3	15.7	252
4	17.5	275
5	11.6	184
6	15.6	268

$\sum X_i = 1367$ $\qquad \bar{X} = 227.83$

$\sum Y_i = 84.6$ $\qquad \bar{Y} = 14.10$

$\sum X_i^2 = 320{,}281$

$\sum Y_i^2 = 1{,}224.76$

$\sum X_i Y_i = 19{,}794.9$

The original selection of service stations was not specified in advance and, therefore, are quite arbitrary. Once they had been selected, however, we could in principle have the same driver repeat the trip several times, each time stopping at the same six service stations. Hence, the six values of X would be the same in all cases and we would generate as many values for Y as there were repetitions of the experiment. The values of X can therefore be considered to be fixed on successive observations.

It would not, of course, be necessary to use the same distances to generate additional values of Y. Indeed, in most economic problems it would be impossible in practice to hold the values of the independent variables fixed in repeated samples. Least squares techniques may be used regardless of whether the X's are considered a set of fixed numbers or a set of observations on a random variable.

There are several sources of random variation in the amount of gasoline used.

1. *Errors of measurement*. The station attendants were not directed to fill the gasoline tank to some accurately marked level but simply to "fill it up with 'regular.'" Hence, any normal variability of behavior among serivce station attendants, including attempts to round off the total charge at some multiple of five or ten cents, would have operated without special restraint.

2. *Differences in driving conditions*. Although all driving was done on good highways, the terrain differed between refills. The degree of traffic congestion and hence of acceleration and retardation differed among the various stretches

due to the differences in the times of day at which they were driven and the proximity to large cities. Speed limits differed among the various states, and driving speeds were adjusted to the speed limits. In addition, differences in temperature and in the type of road surface may have had some effects upon gasoline consumption per mile over the different stretches.

3. *Differences in quality of gasoline.* At each stop, "regular" gasoline was requested but the brand was not specified. Hence, there may have been some differences in the specifications of the different gasolines used which would have affected gasoline consumption per mile.

4. *Possibly* cumulative changes may have occurred during the trip in the carburetor adjustment and/or other characteristics of the automobile which were significant for gasoline consumption. However, it was the same car throughout and the driving was completed within 48 hours elapsed time.

Our basic hypothesis in the subsequent analysis is that there is a straight line relationship between miles and gasoline consumed. It also seems reasonable to expect that the constant term in the least squares straight line will not differ significantly from zero.

A least squares regression line fitted to the six observations gives us

$$\hat{Y} = \underset{(1.382)}{0.682} + \underset{(0.006)}{0.059}X$$

The standard error of estimate $s_{y.x}$ is 0.5620 gallons.[3]

The relationship seems very close, as we expected, and the error of estimate of a little more than half a gallon seems moderate in relation to the average amount of gasoline used between stops which was 14.1 gallons. The constant term of the regression equation is considerably smaller than its standard error, so our hypothesis that the relationship between Y and X is a proportional one is certainly not overturned. The regression slope, 0.059 gallons per mile, is not very different from that of a line drawn from the origin through the point of means, $\bar{Y} = 14.1$ gallons and $\bar{X} = 227.83$ miles traveled. The slope of the latter line would be approximately 0.062 gallons per mile, and its reciprocal would be slightly over 16 miles per gallon.

On the basis of this experiment, we cannot directly predict what our gasoline consumption would be under different driving conditions such as (a) stop-and-start driving in city traffic or (b) driving through mountainous country. However, if we made a similar trip in the same automobile on good roads over similar terrain, stopping for gasoline only every 200 to 300 miles,

[3] It seems reasonable to consider gallons of gas (Y) as the dependent variable in this example. Note, however, that the regression of miles (X) on gallons (Y) is $X = -2.0 + 16.3Y$ and would not give exactly the same results as the regression of gallons on miles except at the point ($\bar{Y}$, $\bar{X}$).

we should be somewhat surprised to find our gasoline mileage averaging less than 15 or more than 17 miles per gallon. If our mileage per gallon on a similar trip is as low as 14 or as high as 18 miles per gallon, it is likely that something about the car is different—perhaps the carburetor adjustment or the total load.

* 9.4 NONLINEAR RELATIONSHIPS

There are times when either economic theory or empirical facts indicate that a nonlinear form should be used to describe the relationship between two variables. In such cases, it is often possible to find a transformation of the data such that the transformed data is approximately linear and standard linear regression techniques may be applied.

As an example, it appears that a nonlinear line such as the one drawn in Figure 9.7 is more likely to fit the data then a straight line. The curve drawn in Figure 9.7 may be represented by a function of the form

$$Z = AW^{\beta} \tag{9.36}$$

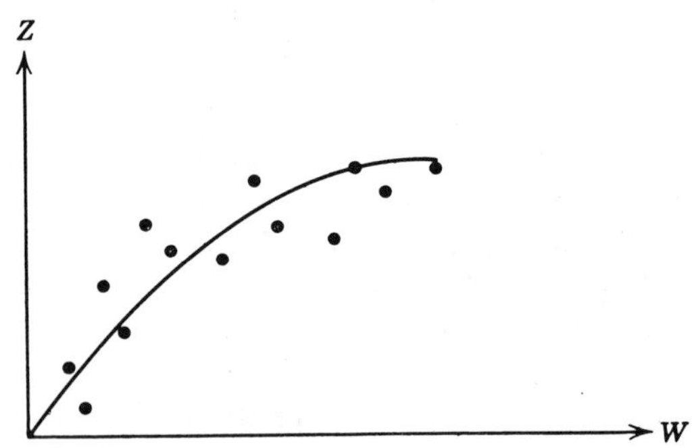

Figure 9.7 A nonlinear relationship.

where A and β are constants. If W is equal to one, then Z is equal to A. If β is positive, Z is equal to zero when W is equal to zero. When β is between zero and one, the slope of the line continually decreases. The effect of changes in the value of the parameter β on the function (9.36) is illustrated in Figure 9.8. If β is greater than one, the slope continually increases as W increases. If β is negative, the function does not pass through the origin at all.

All of the points in the scatter diagram in Figure 9.7 do not fall exactly on the curve $Z = AW^{\beta}$. Thus, the function (9.36) is only an approximation

* This section may be omitted without loss of continuity.

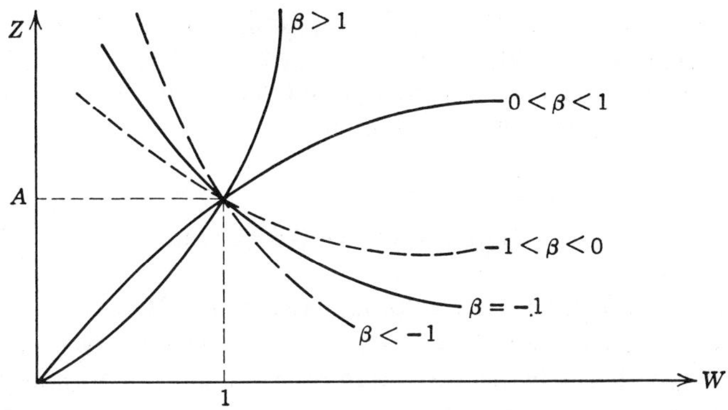

Figure 9.8 The effect of the parameter β.

to the true relationship between the variables W and Z. A more accurate description of the data would involve a disturbance term. If we postulate a multiplicative disturbance term v, the relationship takes the form

(9.37) $$Z_i = AW_i^{\beta} v_i$$

The logarithm of this function is

(9.38) $$\log Z_i = \log A + \beta \log W_i + \log v_i$$

If we let

$$\begin{aligned} Y_i &= \log Z_i \\ \alpha &= \log A \\ \beta &= \beta \\ X_i &= \log W_i \\ u_i &= \log v_i \end{aligned}$$

the function can be rewritten as

$$Y_i = \alpha + \beta X_i + u_i$$

and standard least squares techniques can be used to estimate the parameters α and β.[4]

In some cases, making transformations to keep the calculations in a simple linear framework may lead to a violation of some of the assumptions underlying the linear estimation techniques. For example, the assumption of a

[4] The assumption of a multiplicative disturbance is frequently difficult to support on the basis of a priori information. It is a convenient assumption to work with, however, since the disturbance term is additive in the transformed relation.

multiplicative disturbance term in (9.37) yields an additive disturbance term in the transformed relation (9.38). We may find, however, that the additive disturbance terms are not independently distributed. Fortunately, it is often possible to check from the data on the validity of the basic assumptions. Several such checks will be discussed latter.

9.5 CORRELATION ANALYSIS

Economists frequently use correlation analysis when they want to compare the closeness of association between variables. In this section we shall consider correlation analysis involving two variables. Many of the concepts presented here will be extended to cases where there are more than two variables when we discuss multiple regression analysis in Chapter 10.

The correlation coefficient presented in Chapter 5 was derived on the assumption that the random variables X and Y had a joint (or bivariate) distribution. In regression analysis, on the other hand, we are dealing with a collection of subpopulations of Y's corresponding to fixed values of X. In this case, X is not a random variable. A correlation coefficient, closely related to the one discussed in Chapter 5, can be defined for the regression model presented in equations (9.1) to (9.5).

The interpretation of a correlation coefficient depends on the type of relationship assumed to exist between the variables X and Y. When both X and Y are random variables, the correlation coefficient serves as a measure of the covariability that exists between the variables. In regression analysis, the correlation coefficient provides a measure of the closeness of fit of a regression line to the distribution of observations.

In this section we shall define correlation coefficients for (1) the regression model and (2) for the case in which both X and Y are random variables. We shall also show how to estimate correlation coefficients from sample data and how to use such estimates to test hypotheses about the population correlation coefficient. We begin with the case in which both X and Y are random variables.

9.5.1 The Bivariate Distribution

Suppose that there are 100 students in an economics course and that each student takes two tests during the semester. We shall let X and Y denote the scores obtained on the first and second tests, respectively. The frequency distributions of scores on the tests are presented in Sections *a* and *b* of Figure 9.9. These distributions show the numbers of students who received the various possible scores on each test. We can readily see, for example, that 18 students received a score of 50 on the first test but that only 13 students received a score of 50 on the second test.

The basic information needed to determine the closeness of the association between the scores on the two tests is presented in Table 9.3. The last row in Table 9.3 is the frequency distribution of scores on the first test. The last

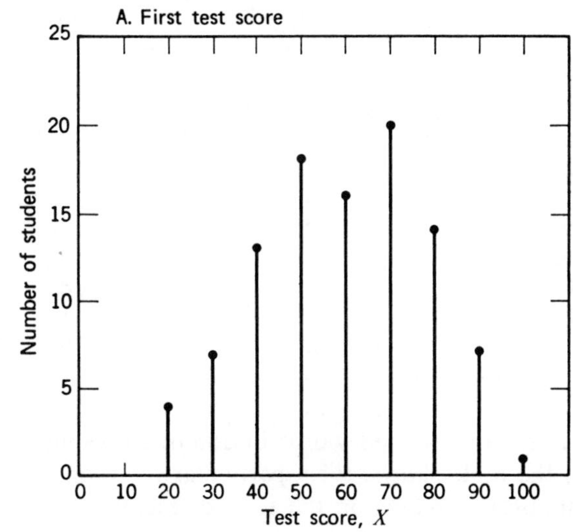

Figure 9.9a Frequency distribution of scores on Test 1.

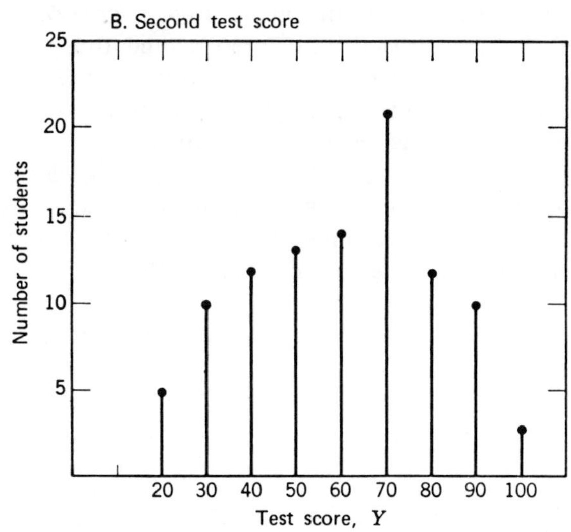

Figure 9.9b Frequency distribution of scores on Test 2

Table 9.3 Joint Frequency Distribution of Test Scores

		Score on the First Test X									
	Y	20	30	40	50	60	70	80	90	100	Total
	20	2	2	1							5
	30	2	2	3	2	1					10
	40		1	2	3	3	2	1			12
Score	50		1	2	4	4	2				13
on the	60		1	2	4	2	4	1			14
Second	70			3	4	5	6	3			21
Test	80				1	1	4	4	2		12
	90						2	4	4		10
	100							1	1	1	3
	Total	4	7	13	18	16	20	14	7	1	100

column in the table is the frequency distribution of scores on the second test. Each column in the table shows, for a particular score on the first test, the frequency distribution of scores on the second test. Each row shows, for a particular score on the second test, the frequency distribution of scores on the first test. For example, there were 14 students who received a score of 60 on the second test. Only one of these students received a score of 30 on the first test, two received 40, four received 50, two received 60, four received 70, and one received 80. As we look across the successive rows of Table 9.3, we see that those students who received low (high) scores on the first test tended to receive low (high) scores on the second test. In other words, X and Y tend to increase together. Although X and Y are obviously dependent, there is no compelling reason to believe that X causes Y or that Y causes X. In fact, if test Y follows test X, it seems reasonable to conclude that, if there is a cause-effect relationship between these variables, it would have to be of the "X causes Y" form.

If there had been a very large number of students and the test scores could vary continuously from 0 to 100, the joint distribution of X and Y would have a frequency surface similar to that shown in Figure 9.10. When this population is sliced at a given value of X, say, X_0, we obtain a frequency distribution for Y given the value X_0. Similarly, we have different frequency distributions for X corresponding to various given values of Y. The correlation coefficient defined in Chapter 5 was based on this type of bivariate distribution.

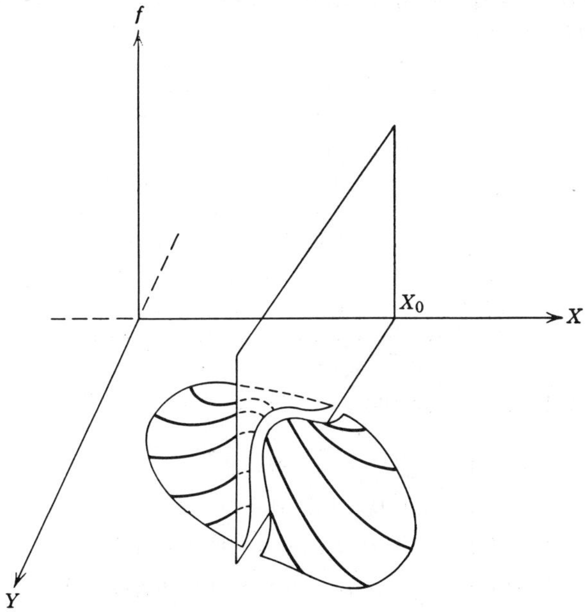

Figure 9.10 Joint distribution of X and Y.

9.5.2 The Correlation Coefficient

Let X and Y be random variables with a bivariate distribution with

$$E(X) = \mu_x, E(Y) = \mu_y, \quad \mathrm{Var}(X) = \sigma_x^2, \quad \mathrm{Var}(Y) = \sigma_y^2$$

and

$$\mathrm{Cov}(X, Y) = E[(X - \mu_x)(Y - \mu_y)]$$

The population correlation coefficient is defined as

$$\rho = \frac{\mathrm{Cov}(X, Y)}{\sigma_x \sigma_y} \tag{9.39}$$

It was shown in Chapter 5 that ρ is the covariance of two standardized variables. The population correlation coefficient is therefore independent of units of measurements, symmetrical with respect to X and Y, and can be either positive or negative.

Suppose that X and Y are discrete random variables with the following probability distribution:

	X		
Y	1	3	$g(y)$
2	0.5	0	0.5
6	0	0.5	0.5
$f(x)$	0.5	0.5	

The expected values and variances of X and Y are

$$\begin{aligned} E(X) &= 0.5(1) + 0.5(3) = 2 \\ E(Y) &= 0.5(2) + 0.5(6) = 4 \\ \mathrm{Var}(X) &= E(X - \mu_x)^2 = 0.5(1) + 0.5(1) = 1 \\ \mathrm{Var}(Y) &= E(Y - \mu_y)^2 = 0.5(4) + 0.5(4) = 4 \end{aligned}$$

The covariance of X and Y is

$$\mathrm{Cov}(X, Y) = E(XY) - E(X)E(Y) = 0.5(2) + 0.5(18) - (2)(4) = 2$$

The population correlation coefficient is therefore equal to

$$\rho = \frac{2}{\sqrt{1}\sqrt{4}} = 1.0$$

In other words, there is perfect covariability between X and Y. Inspection of the probability distribution of X and Y will verify that $Y = 2X$ for all values of X. If the probability distribution has been such that $Y = -2X$, then the population correlation coefficient would have been equal to -1. Although in this particular example the intercept of the relationship between X and Y is zero, perfect correlation occurs whenever there is an exact linear relationship between X and Y regardless of the value of the intercept. To verify this, let $W = 4 + X$, note that $Y = -8 + 2W$, and compute the population correlation coefficient for Y and W.

If X and Y are independent random variables, then $\mathrm{Cov}(X, Y) = 0$, and as a result, the population correlation coefficient is also equal to zero. Recall, however, that $\mathrm{Cov}(X, Y) = 0$ does not necessarily mean that X and Y are independent. To illustrate this, suppose that X and Y have the following probability distribution:

	X			
Y	1	2	3	$g(y)$
0	0	1/3	0	3/9
1	1/9	0	1/9	2/9
2	1/9	0	1/9	2/9
3	1/9	0	1/9	2/9
$f(x)$	3/9	3/9	3/9	

The variables X and Y are not independent random variables. In other words, $f(x, y)$ is not equal to the product $f(x)g(y)$ for all values of X and Y. The expected values of X and Y are

$$E(X) = 2 \qquad \text{and} \qquad E(Y) = \frac{4}{3}$$

The covariance of X and Y is

$$\text{Cov}(X, Y) = E(XY) - E(X)E(Y) = \left(\frac{24}{9}\right) - \left(\frac{24}{9}\right) = 0$$

Thus, the population correlation coefficient is equal to zero even though X and Y are dependent random variables. A three-dimensional graph of X, Y and the probabilities $f(x, y)$ would reveal the lack of any linear relationship between X and Y. The construction of such a graph is left as an exercise for the reader.

In most practical problems, it is necessary to estimate ρ from sample data. Let us assume that a random sample is to be drawn from a bivariate *normal* distribution. It can be shown that the maximum-likelihood estimator of ρ is

$$\text{(9.40)} \qquad \hat{\rho} = r = \frac{\sum (X_i - \bar{X})(Y_i - \bar{Y})}{\sqrt{\sum (X_i - \bar{X})^2}\sqrt{\sum (Y_i - \bar{Y})^2}}$$

$$= \frac{n \sum X_i Y_i - (\sum X_i)(\sum Y_i)}{\sqrt{[n \sum X_i^2 - (\sum X_i)^2][n \sum Y_i^2 - (\sum Y_i)^2]}}$$

The estimator r is called the sample correlation coefficient. In other words, the sample correlation coefficient is equal to the estimator of $\text{Cov}(X, Y)$ divided by the product of the estimators of σ_x and σ_y.

To illustrate the calculation of r, suppose that a random sample of six students is selected from the entering freshman class of a junior college. Each student is given an IQ test on entering the college, and at the end of the year, an achievement test. The scores are presented in Table 9.4

Table 9.4

Student	Score: IQ Test X	Score: Achievement Test Y
1	110	90
2	120	80
3	125	90
4	130	95
5	130	85
6	140	95

We first compute

$$\sum X_i = 755 \qquad \sum Y_i = 535$$
$$\sum X_i^2 = 95{,}525 \qquad \sum Y_i^2 = 47{,}875$$
$$\sum X_i Y_i = 67{,}450$$
$$n\sum X_i Y_i - (\sum X_i)(\sum Y_i) = 775$$
$$n\sum X_i^2 - (\sum X_i)^2 = 3125$$
$$n\sum Y_i^2 - (\sum Y_i)^2 = 1025$$

The sample correlation coefficient is found to be

$$r = \frac{775}{\sqrt{(3{,}125)(1{,}025)}} \doteq \frac{775}{1790} \doteq 0.433$$

On the basis of this sample, it appears that students with higher IQ's tend to have higher achievement test scores at the end of their freshman year at this junior college. The relationship between IQ and achievement test scores does not appear to be especially close, however.

To perform tests of hypotheses about ρ, we need to know the sampling distribution of r. We shall continue to assume that X and Y have a bivariate normal distribution. The sampling distribution of r depends only on the value of ρ and the sample size n.

If we assume that $\rho = 0$, the sampling distribution of r is symmetric around 0 with a variance of

$$\text{Var}(r) = \frac{1 - r^2}{n - 2} \tag{9.41}$$

It can be shown, in this case, that the variable

$$t = \frac{r}{\sqrt{(1 - r^2)/(n - 2)}} \tag{9.42}$$

has a t distribution with $n - 2$ degrees of freedom. Tests of the significance of r are based on (9.42). We shall use the test scores of the six junior college students to illustrate the t test.

The hypothesis to be tested is that there is no relationship between a student's IQ and his achievement test score. The alternative hypothesis is that there is a positive correlation. Thus, $H_0: \rho = 0$ is to be tested against $H_a: \rho > 0$. We shall use the 5 percent level of significance. The region of rejection is in the upper tail of the t distribution, and the critical value of t is 2.1318. The observed value of t is

$$t = \frac{0.433}{\sqrt{(1 - (0.433)^2)/4}} = \frac{0.433}{0.451} = 0.96$$

Since the observed value 0.96 is in the region of acceptance, we cannot reject $H_0: \rho = 0$ at the 5 percent level of significance. In other words, we do not have enough evidence to conclude that there is a positive correlation between IQ and the achievement test score.

If we assume $\rho \neq 0$, the sampling distribution of r is no longer symmetric. It can be shown, however, that when r is transformed by

$$z_r = \frac{1}{2} \log_e \frac{1 + r}{1 - r}$$

the new variable z_r will be approximately normally distributed with mean

$$E(z_r) = z_\rho = \frac{1}{2} \log_e \frac{1 + \rho}{1 - \rho}$$

and variance

$$\text{Var}(z_r) = \sigma_r{}^2 = \frac{1}{n - 3}$$

Tests of significance of r are based on

$$z = \frac{z_r - z_\rho}{\sigma_r}$$

The values of z_r and z_ρ can be obtained directly from Table X in Appendix I.

Suppose that we measure the height and weight of 19 persons selected at random from a well-defined population and find that $r = 0.8$. Is there sufficient evidence to reject the hypothesis H_0: $\rho = 0.5$ in favor of the alternative H_a: $\rho > 0.5$ at the 5 percent level of significance?

In this case, the region of rejection is in the upper tail of the normal distribution and values of z larger than 1.645 will result in the rejection of H_0 in favor of H_a. We begin by computing

$$z_r = \frac{1}{2}\log_e \frac{1+0.8}{1-0.8} = 0.5 \log_e 9 = 1.09861$$

$$z_\rho = \frac{1}{2}\log_e \frac{1+0.5}{1-0.5} = 0.5 \log_e 3 = 0.54931$$

and

$$\sigma_r = \sqrt{\frac{1}{19-3}} = 0.25$$

The values of z_r and z_ρ can be obtained directly from Table X, Appendix I. The observed value of z is found to be

$$z = \frac{1.09861 - 0.54931}{0.25} = \frac{0.54930}{0.25} = 2.197$$

Since $z = 2.197$ is greater than 1.645, we reject H_0: $\rho = 0.5$ in favor of H_a: $\rho > 0.5$.

Confidence intervals for ρ are obtained by using the fact that

$$P[z_r - (z_{\varepsilon/2})\sigma_r < z_\rho < z_r + (z_{\varepsilon/2})\sigma_r] = 1 - \varepsilon$$

Suppose, for example, that a 95 percent interval for ρ is to be obtained from the sample of 19 measurements on height and weight. In this case, $z_{\varepsilon/2} = 1.96$, $\sigma_r = 0.25$, $z_r = 1.09861$, and the 95 percent confidence interval for z_ρ is

$$0.60861 < z_\rho < 1.58861$$

From the body of Table X, Appendix I, we find

z_ρ	ρ
0.60861	0.54
1.58861	0.92

Therefore, the 95 percent confidence interval of ρ is

$$0.54 < \rho < 0.92$$

9.5.3 Linear Correlation

It is explicitly stated in the regression model developed in section 9.1 that the variable X is not a random variable. Thus, a sample correlation coefficient computed from the data generated by a simple regression model does not measure exactly the same thing as a sample correlation coefficient based on data drawn from a bivariate distribution. The relationship between regression analysis and correlation analysis will be illustrated using the scatter diagram Figure 9.11.

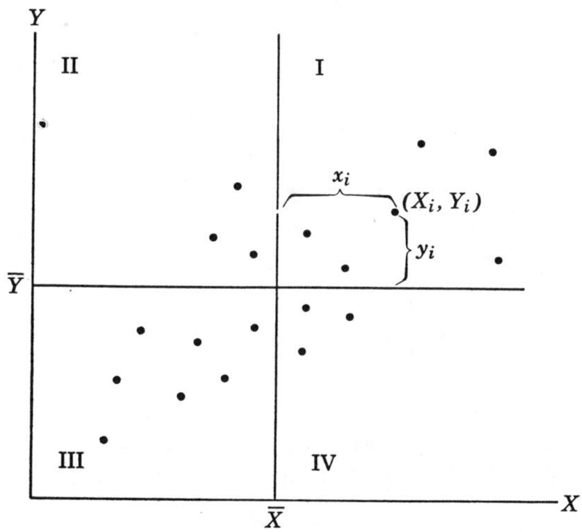

Figure 9.11 Scatter diagram.

The sample observations presented in Figure 9.11 have been divided into four quadrants by erecting perpendiculars to the X and Y axes at $\overline{X}$ and $\overline{Y}$. The distance between any given observation and the perpendicular at $\overline{X}$ is equal to the deviation from $\overline{X}$ and is denoted by $x_i = (X_i - \overline{X})$. The deviations from $\overline{Y}$ are denoted similarly by $y_i = (Y_i - \overline{Y})$. It is clear from inspection of Figure 9.11 that the products $x_i y_i = (X_i - \overline{X})(Y_i - \overline{Y})$ will be positive for all observations in quandrants I and III and negative for all observations in II and IV. Thus, if most of the observations in the sample are in quadrants I and III, the sum of the products $\sum_{i=1}^{n} x_i y_i$ would tend to be positive and would tend to be negative otherwise. The quantity $\sum_{i=1}^{n} x_i y_i$ could serve as a measure of association between X and Y except for the fact that it is influenced by the units of measure of X and Y and can be increased arbitrarily by adding observations. These shortcomings are easily corrected, however, by using the average of the sum of standardized deviations

$$u_i = \frac{x_i}{s_x} = \frac{(X_i - \overline{X})}{s_x} \quad \text{and} \quad v_i = \frac{y_i}{s_y} = \frac{(Y_i - \overline{Y})}{s_y}$$

where

$$s_x = \sqrt{\frac{\sum (X_i - \overline{X})^2}{n}} \quad \text{and} \quad s_y = \sqrt{\frac{\sum (Y_i - \overline{Y})^2}{n}}$$

The average of the sum of products $u_i v_i$ is the sample correlation coefficient

$$r = \frac{\sum u_i v_i}{n} = \frac{\sum (X_i - \bar{X})(Y_i - \bar{Y})}{n s_x s_y} \tag{9.43}$$

$$= \frac{\sum (X_i - \bar{X})(Y_i - \bar{Y})}{\sqrt{\sum (X_i - \bar{X})^2}\sqrt{\sum (Y_i - \bar{Y})^2}}$$

which is identical to the maximum-likelihood estimator of ρ presented in (9.40). The correlation coefficient will be positive if most of the observations are in quadrants I and III and negative otherwise.

There is a close relationship between the correlation coefficient and the slope of the regression line. The least squares estimator for b in the linear equation $Y = a + bX$ is equal to

$$b = \frac{\sum (X_i - \bar{X})(Y_i - \bar{Y})}{\sum (X_i - \bar{X})^2} = \frac{\sum x_i y_i}{\sum x_i^2} = \frac{\sum x_i y_i}{n s_x^2} \tag{9.44}$$

which is equivalent to the computing formula (9.12). Comparing equations (9.43) and (9.44), it is seen that the least squares estimator of b can be written as

$$b = r\left[\frac{s_y}{s_x}\right] \tag{9.45}$$

Both s_y and s_x are always positive. The sign of b is therefore always identical to that of r.

Next, let us investigate the relationship between r and the sample standard deviation of regression $s_{y.x}$ defined by equation (9.17). We begin by noting that the total deviation $y_i = (Y_i - \bar{Y})$ is equal to the sum of the "unexplained error" $e_i = (Y_i - \hat{Y}_i)$ and the "explained error" $\hat{y}_i = (\hat{Y}_i - \bar{Y})$ where $\hat{Y}_i = a + bX_i$ is computed from the least squares regression line. In other words,

$$y_i = (Y_i - \bar{Y}) = (Y_i - \hat{Y}_i) + (\hat{Y}_i - \bar{Y}) = e_i + \hat{y}_i \tag{9.46}$$

Squaring both sides of (9.46) and summing over i, we obtain

$$\sum y_i^2 = \sum e_i^2 + \sum \hat{y}_i^2 + 2\sum e_i \hat{y}_i \tag{9.47}$$

which simplifies to

$$\sum y_i^2 = \sum e_i^2 + \sum \hat{y}_i^2 \tag{9.48}$$

because the term $2\sum e_i \hat{y}_i$ is equal to zero.[5]

[5] This is verified as follows. First note that, by definition, the least squares estimator of the intercept coefficient is equal to $a = \bar{Y} - b\bar{X}$ and therefore

$$\hat{y}_i = \hat{Y}_i - \bar{Y} = a + bX_i - \bar{Y} = \bar{Y} - b\bar{X} + bX_i - \bar{Y} = b(X_i - \bar{X}) = bx_i$$

The square of the correlation coefficient is called the *coefficient of determination* and is equal to the ratio of the "explained" sum of squares $\sum \hat{y}_i^2$ divided by the "total" sum of squares $\sum y_i^2$. This is readily seen as follows:

$$\frac{\sum \hat{y}_i^2}{\sum y_i^2} = \frac{\sum (bx_i)^2}{\sum y_i^2} = b^2 \frac{\sum x_i^2}{\sum y_i^2} = b^2 \left[\frac{s_x^2}{s_y^2}\right] = r^2 \tag{9.49}$$

Dividing both sides of (9.48) by $\sum y_i^2$, substituting r^2 for the ratio

$$(\textstyle\sum \hat{y}_i^2 / \sum y_i^2),$$

and rearranging terms, we obtain

$$r^2 = 1 - \frac{\sum e_i^2}{\sum y_i^2} \tag{9.50}$$

The "unexplained" sum of squares $\sum e_i^2$ is always less than or equal to the "total" sum of square $\sum y_i^2$. The coefficient of determination is therefore always greater than or equal to zero but less than or equal to unity. If $\sum e_i^2$ is equal to zero, the sample standard deviation of regression $s_{y\,.\,x}$ is also equal to zero and r^2 is equal to unity. (It is left to the reader to verify that, if $r^2 = 1$, then s_a and s_b are equal to zero.) Thus, the correlation coefficient computed from data generated by a regression model can be viewed as a measure of the closeness of fit of the data to the regression line. In general, the closer the data fit the regression line, the closer r^2 will be to unity. If all the points fall exactly on the regression line, for example, then r^2 is equal to unity.[6]

We shall illustrate these relationships using the numerical data in Table 9.5. Economic theory suggests that there may be some type of relationship between

[6] This statement is based on the implicit assumption that the Y values are scattered. If Y is always equal to a fixed value m regardless of the value of X, then all points lie on a horizontal line at $Y = m = \bar{Y}$. In this case both $\sum(Y_i - \bar{Y})$ and $\sum(\hat{Y}_i - \bar{Y})$ are equal to zero. Thus r^2 is equal to zero even though all points lie on a straight line. In a sense, there is no regression problem in this case because the Y values have no distribution.

Second, substitute the above result into $\sum e_i \hat{y}_i$ and simplify to obtain

$$\begin{aligned}
\sum e_i \hat{y}_i &= \sum (Y_i - \hat{Y}_i) b(X_i - \bar{X}) \\
&= b \sum x_i (Y_i - \bar{Y} - \hat{Y}_i - \bar{Y}) \quad \text{(add and subtract } \bar{Y}) \\
&= b \sum x_i (y_i - \hat{y}_i) \\
&= b \sum x_i (y_i - bx_i) \quad \text{(substitute in for } \hat{y}_i) \\
&= b \sum x_i y_i - b^2 \sum x_i^2 \\
&= \frac{(\sum x_i y_i)^2}{\sum x_i^2} - \frac{(\sum x_i y_i)^2}{\sum x_i^2} \quad \text{[using (9.44)]} \\
&= 0
\end{aligned}$$

as previously stated.

wage increases and price increases. Suppose that initially we hypothesize that price increases are dependent on and linearly related to wage increases and state this relationship in the form

$$Y_i = \alpha + \beta X_i + u_i \tag{9.51}$$

where Y_i is the percentage increase in prices in industry i and X_i is the corresponding percentage increases in wages during the 1953–1957 period.

Table 9.5 Price and Wage Increases in Selected Industries, 1953–1957

Industry	Percent Price Rise Y	Percent Wage Rise X
Durable manufacturing	13.0	19.3
Electrical machinery	21.9	18.2
Rubber products	20.5	20.2
Contract construction	9.8	21.0
Steelmaking	30.7	26.4
Coal mining	13.4	22.6
Automobiles	14.1	19.2
Chemicals	3.5	22.4

Source. Bureau of Labor Statistics, *Monthly Labor Review.*

$\sum Y_i = 126.9$ $\qquad$ $\sum Y_i^2 = 2498.01$

$\sum X_i = 169.3$ $\qquad$ $\sum X_i^2 = 3630.89$

$\sum X_i Y_i = 2731.82$ $\qquad$ $(\sum Y_i)^2 = 16{,}103.61$

$\sum X_i \sum Y_i = 21{,}484.17$ $\qquad$ $(\sum X_i)^2 = 28{,}662.49$

The least-squares estimates of α and β are

$$b = \frac{n\sum X_i Y_i - \sum X_i \sum Y_i}{n\sum X_i^2 - (\sum X_i)^2} = \frac{21{,}854.56 - 21{,}484.17}{29{,}047.12 - 28{,}662.49}$$

$$= \frac{370.39}{384.63} = 0.963$$

and

$$a = \bar{Y} - b\bar{X} = 15.862 - 0.963(21.162) = -4.517$$

The least-squares regression line is therefore equal to

$$\hat{Y}_i = -4.517 + 0.963X_i \tag{9.52}$$

The sample standard deviations for X and Y are equal to

$$s_x = \sqrt{\frac{384.63}{8}} = 6.928$$

and

$$s_y = \sqrt{\frac{3880.47}{8}} = 22.023$$

Using the computing formula (9.40), the sample correlation coefficient is found to be equal to

$$r = \frac{370.39}{\sqrt{(384.63)(3880.47)}} = \frac{370.39}{1221.70} = 0.3032$$

Checking this result against the relationship (9.45), we see that

$$b = r\left[\frac{s_y}{s_x}\right] = 0.3032\left[\frac{22.023}{6.928}\right] = 0.963$$

as expected.

The low value of r indicates that the data do not fit the least-squares regression line very well. The fact that the regression line does not "explain" much of the variance in the Y values is readily seen by noting that the total sum of squares is equal to

$$\sum y_i^2 = \sum Y_i^2 - \frac{1}{n}\left(\sum Y_i\right)^2 = 2498.01 - 2012.99 = 485.06$$

while the error sum of squares is equal to

$$\begin{aligned}\sum e_i^2 &= \sum Y_i^2 - a\sum Y_i - b\sum X_i Y_i \\ &= 2498.01 + 573.21 - 2630.74 = 440.48\end{aligned}$$

In other words, $100\left[\frac{440.48}{485.06}\right] = 90.8$ percent of the total variance in the Y values is left "unexplained" by the regression model. Another way to arrive at the same result is to compute

$$r^2 = (0.3032)^2 = 1 - \left[\frac{440.48}{485.06}\right] = 0.092$$

which indicates that only 9.2 percent of the total variance of the Y values is "explained" by the regression line.

The initial hypothesis that price increases are dependent on and linearly related to wage increases is not substantiated by the available data. The lack

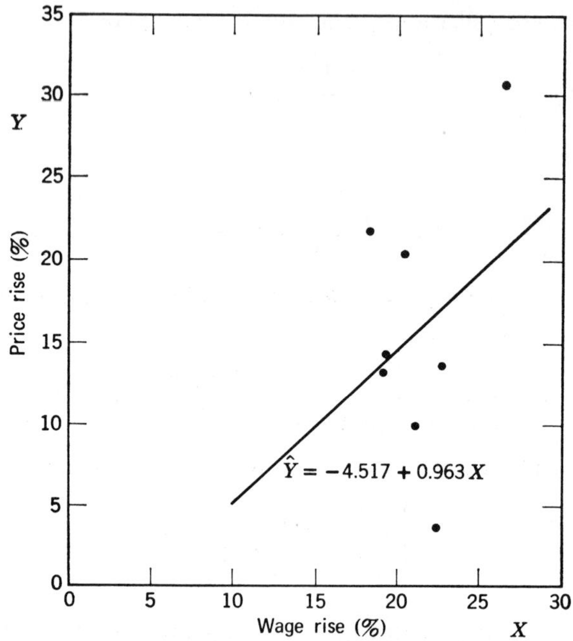

Figure 9.12 Price and wage increases.

of a close linear relationship between the percentage increases in wages and prices is verified by inspection of Figure 9.12. For the industries in the sample, there was a great deal of variation in the percentage increases in prices while the percentage increases in wages are about the same for all industries. From the limited information available, it appears that inflation during the 1953–1957 period tended to be concentrated in the capital-goods industries such as steelmaking and electrical machinery. More detailed information would be required in order to determine if the inflation in these "bottleneck" industries was of the "demand-pull" or "cost-push" variety.

9.6 ECONOMIC APPLICATIONS OF SIMPLE REGRESSION ANALYSIS[7]

Virtually all economic concepts involve relationships between two or more variables. Demand and supply curves, production functions, cost curves, and consumption functions are important in theory and commonly used in

[7] In actual practice, most of the regression relationships discussed in this section would normally involve several independent variables and be part of a multiple equation model. Some readers may prefer to skim (or omit) this section for the moment and return to it after studying the material on multiple regression.

applied economic analysis. Several such relationships estimated by means of regression analysis are presented below.

9.6.1 Consumer Demand Curves

The statistical estimation of demand curves has a history extending back to the early 1900's. In this section, we shall discuss several consumer demand curves estimated by B. F. Stanton in 1961. Stanton was interested in determining whether the aggregate United States demand curves for certain commodities had different slopes and levels in different seasons of the year. He consolidated quarterly data into two periods, summer (including the second and third quarters of the calendar year) and winter (including the fourth quarter of a given year and the first quarter of the succeeding one). He then estimated separate demand curves for the winter and summer seasons. His results for three commodities are shown in Figures 9.13, 9.14, and 9.15.

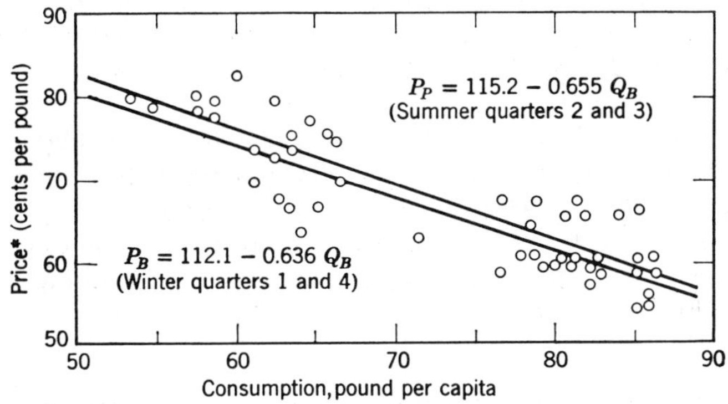

Figure 9.13 Beef: price-consumption relations—quarterly data, 1948–1959. U.S. Department of Agriculture, Neg. 8282–60(12), Agr. Mkt. Service. The asterisk (*) indicates retail price deflated by C.P.I.

In the case of beef, there appears to be no significant difference in the demand curves for the two seasons. However, the demand for pork is much more elastic, in winter than in summer and the winter demand curve stands at a higher level than the summer curve over most of its length. Thus, it appears that during 1953–1959, consumers were willing to pay more than 50 cents a pound in the winter for a pork supply running at an annual rate of 65 pounds, but only a little over 40 cents a pound for the same rate of supply in the summer.

The summer and winter observations for beef are closely intermingled, much as if they might be two samples drawn from the same population. In

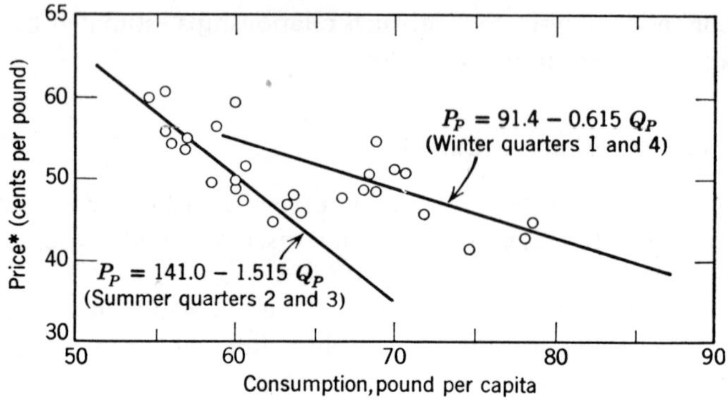

Figure 9.14 Pork: price-consumption relations—quarterly data, 1953–1959. U.S. Department of Agriculture, Neg. 8283–60(12), Agr. Mkt. Service. The asterisk (*) indicates retail price deflated by C.P.I.

the case of pork, there is little intermingling of the summer and winter observations and the general drift or "slope" of the two sets of observations looks quite different, as though they were samples drawn from distinctly different populations.

The broiler case appears to be intermediate between those of beef and pork. Visually, the difference between slopes of the summer and winter demand curves does not appear large in relation to the scatter of the observations about the two regression lines, but it does appear that the levels of the curves over the range of annual consumption rates experienced during 1953–1959 are significantly different.

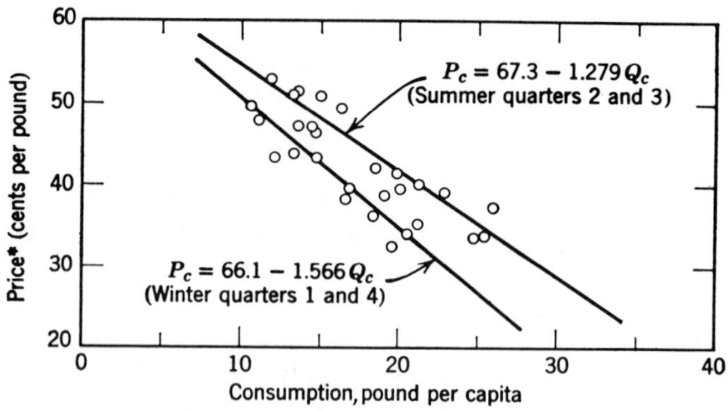

Figure 9.15 Broilers: price consumption relations—quarterly data, 1953–1959. U.S. Department of Agriculture, Neg. 8284–60(12), Agr. Mkt. Service. The asterisk (*) indicates retail price deflated by C.P.I.

Economists are often interested in knowing the degree of responsiveness of quantity demanded to changes in the market price. Such information is useful in planning price support programs or when forecasting price changes. The coefficient of price elasticity of demand is used to measure the degree of responsiveness. By definition, the coefficient of price elasticity of demand is equal to

$$\eta = \frac{\Delta Q}{\Delta P}\frac{P}{Q} = \frac{\text{percentage change in quantity}}{\text{percentage change in price}}$$

where Q is quantity, ΔQ is the change in quantity, P is price, and ΔP is the change in price. If the absolute value of η is between 0 and 1, we say that demand is inelastic, in which case a price increase (decrease) results in an increase (decrease) in total expenditures on the commodity. If, on the other hand, the absolute value of η is greater than 1, demand is said to be elastic and a price increase (decrease) results in a decrease (increase) in total expenditures.

The coefficient of price elasticity of demand is different at each point on an arithmetically linear demand curve. At a particular point (p_1, q_1) on the arithmetically linear demand curve $q = a + bp$, the coefficient is equal to

$$\eta = b\left(\frac{p_1}{q_1}\right)$$

The pork demand curves in Figure 9.14 are written in the form

$$p = -\left(\frac{1}{b}\right)a + \left(\frac{1}{b}\right)q$$

For the points on the demand curve at which $q_1 = 65$ pounds, the values of η are calculated as follows:

Summer

$$\begin{aligned} p_1 &= 141.0 - 1.515q_1 = 141.0 - (1.515)65 \\ &= 141.0 - 98.475 = 42.525 \text{ cents} \end{aligned}$$

$$\eta_s = \left(-\frac{1}{1.515}\right)\frac{42.525}{65.000} = -0.660(0.670) = -0.442$$

Winter

$$\begin{aligned} p_1 &= 91.4 - 0.615q_1 = 91.4 - (0.615)65 \\ &= 91.4 - 39.975 = 51.425 \text{ cents} \end{aligned}$$

$$\eta_w = \left(-\frac{1}{0.615}\right)\frac{51.425}{65.000} = -1.626(0.791) = -1.286$$

If we calculate the elasticities at points nearer the centers of the respective sets of observations, say, 60 pounds and 51.1 cents for summer and 70 pounds and 48.4 cents for winter, the results are as follows:

$$\textit{Summer} \quad \eta_s = -0.562$$
$$\textit{Winter} \quad \eta_w = -1.137$$

At any given price, the winter demand for pork is more elastic than the summer demand. Thus, a price change in the winter has a greater effect on quantity of pork demanded than a similar price change has in the summer.

9.6.2 Aggregate Consumption Functions[8]

Figure 9.16 represents simple regression relationships between personal consumption expenditures and disposable personal income (both in billions of dollars of 1961 purchasing power) for 1929–1961 as a whole and for two subperiods, 1929–1941 and 1949–1961. For any given period, the slope of the regression line is an estimate of the change in personal consumption expenditures that would result from a one dollar increase in disposable personal income.

Table 9.6 Aggregate Consumption Functions for the United States 1929–1961[a]

Equation Number	Years	Regression Equation
1	1929–1941 plus 1949–1961	$X_1 = 9.172 + 0.898X_2$
2	1929–1941	$X_1 = 32.338 + 0.736X_2$
3	1949–1961	$X_1 = 4.882 + 0.913X_2$

[a] X_1 is personal consumption expenditures in billions of dollars per year. X_2 is disposable personal income in billions of dollars per year. Both variables are expressed in constant 1961 dollars.

The change in consumption due to a change in income is commonly called the marginal propensity to consume. This concept figured prominently in Keynes' general theory of employment, interest, and money and is important in formulating policies for economic stability.

[8] The consumption functions presented in this section are frequently referred to as Keynesian consumption functions. This in usually one equation in a set of simultaneous equations representing a Keynesian model of the economy. Simple regression techniques are not always suitable for estimating the relationships in a set of simultaneous equations. In the case of the consumption function, it is somewhat unrealistic to specify income as an independent variable, since the level of consumption partly determines income. The discussion of the empirical results presented in this section is not meant to imply that simple regression techniques are always suitable for estimating simple Keynesian consumption functions.

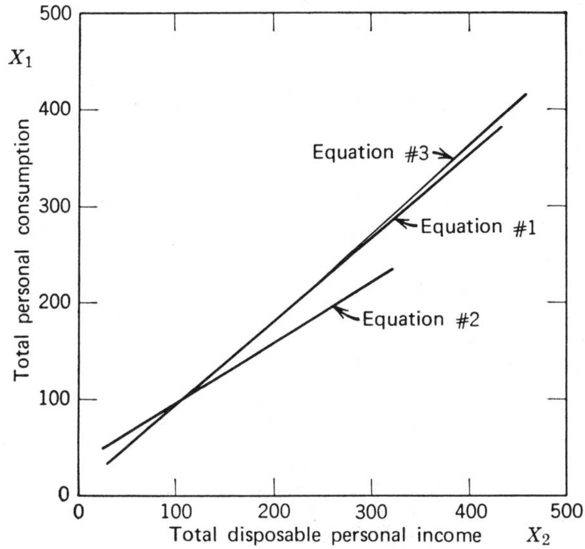

Figure 9.16 Aggregate consumption functions for the United States, 1929–1961.

The regression equations for the several time periods are given in Table 9.6. Equation 2 implies a marginal propensity to consume of approximately 0.74 during 1929–1941. In other words, if total disposable income increased by a billion dollars, total consumption expenditures were expected to increase by approximately 0.74 billion dollars. Equation 3 suggests that the marginal propensity to consume for the United States was somewhat higher after World War II. According to Equation 3, the marginal propensity to consume was 0.91 during the 1949-1961 period. The observations from 1942 to 1948 were not used in the regressions because of the many unusual features of this period—direct price controls, consumer rationing, rationing of raw materials to manufacturers, and many other devices and situations. If we are interested in estimating a consumption function which is applicable to policy or forecasting problems in a peacetime economy, use of the 1942–1948 observations will "contaminate" the regression relationship. In terms of the discussions of frequency distributions and sampling in the preceding chapters, it is clear that a regression relationship based on all the years 1942 to 1961 is equivalent to combining samples drawn from two distinctly different populations.

9.6.3 Production Functions

Production functions have wide application in economic analysis. Much of the theory of resource use within the individual firm revolves around production function concepts. Figure 9.17 shows regression relationships

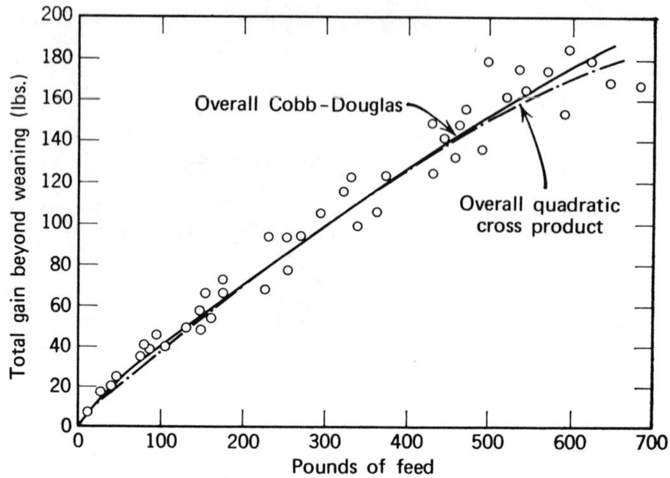

Feed-gain relationships for a 16 percent protein ration with aureomycin

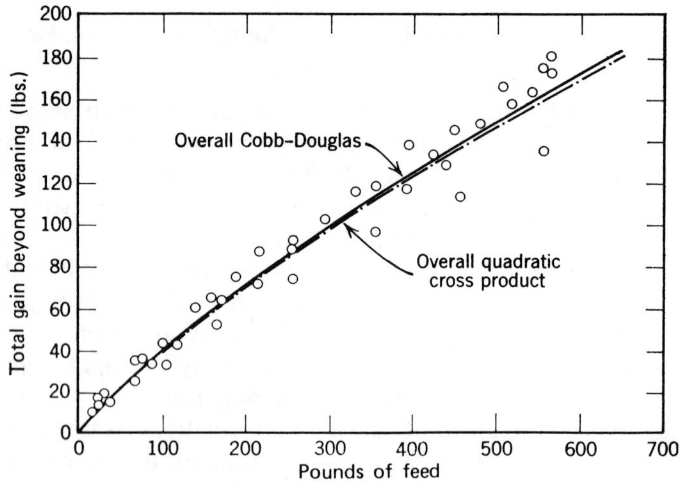

Feed-gain relationships for a 14 percent protein ration with aureomycin

Figure 9.17 Production functions.

fitted to experimental data on amounts of different kinds of protein rations fed to pigs and the associated gains in weight. These regression equations represent simple production functions that show the relationship between input (feed) and output (pork). Two kinds of regression curves have been fitted to the data from each experiment. In the first experiment, a number of pigs were fed on a 16 percent protein ration while, in the second, a number of pigs were fed on a 14 percent protein ration. (The two groups of pigs represent random drawings from the same population, which included a relatively homogeneous lot of pigs in terms of genetic characteristics.)

From an economic standpoint, the investigators were interested in the shapes of the production functions for each group of pigs separately, as well as in possible differences between the production functions obtained with a 14 percent rather than a 16 percent protein ration. Assuming that the early increments of feed produced weight gains which were worth significantly more than the value of the feed, production theory indicates that profit from the feeding operation would be maximized if continued until the next \$1 worth of feed would increase the value of the pig by only \$1. The regression lines in Figure 9.17 do suggest a gradual tapering off of the marginal physical productivity of successive increments of feed.

9.6.4 Cost Functions

Figure 9.18 shows a regression relationship between the total expenses of a busline in the United Kingdom and the millions of car miles covered by its equipment in successive four-week periods. The estimated relationship between total expenses in £100,000, Y, and millions of car miles traveled, X, is equal to

$$\hat{Y} = 0.6558 + 0.4433X$$

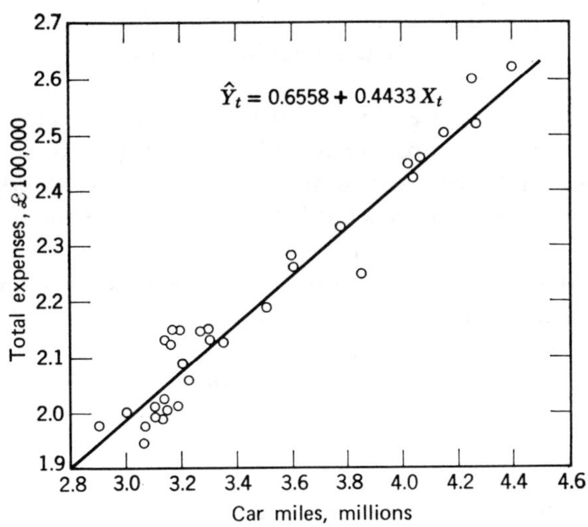

Figure 9.18 Total expenses of a road passenger transport company as a function of total car miles driven, United Kingdom, 1949–1952, by week periods. (*Source*. J. Johnston, *Statistical Cost Analysis* (New York: McGraw-Hill Book Company, Inc. 1960), p. 80.)

The constant term or intercept is an estimate of the expected fixed costs of the company. In other words, the company can expect to incur a cost of £65,580 during each four-week period even if its buses are completely idle. The slope of the regression line indicates that each additional million car miles traveled by the company's buses during a four-week period would increase its total expenses by £44,330. Since a pound sterling was worth about $2.80 during the period indicated, this amounts to about 12 cents a mile. The slope of the regression line is an estimate of marginal cost. The linear regression equation implies that the company's marginal cost was constant over the range from 2.8 to 4.6 million miles during the 1949–1952 period.

The examples presented here are a minute sampling of the tremendous volume of regression analysis, published and unpublished, that is constantly being done by economists and business analysts interested in estimating economic relationships and using them for purposes of policy, prediction, or control.

9.6.5 Nonsense and Spurious Regressions

Many years ago a statistician wrote, "With statistics the unscrupulous deceive the unwary and the innocent deceive themselves." This appraisal is still widely applicable to the interpretation of regression relationships based on time series. There are many devices, tests, and safeguards that a competent economic statistician can use to avoid gross misinterpretation of regressions based on time trends. But the temptation to build causal explanations around the trend elements in possibly unrelated time series is very strong among the public at large.

In the United States since World War II, population has increased steadily. Output per manhour and per manyear has also increased; so have prices of nonfarm goods and services. The numbers of workers in agriculture and in coal mining have persistently declined. Thus, time series data on these variables will show strong trends.

If a scatter diagram is made showing paired observations of two time series which have no known logical relationship but each of which contains a strong trend, the scatter will suggest a high degree of relationship between the two. The average value of one will increase (or decrease) steadily as the other increases over time. In such a case, statistical tests routinely applied will suggest that the relationship could hardly be due to chance. If the two series are logically related, as in the case of the consumption functions in Figure 9.16, we are inclined to take the regression relationship at its face value. But if there is no logical basis for expecting a permanently close association between two series, the relationship has no practical usefulness or safe interpretation.

Regression relationships based on data other than time series data can also be deceiving at times. It should be clear that the various arithmetic operations involved in regression analysis can be applied to any set of numbers regardless of whether these numbers have any practical or scientific meaning. For example, we could randomly select ten numbers out of a telephone directory, tabulate them in a column parallel to ten numbers selected from a table of random digits, plot the observations, and draw some kind of line or curve representing the central tendency or drift of the observations.

In general, if there were no logical reason for expecting a systematic relationship between two sets of numbers, we would expect to find the observations so widely scattered that no systematic upward or downward drift or slope would appear. But if the number of (logically unrelated) observations were small, we would not infrequently find scatter diagrams which *looked* as though there was a definite regression relationship.

At the extreme, it is clear that if each of two series contains only two observations, these define two points through which one and only one straight line can be drawn. It is quite unlikely that the slope of such a line will be exactly zero. Sets of three or four observations (logically unrelated) will also quite frequently suggest to the eye a definite relationship. But as the number of observations in two unrelated series increases, we expect that the great majority of samples will show a wide scatter and that no line will appear to be of much use for estimating values of one series from specified values of the other.

* 9.7 A NOTE ON ECONOMIC MODELS

In recent years, economic models have become important tools in planning, analysis, and forecasting. A model consists of a set of relationships or equations which describe the links between different variables. Most economic models consist of structural relationships that are derived from economic theory and/or the technological facts of production. The structural relationships may be grouped into four categories—behavioral, technical, institutional, and identities.

The *behavioral* relationships include demand curves, supply curves, and Engel curves that reflect the behavior of particular kinds of economic units. Consumers adjust their purchases to changes in commodity prices and in their own incomes; business firms adjust their outputs to changes in the prices of raw materials and factors of production and to the state of market demand for their products.

Technical relationships are based on technological knowledge and, therefore, they include mainly production functions. Some statistical production

* This section may be omitted without loss of continuity.

functions have already been shown in Figure 9.17. Table 9.7 shows two linear production functions or "activities" of the type which is basic to linear programming. They reflect two sets of farming operations, one of which (with average weather) would produce 40 bushels of corn and the other 20 bushels of wheat. The upper part of Table 9.7 shows inputs and expected output on a per acre basis, while the lower part states requirements for land and labor per bushel of expected output. If the production process indicated for corn were replicated on one million acres of land of the appropriate quality, the production process in total would require 0.8 million hours of planting labor, 1.0 million hours of cultivating labor, and 0.8 million hours of harvesting labor. With average weather, these inputs would produce 40 million bushels of corn.

Table 9.7 also presents an example of input-output relationships for the entire United States transportation equipment industry in 1947. The column of input requirements per $10,000 of gross output is formally a *linearly homogeneous production function* of the same type as the production activities for wheat and corn. The input categories in an input-output table are so broad that it is necessary to express them in terms of dollar values. These dollar figures implicitly include 1947 quantities of the many specific input commodities valued at their 1947 prices.

Input-output coefficients imply an exact linear relationship between the amount of total gross output of a given industry and the amount of inputs required from each of the industries which supply it and are, therefore, usually used as though they were exact numbers. In reality, of course, they are subject to various imperfections resulting from inaccuracies in the original data, technological changes over time, and aggregation problems.

Institutional relationships in economics are those specified by law or by regulatory agencies. A clear-cut example would be a schedule of personal income tax rates prescribed by the Congress. In principle, the effect of a change in the tax schedule upon the after-tax income of a specified individual or family is completely predictable, assuming that the size of money income before taxes does not change.

Figure 9.19 illustrates such a relationship based on the United States federal income tax forms for 1961. Curve *A* shows the tax which would be paid by a married taxpayer with a total of four exemptions filing a joint return, under the actual tax rate schedule of 1961. Curve *B* shows the tax payments associated with each specified level of personal income before taxes if the tax rates were reduced by four percentage points in each bracket. Finally, Curve *C* shows the effects of increasing personal exemptions by $400 for each family member. If Curve *A* were a straight line, the tax change reflected in Curve *B* would reduce its slope relative to Curve *A* but it would intersect the horizontal axis—that is, zero tax liability—at the same point. The tax change reflected in

Table 9.7 Examples of Technical Relationships

I. *Microrelations: Production "Activities" on a Particular Farm*

	Corn	Wheat
A. Outputs and Inputs Per Acre		
Yield (output)	40 bu	20 bu
Planting labor requirement	0.8 hr	0.8 hr
Cultivating labor requirement	1.0 hr	0 hr
Harvesting labor requirement	0.8 hr	2.0 hr
B. Inputs Required Per Bushel of Output[a]		
Land (acres)	0.025	0.05
Planting labor (hr)	0.02	0.04
Cultivating labor (hr)	0.025	0
Harvesting labor (hr)	0.02	0.10

II. *Macrorelations: An Input-Output Production Function for Transport Equipment*

	Direct Input Requirements Per \$10,000 Gross Output of Transport Equipment[b]
Supplying Industry	
Transport equipment	\$2,128
Iron and steel	1,389
Machinery	1,069
Rubber products	319
Nonferrous metals	210
Transportation	170
Nonmetallic mineral products	125
Chemicals	92
Other industries	356
Subtotal	\$5,858
"Value added" in transport equipment industry	4,042
Total inputs	\$10,000

[a] Per acre figure divided by yield per acre.

[b] Direct requirements from specified industries to support \$10,000 worth of total output from the transport equipment industry, United States, 1947 (assumes 1947 prices for output of all industries).

Sources. Microrelations from Earl O. Heady, and Wilfred Candler, *Linear Programming Methods*, (Iowa: Iowa State University Press, 1958), p. 35, Table 2.1; and macrorelation from Hollis B. Chenery and Paul G. Clark, *Interindustry Economics* (New York: John Wiley and Sons, Inc., 1959), pp. 222–223.

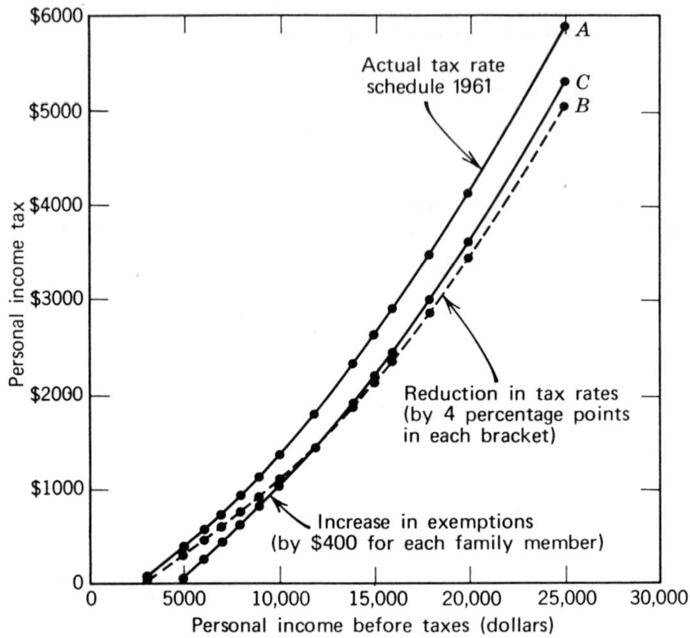

Figure 9.19 Personal income tax and personal income: effects of changes in exemptions and tax rate schedules. Based on rates for married taxpayers with total of four exemptions, filing joint returns, U.S. *Federal Income Tax Forms for 1961*, p. 9, Schedule II.

Curve *C* would shift the zero tax point to the right by $1600 but would leave the slope of Curve *C* identical with that of Curve *A*.

In principle, the effect of a change in the tax rate schedule could be calculated for every actual and potential taxpayer. Hence, tax rate Schedule A would lead to a definite relationship between personal income tax revenues and personal income before taxes in terms of national aggregates. This would also imply a definite relationship at the national level between disposable personal income after taxes and personal income before taxes. Changes from Schedule A to Schedule B or C should influence these relationships among national aggregates approximately in the manner shown in Figure 9.19.

The fourth category of structural economic relationships consists of *identities* or *balance equations* of a bookkeeping nature. For example, the gross national product (GNP) is, by definition, equal to the sum of personal consumption expenditures, gross private capital formation, government purchases of goods and services, and net foreign trade. If each of the four components were to be estimated separately and independently of an estimate for total GNP, only by chance would they add up to the estimated total. Other identities connect, for example, gross farm income, production expenses and net farm income or retail food prices, equivalent farm prices, and food marketing margins.

From a statistical viewpoint, simple regressions involve only two variables. However, most problems of importance in economic theory and in real economic behavior involve more than two simultaneously changing variables, and some involve several distinct equations operating simultaneously or with time lags. Hence, it is particularly important that we see the mechanics and arithmetic of simple regression analysis against a realistic economic background.

Figure 9.20 expresses a hypothesis about the manner in which the produc-

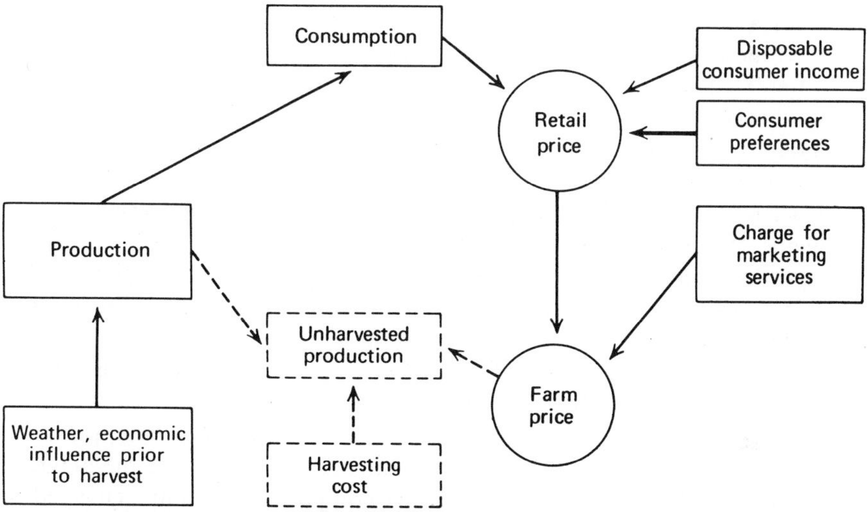

Figure 9.20 Demand and supply structure for potatoes. Arrows show direction of influence. Heavy arrows indicate major paths of influence which account for the bulk of the variation in current prices. Light solid arrows indicate definite but less important paths; dashed arrows indicate paths of negligible, doubtful, or occasional importance. (U.S. Department of Agriculture, Neg. 48930–X, Bureau of Agricultural Economics.)

tion, consumption, retail price, and farm price of potatoes are related to one another and to certain other factors. The directions of the arrows express hypotheses about the direction of influence of one variable upon another. Each variable is measurable in the form of an economic time series except for "consumer preferences." "Weather" can be quantified in terms of joint distributions of rainfall, temperature, and humidity just before and during the growing season. "Economic influences prior to harvest" can also, in principle, be quantified in terms of the price of potatoes, prices of crops which compete for the same kind of land, prices of inputs used in potato production, and changes in employment and other factors affecting consumer demand.

In this model, we might state the following objectives for regression analysis:

(a) Corresponding to each arrow in the diagram, we would like to estimate a coefficient (or a curvilinear relationship) stating the expected change in the variable to which the arrow leads that is associated with a one-unit increase in the variable from which it comes.

(b) We should like to determine each of these coefficients, and hence the entire set of regression coefficients implicit in the diagram, in such a way that, given the time series values of disposable consumer income, weather, and preharvest economic influences actually experienced over a period of years, we could estimate time series values of consumption, retail price, farm price, and unharvested production that would correspond closely to the values of these variables actually observed.

(c) If these estimated or "reconstructed" time series reproduce fairly accurately both the observed trends and the year-to-year changes in the original data, we may wish to use the regression results to predict future values of the dependent or *endogenous* variables.

The pattern of arrows in the diagram also expresses hypotheses as to which of the structural relationships and coefficients can be estimated appropriately by means of simple regression methods. For example, it appears from the diagram that consumption can be estimated solely on the basis of knowledge of changes in production; this implies a two-variable or simple regression relationship. On the other hand, there are at least two measurable variables plus "consumer preferences" affecting the retail price of potatoes. Thus, an explanation of retail price requires multiple regression analysis.

Figure 9.21 expresses a hypothesis concerning the demand-supply-price structure for hogs and pork. It is on the same logical footing as the diagram for potatoes that was already discussed. A competent economist engaged in a serious investigation of the pork and hog economy would try to conceptualize this model in terms of consumption theory and theory of the firm; certain sequences of arrows would be recognized as technical relationships (number of hogs slaughtered as a function of the number of sows bred in the preceding year) and others as identities (pork production is the product of number of hogs slaughtered times pork yield per hog). The investigator could provide points of entry in his conceptual model for "institutional" changes such as an increase in sales taxes on retail pork (food) sales, the adoption of a federal subsidy to meat packers to reduce the margin between retail pork prices and the farm price of hogs, and other conceivable types of public intervention.

If one variable, such as the retail price of pork, is in fact strongly influenced by two or more other variables (such as disposable consumer income, supplies of other meats and poultry, and pork consumption), it is unreasonable to expect that the simple regression of retail prices upon any one of the two or more associated variables will closely fit the observations. The scatter of

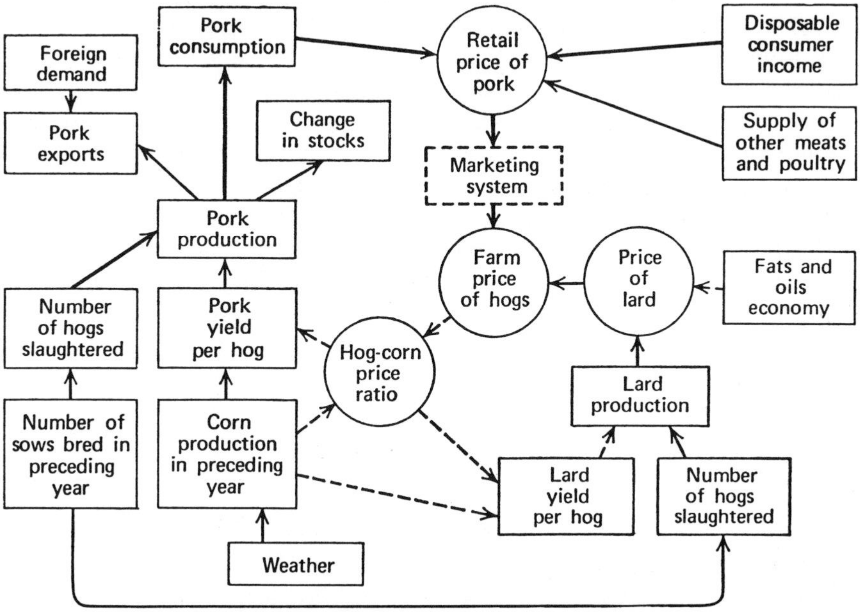

Figure 9.21 The demand and supply structure for pork. Arrows show direction of influence. Heavy arrows indicate major paths of influence which account for the bulk of the variation in current prices. Light solid arrows indicate definite but less important paths. Dashed arrows indicate paths of negligible, doubtful, or occasional importance. (U.S. Department of Agriculture, Neg. 48936-X, Bureau of Agricultural Economics.

observations around a simple regression between retail price of pork and pork consumption is affected by variations in disposable income, the supply of competing meats, and changes in consumer attitudes toward pork and competing foods. Hence, we should expect that a simple regression relationship between pork production and the farm price of hogs would be vulnerable to errors of measurement in pork consumption and the retail price of pork as well as in the two variables specifically included in the regression. The relationship between pork production and the farm price of hogs would be subject to disturbances resulting from variations in disposable consumer income and the supply of other meats and poultry; it would also be vulnerable to changes in unit costs of labor and in prices of intermediate goods and services used in the marketing process.

It should be clear from Figure 9.21 that we cannot hope to use a simple regression between hog prices and pork production safely *unless* we are aware of the complete demand-supply-price structure into which these two variables fit. From a structural standpoint, this particular regression is equivalent to drawing a fictitious arrow from pork production to the farm price of hogs over a path where no real arrow exists. It puts us in the position of the English roofer who during a London "pea-souper" inadvertently shingled nine feet of fog! When the fog lifted, his shingles fell clattering to the ground.

In summary, the arrows in diagrams such as Figure 9.21 represent paths over which economic influences can be expected to travel. Regression analysis can be used appropriately to estimate or map out such paths of influence and relationship as exist in the real economy. Neither statistics nor economics can save us from the results of our folly if we treat nonstructural regression relationships as if they were real economic thoroughfares. In practical research applications, we must use economic theory, common sense, and knowledge of production and marketing processes to avoid the dead ends that have characterized many earlier applications of regression analysis in economics and business.

EXERCISES

Section 9.1

1. Suppose that the *population* consists of six points:

$$(1, 2), (4, 6), (2, 4), (2, 3), (3, 5), (5, 10)$$

(a) Plot the points on a scatter diagram.
(b) Compute the regression function $Y = \alpha + \beta X$.
(c) Do the error terms satisfy the condition $E(u) = 0$?

2. Suppose that a straight line passes through the points (X_1, Y_1) and (X_2, Y_2). Show that the slope of the line is equal to

$$m = \frac{(Y_2 - Y_1)}{(X_2 - X_1)}$$

Section 9.2

3. Two different college entrance examinations were given to a selected group of freshmen. The following scores were recorded.

Student	Test Score		Student	Test Score	
i	Exam Y	Exam X	i	Exam Y	Exam X
1	55	58	11	43	67
2	58	70	12	54	67
3	62	58	13	26	52
4	59	59	14	43	74
5	52	61	15	40	56
6	40	63	16	55	67
7	43	70	17	56	71
8	56	67	18	50	56
9	34	51	19	50	68
10	64	70			

(a) Estimate the regression function $Y = \alpha + \beta X$.

(b) Plot the scores in a scatter diagram. Does the assumption of a linear relationship between X and Y appear justified on the basis of a visual inspection of the scatter diagram?

(c) Is it meaningful to define the score on Exam Y as a dependent variable in this case? Explain.

4. Prove that the least squares regression line $\hat{Y} = a + bX$ always passes through the point of means $(\bar{X}, \bar{Y})$.

5. Using the following data:

W	1	2	3	4	5
Y	14	17	24	22	30

(a) Compute the least squares regression line $\hat{Y} = a + bW$.

(b) Let $X = (W - 3)$. Compute the least squares regression line $\hat{Y} = a + bX$.

6. Using the following data:

X	2	3	3	4	5	5
Y	18	26	22	28	38	36

(a) Compute the least squares regression line $\hat{Y} = a + bX$.

(b) Compute the least squares regression line $\hat{X} = c + dY$.

(c) Plot the data in a scatter diagram and draw in the regression lines obtained in parts (a) and (b) above.

(d) Explain why the regression of Y on X is not identical to the regression of X on Y.

Section 9.3

7. Data from a random sample of 22 families yield the consumption function

$$\hat{C} = 12 + \underset{(0.05)}{0.90}\,Y$$

where C is consumption and Y is income measured in dollars.

(a) Test the hypothesis that the marginal propensity to consume is equal to 0.83. Use a 5 percent level of significance.

(b) What are the 95 percent confidence limits for the regression coefficient β?

8. Suppose the basic regression equation is assumed to be $Y = \beta X + u$.

(a) Specify the conditions you would use to define the complete regression model.

(b) Derive the least squares estimator b of β.

(c) Using the following data:

Investment (Y)	65	57	57	54	66
Change in output (X)	26	13	16	−7	27

(i) Compute the regression of Y on X and plot the result on a scatter diagram.
(ii) Test the hypothesis that $\beta = 0$ against the alternative $\beta > 0$ at the 5 percent level of significance. Explain why there are $(n-1)$ degrees of freedom in this case.
(iii) Under what conditions is the test in part (ii) valid?

9. Suppose the random variable Y has a normal distribution with mean $\alpha + \beta X$ and variance σ^2 where X can take on only the values zero and one.
(a) Set up a test of the hypothesis that β is equal to zero.
(b) Compare the test derived in part (a) above with the test of the equality of means of two normal distributions having the same variance.

10. Use the following data on rice prices and consumption to estimate a demand function of the form $\hat{Q} = a + bP$ where Q is consumption of rice in 1000's of metric tons and P is the retail price of rice in dollars per metric ton.

Year	*Consumption* Q	*Price* P
1957	178	105
1958	224	105
1959	160	130
1960	315	130
1961	229	130
1962	250	150
1963	181	150
1964	306	170
1965	237	170
1966	300	180

11. The following questions are to be answered using the data and results from Exercise 10.
(a) What is the economic interpretation of the coefficients a and b of the demand function $\hat{Q} = a + bP$?
(b) Are the values of a and b consistent with what you would expect on the basis of economic theory?
(c) Suggest several alternative models that you believe would provide a better estimate of the effect of the retail price on the quantity of rice consumed.
(d) Using the estimated demand function, project consumption if the retail price were to increase to $200 per metric ton.
(e) Compute a 90 percent confidence interval for the projection obtained in part (d) above.

Section 9.4

12. Using the following data:

W	2	4	6	5	4
Z	8	28	70	52	32

(a) Estimate the parameters A and β of the regression function

$$Z = AW^{B}v$$

(b) Graph the data and the estimated regression line.

13. Explain how to derive the least squares estimators of α, β, and γ for the nonlinear function

$$Y_i = \alpha + \beta X_i + \gamma X_i^{\,2} + u_i$$

14. Derive the least squares estimators of α, β, and γ for the nonlinear function

$$Y_i = \alpha + \beta X_i + \gamma X_i^{\,2} + u_i$$

under the assumption that the sum of the X's is equal to zero.

Section 9.5

For the following three questions, select the correct answer and explain why it is correct.

15. If the standard error of estimate is zero and the standard deviation of the dependent variable is not zero, then the sample correlation coefficient is:
(a) 1 or -1.
(b) Between 1 and 0.
(c) Zero.
(d) Between -1 and 0.
(e) Further information is needed to answer this question.

16. If the regression coefficient is zero and the standard deviation of the dependent variable is not zero, then the sample correlation coefficient is:
(a) 1 or -1.
(b) Between 1 and 0.
(c) Zero.
(d) Between -1 and 0.
(e) Further information is needed to answer this question.

17. If the sample correlation coefficient between two variables is -0.90, then:
(a) 90 percent of the observations lie on the regression line.
(b) 81 percent of the observations lie on the regression line.
(c) The average change in the dependent variable which accompanies a unit change in the independent variable is -0.9 units.
(d) None of the above.

18. "If no relationship exists, the correlation coefficient will be zero, but a correlation coefficient of zero does not mean that no relationship exists." Do you agree or disagree? Explain.

19. A least-squares regression study of the linear relation between the price of wheat per bushel and the number of bushels sold during a given period of time yields the following results:
(a) If the price of wheat is \$2 per bushel, we may be 80 percent confident that more than 1,976,000 bushels but less than 4,024,000 bushels of wheat were sold.

(b) 85 percent of the variation in wheat sales is linearly associated with price changes.

(c) For all those periods in which the price is \$3 per bushel, the predicted average level of sales is 2,500,000 bushels.

In view of these results, what were the numerical values obtained for each of the following statistics?

(i) The regression coefficient b.

(ii) The regression constant a.

(iii) The simple correlation coefficient r.

20. A study of the relationship between smoking and lung cancer is summarized in the following estimated regression equation relating the number of years a person has been smoking (X) and the percentage-incidence of lung cancer within each group (Y):

$$\hat{Y} = -2 + 1.7X \qquad r = 0.80$$

(a) Explain the meaning of the estimates "-2" and "1.7" in the regression equation.

(b) What is the expected rate of incidence of lung cancer for persons who have been smoking for 20 years?

(c) If r had been equal to 1.00, could we conclude that smoking was the single cause of lung cancer?

(d) Suppose that the estimated regression equation was obtained from a random sample of 100 smokers. Test the hypothesis that the correlation coefficient is equal to zero. Use the 1 percent level of significance.

21. Suppose that an investigator interested in the relationship between advertising expenditures and sales decides to make a study of large United States manufacturers with annual gross sales of at least \$250,000. Based on a random sample of such manufacturers, the investigator computes the regression line $\hat{Y} = 40{,}000 + 6X$ where Y is gross annual sales and X is annual advertising expenditures.

(a) In what sense is the least-squares regression line the line that fits the data "best"?

(b) Because the regression function "explained" so much of the variance of gross annual sales, the investigator who undertook this study concluded that greater advertising expenditures lead to greater sales. Do you agree with this conclusion? Why or why not?

22. Explain briefiy why you agree or disagree with the following statements.

(a) A correlation coefficient of $+1.0$ between X and Y means that X causes Y, but a correlation coefficient of -1.0 means that X does not cause Y.

(b) If the regression coefficient is zero, the correlation coefficient is zero.

(c) If the slope of the regression line is 1, this means that there is perfect correlation between X and Y.

(d) It is possible that the sample correlation coefficient is positive when, in fact, there is no correlation between X and Y.

(e) Least-squares regression techniques cannot be used if the basic relationship between X and Y is nonlinear.

(f) The standard error of estimate measures the dispersion of sampling distribution of the estimate around the true mean.

23. Explain the relationship between the joint frequency distribution of two variables and a simple regression analysis showing the interdependence between the same two variables.

24. A sample correlation coefficient of 0.42 is obtained from a random sample of 19 observations on X and Y. Test the hypothesis that the true value of the correlation coefficient is zero using a 10 percent level of significance.

Section 9.6

25. A study of two crops gave the following information:

Crop A	*Crop B*
$\hat{Y} = 200 + 0.8X$	$\hat{Y} = 50 + 1.2X$
$s_{y.x} = 30$	$s_{y.x} = 20$
$r = 0.7$	$r = 0.9$

where Y is output measured in bushels and X is the amount of rainfall measured in inches per crop season.

(a) If there were no rainfall, could you use these regression equations to predict the amounts of output of the two crops? Why or why not?
(b) Which of the two crops gains more by an increase in rainfall? Briefly state the basis for your answer.
(c) For which of the two crops can you predict crop output more accurately?

26. The following data were collected from five different plants in a certain industry:

Total cost (Y)	80	44	51	70	61
Production (X)	12	4	6	11	8

(a) Estimate a linear total cost function of the form $\hat{Y} = a + bX$ for this industry.
(b) What is the economic significance of the estimates a and b?
(c) Test the hypothesis that the marginal cost of producing the item in question is equal to 4 against the alternative that the marginal cost is greater than 4. Use the 10 percent level of significance.

27. It was found by regression analysis that the weight of chickens was directly related to the amount of vitamins placed in their feed. The equation was $\hat{W} = 3 + 0.1V$ where W is the weight of the chicken in pounds and V is units of vitamins. The equation was based on a random sample of 248 chickens. The correlation coefficient was found to be $r = 0.85$.

(a) Write a short report on the statistical and economic meaning of the regression equation, and discuss the advisability of adding vitamins to chicken feed.

(b) If the price of chickens is 15 cents a pound and the price of vitamins is 1 cent per unit, how many units of vitamins should be added to the feed for maximum profit?

(c) If the cost of raising a chicken is 40 cents for everything other than vitamins which are 1 cent per unit, how many chickens should the firm raise if chickens are selling for 15 cents per pound?

(d) Propose a revised experiment for the "experts" at the chicken lab.

28. Using the data on aggregate consumption functions presented in Section 9.6.2:

(a) What was the marginal propensity to consume during the 1949–1961 period?

(b) Develop a multiple equation model that shows how the consumption function number 3 in Table 9.6 is related to such items as the supply of money and total investment.

(c) Does it seem entirely realistic to consider disposable personal income (X_2) as an independent variable? Discuss.

29. Graph the following cost curves for values of X between 0 and 5 using the basic cost function presented in Section 9.6.4:

(a) Total fixed costs.

(b) Total variable costs.

(c) Marginal costs.

(d) Average variable costs.

Section 9.7

30. Relationships in economic models can be classified as behavioral, technical, institutional and identities. How could you classify the demand, consumption, and cost functions discussed in Section 9.6?

31. Develop a simple block digaram such as that shown in Figure 9.21 showing how various factors are related to the retail price of beef.

SELECTED REFERENCES

Ezekiel, Mordecai and K. A. Fox, *Methods of Correlation and Regression Analysis*, Third Edition, John Wiley and Sons, New York, 1959.

Fisher, R. A., *Statistical Methods for Research Workers*, Thirteenth Revised Edition, Hafner Publishing Co., New York, 1963.

Fox, K. A., *Intermediate Economic Statistics*, John Wiley and Sons, Inc., New York, 1968.

Goldberger, A. S., *Topics in Regression Analysis*, The Macmillan Company, New York, 1968.

Kane, E. J., *Economic Statistics and Econometrics: An Introduction to Quantitative Economics*, Harper and Row, Publishers, New York, 1968.

CHAPTER 10

Multiple Regression Analysis

All of the basic logical concepts of regression analysis have been introduced in Chapter 9. We shall now extend these basic concepts to regression equations involving more than two variables. Matrix notation is used throughout this chapter. Those persons unfamiliar with matrix algebra should study Section IV in Appendix II before starting this chapter. There are several reasons for turning to matrix notation at this point. First, we believe that the multiple regression model can be understood more easily when presented compactly in matrix notation. Second, those students who plan to study economic statistics further will eventually need to be familiar with the basic concepts of matrix algebra because they are commonly used in econometrics textbooks. And third, we want to emphasize the concepts of multiple regression analysis rather than present computing techniques by which to estimate the parameters of the multiple regression model.

The plan of this chapter is as follows. We shall first present the basic multiple regression model and methods for measuring the closeness of fit of the regression line and testing hypotheses about the regression coefficients. We shall then generalize the basic model and consider several special problems frequently encountered in the analysis of economic data. Several examples of multiple regression problems will be presented in the final section of the chapter.

Our intention in this chapter is to introduce some of the material usually encountered in more advanced courses in econometrics and to familiarize the reader with the notation commonly employed in econometric studies. This is a difficult chapter, and it will frequently be necessary for the reader to write out the matrix notation in detail in order to understand how certain results are obtained.

10.1 THE MULTIPLE REGRESSION MODEL

Let us assume that a linear relationship exists between a variable Y, called the *dependent variable*, and k variables $X_1, X_2, \ldots, X_k$ called *independent variables*, plus a disturbance term u that represents the *error in the equation.* If we have a sample of n observations on the $k + 1$ variables $Y, X_1, X_2, \ldots, X_k$, the data may be presented as a set of linear equations:

$$\begin{aligned} Y_1 &= \beta_1 X_{11} + \beta_2 X_{21} + \cdots + \beta_k X_{k1} + u_1 \\ Y_2 &= \beta_1 X_{12} + \beta_2 X_{22} + \cdots + \beta_k X_{k2} + u_2 \\ \cdots & \quad \cdots \quad \cdots \quad \cdots \quad \cdots \quad \cdots \\ Y_n &= \beta_1 X_{1n} + \beta_2 X_{2n} + \cdots + \beta_k X_{kn} + u_n \end{aligned}$$

or simply

$$Y_i = \beta_1 X_{1i} + \beta_2 X_{2i} + \cdots + \beta_k X_{ki} + u_i \qquad \text{for } i = 1, 2, \ldots, n \tag{10.1}$$

where the β coefficients are unknown constants. The n equations in (10.1) can be written compactly in matrix notation as

$$Y = X\beta + U$$

where

$$Y = \begin{bmatrix} Y_1 \\ Y_2 \\ \vdots \\ Y_n \end{bmatrix} \qquad \beta = \begin{bmatrix} \beta_1 \\ \beta_2 \\ \vdots \\ \beta_k \end{bmatrix} \qquad U = \begin{bmatrix} u_1 \\ u_2 \\ \vdots \\ u_n \end{bmatrix}$$

and

$$X = \begin{bmatrix} X_{11} & X_{21} & \cdots & X_{k1} \\ X_{12} & X_{22} & \cdots & X_{k2} \\ \vdots & \vdots & \cdots & \vdots \\ X_{1n} & X_{2n} & \cdots & X_{kn} \end{bmatrix}$$

The β coefficients and the parameters of the distribution of the error terms are to be estimated from sample data. The estimation procedures used depend, to a large extent, on the set of assumptions made about the variables X and the disturbance term U. The simplest set of assumptions is:

1. The expected values of the disturbance terms are equal to zero. That is, $E(u_i) = 0$ for all i. This is written in matrix notation as

$$E(U) = \begin{bmatrix} E(u_1) \\ E(u_2) \\ \vdots \\ E(u_n) \end{bmatrix} = \begin{bmatrix} 0 \\ 0 \\ \vdots \\ 0 \end{bmatrix} = 0$$

Notice that the operation of taking the expected value of a vector consists of taking the expected value of each element in the vector (or matrix).

2. It is assumed that the disturbance terms are not correlated with one another and that they have the same variance σ^2 for all possible sets of values of the variables X. That is,

$$E(u_i u_j) = 0 \qquad \text{for } i \neq j$$

and

$$E(u_i^2) = \sigma^2 \qquad \text{for all } i$$

This assumption is written in matrix notation as

$$E(UU') = \begin{bmatrix} E(u_1^2) & E(u_1u_2) & \cdots & E(u_1u_n) \\ E(u_1u_2) & E(u_2^2) & \cdots & E(u_2u_n) \\ \vdots & \vdots & \vdots & \vdots \\ E(u_1u_n) & E(u_nu_2) & \cdots & E(u_n^2) \end{bmatrix} = \begin{bmatrix} \sigma^2 & 0 & \cdots & 0 \\ 0 & \sigma^2 & \cdots & 0 \\ \vdots & \vdots & \vdots & \vdots \\ 0 & 0 & \cdots & \sigma^2 \end{bmatrix} = \sigma^2 I_n$$

where U' is a 1 by n row vector $(u_1 u_2 u_3 \cdots u_n)$ and I_n is an n by n matrix with ones in the principal diagonal and zeros elsewhere. The property that the diagonal terms are all the same is referred to as homoscedasticity.

3. The numbers $X_{1i}, X_{2i}, \ldots, X_{ki}$ are real numbers containing no random errors or disturbances. In other words, the matrix X is a set of fixed numbers in the sense that they are not subject to random variation. A second sample may, of course, yield a new set of real numbers and thus a different matrix X.

4. The matrix X has $k < n$ linearly independent columns. In other words, X has rank $k < n$. This condition is satisfied if the number of observations n exceeds the number k of β coefficients to be estimated and no exact linear relationships exist between any of the X variables.

10.1.1 Point Estimation

When the above four conditions are satisfied, use of the method of least squares provides the best linear, unbiased estimators of the β coefficients. This result has already been discussed for the simple regression case involving one dependent variable. The discussion of point estimation in this section is identical to the discussion in Chapter 9 except that we are now dealing with k independent variables and, therefore, use matrix notation.

Let b denote the k by 1 vector of least squares estimators. That is,

$$b = \begin{bmatrix} b_1 \\ b_2 \\ \vdots \\ b_k \end{bmatrix}$$

The residuals from the fitted regression function are used to derive the least squares estimator b. The residuals are equal to

$$\begin{bmatrix} e_1 \\ e_2 \\ \vdots \\ e_n \end{bmatrix} = \begin{bmatrix} Y_1 \\ Y_2 \\ \vdots \\ Y_n \end{bmatrix} - \begin{bmatrix} X_{11} & X_{21} & \cdots & X_{k1} \\ X_{12} & X_{12} & \cdots & X_{k2} \\ \vdots & \vdots & \vdots & \vdots \\ X_{1n} & X_{2n} & \cdots & X_{kn} \end{bmatrix} \begin{bmatrix} b_1 \\ b_2 \\ \vdots \\ b_k \end{bmatrix}$$

and can be written compactly as

$$e = Y - Xb$$

The sum of squared residuals, say, ϕ, is written as

$$\begin{aligned} \text{(10.2)} \qquad \phi &= \sum e_i^2 = \sum (Y_i - b_1 X_{1i} - b_2 X_{2i} - \cdots - b_k X_{ki})^2 \\ &= e'e = (Y - Xb)'(Y - Xb) \\ &= Y'Y - 2b'X'Y + b'X'Xb \end{aligned}$$

(Note that $b'X'Y$ is a scalar and, therefore, is equal to its transpose $Y'Xb$.)

The least squares estimator of β is the vector b that minimizes the sum of squared residuals ϕ. Differentiating ϕ with respect to each component of b and setting the resulting equations equal to zero provides a set of normal equations:

$$\frac{\partial \phi}{\partial b_1} = 2\left(-\sum X_{1i} Y_i + b_1 \sum X_{1i}^2 + b_2 \sum X_{1i} X_{2i} + \cdots + b_k \sum X_{1i} X_{ki}\right) = 0$$

$$\frac{\partial \phi}{\partial b_2} = 2\left(-\sum X_{2i} Y_i + b_1 \sum X_{2i} X_{1i} + b_2 \sum X_{2i}^2 + \cdots + b_k \sum X_{2i} X_{ki}\right) = 0$$

$$\vdots \qquad\qquad \vdots$$

$$\frac{\partial \phi}{\partial b_k} = 2\left(-\sum X_{ki} Y_i + b_1 \sum X_{ki} X_{1i} + b_2 \sum X_{ki} X_{2i} + \cdots + b_k \sum X_{ki}^2\right) = 0$$

This set of normal equations is written in matrix form as

$$\frac{\partial \phi}{\partial b} = -2X'Y + 2X'Xb = 0$$

which is equivalent to

$$\text{(10.3)} \qquad X'Xb = X'Y$$

When X has rank k, the normal equation (10.3) has a unique solution and the least squares estimator b is equal to

$$\text{(10.4)} \qquad b = (X'X)^{-1}X'Y$$

Suppose, for example, that we want to estimate the simple regression equation

$Y = \beta_1 + \beta_2 X_2 + u$. In this case, $k = 2$ and the variable X_{1i} is equal to 1 for all i. The normal equation (10.3) is therefore equal to

$$\begin{bmatrix} 1 & 1 & \cdots & 1 \\ X_{21} & X_{22} & \cdots & X_{2n} \end{bmatrix} \begin{bmatrix} 1 & X_{21} \\ 1 & X_{22} \\ 1 & X_{23} \\ \vdots & \vdots \\ 1 & X_{2n} \end{bmatrix} \begin{bmatrix} b_1 \\ b_2 \end{bmatrix} = \begin{bmatrix} 1 & 1 & \cdots & 1 \\ X_{21} & X_{22} & \cdots & X_{2n} \end{bmatrix} \begin{bmatrix} Y_1 \\ Y_2 \\ Y_3 \\ \vdots \\ Y_n \end{bmatrix}$$

After multiplying X' by X and X' by Y, we obtain

$$\begin{bmatrix} n & \sum X_{2i} \\ \sum X_{2i} & \sum X_{2i}^2 \end{bmatrix} \begin{bmatrix} b_1 \\ b_2 \end{bmatrix} = \begin{bmatrix} \sum Y_i \\ \sum Y_i X_{2i} \end{bmatrix}$$

Solving for b, we have

$$\begin{bmatrix} b_1 \\ b_2 \end{bmatrix} = \begin{bmatrix} n & \sum X_{2i} \\ \sum X_{2i} & \sum X_{2i}^2 \end{bmatrix}^{-1} \begin{bmatrix} \sum Y_i \\ \sum Y_i X_{2i} \end{bmatrix} \tag{10.5}$$

which is equivalent to

$$\begin{bmatrix} b_1 \\ b_2 \end{bmatrix} = \frac{1}{|X'X|} \begin{pmatrix} \sum X_{2i}^2 & -\sum X_{2i} \\ -\sum X_{2i} & n \end{pmatrix} \begin{pmatrix} \sum Y_i \\ \sum Y_i X_{2i} \end{pmatrix}$$

where the determinant $|X'X|$ is equal to

$$|X'X| = n \sum (X_{2i} - \bar{X}_2)^2 = n \sum X_{2i}^2 - \left(\sum X_{2i}\right)^2$$

Multiplying the matrices on the right-hand side of (10.5) and writing b in its simplest form, we obtain

$$\begin{bmatrix} b_1 \\ b_2 \end{bmatrix} = \begin{bmatrix} \dfrac{\sum Y_i \sum X_{2i}^2 - \sum X_{2i} \sum Y_i X_{2i}}{n \sum X_{2i}^2 - (\sum X_{2i})^2} \\ \dfrac{n \sum Y_i X_{2i} - \sum Y_i \sum X_{2i}}{n \sum X_{2i}^2 - (\sum X_{2i})^2} \end{bmatrix} \tag{10.6}$$

The term for b_1 on the right-hand side of (10.6) is equivalent to $\bar{Y} - b_2 \bar{X}$. The term for b_2 is seen to be identical to the least squares estimator of the slope coefficient derived in Chapter 9.

The expected value of b is equal to β. The result is obtained by substituting $X\beta + U$ for Y in equation (10.4) before taking the expected value of b. Making this substitution, we obtain

$$\begin{aligned} b &= (X'X)^{-1} X'(X\beta + U) \\ &= (X'X)^{-1} X'X\beta + (X'X)^{-1} X'U \\ &= \beta + (X'X)^{-1} X'U \end{aligned} \tag{10.7}$$

because $(X'X)^{-1}(X'X) = I_k$ is a k by k identity matrix and $I_k\beta = \beta$. The expected value of b is therefore equal to

$$(10.8) \qquad \begin{aligned} E(b) &= E(\beta) + E[(X'X)^{-1}X'U] \\ &= \beta + (X'X)^{-1}X'E(U) \\ &= \beta \end{aligned}$$

because X is fixed and $E(U) = 0$.

The variance of b is $\sigma^2\ (X'X)^{-1}$. This result is slightly more difficult to obtain. By definition, the variance of b is equal to

$$E[(b-\beta)(b-\beta)']$$

$$= \begin{bmatrix} E(b_1-\beta_1)^2 & E[(b_1-\beta_1)(b_2-\beta_2)] & \cdots & E[(b_1-\beta_1)(b_k-\beta_k)] \\ E[(b_1-\beta_1)(b_2-\beta_2)] & E(b_2-\beta_2)^2 & \cdots & E[(b_2-\beta_2)(b_k-\beta_k)] \\ \vdots & \vdots & & \vdots \\ E[(b_1-\beta_1)(b_k-\beta_k)] & E[(b_k-\beta_k)(b_2-\beta_2)] & \cdots & E(b_k-\beta_k)^2 \end{bmatrix}$$

From (10.7) we see that $(b - \beta) = (X'X)^{-1}X'U$ and therefore

$$(10.9) \qquad \text{Var}(b) = E[(b-\beta)(b-\beta)'] = E[(X'X)^{-1}X'U'X(X'X)^{-1}]$$

This result follows from the rule that $(AB)' = B'A'$ and the fact that $(X'X)^{-1}$ is a symmetric matrix and therefore $(X'X)^{-1} = [(X'X)^{-1}]'$. To simplify (10.9), we use the assumption that X is fixed to obtain

$$\text{Var}(b) = (X'X)^{-1}X'E(UU')X(X'X)^{-1}$$

But $E(UU') = \sigma^2 I_n$, and thus

$$\text{Var}(b) = \sigma^2(X'X)^{-1}X'I_nX(X'X)^{-1}$$

since σ^2 is a scalar and can therefore be moved to the front of the matrix.

Next, note that $X'I_n = X'$ from which it follows that

$$\text{Var}(b) = \sigma^2(X'X)^{-1}X'X(X'X)^{-1}$$

Finally, we see that $(X'X)(X'X)^{-1} = I_k$ and $(X'X)^{-1}I_k = (X'X)^{-1}$ from which it follows that

$$(10.10) \qquad \text{Var(b)} = \sigma^2(\text{X}'\text{X})^{-1}$$

In the case of simple regression where $Y = \beta_1 + \beta_2 X_2 + u$, we have

$$\text{Var}(b) = \begin{bmatrix} \text{Var}(b_1) & \text{Cov}(b_1, b_2) \\ \text{Cov}(b_1, b_2) & \text{Var}(b_2) \end{bmatrix}$$

$$= \begin{bmatrix} \dfrac{\sigma^2 \sum X_{2i}{}^2}{n\sum(X_{2i}-\bar{X}_2)^2} & \dfrac{-\sigma^2\sum X_{2i}{}^2}{n\sum(X_{2i}-\bar{X}_2)^2} \\ \dfrac{-\sigma^2\sum X_{2i}}{n\sum(X_{2i}-\bar{X}_2)^2} & \dfrac{\sigma^2}{\sum(X_{2i}-\bar{X}_2)^2} \end{bmatrix}$$

For the linear regression model defined by the assumptions (1)–(4), the least squares estimators of the β coefficients are linear unbiased estimators and possess a smaller variance than any other linear unbiased estimators. That b is unbiased is shown by (10.8). The proof that Var(b) is smaller than the variance of any other linear unbiased estimator of β will not be presented here but can be found in most advanced text books on economic statistics.[1]

Our discussion of point estimation is not complete until we obtain an estimator, say, s^2, of σ^2. As in the case of simple regression, this estimator is based on the sum of squared residuals from the fitted regression line. You will recall that the observed residuals from the fitted regression line are equal to

$$e = Y - Xb \tag{10.11}$$

Since $Y = X\beta + U$ and $b = (X'X)^{-1}X'Y$, the observed residuals can be written as

$$\begin{aligned} e &= X\beta + U - X[(X'X)^{-1}X'(X\beta + U)] \\ &= U - X(X'X)^{-1}X'U \\ &= (I_n - X(X'X)^{-1}X')U \\ &= MU \end{aligned}$$

where the matrix M is defined as $M = I_n - X(X'X)^{-1}X'$.

The matrix M has two very useful properties. First, M is a symmetric matrix. That is, $M' = I_n + X(X'X)X'^{-1}X' = M$. Second, M remains unchanged when multiplied by itself. That is,

$$\begin{aligned} M^2 &= (I_n - X(X'X)^{-1}X')(I_n - X(X'X)^{-1}X') \\ &= I_n - 2X(X'X)^{-1}X' + X(X'X)^{-1}(X'X)(X'X)^{-1}X' \\ &= I_n - X(X'X)^{-1}X' \end{aligned}$$

The sum of squared residuals is therefore equal to

$$e'e = U'M'MU = U'MU$$

because $M'M = M^2 = M$. Writing $e'e$ out in detail, we have

$$U'MU = (u_1 u_2 \cdots u_n)\begin{bmatrix} m_{11} & m_{12} & \cdots & m_{1n} \\ m_{21} & m_{22} & \cdots & m_{2n} \\ \vdots & \vdots & & \vdots \\ m_{n1} & m_{n2} & & m_{nn} \end{bmatrix}\begin{bmatrix} u_1 \\ u_2 \\ \vdots \\ u_n \end{bmatrix}$$

or

$$U'MU = \sum_i \sum_j m_{ij} u_i u_j$$

[1] J. Johnston, *Econometric Methods*, (New York: McGraw-Hill Book Company, Inc., 1960), Chapter 4.

Next, recall that by assumption (2), we have $E(u_i^2) = \sigma^2$ and $E(u_i u_j) = 0$ for $i \neq j$. Therefore,

$$E(e'e) = E(U'MU) = \sum_i \sum_j m_{ij} E(u_i u_j) = \sigma^2 \sum_i m_{ii}$$

because all terms with $i \neq j$ are equal to zero. We now have to evaluate the quantity $\sum_i m_{ii}$.

The sum of the elements in the principal diagonal of a square matrix A is called the *trace* of the matrix and denoted by trA. Furthermore, it can be shown that:

(a) The trace of the sum of two matrices is the sum of the traces of each matrix.

(b) If A and B are any two matrices such that AB and BA exist, then the trace of AB is the same as the trace of BA.

Therefore,

$$\begin{aligned}\sum_i m_{ii} &= trM = tr(I_n - X(X'X)^{-1}X') \\ &= trI_n - tr(X(X'X)^{-1}X') \\ &= n - k\end{aligned}$$

because

$$tr(X(X'X)^{-1}X') = tr(X'X(X'X)^{-1}) = trI_k = k$$

An unbiased estimator of σ^2 is obtained by letting

$$s^2 = \frac{e'e}{n-k} \tag{10.12}$$

The calculation of s^2 is simplified by using the fact that $e'e$ is equal to

$$\begin{aligned}e'e &= (Y - Xb)'(Y - Xb) \\ &= Y'Y - 2b'X'Y + b'X'Xb \\ &= Y'Y - 2b'X'Y + b'X'X(X'X)^{-1}X'Y \\ &= Y'Y - 2b'X'Y + b'X'Y \\ &= Y'Y - b'X'Y\end{aligned}$$

which, when written out in detail, becomes

$$\sum e_i^2 = \sum Y_i^2 - b_1 \sum X_{1i} Y_i - b_2 \sum X_{2i} Y_i - \cdots - b_k \sum X_{ki} Y_i$$

In the case of simple regression with $Y = \beta_1 + \beta_2 X_2 + u$, we have

$$\sum e_i^2 = \sum Y_i^2 - b_1 \sum Y_i - b_2 \sum X_{2i} Y_i$$

and

$$s^2 = \frac{\sum e_i^2}{n-2} = s_{y \cdot x}^2$$

Compare this result with formula (9.18) in Chapter 9.

10.1.2 A Numerical Example

The basic ideas of multiple regression are relatively simple. Computation of the regression coefficients can be rather troublesome, however, without the help of modern computing facilities. We shall use a simplified example to illustrate one way to compute the regression coefficients when such facilities are not available.

Let us assume that the amount of a particular good that is consumed by a family depends on the price of the good and the income of the family. To estimate the demand function for this good, we could survey a number of families and determine:

Y_i = the number of units consumed by family i

X_{2i} = the price (in dollars) paid per unit by family i

X_{3i} = the annual income (in 1000's of dollars) of family i

Hypothetical survey data for 10 families are presented in Table 10.1.

Let us assume that the demand curve is a linear function of the form

$$Y_i = \beta_1 + \beta_2 X_{2i} + \beta_3 X_{3i} + u_i$$

for all i families.

The normal equations for the least squares estimators of β_1, β_2, and β_3 are

$$\begin{aligned} nb_1 + b_2 \sum X_{2i} + b_3 \sum X_{3i} &= \sum Y_i \\ b_1 \sum X_{2i} + b_2 \sum X_{2i}^2 + b_3 \sum X_{2i} X_{3i} &= \sum X_{2i} Y_i \\ b_1 \sum X_{3i} + b_2 \sum X_{2i} X_{3i} + b_3 \sum X_{3i}^2 &= \sum X_{3i} Y_i \end{aligned}$$

Using the data in Table 10.1, we obtain

$$\begin{aligned} 10b_1 + 40b_2 + 47b_3 &= 74 \\ 40b_1 + 180b_2 + 192b_3 &= 282 \\ 47b_1 + 192b_2 + 237b_3 &= 375 \end{aligned}$$

This set of equations is written in matrix form as:

$$X'Xb = \begin{bmatrix} 10 & 40 & 47 \\ 40 & 180 & 192 \\ 47 & 192 & 237 \end{bmatrix} \begin{bmatrix} b_1 \\ b_2 \\ b_3 \end{bmatrix} = \begin{bmatrix} 74 \\ 282 \\ 375 \end{bmatrix} = X'Y$$

Table 10.1 Survey Data on Family Consumption

Family i	Consumption Y	Price X_2	Income X_3
1	5	2	3
2	8	3	4
3	8	5	6
4	9	4	5
5	9	6	7
6	13	2	6
7	6	3	4
8	9	4	5
9	4	5	4
10	3	6	3

$\sum Y_i^2 = 626$ $\qquad \sum Y_i = 74$

$\sum X_{2i} Y_i = 282$ $\qquad \sum X_{3i} Y_i = 375$

$\sum X_{2i}^2 = 180$ $\qquad \sum X_{2i} = 40$

$\sum X_{3i}^2 = 237$ $\qquad \sum X_{3i} = 47$

$\sum X_{2i} X_{3i} = 192$

$\sum (X_{2i} - \bar{X}_2)^2 = 180 - 160 = 20$

$\sum (X_{3i} - \bar{X}_3)^2 = 237 - 221 = 16$

$\sum (X_{2i} - \bar{X}_2)(X_{3i} - \bar{X}_3) = 192 - 188 = 4$

$\sum (Y_i - \bar{Y})(X_{2i} - \bar{X}_2) = 282 - 296 = -14$

$\sum (Y_i - \bar{Y})(X_{3i} - \bar{X}_3) = 375 - 348 = 27$

Since $b = (X'X)^{-1}X'Y$, we shall compute the inverse matrix $(X'X)^{-1}$ first and then multiply it by $X'Y$ to find b. Once $(X'X)^{-1}$ is known, we can estimate σ^2 and then compute the variances of b using the relationship $\text{Var}(b) = \sigma^2(X'X)^{-1}$.

There are a number of different ways to compute the inverse of a matrix. The procedure used below is to perform a sequence of row operations on $(X'X)$ that will reduce it to the identity matrix I. The inverse matrix $(X'X)^{-1}$ is obtained by performing the same sequence of operations on the identity matrix I.

As a numerical example, let us take

$$A = \begin{pmatrix} 2 & 4 \\ 2 & 1 \end{pmatrix}, \qquad I = \begin{pmatrix} 1 & 0 \\ 0 & 1 \end{pmatrix}$$

Divide the first row of A and I by 2:

$$\begin{pmatrix} 1 & 2 \\ 2 & 1 \end{pmatrix} \qquad \begin{pmatrix} \frac{1}{2} & 0 \\ 0 & 1 \end{pmatrix}$$

Subtract 2 times row 1 from row 2:

$$\begin{pmatrix} 1 & 2 \\ 0 & -3 \end{pmatrix} \qquad \begin{pmatrix} \frac{1}{2} & 0 \\ -1 & 1 \end{pmatrix}$$

Divide row 2 by -3:

$$\begin{pmatrix} 1 & 2 \\ 0 & 1 \end{pmatrix} \qquad \begin{pmatrix} \frac{1}{2} & 0 \\ \frac{1}{3} & -\frac{1}{3} \end{pmatrix}$$

Subtract 2 times row 2 from row 1 to obtain

$$I = \begin{pmatrix} 1 & 0 \\ 0 & 1 \end{pmatrix} \qquad A^{-1} = \begin{pmatrix} -\frac{1}{6} & \frac{4}{6} \\ \frac{2}{6} & -\frac{2}{6} \end{pmatrix}$$

To check this result, we note that, if A^{-1} is the inverse of A, then $A^{-1}A = I$. Multiplying A and A^{-1}, we obtain

$$\begin{pmatrix} -\frac{1}{6} & \frac{4}{6} \\ \frac{2}{6} & -\frac{2}{6} \end{pmatrix} \begin{pmatrix} 2 & 4 \\ 2 & 1 \end{pmatrix} = \begin{pmatrix} 1 & 0 \\ 0 & 1 \end{pmatrix}$$

The steps in the inversion of $X'X$ are given in Table 10.2. The matrices $X'X$ and I are presented in Rows (a), (b), and (c). The operations required to obtain $(X'X)^{-1}$ are shown in Rows (d)–(r). The resulting inverse is presented in Columns (5), (6), and (7) of Rows (p), (q), and (r). That is,

$$(X'X)^{-1} = \begin{bmatrix} 1.89411768 & -0.14901975 & -0.25490185 \\ -0.14901961 & 0.05261437 & -0.01307189 \\ -0.25490197 & -0.01307186 & 0.06535945 \end{bmatrix}$$

The sums of the numbers in Rows (a), (b), and (c) are presented in the "row sum operations" column and in the "check sum" column thereafter. These sums provide a means of checking the arithmetic at each step of the inversion process. The operation applied to the number in each row is also applied to the "row sum operations" column, and the sum of the numbers in each row should equal the corresponding figure in that column. For example, the sum of the numbers in Row (a) is 98.0. The first operation is to divide

Table 10.2 The inversion of $(X'X)$.

Operation	Row	Column (1)	(2)	(3)	(4)	(5)	(6)	Row Sum Operations	Check Sum
—	(a)	10.0	40.0	47.0	1.0	0.0	0.0	98.0	—
—	(b)	40.0	180.0	192.0	0.0	1.0	0.0	413.0	—
—	(c)	47.0	192.0	237.0	0.0	0.0	1.0	477.0	—
(a) ÷ 10	(d)	1.0	4.0	4.7	0.1	0.0	0.0	9.80	9.80
(b) ÷ 40	(e)	1.0	4.5	4.8	0.0	0.025	0.0	10.325	10.325
(c) ÷ 47	(f)	1.0	4.08510638	5.04255319	0.0	0.0	0.02127659	10.14893617	10.14893616
(d)	(g)	1.0	4.0	4.7	0.1	0.0	0.0	9.80	9.80
(e) − (d)	(h)	0.0	0.5	0.1	−0.1	0.025	0.0	0.525	0.525
(f) − (d)	(i)	0.0	0.08510638	0.34255319	−0.1	0.0	0.02127659	0.34893617	0.34893616
(g)	(j)	1.0	4.0	4.7	0.1	0.0	0.0	9.80	9.80
(h) ÷ 0.5	(k)	0.0	1.0	0.2	−0.2	0.05	0.0	1.05	1.05
(i) − 0.08510638(k)	(l)	0.0	0.0	0.32553192	−0.08297873	−0.00425531	0.02127659	0.25957448	0.25957447
(j) − 4(k)	(m)	1.0	0.0	3.9	0.9	−0.2	0.0	5.60	5.60
(k)	(n)	0.0	1.0	0.2	−0.2	0.05	0.0	1.05	1.05
(l) ÷ 0.32553192	(o)	0.0	0.0	1.0	−0.25490197	−0.01307186	0.06535945	0.79738564	0.79738562
(m) − 3.9(o)	(p)	1.0	0.0	0.0	1.89411768	−0.14901975	−0.25490185	2.49019601	2.49019608
(n) − 0.2(o)	(q)	0.0	1.0	0.0	−0.14901961	0.05261437	−0.01307189	0.89052288	0.89052288
(o)	(r)	0.0	0.0	1.0	−0.25490197	−0.01307186	0.06535945	0.79738564	0.79738562

Row (a) by 10. Dividing 98.0 by 10, we obtain 9.80 and record the result in the "row sum operations" column of Row (d). If all of the divisions are carried out correctly, the sum of the numbers in Columns (1)–(6) should equal 9.80, which it does.

The least squares estimates are obtained by multiplying $(X'X)^{-1}$ by $X'Y$. The resulting estimates are:

$$b_1 = 1.89411768(74) - 0.14901975(282) - 0.25490185(375) = 2.5529$$
$$b_2 = -0.14901961(74) + 0.05261437(282) - 0.01307189(375) = -1.0921$$
$$b_3 = -0.25490197(74) - 0.01307186(282) + 0.06535945(375) = 1.9608$$

The estimated regression equation is therefore equal to

$$\hat{Y} = 2.5529 - 1.0921\ X_2 + 1.9608\ X_3$$

To estimate Var(b), we must first estimate σ^2. An unbiased estimate of σ^2 can be obtained from (10.12):

$$s^2 = \frac{e'e}{n-k} = \frac{1}{n-k}(Y'Y - b'X'Y)$$

Carrying out the computations, we obtain

$$Y'Y = \sum Y_i^2 = 626$$

$$b'X'Y = 2.5529(74) - 1.0921(282) + 1.9608(375) = 616.24$$

$$e'e = 626.00 - 616.24 = 9.76$$

and

$$s^2 = \frac{9.76}{7} = 1.39$$

The estimated variances of the regression coefficients are obtained by multiplying s^2 by the elements of the principle diagonal of $(X'X)^{-1}$ to obtain

$$s_{b_1}{}^2 = 1.39(1.89411768) = 2.633$$
$$s_{b_2}{}^2 = 1.39(0.05261437) = 0.073$$
$$s_{b_3}{}^2 = 1.39(0.06535945) = 0.091$$

These estimates will be used in later sections of this chapter.

10.1.3 Interpretation of the Regression Coefficients

The coefficients of a multiple regression equation are called *partial regression coefficients*. The coefficient β_i, for example, shows the average change in the dependent variable Y when there is a unit change in the independent variable X_i while the other independent variables are held constant.

The demand function estimated in Section 10.1.2 will serve to illustrate how multiple regression coefficients are interpreted. The coefficient $b_2 = -1.0821$ indicates that, on the average, the amount of the item purchased will decline by 1.0821 units for each $1 increase in price while family income, X_3, is held constant. The income coefficient, b_3, shows the change in the numbers of units purchased if income increases by $1000 while the price is held constant. The coefficient b_1 has the same units as the dependent variable Y and shows the amount of the item that would be purchased if both price and income were zero.

In some cases we may want to compare the relative importance of the independent variables. If the partial regression coefficients have the same units, they can be compared directly. Otherwise, we shall have to convert them into units of standard deviations. Suppose, for example, that the regression function is

$$Y = \beta_1 + \beta_2 X_2 + \beta_3 X_3 \tag{10.13}$$

This function is identical to

$$\frac{Y}{\sigma_y} = \frac{\beta_1}{\sigma_y} + \frac{\beta_2 \sigma_2}{\sigma_y}\frac{X_2}{\sigma_2} + \frac{\beta_3 \sigma_3}{\sigma_y}\frac{X_3}{\sigma_3} \tag{10.14}$$

$$= \beta_1{}^* + \beta_2{}^* \frac{X_2}{\sigma_2} + \beta_3{}^* \frac{X_3}{\sigma_3}$$

where σ_y is the standard deviation of Y while σ_2 and σ_3 are the standard deviations of X_2 and X_3, respectively. The coefficients $\beta_1{}^*$, $\beta_2{}^*$, and $\beta_3{}^*$ are called *beta coefficients.*

Estimators for the standard partial regression coefficients can be constructed from the least squares estimators of β_1, β_2, β_3 and the usual estimators for the standard deviations of Y, X_2, and X_3. An estimator for $\beta_2{}^*$, for example, would be

$$b_2{}^* = b_2 \frac{s_2}{s_y} = b_2 \sqrt{\frac{\sum (X_{2i} - \bar{X}_2)^2}{\sum (Y_i - \bar{Y})^2}}$$

where b_2 is the least squares estimator of β_2 in (10.13).

As an example, suppose that an educational testing laboratory estimates that

$$Y = 10 + 4X_2 + 0.5X_3$$

where Y is the score received on a certain test

X_2 is the number of hours spent studying for the test

X_3 is the IQ of the person taking the test

In order to compare the relative importance of study-time and *IQ* first, we compute $s_y = 8$, $s_2 = 4$, and $s_3 = 10$. The estimated beta coefficients are therefore equal to

$$b_1{}^* = \left(\frac{10}{8}\right) = 1.25$$

$$b_2{}^* = 4\left(\frac{4}{8}\right) = 2.00$$

$$b_3{}^* = 0.5\left(\frac{10}{8}\right) = 0.625$$

The standardized regression equation is written as

$$\frac{Y}{8} = 1.25 + 2.0\left(\frac{X_2}{4}\right) + 0.625\left(\frac{X_3}{10}\right)$$

The interpretation of this equation is that, on the average, a one-standard deviation change in X_2 will result in a two-standard deviation change in Y and a one-standard deviation change in X_3 will result in a 0.625 standard deviation change in Y. Thus, it appears that study-time was relatively more important than *IQ* in determining the test score on the particular test under study.

10.2 THE COEFFICIENT OF MULTIPLE DETERMINATION

Once a multiple regression equation has been estimated, the investigator often wants to determine the closeness of fit of the regression plane to the actual observations. Let us assume that the estimated regression equation is

$$\hat{Y} = b_1 + b_2 X_2 + b_3 X_3 \tag{10.15}$$

The difference between an observed value of Y, say, Y_i, and the mean value of the Y's can be divided into two parts. We have

$$Y_i - \bar{Y} = (Y_i - \hat{Y}_i) + (\hat{Y}_i - \bar{Y}) \tag{10.16}$$

where $\hat{Y}_i$ is the value of Y predicted from the estimated regression function. Squaring both sides of (10.16) and summing overall values of Y, we obtain

$$\sum (Y_i - \bar{Y})^2 = \sum (Y_i - \hat{Y}_i)^2 + \sum (\hat{Y}_i - \bar{Y})^2 \tag{10.17}$$

The quantity $\sum (Y_i - \bar{Y})^2$ is frequently called the total sum of squares and denoted by S_{YY}. The quantity $\sum (Y_i - \hat{Y}_i)^2$ is called the unexplained sum of squares or error sum of squares and denoted by S_E. The quantity $\sum (\hat{Y}_i - \bar{Y})^2$

is the explained sum of squares and is denoted by S_R. The larger the value of S_R, the closer the observed points are to the regression plane. The ratio of the explained sum of squares to the total sum of squares

$$R^2 = \frac{\sum (\hat{Y}_i - \bar{Y})^2}{\sum (Y_i - \bar{Y})^2} = \frac{S_R}{S_{YY}} \tag{10.18}$$

is called the *coefficient of multiple determination*. The square root of R^2 is called the *multiple correlation coefficient*.

The multiple correlation coefficient R cannot be considered as a measure of the correlation between the X's and Y because the X's in the regression model are fixed variables. Even so, it is a useful summary statistic to measure the goodness of fit of the observed points to the regression plane.

To derive a computational formula for R^2, we note that

$$S_E = e'e = Y'Y - b'X'Y = \sum Y_i^2 - \sum_i \sum_j b_j X_{ji} Y_i$$

and

$$S_{YY} = Y'Y - \frac{1}{n}(\sum Y_i)^2 = \sum Y_i^2 - \frac{1}{n}(\sum Y_i)^2$$

Subtracting the unexplained sum of squares from both sides of (10.17), we obtain

$$S_R = S_{YY} - S_E = \sum_i \sum_j b_j X_{ji} Y_i - \frac{1}{n}(\sum Y_i)^2$$

The coefficient of multiple determination for the regression equation (10.15) is therefore equal to

$$R^2 = \frac{b_1 \sum Y_i + b_2 \sum X_{2i} Y_i + b_3 \sum X_{3i} Y_i - \frac{1}{n}(\sum Y_i)^2}{\sum Y_i^2 - \frac{1}{n}(\sum Y_i)^2} \tag{10.19}$$

The numerical example presented in Section 10.1.2 will be used to illustrate the computation of R^2. The total sum of squares is equal to

$$S_{YY} = 626.0 - 547.6 = 78.4$$

The error sum of squares is $S_E = e'e = 626.0 - 619.06 = 6.94$. The explained sum of squares, as found by subtraction, is

$$S_R = 78.40 - 6.94 = 71.46$$

The coefficient of multiple determination is therefore

$$R^2 = \frac{S_R}{S_{YY}} = \frac{71.46}{78.40} = 0.9114$$

and the multiple correlation coefficient is

$$R = \sqrt{R^2} = \sqrt{0.9114} = 0.9547$$

This result is frequently summarized by saying that 91.14 percent of the total deviation has been explained by the regression line. Note that by selecting b so as to minimize S_E, we have at the same time maximized S_R and the percent of the total deviation explained by the regression line.

When both the X's and Y are stochastic variables, the coefficient of multiple determination can be considered as a measure of the correlation between Y and its best linear estimate $\hat{Y}$.

The estimator R^2 given by (10.18) tends to overestimate the true value of the coefficient of multiple determination. The adjusted estimator is

$$\hat{R}^2 = 1 - \frac{S_E}{S_{YY}}\left[\frac{n-1}{n-k-1}\right] \tag{10.20}$$

where n is the number of observations used when estimating the regression function and k is the number of coefficients in the estimated regression function.

For the numerical example above, we have

$$\hat{R}^2 = 1 - \frac{6.94}{78.4}\left[\frac{10-1}{10-3-1}\right] = 0.8673$$

Since $\left[\frac{n-1}{n-k-1}\right]$ is always greater than one, the adjusted coefficient of multiple determination is always less than the unadjusted R^2.

10.2.1 Partial Correlation

The technique of partial correlation is frequently used in order to measure the separate or net effect of an independent variable when other specified variables are held constant. To distinguish one partial-correlation coefficient from another, we need to indicate which other variables are being held constant. This is done by means of secondary subscripts. The partial-correlation coefficient between X_1 and Y given X_2, X_3, and X_4, for example, would be denoted by $r_{1Y\cdot 234}$. The number of secondary variables listed after the subscripted dot determines the order of the partial-correlation coefficient. Thus $r_{1Y\cdot 234}$ is a third-order partial-correlation coefficient while $r_{1Y\cdot 23}$ is of second order. Using this notation, the coefficient of multiple determination R^2 may be written as $R^2_{Y\cdot 123\cdots k}$ where the subscripts following the dot serve to indicate which explanatory variables were included in the regression equation.

Partial-correlation coefficients of first and higher order can be expressed in terms of partial-correlation coefficients of the next-lower order. The partial-correlation coefficient $r_{1Y\cdot 23}$, for example, is equal to

$$r_{1Y\cdot 23} = \frac{r_{1Y\cdot 3} - r_{2Y\cdot 3}r_{12\cdot 3}}{\sqrt{(1 - r_{2Y\cdot 3}^2)(1 - r_{12\cdot 3}^2)}} \tag{10.21}$$

while $r_{1Y\cdot 3}$ is equal to

$$r_{1Y\cdot 3} = \frac{r_{1Y} - r_{3Y}r_{13}}{\sqrt{(1 - r_{3Y}^2)(1 - r_{13}^2)}} \tag{10.22}$$

and

$$r_{1Y} = \frac{\sum (X_{1i} - \bar{X}_1)(Y_i - \bar{Y})}{\sqrt{\sum (X_{1i} - \bar{X}_1)^2}\sqrt{\sum (Y_i - \bar{Y})^2}} \tag{10.23}$$

An iterative procedure is used to compute the value of $r_{1Y\cdot 23}$. First, find the zero-order coefficients r_{1Y}, r_{2Y}, r_{3Y}, r_{12}, r_{13}, and r_{23}. Then compute $r_{1Y\cdot 3}$, $r_{2Y\cdot 3}$, and $r_{12\cdot 3}$ and use the results to obtain $r_{1Y\cdot 23}$. Using a somewhat simpler example, suppose that the zero-order correlation coefficients associated with the regression equation

$$Y = b_{Y\cdot 12} + b_{Y1\cdot 2}X_1 + b_{Y2\cdot 1}X_2$$

are $r_{1Y} = 0.8$, $r_{2Y} = 0.4$, and $r_{12} = 0.2$. Using equation (10.22), we find the partial-correlation coefficients $r_{1Y\cdot 2}$ and $r_{2Y\cdot 1}$ to be equal to

$$r_{1Y\cdot 2} = \frac{0.8 - (0.4)(0.2)}{\sqrt{(1 - 0.16)(1 - 0.04)}} = \frac{0.72}{\sqrt{0.8064}} = 0.80$$

and

$$r_{2Y\cdot 1} = \frac{0.4 - (0.8)(0.2)}{\sqrt{(1 - 0.64)(1 - 0.04)}} = \frac{0.24}{\sqrt{0.3456}} = 0.408$$

The square of the partial-correlation coefficient between, say, X_1 and Y indicates what percentage of the variance in Y previously left unexplained by the other variables can be accounted for by X_1. Thus, if $r_{1Y\cdot 23}^2 = 0.60$, we would conclude that the variable X_1 accounts for 60 percent of the variance in Y which is left unexplained by X_2 and X_3.

10.2.2 Further Comments on Partial Correlation

The terminology of correlation and regression analysis was first developed in the late nineteenth century by men who were concerned with biological phenomena. They measured variables which were inherently continuous, such as heights and weights of individuals drawn at random from a population of plants or animals of a given species. Often the frequency distribution of measurements on each variable approximated a normal curve. Two such variables would form a *bivariate normal* distribution; three or more such variables would form a *multivariate normal* distribution. As the random samples involved were drawn directly from natural populations, the coefficients of simple, multiple, and partial correlation computed from samples were regarded as estimates of the corresponding population parameters.

Later, scientists conducting agricultural experiments to determine (for example) the effects of fertilizer, irrigation water, and other inputs or practices upon crop yields recognized that bivariate or multivariate normal distributions of their independent variables would be extremely inefficient. Nature does not provide us with a normal population of chemical fertilizer applications from which to sample. The scientists were trying to answer cause-and-effect questions such as: What effect will 80 pounds of nitrogen fertilizer per acre have upon corn yields? Will 160 pounds of nitrogen per acre raise corn yields significantly relative to the effects of 80 pounds of nitrogen per acre? Are the effects of nitrogen and phosphate fertilizers independent of one another, or does nitrogen have different effects on corn yields depending upon the amount of phosphate we apply?

In a pragmatic sense, to ask these questions is to answer them. If we wish to explore the effects of nitrogen (N) and phosphate (P) upon corn yields (Y), we can choose any set of paired values of N and P we like, apply each (N, P) pair to a separate experimental plot, measure the corresponding values of Y at harvest time, and fit a multiple regression equation, $Y = f(N, P)$, to the observations. Viewed in this light, multiple regression analysis becomes a remarkably versatile tool. At the same time, however, coefficients of correlation, whether simple, multiple, or partial, lose their status as estimates of population parameters except in a highly artificial sense.

As a matter of fact, the coefficients of simple correlation between pairs of independent variables in a controlled experiment can be set at any levels we choose. The coefficient of multiple correlation (and its square, the coefficient of multiple determination) can be increased by increasing the standard deviations of the independent variables. In principle, values of the independent variables should be chosen to estimate *regression* relationships as accurately as possible for any given number of observations. (This is only an approximate statement, but to elaborate on it would lead us too far into the specialized field of experimental design.)

Consider the following three-variable multiple regression equation:

$$X_0 = b_{0 \cdot 12} + b_{01 \cdot 2} X_1 + b_{02 \cdot 1} X_2 + u_{0 \cdot 12} \tag{10.24}$$

The simple correlation coefficients among the three possible pairs of variables are r_{01}, r_{02}, and r_{12}, and the coefficient of multiple determination of X_0 by X_1 and X_2 is $R^2_{0 \cdot 12}$.

The coefficient of multiple determination may be written as

$$R^2_{0\cdot 12} = \frac{r_{01}^2 + r_{02}^2 - 2r_{01}r_{02}r_{12}}{1 - r_{12}^2} \tag{10.25}$$

In the special case in which $r_{12} = 0$, this becomes simply

$$R^2_{0\cdot 12} = \frac{r_{01}^2 + r_{02}^2}{1} = r_{01}^2 + r_{02}^2 \tag{10.26}$$

If r_{12} is set equal to zero by an appropriate choice of paired values of X_1 and X_2 in an experimental design, the partial regression coefficients $b_{01\cdot 2}$ and $b_{02\cdot 1}$ can be estimated more accurately than if r_{12} is set at any nonzero value. For the standard error of (say) $b_{02\cdot 1}$ can be written as

$$s_{b_{02\cdot 1}} = \frac{s_0}{s_2}\left[\frac{1 - R^2_{0\cdot 12}}{(1 - r_{12}^2)(n-3)}\right]^{1/2} = \frac{\bar{S}_{0\cdot 12}}{s_2\sqrt{n(1 - r_{12}^2)}} \tag{10.27}$$

$\bar{S}_{0\cdot 12}$ is the standard error of estimate of X_0 from equation (10.24) and s_0 and s_2 are the standard deviations of X_0 and X_2, respectively. If $r_{12} = 0$, (10.27) becomes

$$s_{b_{02\cdot 1}} = \frac{\bar{S}_{0\cdot 12}}{s_2\sqrt{n}} \tag{10.28}$$

Any nonzero value of r_{12} will reduce the denominator of (10.27) and increase the standard error, $s_{b_{02\cdot 1}}$, relative to that in (10.28). The standard error of estimate squared, $\bar{S}^2_{0\cdot 12}$, is an estimate of the population variance $\sigma^2_{0\cdot 12}$ attributable to random disturbances not correlated with X_1 and X_2. We assume that $\sigma^2_{0\cdot 12}$ is the same for all possible pairs of values of X_1 and X_2. Hence, $\bar{S}_{0\cdot 12}$ is independent of s_1 and s_2, and it may be possible to increase s_2 in (10.27) and (10.28) relative to its value in some tentative experimental design, thereby reducing $s_{b_{02\cdot 1}}$. For example, as compared with a normal distribution of X_2, we could select far fewer observations near the center of the experimental range and far more near the extremes. For any final value of s_2 we decided upon, $s_{b_{02\cdot 1}}$ could be reduced still further by increasing the number of observations, n.

Given equation (10.24), we could set X_1 equal to any arbitrary constant and calculate the expected values of X_0 associated with specified values of X_2:

$$\hat{X}_0 = (b_{0\cdot 12} + b_{01\cdot 2}X_1) + b_{02\cdot 1}X_2 \tag{10.29}$$

If we set $X_1 = 0$, 80, 160, and 240, respectively, equation (10.29) would give us four parallel lines with the slope $b_{02\cdot 1}$ and with equally spaced intercepts differing by successive increments of 80 $b_{01\cdot 2}$. Each of these lines would represent the partial regression of X_0 upon X_2 for the specified value of X_1. The partial regression of X_0 upon X_1 would be

$$\hat{X}_0 = (b_{0\cdot 12} + b_{02\cdot 1} X_2) + b_{01\cdot 2} X_1 \tag{10.30}$$

for any specified value of X_2.

The idea of partial regression can be extended to nonlinear cases. For example, what is the relation between corn yields and phosphate when nitrogen is held constant at $N = 0$? The answer can be obtained by applying phosphate at several different levels on different plots and using no nitrogen at all. The relation between Y and P when $N = 80$ can be approached in an equally direct fashion. In principle we can estimate a relationship such as

$$E(Y \mid N = 80) = f(P) \tag{10.31}$$

as closely as we like by increasing the number of observations.

Table 10.3 Joint Frequency Distribution of Independent Variables, Nitrogen and Potash, in Corn Yield Experiment

Phosphate (Pounds per Acre)	Nitrogen (Pounds per Acre) 0	80	160	240
	Number of Observations			
0	2	2	2	2
80	2	2	2	2
160	2	2	2	2
240	2	2	2	2

Table 10.3 will provide some insights into the nature of the multiple regression analysis of experimental data. The 32 experimental plots are given a total of 16 different "treatments." Each "treatment" is a distinct combination of N and P. The joint frequency distribution of N and P is rectangular, and their simple correlation coefficient is zero. For we may write

$$\begin{aligned} E(N \mid P = 0) &= \frac{2}{8}(0 + 80 + 160 + 240) = 120 \\ E(N \mid P = 80) &= \frac{2}{8}(0 + 80 + 160 + 240) = 120 \end{aligned} \tag{10.32}$$

and so on for $P = 160$ and $P = 240$. Also, $E(P \mid N = 0) = (2/8)(0 + 80 + 160 + 240) = 120$, and $E(P)$ is the same for $N = 80$, 160, and 240. The values of

N and P are specifically chosen to achieve zero correlation. This controlled result is more precise than we could expect if N and P were random drawings from a *population* in which N and P were independently distributed, for although $E(r_{np}) = \rho_{np} = 0$ in that case, $r_{np} \neq 0$ in any particular sample.

Table 10.4 shows the corn yields, Y, obtained on the 32 plots. Differences between the two yields obtained for the same combination of N and P on

Table 10.4 Corn Yields Related to Amounts of Phosphorus and Nitrogen Applied, Iowa, 1952[a]

Phosphate (Pounds per Acre)	Nitrogen (Pounds per Acre)			
	0	80	160	240
	Corn Yields (Bushels per Acre)			
0	24.5	28.7	17.3	16.2
	6.2	6.4	4.2	6.8
80	22.1	99.5	115.9	112.4
	30.6	115.4	72.6	125.6
160	12.0	102.2	129.7	130.5
	34.0	108.5	116.3	124.3
240	38.0	97.2	127.6	121.1
	35.0	107.8	125.8	114.2

Source. Earl O. Heady, John T. Pesek, and William G. Brown, "Crop Response Surfaces and Economic Optima in Fertilizer Use," *Iowa State College Agricultural Experiment Station Research Bulletin* 424, March 1955.

[a] Two figures are shown in each cell since the treatments were replicated.

two different plots obviously cannot be attributed to differences in these variables; they provide an estimate of experimental error, uncorrelated with N and P.

Figure 10.1 shows the mean values of each of the 16 pairs of corn yields. The four mean values corresponding to a particular value of N are connected to show the partial relation between Y and P when N is held constant at the specified level. The relations are somewhat erratic, as each mean is based on only two observations. However, the overall impression gained from Figure 10.1 is that the relation between Y and P is nonlinear, reaching a maximum somewhere between $P = 160$ and $P = 240$ for any given level of N. The *level* of the partial relation between Y and P rises sharply between $N = 0$ and $N = 80$ and increases only moderately as N moves from 80 to 160 and from 160 to 240. For $P = 160$ to 240, the levels of the partial relations are almost

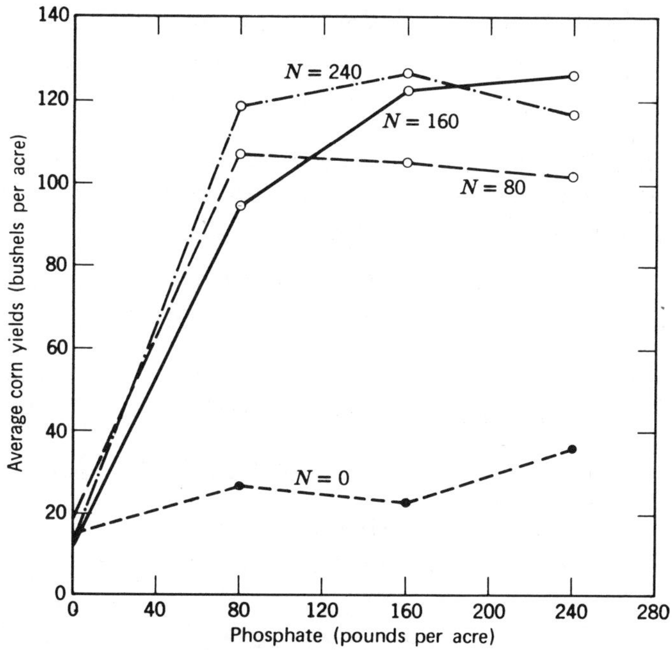

Figure 10.1 Relations between means of 16 pairs of corn yields in Table 10.4 and amounts of phosphate applied, for each of four levels of Nitrogen ($N = 0$, 80, 160, and 240 pounds per acre).

identical for $N = 160$ to 240, suggesting that Y also reaches a maximum with respect to N in that range.

A priori considerations and other experiments lead us to assume that the relation between Y, P, and N is continuous, that Y shows diminishing returns with respect to P and N, and that Y reaches a maximum value beyond which additional amounts of P and N no longer increase Y and may actually reduce it. There is some suggestion in Figure 10.1 that increases in N not only raise the level of the partial relation of Y to P but also change its shape. The last feature implies that the multiple regression surface, $Y = f(P, N)$, probably involves a multiplicative term in P and N in addition to additive terms.

One multiple regression equation embodying these features is

$$\hat{Y} = 7.3652 + \underset{(0.1490)}{0.8177P} + \underset{(0.1490)}{0.58448N} - \underset{(0.000556)}{0.002806P^2} - \underset{(0.000556)}{0.002059N^2} + \underset{(0.000444)}{0.001444NP} \tag{10.33}$$

$$\bar{S} = 20.1169$$

$$R^2 = 0.8584$$

If we set $N = 0$ in equation (10.33), the partial regression of Y upon P becomes

$$\hat{Y} = 7.3652 + 0.8177P - 0.002806P^2 \tag{10.34}$$

If we set $N = 100$ (although we had no observations precisely at $N = 100$, we feel justified in assuming that Y is a continuous function of N), the partial regression of Y upon P becomes

$$\begin{aligned} \hat{Y} = {} & 7.3652 + 58.448 - 20.59 \\ & + 0.8177P + 0.1444P - 0.002806P^2 \end{aligned} \tag{10.35}$$

or

$$\hat{Y} = 45.2232 + 0.9621P - 0.002806P^2$$

If we set $N = 200$, the partial regression of Y upon P becomes

$$\begin{aligned} \hat{Y} = {} & 7.3652 + 116.896 - 82.36 \\ & + 0.8177P + 0.2888P - 0.002806P^2 \end{aligned} \tag{10.36}$$

or

$$\hat{Y} = 41.9012 + 1.1065P - 0.002806P^2$$

Perhaps this discussion is sufficient for illustrating partial regression concepts. However, we would like to do a little more justice to the power and elegance of multiple regression analysis based on well-designed experiments. Table 10.5 shows the complete design of the corn yield and fertilizer experiment from which we selected our 32 observations. The complete design included 114 observations. The rectangular array of values of N and P (actually $P_2 0_5$) evidently achieves either zero correlation or very low correlation between N and P. Also, the observations on N and P are spaced only 40 pounds apart instead of 80. The 57 different treatments, each replicated on two plots, provide 57 degrees of freedom for estimating the level of experimental error.

The regression equation chosen (by the team of economists and agronomists who performed this experiment) as best fitting the data and also meeting a priori considerations was

$$\begin{aligned} \hat{Y} = {} & -5.682 - \underset{(0.040)}{0.316N} - \underset{(0.040)}{0.417P} + \underset{(0.8676)}{6.3512\sqrt{N}} \\ & + \underset{(0.8680)}{8.5155\sqrt{P}} + \underset{(0.0385)}{0.3410\sqrt{NP}} \end{aligned} \tag{10.37}$$

Table 10.5 Experimental Yields of Corn for Varying Levels of Fertilizer Application on Calcareous Ida Silt-Loam Soil in Western Iowa in 1952 (Yields are in Bushels per Acre)[a]

P_2O_5, lb	Nitrogen, lb								
	0	40	80	120	160	200	240	280	320
0	24.5	23.9	28.7	25.1	17.3	7.3	16.2	26.8	25.1
	6.2	11.8	6.4	24.5	4.2	10.0	6.8	7.7	19.0
40	26.7	60.2			96.0	95.4			81.9
	29.6	82.5			107.0	95.4			76.4
80	22.1		99.5		115.9		112.4		129.0
	30.6		115.4		72.6		125.6		82.0
120	44.2			119.4	113.6			114.9	124.6
	21.9			97.3	102.1			129.2	83.0
160	12.0	96.2	102.2	133.3	129.7	105.7	130.5	123.6	135.6
	34.0	80.7	108.5	124.4	116.3	115.5	124.3	142.5	122.7
200	37.7	81.1			128.7	140.3			136.0
	34.2	51.0			109.3	142.2			118.2
240	38.0		97.2		127.6		121.1		130.9
	35.0		107.8		125.8		114.2		144.9
280	32.4			129.5	134.4			130.0	124.8
	27.4			125.2	127.6			141.9	114.1
320	5.3	79.5	116.9	135.7	122.9	138.7	127.3	131.8	127.9
	17.9	39.7	83.6	121.5	122.7	126.1	139.5	111.9	118.8

Source. Earl O. Heady, John T. Pesek, and William G. Brown, "Crop Response Surfaces and Economic Optima in Fertilizer Use," *Iowa State College Agricultural Experiment Station Research Bulletin 424*, March 1955, p. 330.

[a] Two figures are shown in each cell, since the treatments were replicated (that is, two plots received the same fertilizer quantities and ratios).

The partial regressions of Y on P for $N = 0$, 100, and 200, respectively, are

$$Y = -5.682 - 0.417P + 8.5155\sqrt{P}; \quad (N = 0)$$
$$Y = 26.230 - 0.417P + 11.9255\sqrt{P}; \quad (N = 100)$$
$$Y = 20.924 - 0.417P + 13.3372\sqrt{P}; \quad (N = 200)$$

10.3 TESTS OF HYPOTHESES INVOLVING PARTIAL REGRESSION COEFFICIENTS

In many situations, an investigator wants to test a hypothesis involving one, or several, of the β_1. Let us begin with an arbitrary linear function of the least squares estimators, say,

$$A'b = \sum_{i=1}^{k} a_i b_i$$

where the coefficients a_i are constants.

In order to derive significance tests about $A'b$, we begin by adding the assumption that the disturbance terms u_i are normally distributed. Written compactly, the regression model becomes

$$Y = X\beta + U \tag{10.38}$$

where

U is $N(0, \sigma^2 I_n)$

X is a set of fixed numbers such that $(X'X)^{-1}$ exists

Y, X, β, U and I_n are matrices defined as in Section 10.1

The statement U is $N(0, \sigma^2 I_n)$ is read "the vector of disturbance terms U has a multivariate normal distribution with mean 0 and variance $\sigma^2 I_n$." In this case, the least squares estimators of β are identical to the maximum likelihood estimators.

From (10.7), it is seen that the least squares estimator b is equal to β plus a linear function of U. It can be shown that, if U is $N(0, \sigma^2 I_n)$, then b has a multivariate normal distribution with mean β and variance $\sigma^2(X'X)^{-1}$. The quantity $A'b$ is therefore a linear function of joint normally distributed variables and is, in turn, normally distributed.

The expected value of $A'b$ is

$$E(A'b) = A'E(b) = A'\beta \tag{10.39}$$

Suppose, for example, that the regression function is

$$Y = \beta_1 + \beta_2 X_2 + \beta_3 X_3 + u \tag{10.40}$$

and the estimated regression equation is equal to

(10.41) $$\hat{Y} = b_1 + b_2 X_2 + b_3 X_3$$

If $A' = (0\ 1\ 1)$, then

$$A'b = (0\ 1\ 1)\begin{pmatrix} b_1 \\ b_2 \\ b_3 \end{pmatrix} = b_2 + b_3$$

and

$$E(A'b) = E(b_2 + b_3) = E(b_2) + E(b_3) = \beta_2 + \beta_3$$

The variance of $A'b$ is equal to

$$\text{Var}(A'b) = \sigma^2 A'(X'X)^{-1}A$$

Using (10.40), (10.41) and $A' = (0\ 1\ 1)$, we find

$$\text{Var}(A'b) = \sigma^2\left[(0\ 1\ 1)\begin{pmatrix} c_{11} & c_{12} & c_{13} \\ c_{21} & c_{22} & c_{23} \\ c_{31} & c_{32} & c_{33} \end{pmatrix}\begin{pmatrix} 0 \\ 1 \\ 1 \end{pmatrix}\right]$$
$$= \sigma^2(c_{22} + c_{32} + c_{23} + c_{33})$$

where the c_{ij} represent the elements in the matrix $(X'X)^{-1}$.

A standard normal variable z^* can be formed by subtracting $E(A'b)$ from $A'b$ and dividing by $\sqrt{\text{Var}(A'b)}$. Doing this, we obtain

(10.42) $$z^* = \frac{A'b - A'\beta}{\sigma\sqrt{A'(X'X)^{-1}A}}$$

It can be shown that $A'b$ is distributed independently of $e'e = \sum e_i^2$ and that $e'e/\sigma^2$ has an χ^2 distribution with $n - k$ degrees of freedom. A t-statistic can therefore be constructed using the formula (6.18). The normal variable z^* is divided by the square root of $e'e/\sigma^2(n - k)$ to obtain

(10.43) $$t^* = \frac{(A'b - A'\beta)/\sigma\sqrt{A'(X'X)^{-1}A}}{(1/\sigma)\sqrt{e'e/(n-k)}} = \frac{A'b - A'\beta}{\sqrt{A'(X'X)^{-1}A(e'e)/(n-k)}}$$

where

$$s_{A'b} = \sqrt{\frac{A'(X'X)^{-1}A(e'e)}{n-k}}$$

is the standard error of $A'b$ and t^* has a t distribution with $n - k$ degrees of freedom.

To test a hypothesis about a single regression parameter, say, β_i, we let $a_i = 1$ and set all of the other elements of the vector A equal to zero. Then

$A'b = b_i$ and $A'(X'X)^{-1}A = c_{ii}$ where c_{ii} is the ith element in the principal diagonal of $(X'X)^{-1}$. The appropriate t-statistic is

$$t^* = \frac{b_i - \beta_i}{s_{b_i}} = \frac{b_i - \beta_i}{\sqrt{c_{ii}\sum e_i^2/(n-k)}} \tag{10.44}$$

To test any particular hypothesis about β_i, we substitute the hypothetical value of β_i in (10.44) and reject the hypothesis if t^* lies in a critical region of the test.

The numerical data in Section 10.1.2 will be used to illustrate the test procedure. The estimated regression function was

$$\hat{Y} = 2.5529 - 1.0821X_2 + 1.9608X_3$$

Suppose we wish to test the hypothesis

$$H_0\colon \beta_2 = 0$$

against the alternative

$$H_a\colon \beta_2 < 0$$

at the 5 percent level of significance. In this case, there are $n - k = 7$ degrees of freedom and the hypothesis H_0 is rejected if the observed value of t is less than -1.8946.

The standard error of b_2 is

$$\sqrt{s_{b_2}{}^2} = \sqrt{s^2(c_{22})} = \sqrt{1.39(0.05261)} = 0.27$$

The computed value of t is therefore equal to

$$t = \frac{b_2}{\sqrt{s_{b_2}{}^2}} = \frac{-1.0921}{0.27} = -4.045$$

which leads to the rejection of H_0. In other words, the price appears to have a significance effect on the quantity purchased.

As a second example, suppose that a production function is estimated from a random sample of 19 plants and found to be

$$\hat{Q} = 0.25 + \underset{(0.60)}{2.40L} + \underset{(0.80)}{1.60K} \qquad R^2 = 0.81$$

where

Q is quantity produced
L is quantity of labor used
K is a measure of the capital input

The numbers in parenthesis below the regression coefficients are the standard errors of the regression coefficients. That is, $s_{b_2} = 0.60$ and $s_{b_3} = 0.80$. To test $H_0\colon \beta_3 = 0$ against the alternative $H_a\colon \beta_3 > 0$, we compute

$$t = \frac{1.60}{0.80} = 2.0$$

The tabled value of t for a one-tail test with $n - k = 16$ degrees of freedom at the 5 percent level of significance is 1.75. The critical region of the test consists of all values of t greater than 1.75. The observed value of 2.0 is in the critical region and leads to the rejection of H_0. We conclude that there is a positive relationship between output Q and the measure of capital input K.

In some cases, an investigator wants an overall test of whether or not the regression relationship is significant. The hypothesis tested in this case is

$$H_0\colon \beta_1 = \beta_2 = \cdots = \beta_k = 0 \tag{10.45}$$

The appropriate test statistic is constructed as follows. We have seen in Section 10.2 that the total sum of squared observations for the dependent variable, $Y'Y$, can be divided into two parts: the sum of squared residuals $e'e$ and a part that measures the reduction in $Y'Y$ due to the explanatory variables, $b'X'Y$. It can be shown that $b'X'Y/\sigma^2$ is distributed as χ^2 with k degrees of freedom and is independent of $e'e/\sigma^2$ which is distributed as χ^2 with $n - k$ degrees of freedom. The ratio

$$F = \frac{b'X'Y/\sigma^2 k}{e'e/\sigma^2(n-k)} = \frac{(n-k)b'X'Y}{k(e'e)} = \frac{(n-k)\sum_j \sum_j b_j X_{ji} Y_i}{k \sum_i e_i^2} \tag{10.46}$$

therefore has an F distribution with k and $n - k$ degrees of freedom. If $b'X'Y$ is small relative to $e'e$, then there has been relatively little reduction in $Y'Y$ due to the explanatory variables and the regression cannot be considered significant. Thus, small values of F lead to the acceptance of the hypothesis (10.45) that the regression relationship is not significant.

10.4 THE GENERAL LINEAR REGRESSION MODEL

In the preceding sections, it was assumed that the variance-covariance matrix of errors is given by

$$E(UU') = \sigma^2 I_n$$

This assumption may be replaced by the more general assumption that

$$E(UU') = \sigma^2 V \tag{10.47}$$

where V is an n by n positive definite matrix. The general linear regression model can then be written as

$$Y = X\beta + U \tag{10.48}$$

where

(i) $E(U) = 0$
(ii) $E(UU') = \sigma^2 V$
(iii) X is a fixed matrix of rank k

It can be shown that, if $E(UU')$ is not equal to $\sigma^2 I_n$, then the ordinary least squares estimators derived in Section 10.1 are not minimum variance estimators of β.

If V is positive definite, however, there exists a nonsingular matrix T of order n such that $TVT' = I_n$ and $T'T = V^{-1}$. Thus, by multiplying both sides of the regression equation (10.48) by the matrix T, we obtain

$$TY = TX\beta + TU = TX\beta + Z \tag{10.49}$$

where Z is such that $E(Z) = E(TU) = TE(U) = 0$ and $E(ZZ') = E(TUU'T') = TE(UU')T' = \sigma^2 TVT' = \sigma^2 I_n$. Since T is a nonsingular matrix of order n, the matrix TX is a fixed matrix with rank k; the rank of X.

The generalized least squares estimator of β is equal to

$$b^* = (X'T'TX)^{-1}X'T'TY = (X'V^{-1}X)^{-1}X'V^{-1}Y \tag{10.50}$$

The estimator b^* is unbiased. To verify this we note that

$$E(b^*) = E[(X'V^{-1}X)^{-1}X'V^{-1}Y] = E[(X'V^{-1}X)^{-1}X'V^{-1}(X\beta + U)] = \beta$$

since $(X'V^{-1}X)^{-1}X'V^{-1}X = I$ and $E(U) = 0$.

The variance of the generalized least squares estimator is given by

$$\text{Var}(b^*) = \sigma^2(X'T'TX)^{-1} = \sigma^2(X'V^{-1}X)^{-1} \tag{10.51}$$

It can be shown that for the individual coefficients

$$\text{Var}(b_i^*) < \text{Var}(b_i) \qquad \text{for } i = 1, 2, \ldots, k$$

In other words, that the generalized least squares estimators are more efficient estimators of the regression coefficients β_i for the general regression model than the ordinary regression coefficients derived in Section 10.1.

The unbiased estimator of σ^2 obtained using the residuals from the regression fitted to the transformed model is equal to

$$s^2 = \frac{e'T'Te}{n-k} = \frac{e'V^{-1}e}{n-k} \tag{10.52}$$

where

$$Te = TY - TXb^*$$

and b^* is the generalized least squares estimator of β.

Tests of hypotheses about the vector β may be constructed on the assumption that U is normally distributed with mean zero and variance $\sigma^2 V$. In this case, the generalized least squares estimator b^* is a linear function of U and is itself normally distributed with mean β and variance $\sigma^2(X'V^{-1}X)^{-1}$. Furthermore, it can be shown that b^* and the statistic s^2 given by equation (10.52) are distributed independently. In this case, it is possible to construct a *t-test* of the hypothesis that a single element of β, say, β_i, is equal to a particular value, say, β_i^0. The appropriate t-statistic is

$$t_{b_i*} = \frac{b_i^* - \beta_i}{\sqrt{(e'V^{-1}e)(X'V^{-1}X)_{ii}^{-1}/(n-k)}} \tag{10.53}$$

where $(X'V^{-1}X)_{ii}{}^{-1}$ is the diagonal element of $(X'V^{-1}X)^{-1}$ corresponding to Row i and Column i. The statistic t_{b_i*} has a t-distribution with $n-k$ degrees of freedom.

The appropriate F-statistic for a goodness of fit test for the regression as a whole (that is, a test of the hypothesis that $\beta = 0$) is

$$F = \left[\frac{b^{*\prime}X'V^{-1}Y}{e'V^{-1}e}\right]\left[\frac{n-k}{k}\right] \tag{10.54}$$

This F-statistic has an F distribution with k and $n-k$ degrees of freedom.

Several specific examples of the general linear regression model will be presented in the following sections.

10.5 AUTOCORRELATION

In the regression model presented at the beginning of this chapter, it was assumed that the error term u_i was an independent random variable. In other words, that successive error terms are drawn independently of previous values so that

$$E(u_i u_{i+s}) = 0 \qquad \text{for all } i \text{ and all } s \neq 0$$

Many difficulties in the analysis of economic data arise when this assumption is not justified. In this section, we shall consider the case where the error terms are serially dependent or *autocorrelated.*

Let us begin with a very simple form of autocorrelation. Suppose the regression equation is of the form

$$Y_i = \alpha + \beta X_i + u_i \tag{10.55}$$

where

$$u_i = pu_{i-1} + \varepsilon_i$$
$$|p| < 1$$

and

$$E(\varepsilon_i) = 0$$
$$\left.\begin{aligned} E(\varepsilon_i \varepsilon_{i+s}) &= \sigma_\varepsilon^2 \qquad \text{for } s = 0 \\ &= 0 \qquad \text{for } s \neq 0 \end{aligned}\right\} \quad \text{for all } i$$

The error term u_i is said to follow a first-order autoregression scheme.

To find the expected value of the error term u_i, we note that

$$\begin{aligned} u_i &= pu_{i-1} + \varepsilon_i = p(pu_{i-2} + \varepsilon_{i-1}) + \varepsilon_i \\ &= \varepsilon_i + p\varepsilon_{i-1} + p^2\varepsilon_{i-2} + \cdots \\ &= \sum_{r=0}^{\infty} p^r \varepsilon_{i-r} \end{aligned} \tag{10.56}$$

Therefore

$$E(u_i) = \sum_{r=0}^{\infty} p^r E(\varepsilon_{i-r}) = 0$$

The variance-covariance matrix of the error terms is

$$V = E(UU') = \begin{bmatrix} E(u_1^2) & E(u_1 u_2) & \cdots & E(u_1 u_n) \\ E(u_2 u_1) & E(u_2^2) & \cdots & E(u_2 u_n) \\ \vdots & \vdots & & \vdots \\ E(u_n u_1) & E(u_n u_2) & \cdots & E(u_n^2) \end{bmatrix} \tag{10.57}$$

But[2]

$$\begin{aligned} E(u_i^2) &= E(\varepsilon_i^2) + p^2 E(\varepsilon_{i-1}^2) + p^4 E(\varepsilon_{i-2}^2) + \cdots \\ &= (1 + p^2 + p^4 + \cdots)\sigma_\varepsilon^2 \\ &= \frac{\sigma_\varepsilon^2}{1 - p^2} = \sigma_u^2 \qquad \text{for all } i \end{aligned} \tag{10.58}$$

and it can be shown that

$$E(u_i u_{i-s}) = p^s \sigma_u^2 \qquad \text{for } s \neq 0$$

Therefore

$$E(UU') = \sigma_u^2 \begin{bmatrix} 1 & p & p^2 & \cdots & p^{n-1} \\ p & 1 & p & \cdots & p^{n-2} \\ \vdots & \vdots & \vdots & & \vdots \\ p^{n-1} & p^{n-2} & p^{n-3} & \cdots & 1 \end{bmatrix}$$

[2] Let $S = 1 + p^2 + p^4 + \cdots$, then $p^2 S = p^2 + p^4 + p^8 + \cdots$ and $S - p^2 S = 1$. Thus $S = 1/(1 - p^2)$.

Since $E(UU')$ is not equal to $\sigma^2 I$, the ordinary least squares estimators of α and β derived in Section 10.1 are no longer the best, linear, unbiased estimators. Furthermore, the significance tests based on the t and F distributions are no longer valid when the error terms are autocorrelated.

Unfortunately, the generalized-least squares estimator b^* cannot be computed unless the variance-covariance matrix V is known. Exact knowledge of V is seldom available. In some cases, it is possible to approximate b^* by using an estimate of V. This approach would usually be considered better than simply ignoring the problem of auto-correlation. Alternative approaches are available for estimating β if a suitable estimate of V cannot be obtained.

One alternative is to transform the original observations in such a way as to eliminate the autocorrelation of the error terms. Such an approach requires that we be able to test for the presence of autocorrelated error terms. The Durbin-Watson d statistic provides such a test.

10.5.1 The Durbin-Watson d Statistic

Let e_i for $i = 1, 2, \ldots, n$ denote the residuals from a simple least squares regression line. That is,

$$e_i = Y_i - a - \sum_{j=1}^{k} b_j X_{ji}$$

where a, the intercept, and the b_j are the simple least squares estimators. The Dubin-Watson statistic is defined as

$$d = \frac{\sum_{i=2}^{n} (e_i - e_{i-1})^2}{\sum_{i=1}^{n} e_i^2} \tag{10.59}$$

Durbin and Watson have tabulated upper and lower bounds d_U and d_L for various values of n (the sample size) and k (the number of independent variables). The 1 and 5 percent significance levels of d_L and d_U for $k = 1, \ldots, 5$ and various values of n between 15 and 100 are found in Table IX of Appendix I. The Durbin-Watson statistic is frequently denoted by D-W.

To test the hypothesis:

H_0: there is no positive serial correlation

against the alternative

H_1: there is positive serial correlation

we compute d and compare it with the critical value d_L. If there is positive serial correlation (i.e., if $0 < p < 1$), then d will tend to be small. Thus, if

d is less than d_L, we reject H_0 and conclude that there is positive serial correlation. If d is greater than d_U, we accept the hypothesis that there is no positive serial correlation. If $d_L < d < d_U$, then the test is inconclusive and further observations would be required in order to determine if positive serial correlation is present.

To test for negative autocorrelation, we use $4 - d$ instead of d. If $4 - d < d_L$, we reject the hypothesis of no negative auto-correlation in favor of the alternative hypothesis that there is negative autocorrelation. If $4 - d > d_U$, we accept the hypothesis that there is no negative autocorrelation. If $d_L < 4 - d < d_U$, the test is inconclusive.

For a two-tail test, we conclude that d is significant and there is autocorrelation if $d < d_L$ or $4 - d < d_L$. If, however, $d > d_U$ or $4 - d > d_U$, then d is not significant and there is no autocorrelation. Otherwise, the test is inconclusive. When the 5 percent Durbin-Watson table is used to find d_U and d_L, the level of significance of the two-tail test becomes 10 percent.

A numerical example will serve to clarify the computational procedure. The values of Y presented in Table 10.6 were obtained using the model

$$Y_i = 20 + 2X_i + u_i$$

where $u_i = 0.5u_{i-1} + \varepsilon_i$ and the error terms ε_i are distributed independently with $E(\varepsilon_i) = 0$. The ordinary least squares regression line is

$$\hat{Y} = 20.858 + 1.941X$$

The Durbin-Watson d-statistic is found to be equal to

$$d = \frac{23.491}{42.140} = 0.557$$

In this case, we have $n = 17$ observations and $k = 1$ independent variables. Using a 5 percent level of significance, we find from Table IX in Appendix I that $d_L = 1.13$ and $d_U = 1.38$. Since the observed value $d = 0.557$ is less than $d_L = 1.13$, we accept the alternative hypothesis that there is positive serial correlation.

Iterative procedures are frequently used to eliminate autocorrelation. The regression model (10.55) presented at the beginning of this section will be used to illustrate one such approach. Since $u_i = pu_{i-1} + \varepsilon_i$, the regression equation can be written as

$$Y_i = \alpha + \beta X_i + pu_{i-1} + \varepsilon_i$$

But $u_{i-1} = Y_{i-1} - \alpha - \beta X_{i-1}$, and therefore

$$Y_i = \alpha + \beta X_i + p(Y_{i-1} - \alpha - \beta X_{i-1}) + \varepsilon_i$$

Table 10.6 Original Data

i	X_i	Y_i	$\hat{Y}_i$	$(Y_i - \hat{Y}_i) = e_i$	$(e_i - e_{i-1})$
1	2	25.5	24.740	0.760	—
2	3	27.8	26.681	1.119	0.359
3	4	30.4	28.622	1.778	0.659
4	5	33.2	30.563	2.637	0.859
5	6	34.6	32.504	2.096	−0.541
6	7	35.3	34.445	0.855	−1.241
7	8	38.6	36.386	2.214	1.359
8	4	28.3	28.622	−0.322	−2.536
9	5	28.7	30.563	−1.863	−1.541
10	6	30.7	32.504	−1.804	0.059
11	7	33.3	34.445	−1.145	0.659
12	3	26.2	26.681	−0.481	0.664
13	4	29.1	28.622	0.478	0.959
14	5	28.5	30.563	−2.063	−2.541
15	8	34.3	36.386	−2.086	−0.023
16	2	23.1	24.740	−1.640	0.446
17	4	28.1	28.622	−0.522	1.118

$$\sum_{i=1}^{n} X_i = 83 \qquad \sum_{i=1}^{n} Y_i = 515.7$$

$$\sum_{i=1}^{n} X_i^2 = 463 \qquad \sum_{i=1}^{n} X_i Y_i = 2630$$

$$\sum_{i=1}^{n} e_i^2 = 42.140 \qquad \sum_{i=2}^{n} (e_i - e_{i-1})^2 = 23.491$$

$$\sum_{i=2}^{n} e_{i-1}^2 = 41.86 \qquad \sum_{i=2}^{n} e_i e_{i-1} = 29969$$

Rearranging terms, we obtain

$$(Y_i - pY_{i-1}) = (\alpha - p\alpha) + \beta(X_i - pX_{i-1}) + \varepsilon_i \tag{10.60}$$

This relationship can be written as

$$Y_i' = \alpha' + \beta X_i' + \varepsilon_i \qquad \text{for } i = 2, 3, \ldots, n \tag{10.61}$$

where the error terms ε_i are such that

$$E(\varepsilon_i^2) = \sigma_\varepsilon^2 \text{ and } E(\varepsilon_i \varepsilon_{i-s}) = 0$$

for all i and all $s \neq 0$. Thus, if Y' and X' were known, simple least squares regression would provide the best, linear, unbiased estimators for α' and β.

To estimate Y' and X', we first obtain simple least squares estimates of α and β using the original Y and X values. These estimates are then used to compute the residuals

$$e_i = Y_i - a - bX_i$$

The residuals are then used to estimate the autocorrelation p for a first-order autoregressive scheme of the form

$$u_i = pu_{i-1} + \varepsilon_i$$

The least squares estimator of p is

$$r = \frac{\sum_{i=2}^{n} e_i e_{i-1}}{\sum_{i=2}^{n} e_{i-1}^2} \tag{10.62}$$

The coefficient r is then used to compute the transformed variables:

$$\left.\begin{aligned} Y_i' &= (Y_i - rY_{i-1}) \\ \text{and} \qquad X_i' &= (X_i - rX_{i-1}) \end{aligned}\right\} \text{for } i = 2, 3, \ldots, n$$

Simple least squares is then used to estimate α' and β in (10.61). If the residuals obtained from the new relationship (10.61) are significantly autocorrelated, they can be used to provide a new estimate of p which, in turn, is used to obtain a new set of transformed variables from the original observations. This iterative procedure can be continued until the autocorrelation is eliminated. The procedure can be extended directly to the case of more than one independent variable.

The data presented in Table 10.6 will be used to illustrate this procedure. The least squares estimate of p is

$$r = \frac{29.969}{41.860} = 0.716$$

This estimate of p is then used to obtain the transformed values of X and Y presented in Table 10.7. The least squares regression line for the transformed values of X and Y is

$$\hat{Y}' = 5.737 + 2.017X'$$

An estimate of α can be obtained directly from the least squares estimate of α' by noting that $\alpha' = \alpha - p\alpha$ and thus

$$\alpha = \frac{\alpha'}{1 - p}$$

Table 10.7 Transformed Data

i	X_i'	Y_i'	$(Y_i' - \hat{Y}_i') = e_i'$	$(e_i' - e_{i-1}')$
1	—	—	—	—
2	1.568	9.542	0.643	—
3	1.852	10.495	1.023	0.380
4	2.136	11.434	1.389	0.366
5	2.420	10.829	0.211	−1.178
6	2.704	10.526	−0.665	−0.876
7	2.988	13.325	1.561	2.226
8	−1.728	0.663	−1.589	−3.150
9	2.136	8.437	−1.608	−0.019
10	2.420	10.151	−0.467	1.141
11	2.704	11.319	0.128	0.595
12	−2.012	2.357	0.678	0.550
13	1.852	10.341	0.869	0.191
14	2.136	7.665	−2.380	−3.249
15	4.420	13.894	−0.758	1.622
16	−3.728	−1.459	0.323	1.081
17	2.568	11.560	0.644	0.321

$$\sum_{i=2}^{n} X_i' = 24.436 \qquad \sum_{i=2}^{n} Y_i' = 141.079$$

$$\sum_{i=2}^{n} X_i'^2 = 105.333 \qquad \sum_{i=2}^{n} X_i' Y_i' = 352.671$$

$$\sum_{i=2}^{n} e_i'^2 = 19.630 \qquad \sum_{i=3}^{n} (e_i' - e_{i-1}')^2 = 33.765$$

The least squares estimate of α is therefore equal to

$$a = \frac{5.737}{1 - 0.716} = 20.201$$

Thus the new regression line for the original data can written as

$$\hat{Y} = 20.201 + 2.017X$$

The Durbin-Watson d-statistic for the transformed data is

$$d' = \frac{33.765}{19.630} = 1.72$$

We now have only 16 observations. The significance points for $k = 1$ and the 5 percent level of significance are $d_L = 1.10$ and $d_U = 1.37$. Since the observed value $d' = 1.72$ is greater than d_U, we accept the hypothesis that there is no positive serial correlation when using the transformed data.[3]

10.5.2 The Use of First Differences to Reduce Autocorrelation

When we discuss autocorrelation in the abstract, our message tends to be simple and mechanistic: "Help stamp out autocorrelation!" In seriously intended studies of real economic time series, however, proper specification of the model by which the data were generated will often eliminate significant autocorrelation in the residuals from simple or multiple regression equations.

Various specification errors may lead to autocorrelation in such residuals. For example, if the true relation between X_1 and X_2 is a second-degree parabola but we inadvertently fit a straight line, the residuals corresponding to extreme values of X_2 will tend to be opposite in sign from the residuals corresponding to more central values of X_2. If a second-degree parabola is fitted in this case, autocorrelation may be eliminated.

Alternatively, if the true model generates X_1 as a linear function of X_2 and X_3 but we erroneously express X_1 as a function of X_2 only, any trend and/or cyclical components in X_3 which are not correlated with X_2 will tend to produce autocorrelation in the residuals from

$$X_1 = a_{1\cdot 2} + b_{12} X_2 + \mu_{1\cdot 2} \tag{10.63}$$

In this case, we should find that the residuals $\mu_{1\cdot 2}$ are significantly correlated with the omitted variable X_3. When the correct equation

$$X_1 = a_{1\cdot 23} + b_{12\cdot 3} X_2 + b_{13\cdot 2} X_3 + \mu_{1\cdot 23} \tag{10.64}$$

is fitted, the residuals $\mu_{1\cdot 23}$ should (under our present assumptions) show no significant autocorrelation. But if the true model contained still another variable X_4 with some trend and/or cyclical components not correlated with X_2 and X_3, $\mu_{1\cdot 23}$ should show (1) significant autocorrelation and (2) signi-

[3] An alternative test statistic used to check for the presence of autocorrelation is known as the ratio of mean square successive difference to the variance and is frequently referred to as the von Neumann ratio. It is usually symbolized by δ^2/S^2 and for a series of n observations $x_1, x_2, \ldots x_n$ it is defined as

$$\frac{\delta^2}{S^2} = \frac{(1/(n-1)) \sum_{i=2}^{n} (x_i - x_{i-1})^2}{(1/n) \sum_{i=1}^{n} (x_i - \bar{x})^2}$$

For a detailed discussion of the use of δ^2/S^2, see Carl F. Christ, *Econometric Models and Methods*, (New York: John Wiley & Sons, Inc., 1966), Chapter X.

ficant correlation with X_4; the residuals $\mu_{1\cdot 234}$ from

(10.65) $X_1 = a_{1\cdot 234} + b_{12\cdot 34}X_2 + b_{13\cdot 24}X_3 + b_{14\cdot 23}X_4 + \mu_{1\cdot 234}$

should not show significant autocorrelation.

The property of significant autocorrelation in residuals can be used in forecasting future values of a variable *if and only if* we have strong reasons for believing that the factors responsible for autocorrelation during the sample period will continue to operate in the same fashion during the forecast period. These reasons may be unconvincing unless we can express them in terms of a model and specify the data that would be required to test its validity. This approach leads us forward to an expanded model which (if fitted to the appropriate data) should result in random residuals.

Some economic questions can be stated about as well in terms of year-to-year changes as in terms of the absolute values of the variables. For example, changes in the price of a perishable food product from one year to the next may be caused primarily by year-to-year changes in (1) the quantity of the product available for consumption and (2) the disposable income of consumers. In addition to these short-term factors, there may be longer-term factors such as changes in the age distribution of the consuming population or changes in the nation's transportation and communications systems which take many years to run their course. The latter factors may change smoothly and continuously and thus not obscure the year-to-year action of the short-term factors. If our interest focuses on the short-term factors, our specification problem is to formulate a model which is complete and accurate in accounting for year-to-year changes and which requires of the omitted variables only that they change slowly over periods of several years' duration.

An example of first difference transformations. Table 10.8 contains annual data on the retail price and per capita consumption of pork and on the per capita disposable income of consumers in the United States during 1922–1941. In terms of the original values of the variables, we may compute the following sequence of equations:

(10.66) $$X_1 = \underset{(11.6897)}{36.1580} - \underset{(0.1768)}{0.1646}X_2 + \mu_{1\cdot 2};$$

$$r_{12}{}^2 = 0.0459$$
$$\text{D-W} = 0.4178$$

(10.67) $$\mu_{1.2} = \underset{(3.7024)}{-24.9890} + \underset{(0.00654)}{0.04478}X_3 + \varepsilon_{(\mu_1 . 2)(3)}$$

$$r^2_{(\mu_1 . 2)(3)} = 0.7225$$

(10.68) $$X_1 = \underset{(5.6304)}{22.8730} - \underset{(0.0858)}{0.3826}X_2 + \underset{(0.00602)}{0.04951}X_3 + \mu_{1.23}$$

$$R^2_{1\cdot 23} = 0.8082$$
$$\text{D-W} = 0.2602$$

Table 10.8 Pork: Basic Data on Retail Price, Consumption, and Disposable Personal Income, United States, 1922–1941

	Original Values			First Differences			Logarithms			First Differences of Logarithms		
Year	X_1 Retail Price of Pork (Cents per lb)	X_2 Pork Consumption (lb per Capita)	X_3 Disposable Personal Income (Dollars) per Capita)	X_4 $= \Delta X_1$	X_5 $= \Delta X_2$	X_6 $= \Delta X_3$	X_7 $= \log X_1$ Price	X_8 $= \log X_2$ Consumption	X_9 $= \log X_3$ Income	X_{10} $= \Delta \log X_1$ Price	X_{11} $= \Delta \log X_2$ Consumption	X_{12} $= \Delta \log X_3$ Income
1922	26.8	65.7	541	—	—	—	1.428	1.818	2.733	—	—	—
1923	25.3	74.2	616	−1.5	8.5	75	1.403	1.870	2.790	−0.025	0.052	0.057
1924	25.3	74.0	610	0	−0.2	−6	1.403	1.869	2.785	0.000	−0.001	−0.005
1925	31.1	66.8	636	5.8	−7.2	26	1.493	1.825	2.803	0.090	−0.044	0.018
1926	33.3	64.1	651	2.2	−2.7	15	1.522	1.807	2.814	0.029	−0.018	0.011
1927	31.2	67.7	645	−2.1	3.6	−6	1.494	1.831	2.810	−0.028	0.024	−0.004
1928	29.5	70.9	653	−1.7	3.2	8	1.470	1.851	2.815	−0.024	0.020	0.005
1929	30.3	69.6	682	0.8	−1.3	29	1.481	1.843	2.834	0.011	−0.008	0.019
1930	29.1	67.0	604	−1.2	−2.6	−78	1.464	1.826	2.781	−0.017	−0.017	−0.053
1931	23.7	68.4	515	−5.4	1.4	−89	1.375	1.835	2.712	−0.089	0.009	−0.069
1932	15.6	70.7	390	−8.1	2.3	−125	1.193	1.849	2.591	−0.182	0.014	−0.121
1933	13.9	69.6	364	−1.7	−1.1	−26	1.143	1.843	2.561	−0.050	−0.006	−0.030
1934	18.8	63.1	411	4.9	−6.5	47	1.274	1.800	2.614	0.131	−0.043	0.053
1935	27.4	48.4	459	8.6	−14.7	48	1.438	1.685	2.662	0.164	−0.115	0.048
1936	26.9	55.1	517	−0.5	6.7	58	1.430	1.741	2.713	−0.008	0.056	0.051
1937	27.7	55.8	551	0.8	0.7	34	1.442	1.747	2.741	0.012	0.006	0.028
1938	24.5	58.2	506	−3.2	2.4	−45	1.389	1.765	2.704	−0.053	0.018	−0.037
1939	22.2	64.7	538	−2.3	6.5	32	1.346	1.811	2.731	−0.043	0.046	0.027
1940	19.3	73.5	576	−2.9	8.8	38	1.286	1.866	2.760	−0.060	0.055	0.029
1941	24.7	68.4	697	5.4	−5.1	121	1.393	1.835	2.843	0.107	−0.031	0.083

Equation (10.66) expresses price (X_1) as a function only of per capita consumption (X_2). The coefficient of determination is very small, and the Durbin-Watson statistic value of 0.4178 indicates high positive correlation in the residuals. The omitted variable X_3, per capita income, had a strong cyclical pattern and was correlated quite closely [$r^2_{(\mu_1 . 2)(3)} = 0.7225$] with the residuals from the price-consumption equation. Adding X_3 to equation (10.66) raises the coefficient of multiple determination to $R^2_{1 \cdot 23} = 0.8082$; each of the partial regression coefficients in equation (10.68) is highly significant. But the Durbin-Watson statistic, 0.2602, indicates that high positive autocorrelation persists.

Columns (4), (5), and (6) of Table 10.8 show the arithmetic first differences of variables X_1, X_2, and X_3, respectively; we will call these variables X_4, X_5, and X_6 or ΔX_1, ΔX_2, and ΔX_3 interchangeably. We compute the following sequence of equations:

(10.69)
$$\Delta X_1 = \underset{(0.61819)}{-0.03517} - \underset{(0.1076)}{0.5303}\Delta X_2 + \mu_{4 \cdot 5}$$
$$r_{45}{}^2 = 0.5881 \qquad D\text{-}W = 0.7895$$

(10.70)
$$\mu_{4 \cdot 5} = \underset{(0.2009)}{-0.3394} + \underset{(0.00341)}{0.04134}\Delta X_3 + \varepsilon_{(\mu_4 . 5)(6)}$$
$$r^2_{(\mu_4 . 5)(6)} = 0.8964$$

(10.71)
$$\Delta X_1 = \underset{(0.1988)}{-0.3835} - \underset{(0.0344)}{0.4899}\Delta X_2 + \underset{(0.00339)}{0.04172}\Delta X_3 + \mu_{4 \cdot 56}$$
$$R^2_{4.56} = 0.9607 \qquad D\text{-}W = 1.4229$$

Equation (10.69) shows a substantial correlation ($r_{45}{}^2 = 0.5881$) between year-to-year changes in price and consumption. The Durbin-Watson statistic, 0.7895, still shows high positive autocorrelation though less than in equations (10.66) and (10.68). The residuals $\mu_{4 \cdot 5}$ are very highly correlated with year-to year changes in consumers' income, ΔX_3.

Equation (10.71) expresses year-to-year changes in price as a function of year-to-year changes in consumption and income. The coefficient of multiple determination is $R^2_{4 \cdot 56} = 0.9607$; the partial regression coefficients are 12 to 14 times as large as their standard errors; and the Durbin-Watson statistic, 1.4229, is substantially larger than in any of the preceding equation. Entering Table IX, Appendix I, with $n = 19$ observations and $k' = 2$ independent variables, we find that $d_L = 1.08$ and $d_U = 1.53$. The test for positive autocorrelation at the 5 percent level of probability is inconclusive.

From other information about the United States economy, we know that the level of prices paid by consumers fell substantially from the 1920's to the early 1930's and rose moderately during the late 1930's. This movement in consumer prices is reflected in our retail pork price and consumer income variables, but the level of pork consumption per capita was about as high during 1930–1934 as during the 1920's. The theory of consumer demand implies that equal percentage changes in all prices and in the money income of a consumer should leave his purchases (in terms of quantities) unaffected. This requirement could be met approximately if we divided X_1 and X_3 by an index number of consumer prices, but other data adjustments and additional variables would be needed if we wished to apply consumption theory in a sophisticated way to the facts of 1922–1941 price and income changes.

An intuitive shortcut would be to assume that percentage or logarithmic relationships (price and income elasticities of demand) are more stable than arithmetic relationships in consumer demand when there are major changes in the general price level. Transforming X_1, X_2, and X_3 into logarithms gives us

$$\log X_1 = \underset{(0.4207)}{0.0540} - \underset{(0.2026)}{1.0115} \log X_2 + \underset{(0.1198)}{1.1592} \log X_3 + \mu_{7\cdot 89} \tag{10.72}$$

$$R^2_{7\cdot 89} = 0.8541$$
$$\text{D-W} = 0.2902$$

Our notation for R^2 and μ implies $X_7 = \log X_1$, $X_8 = \log X_2$, and $X_9 = \log X_3$. The logarithmic transformation of the original variables leaves us with high positive autocorrelation in the residuals just as did equation (10.68).

We next take *first differences of the logarithms* and fit the equation

$$\Delta \log X_1 = -\underset{(0.00417)}{0.00696} - \underset{(0.1024)}{1.2486}\, \Delta \log X_2 + \underset{(0.0858)}{1.0782}\, \Delta \log X_3 + \mu_{10\cdot 11\cdot 12} \tag{10.73}$$

$$R^2_{10\cdot 11,12} = 0.9562$$
$$\text{D-W} = 2.3264$$

Our notation for R^2 and μ implies $X_{10} = \Delta \log X_1$, $X_{11} = \Delta \log X_2$, and $X_{12} = \Delta \log X_3$. The t-ratios of the partial regression coefficients and the coefficient of multiple determination are about the same as in equation (10.71), but the Durbin-Watson statistic, 2.3264, shows no evidence of autocorrelation. For $n = 19$ and $k' = 2$, $d_L = 1.08$ and $d_U = 1.53$; these are the lower and upper bounds to be used when testing for positive autocorrelation at the 5 percent level.

10.6 HETEROSCEDASTICITY

If the variance-covariance matrix of errors is given by

$$E(UU') = \sigma^2 V = \sigma^2 \begin{bmatrix} v_{11} & 0 & \cdots & 0 \\ 0 & v_{22} & \cdots & 0 \\ \vdots & \vdots & & \vdots \\ 0 & 0 & \cdots & v_{nn} \end{bmatrix}$$

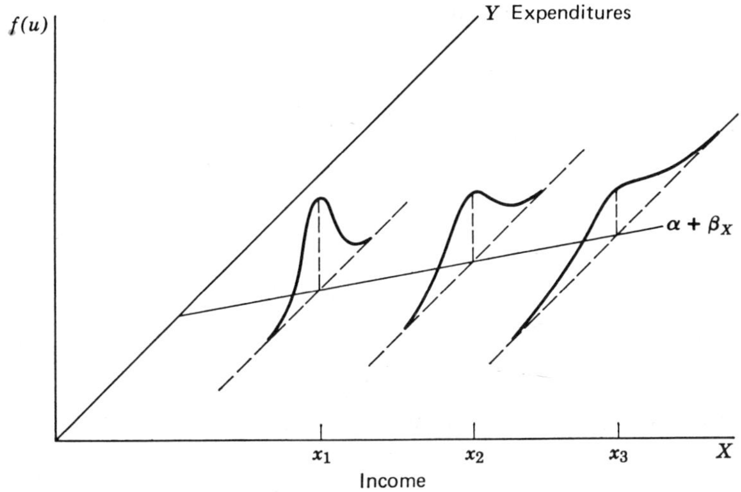

Figure 10.2 Heteroscedasticity.

where some of the positive diagonal elements v_{ii} are not equal to one, the error terms are said to be *heteroscedastic*. In other words, the error terms are independently distributed variables but there are differences in the variances of the distributions associated with different fixed values of the independent variables. One such situation is pictured in Figure 10.2 where Y is family expenditures and X is family income. An increase in the variance of expenditures as income increases is frequently expected in family budget studies.

For concreteness, we shall restrict the discussion of heteroscedastic errors to simple regression and assume that the variance of the error terms is proportional to X^2; that is,

$$E(u_i^2) = \sigma^2 X_i^2$$

The regression model in this case is

$$Y = X\beta + U \tag{10.74}$$

where

(i) $E(U) = 0$

(ii) $E(UU') = \sigma^2 \begin{bmatrix} X_1^2 & 0 & \cdots & 0 \\ 0 & X_2^2 & \cdots & 0 \\ \vdots & \vdots & & \vdots \\ 0 & 0 & & X_n^2 \end{bmatrix}$

(iii) X is a fixed matrix

In order to compute the generalized least-squares estimators, we must first find a matrix T such that

$$E(TUU'T') = \sigma^2 I_n$$

The matrix T in this case is equal to

$$T = \begin{bmatrix} \frac{1}{X_1} & 0 & \cdots & 0 \\ 0 & \frac{1}{X_2} & \cdots & 0 \\ \vdots & \vdots & & \vdots \\ 0 & 0 & & \frac{1}{X_n} \end{bmatrix} = T'$$

Premultiplying (10.74) by T, we have

$$TY = TX\beta + TU \tag{10.75}$$

The generalized least-squares estimators in this case are equal to

$$b^* = (X'T'TX)^{-1}X'T'TY$$

When dealing with the simple regression model $Y = \beta_1 + \beta_2 X$, the generalized least-squares b^* in this case becomes

$$\begin{bmatrix} b_i^* \\ b_2^* \end{bmatrix} = \begin{bmatrix} \dfrac{n\sum\left(\frac{1}{X_i^2}\right)Y_i - \sum\frac{1}{X_i}\sum\frac{Y_i}{X_i}}{n\sum\frac{1}{X_i^2} - \left(\sum\frac{1}{X_i}\right)^2} \\ \dfrac{\sum\frac{1}{X_i^2}\sum\frac{Y_i}{X_i} - \sum\frac{1}{X_i}\sum\frac{Y_i}{X_i^2}}{n\sum\frac{1}{X_i^2} - \left(\sum\frac{1}{X_i}\right)^2} \end{bmatrix} \tag{10.76}$$

Another way to obtain the least squares estimators in (10.76) is to note that the transformed model (10.75) is equivalent to

$$\frac{Y_i}{X_i} = \beta_1 \frac{1}{X_i} + \beta_2 + \varepsilon_i$$

where $\varepsilon_i = (u_i/X_i)$. Thus the estimator b_2^* is obtained by computing the intercept of the regression of Y_i/X_i on $1/X_i$.

A numerical example based on the hypothetical data in Table 10.9 will serve to illustrate the computations involved. The regression coefficients are equal to

$$b_1^* = \frac{6(32.15) - (3.3)(45.1)}{6(2.55) - (3.3)^2} = \frac{44.07}{4.41} = 9.99$$

Table 10.9 Hypothetical Data for the Heteroscedastic Case

i	X_i	$\frac{1}{X_i}$	$\frac{1}{X_i^2}$	Y_i	$\frac{Y_i}{X_i}$	$\frac{Y_i}{X_i^2}$
1	1	1.0	1.00	12	12.0	12.00
2	2	0.5	0.25	15	7.5	3.75
3	10	0.1	0.01	27	2.7	0.27
4	5	0.2	0.04	22	4.4	0.88
5	1	1.0	1.00	12	12.0	12.00
6	2	0.5	0.25	13	6.5	3.25
Totals	21	3.3	2.55	101	45.1	32.15

	$\overline{Y}_i$	$(Y_i - \overline{Y}_i) = e_i$	e_i^2	$\frac{1}{X_i^2}(e_i^2)$
1	12.01	−0.01	0.0001	0.0001
2	14.03	0.97	0.9409	0.2352
3	30.19	−3.19	10.1761	0.1018
4	20.09	1.91	3.6481	0.1459
5	12.01	−0.01	0.0001	0.0001
6	14.03	−1.03	1.0609	0.2652
Totals	—	—	—	0.7483

and

$$b_2^* = \frac{(2.55)(45.1) - (3.3)(32.15)}{6(2.55) - (3.3)^2} = \frac{8.91}{4.41} = 2.02$$

Thus, we have

$$\hat{Y} = 9.99 + 2.02X$$

The variance-covariance matrix of b^* in this case is equal to

$$\text{Var}(b^*) = \sigma^2(X'T'TX)^{-1} = \sigma^2(X'T^2X)^{-1} = \begin{bmatrix} \text{Var}(b_1^*) & \text{Cov}(b_1^*b_2^*) \\ \text{Cov}(b_1^*b_2^*) & \text{Var}(b_2^*) \end{bmatrix}$$

$$= \frac{\sigma^2}{n\sum\frac{1}{X_i^2} - \left(\sum\frac{1}{X_i}\right)^2} \begin{bmatrix} n & -\sum\frac{1}{X_i} \\ -\sum\frac{1}{X_i} & \sum\frac{1}{X_i^2} \end{bmatrix}$$

Using the data in Table 10.9, we find the variance of b_2^* is equal to

$$\text{Var}(b_2^*) = \frac{\sigma^2(2.55)}{6(2.55) - (3.3)^2} = 0.58\sigma^2$$

An estimator of $\mathrm{Var}(b_2{}^*)$ is obtained by replacing σ^2 with

$$s^2 = \frac{e'T'Te}{n-k} = \frac{(Y - Xb^*)'T^2(Y - Xb^*)}{n-k}$$

In the present case with $k = 2$, we have

$$(n-2)s^2 = \sum \frac{1}{X_i{}^2}(Y_i - b_1{}^* - b_2{}^*X_i)^2 = \sum \frac{1}{X_i{}^2}(e_i{}^2)$$

Using the data in Table 10.9, our estimate of the variance of $b_2{}^*$ is

$$s_{b_2*}^2 = 0.58s^2 = (0.58)(0.1871) = 0.1085$$

To test the hypothesis H_0: $\beta_2 = 0$ against the alternative H_a: $\beta_2 \neq 0$ at the 5 percent level of significance, we first compute

$$t_{b_2*} = \frac{b_2{}^*}{s_{b_2*}} = \frac{2.02}{0.329} = 6.14$$

Since the observed value of t_{b_2*} is greater than the critical value $t = 2.7764$ with four degrees of freedom, we reject the hypothesis that β_2 is equal to zero.

It is left to the reader as an exercise to derive the least squares estimators of β and $\mathrm{Var}(\beta)$ if $k = 2$ and $E(u_i{}^2) = \sigma^2 X_i$.

10.7 MULTICOLLINEARITY

High intercorrelation between independent variables is one of the most vexing and perplexing problems in the multiple regression analysis of nonexperimental observations. It is particularly acute in the case of economic time series. The problem of high intercorrelation between independent variables is frequently referred to as the problem of *multicollinearity*. In the extreme case, two or more of the independent variables may be so highly correlated that their separate effects upon the dependent variable cannot be distinguished. Consider, for example, the linear relationship

(10.77) $$Y_i = \beta_1 + \beta_2 X_{2i} + \beta_3 X_{3i} + \mu_i$$

Suppose that the independent variables are related by an exact linear relationship

$$X_{3i} = cX_{2i}$$

and that we have n observations on Y, X_2, and X_3 with which to estimate the coefficients β_1, β_2, and β_3. The method of least squares breaks down in the present example because the rank of $X'X$ is less than $k = 3$. The determinant $|X'X|$ is therefore equal to zero, and it is impossible to compute

$b = (X'X)^{-1}X'Y$ since $(X'X)^{-1}$ does not exist. To verify this, we note that the matrix $(X'X)$ is equal to

$$(X'X) = \begin{bmatrix} n & \sum X_{2i} & \sum X_{3i} \\ \sum X_{2i} & \sum X_{2i}^2 & \sum X_{2i}X_{3i} \\ \sum X_{3i} & \sum X_{2i}X_{3i} & \sum X_{3i}^2 \end{bmatrix}$$

Substituting cX_{2i} for X_{3i}, we have

$$(X'X) = \begin{bmatrix} n & \sum X_{2i} & c\sum X_{2i} \\ \sum X_{2i} & \sum X_{2i}^2 & c\sum X_{2i}^2 \\ c\sum X_{2i} & c\sum X_{2i}^2 & c^2\sum X_{2i}^2 \end{bmatrix}$$

Next, recall that the value of a determinant is unaltered if a constant multiple of a row or column is subtracted from any other row or column. If we multiply the second row of $(X'X)$ by c and subtract the result from row three, we have

$$A = \begin{bmatrix} n & \sum X_{2i} & c\sum X_{2i} \\ \sum X_{2i} & \sum X_{2i}^2 & c\sum X_{2i}^2 \\ 0 & 0 & 0 \end{bmatrix}$$

Since $|X'X| = |A| = 0$, the matrix $(X'X)$ is singular and $(X'X)^{-1}$ does not exist.

In practice, exact linear relationships seldom occur. The general interdependence of economic phenomena often leads to approximate linear relationships between economic variables, however. Consider, for example, the possibility of including both population and gross national product (GNP) as independent variables in a demand function for rice. In most countries, GNP and population both tend to increase over time and often appear to be linearly related. The usual effect of such multicollinearity is to increase the standard errors of the regression coefficients. Thus our estimates of the separate effects of population and gross national product on demand cannot be expected to be highly reliable.

Awareness of the possible effects of highly correlated independent variables would undoubtedly lead the investigator to search for a way to avoid the problem. One possibility in the present example would be to use per capita income as an independent variable instead of both population and GNP. In other cases it may be helpful to use a priori or "extraneous" information.

The way in which a priori or extraneous information is used depends, among other things, on the type of information available, the purpose of the analysis, and the imagination of the investigator. Although there is a rather long history of the use of a priori information in connection with regression analysis, there are still no fixed rules as to how such information is to be used.

Ideally, in carefully designed experiments, intercorrelations could be set at zero, the observations on each independent variable could be spread over a wide range to reduce the standard errors of the regression coefficients, and clear unidirectional causal dependence could be specified between the independent variables and the dependent variable. If economists could base their regression analysis on data from such experiments, it would be tempting to emphasize tests of significance and to consider the results of each experiment in isolation without drawing upon prior experience or outside information of any kind. A sound interpretation of data from even the best of experiments can seldom be based entirely on the results of tests of significance, however.

Since economists usually deal with nonexperimental observations, the appropriate stance might be, "Don't trust an observation unless it is confirmed by a good theory." And, don't trust a multiple regression analysis of economic time series unless the results accord with prior or outside information!

We are confronted here with an area of permissiveness which is more likely to be reduced by advances in economic knowledge than by advances in statistical technique. At one extreme, we might elect to stand or fall on the results of purely statistical estimates based on a set of time series observations—despite multicollinearity, latent variables, errors of measurement, and aggregation problems, not to mention changes in technology and consumer preferences. At the other extreme, we might ignore statistical estimation entirely and impose all coefficients in a model on the basis of (1) restrictions supplied by economic theory, accounting identities, and technical coefficients (five new tires sold with each new automobile, for example) and (2) judgments based on all kinds of information, collated intuitively rather than rigorously.

As economists, we would be most unwise to cast aside the wealth of theory and empirical observation that has been painstakingly built up by brilliant and dedicated scholars and base our scientific conclusions exclusively on statistical analyses of particular time series samples. So, we must use "prior information" to some extent to supplement our statistical estimation procedures.

But note that if we impose our economics on our statistics, we are inviting the judgments of economists on the cogency of our economic reasoning and the validity of any facts or relationships we adduce from empirical economic studies. "A priori information" is of varying qualities. If we use purely arbitrary coefficients to get around a statistical impasse, we deserve criticism from both economists and statisticians. Bad economics will not rescue us from ambiguous statistics; good economics may.

Like Monsieur Jourdain, who was surprised to learn that he had been speaking prose all his life, we may be surprised to realize the extent to which "prior information" is used in setting up almost any regression analysis.

Consider, for example, a demand function of the type

$$P_i = \beta_1 + \beta_2 Q_i + \beta_3 Y_i + u_i \tag{10.78}$$

where P is the price of pork, Q is the per capita consumption of pork, and Y is disposable personal income per capita. Implicitly, we have assigned a regression coefficient to all variables not included in equation (10.78). Such a decision may be based on a theory which suggests that Q and Y are the most important variables or on previous studies that may have indicated that variables such as the production of beef or the price of lamb have very little effect on the demand for pork. Also, we might find after fitting equation (10.78) to some time series observations that Y had shown little variation other than a linear upward trend. We could not tell from the time series whether the true model might not be

$$P_i = \beta_1 + \beta_2 Q_i + \beta_4 t_i + u_i \tag{10.79}$$

where t is "time" in years. But if we accept equation (10.79) as the true model, we are assigning an a priori value of zero to the coefficient of Y.

If decisions as to the inclusion or exclusion of variables are regarded as the a priori assignment of zero values to certain regression coefficients, it is clear that any regression analysis is an island surrounded by a sea of a priori information.

In the gasoline consumption example of Section 9.3.3, the constant term in the regression equation

$$Y = \alpha + \beta X + u$$

where Y is gasoline used and X is miles traveled, was not significantly different from zero. We would probably have been justified on a priori grounds in setting α equal to zero. If we stipulate that α is equal to zero, then the least squares estimator of β becomes

$$b = \frac{\sum Y_i X_i}{\sum X_i^2}$$

Assuming that α is equal to zero and using the data in Table 10.2, the new estimate of β is found to be

$$b = \frac{19{,}794}{320{,}281} = 0.0618 \text{ gallons of gas per mile}$$

which is equivalent to 16.2 miles per gallon. Our previous estimate of β was 0.059 gallons of gas per mile which is equivalent to nearly 17 miles per gallon. Our new estimate of β based on a priori information about α probably provides us with the best estimate of miles traveled per gallon of gas.

As a final example, suppose we are dealing with a multiple regression relationship

$$Y = \beta_1 + \beta_2 X_2 + \beta_3 X_3 + \beta_4 X_4 + u$$

and find that X_3 and X_4 are highly correlated. If, on the basis of extraneous information, we decide that β_4 can be taken equal to $0.6\beta_3$, then the relationship can be reduced to

$$\begin{aligned} Y &= \beta_1 + \beta_2 X_2 + \beta_3(X_3 + 0.6X_4) + u \\ &= \beta_1 + \beta_2 X_2 + \beta_3 X_5 + u \end{aligned}$$

where X_5 is equal to $(X_3 + 0.6X_4)$. The least squares estimates of β_1, β_2, and β_3 can then be obtained by the usual procedures using X_2 and X_5 as the independent variables.

There are a considerable variety of other a priori restrictions that can be imposed upon either single-equation or multiple-equation models by building the restrictions into the structures of the models so that estimates of other coefficients and minimization of sums of squared residuals are secured subject to the restrictions. Once an investigator recognizes these possibilities, he must also recognize that the quality of the statistically estimated coefficients is directly dependent upon the validity and accuracy of the restrictions. Such a priori restrictions are useful in dealing with problems of multicollinearity only if the restrictions can be justified. If not, then other means must be found for dealing with multicollinearity.

The seriousness of multicollinearity depends on the practical use that will be made of the results of the statistical estimates. If the intention is to use the estimated regression equation to predict future values of the dependent variable, then linear relationships between independent variables may not be considered a very serious problem provided that past relationships are expected to hold during the period of prediction. If, on the other hand, the purpose of the regression analysis is to measure the effect of each independent variable on Y, then multicollinearity may make it impossible to estimate individual regression coefficients with sufficient accuracy.

10.8 ECONOMIC MODELS INVOLVING MULTIPLE REGRESSION ANALYSIS

Several examples of economic problems involving simple regression analysis were given in Chapter 9. Multiple regression analysis is very frequently involved in the empirical derivation of demand functions, production functions, Engel curves, and various macroeconomic relationships. A few illustrations should suffice.

10.8.1 Consumer Demand Functions

The theory of consumer demand states that an individual's purchases of a particular commodity Q will be influenced by the price of the given commodity P_1, the prices of closely related (usually competitive) commodities P_2, and the individual's income Y. The least complex representation of these elements in an empirical demand function would be

$$Q = \beta_1 + \beta_2 P_1 + \beta_3 P_2 + \beta_4 Y + u \tag{10.80}$$

The most general statement of demand theory would make Q depend on the prices of *all* consumer goods and services. If P_2 is an index of prices of all consumer goods and services other than commodity 1, equation (10.80) takes at least formal cognizance of this principle.

The precision of demand analysis based on time series is usually limited by relatively small numbers of observations. In addition, research and development in the consumer goods and services industries tends to modify products and the competitive relationships between them, so that regression relationships calculated for one decade may suffer at least a gradual erosion of their inferential value during the succeeding one. For these reasons, it is seldom possible to demonstrate statistically significant relationships between the consumption of a given commodity and the prices of more than two or three distinct competing commodities or commodity groups.

Several demand functions for food livestock products, all meat and individual meats in the United States are presented in Table 10.10. All variables are expressed in terms of first differences of logarithms; the constant terms are omitted from the table. The net effects in the first column under retail price are estimates of the elasticity of demand for the commodity with respect to its own price.

It will be noted that the own-price elasticities are larger (in absolute value) for individual meats than for all meats as a group, and larger for all meat than for all food livestock products (including meats, poultry and eggs, and dairy products). This progression from greater to smaller elasticities is to be expected when groups of competing commodities are aggregated. An increase in the prices of all other consumer goods and services evidently made livestock products and meat more attractive to consumers at any given level of food livestock and meat prices. The equations for beef and lamb indicate that an increase in the supply of competing meats tended to reduce the consumption of the given meat. The most important mechanism was probably the following: (1) An increase in the supply of pork would reduce the price of pork; (2) this would make beef and lamb relatively less attractive to consumers at the prices of beef and lamb previously existing, so that purchases of beef and lamb would fall *if* their prices were held constant.

Table 10.10 Food Livestock Products: Factors Affecting Year-to-Year Changes in per Capita Consumption, United States, 1922–1941

Commodity or Group	Coefficient of Determination	Effects of 1 Percent Changes in: Retail Price		Price of all other Commodities		Disposable Income[a]		Supply of Competing Commodities[a]	
		Net Effect	Standard Error	Net Effect	Standard Error	Net Effect	Standard Error	Net Effect	Standard Error
		Percent[b]		Percent[b]		Percent[b]		Percent[b]	
All food livestock products	Multiple								
Deflated income	0.95	−0.52	(0.03)	0.70[c]	(0.10)	0.40[d]	(0.03)		
All meat									
Deflated income	0.96	−0.62	(0.04)	0.69[e]	(0.15)	0.51[d]	(0.05)		
Individual meats									
Pork	0.94	−0.81	(0.05)			0.72	(0.07)		
Beef	0.86	−0.79	(0.09)			0.73	(0.08)	−0.41[f]	(0.09)
Lamb	0.59	−0.91*	(0.26)			0.65	(0.23)	−0.83[f]	(0.20)

Source. Karl A. Fox, *Econometric Analysis for Public Policy* (Ames: Iowa State University Press, 1958), p. 116.

[a] Per capita basis.
[b] Coefficients based on first differences of logarithms.
[c] Special index, retail prices other than food livestock products.
* Probably understates the true effect of price on consumption.
[d] Disposable income deflated by retail price index.
[e] Special index, retail prices other than meat.
[f] Consumption per capita, other meats.

Actually, the three meats listed, along with broilers and turkeys, form an interrelated system which can be treated in more consistent and elegant ways than those shown in Table 10.10. However, the individual coefficients and equations in Table 10.10 have been quite useful for economic forecasting and for analyzing some of the impacts of proposed farm policies.

10.8.2 Engel Curves or Family Expenditure Functions

Houthakker (1957) fitted logarithmic multiple regression equations to data from family budget studies for a number of different countries. The equations were of the form

$$\log Y^k = \beta_1{}^k + \beta_2{}^k \log X_2 + \beta_3{}^k \log X_3 + u^k \tag{10.81}$$

where Y^k is expenditure on the kth group of items, X_2 is total expenditure, X_3 is family size, u^k is a disturbance term, and $\beta_1{}^k$, $\beta_2{}^k$, and $\beta_3{}^k$ are constants to be estimated. The parameters were estimated by means of least-squares regression; however, the observations, being averages of incomes and expenditures for groups of families within specified intervals of family income, were weighted according to the number of households represented in each group average. The multiple regression coefficients, $\beta_2{}^k$ and $\beta_3{}^k$, are the partial elasticities of the kth group of items with respect to total expenditure and family size, respectively.[4]

A specific example of Houthakker's results for large cities in the northern United States in 1950 is shown in Table 10.11 (omitting the constant terms). An interesting feature of these four equations is that, although they are fitted independently by least squares, the four b_2 coefficients, each weighted

Table 10.11

			$\beta_1{}^k$		$\beta_2{}^k$		$\beta_3{}^k$
Food	$\log Y^1$	=	$b_1{}^1$	+	$0.693 \log X_2$	+	$0.224 \log X_3$
					(0.017)		(0.016)
Clothing	$\log Y^2$	=	$b_1{}^2$	+	$1.399 \log X_2$	+	$0.016 \log X_3$
					(0.059)		(0.054)
Housing	$\log Y^3$	=	$b_1{}^3$	+	$0.764 \log X_2$	−	$0.155 \log X_3$
					(0.011)		(0.010)
Miscellaneous	$\log Y^4$	=	$b_1{}^4$	+	$1.367 \log X_2$	−	$0.111 \log X_3$
					(0.011)		(0.010)

[4] Hendrik S. Houthakker, "An International Comparison of Household Expenditure Patterns, Commemorating the Centenary of Engel's law," *Econometrica*, October 1957, p. 539.

by the proportion of total expenditure spent on the commodity concerned, must sum to unity, as the four expenditure categories add up precisely to X_2, total consumption expediture. Similarly, the weighted average of the b_3 coefficients is zero, the weights again being the proportions of total expenditure going for each of the four categories. It is clear from these constraints that, if some income elasticities are less than one, others must be greater than one; also that, if (at some constant level of total expenditure) the expenditure for food increases with family size, the expenditures for one or more other categories must decrease as family size increases.

10.8.3 Macroeconomic Models

Since the end of World War II, economists have shown much interest in "models" or sets of equations to express the main features of the operation of a national economy. "Macroeconomic" thinking received a great impulse from Keynes' *General Theory of Employment, Income and Money* (1936) and from the work of several outstanding economists in Sweden. A Dutch economist, Tinbergen, developed an empirical model of the United States economy from 1919–1932 data with the object of testing Keynes' and other theories of the business cycle. The variables in Tinbergen's model were, of course, economic time series; they were highly aggregative; and each of the individual equations was estimated by means of multiple regression.

During the 1950's and 1960's, a number of models have been designed to throw light upon particular mechanisms or sectors of the economy within the framework of total economic activity. Two equations developed by Duesenberry, Eckstein, and Fromm will illustrate this kind of model.[5]

The first equation expresses the net change in inventories (ΔI_t) for the national economy as a function of the current level of sales (S_t), the change in sales since the preceding quarter (ΔS_t), the change in new orders since the preceding quarter ($\Delta 0_t$), the rate of net inventory accumulation in the preceding quarter (ΔI_{t-1}), and the actual levels of inventories (I_{t-1}) and new orders (0_{t-1}) during the preceding quarter:

$$\begin{aligned}(10.82)\quad \Delta I_t = {} & \underset{(0.078)}{0.295S_t} - \underset{(0.267)}{0.947I_{t-1}} - \underset{(0.107)}{0.333\Delta S_t} + \underset{(0.139)}{0.771\Delta 0_{t-1}} + \underset{(0.054)}{0.1150_{t-1}} \\ & + \underset{(0.115)}{0.341\Delta I_{t-1}} + 8.508\end{aligned}$$

The standard error of estimate was \$1.74 billion and $R^2 = 0.81$.

[5] The equations are taken from pages 798 and 804 of J. S. Duesenberry, Otto Eckstein, and Gary Fromm "A Simulation of the United States Economy in Recession," *Econometrica*, October 1960, **28**(4), pp. 749–809.

A second equation from the same model expresses changes in personal consumption expenditures (seasonally adjusted in 1947–1949 dollars per capita) as a function of consumption and income variables for earlier periods:

$$\frac{c_t}{y_{t-1}} = 0.5353 - \underset{(0.0217)}{0.1366}\frac{y_{t-1}}{y_{0_{t-1}}} + \underset{(0.0457)}{0.5704}\frac{c_{t-1}}{y_{t-2}} \tag{10.83}$$

The standard error of estimate was 0.0168 (or 1.68 percentage points) and $R^2 = 0.83$. There was no evidence of autocorrelation in residuals from the equation.

A number of points in these equations are interesting:

1. Personal consumption expenditures in the current quarter (c_t) are expressed as a ratio to disposable personal income during the preceding quarter (y_{t-1}) implying a three-month average lag on the part of consumers in adjusting their expenditure behavior to changes in their incomes.
2. The variable y_0 stands for the highest level of disposable income per capita actually realized before the current period, the theory being that consumers will tend to maintain their expenditures (hence, their standards of living) at the highest level previously attained in the event that current income drops below the previous peak.
3. The inertia in consumer spending was further expressed by including as the second independent variable the ratio of consumption expenditures in the previous quarter (c_{t-1}) to disposable personal income in the quarter preceding that (y_{t-2}). All variables are on a per capita basis.

10.8.4 Prediction Functions

Economists doing applied research frequently use regression analysis for prediction purposes or to check their judgments about relationships between variables rather than to test formal economic theories. The following equation by Fox and Boles (1961) is based on observations from 99 Iowa counties in the year 1955:

$$Y = 22.98 + \underset{(0.1459)}{1.4986X_2} - \underset{(0.0200)}{0.3912X_3} \qquad \bar{S} = \$2.59 \quad R^2 = 0.814 \tag{10.84}$$

where Y equals expenditures on Old Age and Survivors Insurance (OASI) annuities per capita of the total county population, X_2 equals percent of population 65 years of age and older in 1955, and X_3 equals percent of rural farm population in 1950.

OASI payments are a matter of right for eligible individuals, and the benefit payments to an individual are calculated on the basis of definite formulas which include his average earnings in employment covered by the program. Prior to 1955, farm operators and other self-employed persons were not required to participate and many stayed out of it. In the earlier years of the program, employees of very small firms, such as are found more frequently in rural rather than urban areas, were not required to participate. Thus, through 1955, the OASI program primarily affected urban workers.

Hence, a larger than average percentage of rural farm population in a county implies that a smaller proportion of the people over age 65 were eligible for OASI and also that some of those who were eligible would have established lower annuity rights than average workers of their age group in more urban and industrialized counties. The larger the percentage of elderly workers in the population, the larger would be OASI payments per capita of the total population.

It could be argued that the standard errors of the regression coefficients in equation (10.84) are inappropriate because there are only 99 counties in Iowa; there was only one set of financial records per county in the year 1955; and hence, our sample coincides with the population.

An alternative viewpoint is as follows. The residual variance in Y is accounted for by a variety of factors not correlated with X_2 and X_3; these factors might have assumed any number of different combinations of values in 1955; hence, although the 99 values each of X_2 and X_3 *might* be regarded as exhausting the population of those values, still the residuals $(Y - \hat{Y})$ may be regarded as coming from one of an infinite number of possible random samples of Y *given* the specified combination of values of X_2 and X_3.

One more equation from the same study by Fox and Boles may be included:

$$(10.85)\quad Y = -11.3758 + \underset{(0.0124)}{0.1656X_2} + \underset{(0.0311)}{0.0972X_3} \qquad \bar{S} = 14.46 \text{ miles} \quad R^2 = 0.6706$$

where Y is the mileage of primary roads in a given country, X_2 is the area of the county in square miles, and X_3 is the density of the total population of the county in persons per square mile. Road mileages relate to December 31, 1957 and population density to the 1950 Census of Population. An average county in Iowa includes about 576 square miles and contained (in 1957) about 87 miles of primary roads. Thus, the coefficient of X_2 is consistent with an average increase of one-sixth of a mile of primary road mileage for each increase of one square mile in county area above the average for the 99 counties. As the main function of primary roads is to connect towns and cities of 1000 population or larger, total population density within a state with a relatively homogeneous agricultural base will vary almost exclusively on account of variations in urban population. A larger than average urban population in a given county is likely to mean a larger than average number of towns of 1000 population or more.

Although the values of county area and of total population density for each of 99 counties exhausted the population of such values in Iowa, we could still argue that many different values of Y might have emerged in any county even with a specified pair of values of area and total population density. One could, of course, regard the equation for Iowa as throwing some light on a larger population which includes similar variables for several or many other states.

10.8.5 Multiple Regression Analysis with Extremely Large Numbers of Observations

A great deal of statistical analysis of economic data has involved relatively short time series of annual observations of national aggregates or averages. It is refreshing, on occasion, to encounter microeconomic data involving very large numbers of observations, so that degrees of freedom are not a major limitation.

This section is based on a multiple regression analysis of factors affecting salaries of scientists included in the (United States) 1966 National Register of Scientific and Technical Personnel, performed in 1968 by Dr. Emanuel Melichar.[6] Melichar's study is by far the most definitive analysis available as of 1968 which permits meaningful comparisons of the salaries of scientists in different disciplines, holding all other major variables constant. The 1966 National Register includes 242,763 registrants. For purposes of his analysis, Melichar excluded persons not employed full time, those working for the military services or the Public Health Service, members of religious orders, and persons who did not report salary. With these deletions, there were 196,428 full-time salary observations available for analysis.

To these observations, Melichar fitted a multiple regression equation including 58 independent variables, all of the zero-or-one type. The original salaries were converted into logarithms prior to running the regression, as distributions of salaries (and of errors or residuals in salaries) often approach normality in terms of logarithms whereas the arithmetical values are highly skewed. For example, a frequency distribution of scientists' salaries having a median of \$15,000 would have a large concentration between \$10,000 and \$15,000 and very few observations below \$10,000; on the other hand, there would be some salaries in excess of \$40,000 or even \$50,000 in private industry.

Melichar's regression equation explained 54 percent of the variance of the logarithms of salaries around their mean logarithm ($R^2 = 0.54$). Table

[6] Emanuel Melichar, "Factors Affecting 1966 Basic Salries in the National Register Professions," *American Economic Review*, Vol. LVII, No. 5, Part 2 (Supplement), December 1968, pp. 59–69.

Table 10.12 Importance of Selected Characteristics in Explaining Variation in Professional Salaries: All Professions, 1966 National Register

Characteristic	Net Relationship: Partial r^2	Net Relationship: F-Ratio	Number of Variables Used
Highest academic degree	0.159	9260	4
Years of experience and type of employer	0.149	1185	29
Profession	0.066	1062	13
Primary work activity	0.065	3417	4
Sex	0.024	4923	1
Age	0.019	555	7

Source. Melichar, *op. cit.*, page 60.
Note. All relationships are significant at the 0.01 probability level.

10.12 indicates the nature of the explanatory variables used. The highest academic degree was most important; a combination of years of experience and type of employer was almost equally important; profession (to be defined later) made a substantial contribution, as did the primary work activity; sex and age made smaller contributions to explaining the total variance of scientists' salaries. (Age is, of course, highly correlated with years of experience; the partial r^2 of 0.019 represents the contribution of age over and above the estimated contribution of years of experience and the other variables.)

The F-ratios in Table 10.12 are astronomical, owing to the very large number of degrees of freedom available. To gain some insight, we may note that the formula for the standard error of a partial regression coefficient in the multiple regression equation

$$X_0 = b_{0\cdot 123\cdots m} + b_{01\cdot 23\cdots m}X_1 + b_{02\cdot 13\cdots m}X_2 + \cdots + b_{0m\cdot 12\cdots m-1}X_m + u \tag{10.86}$$

may be written as

$$s_{b_{01\cdot 23\cdots m}} = \frac{s_0}{s_1}\left[\frac{1 - R^2_{0\cdot 123\cdots m}}{(1 - R^2_{0\cdot 23\cdots m})(n-m-1)}\right]^{1/2} \tag{10.87}$$

or, in squared (variance) form,

$$s^2_{b_{01\cdot 23\cdots m}} = \frac{\bar{S}^2_{0\cdot 123\cdots m}}{s_1(1 - R^2_{0\cdot 23\cdots m})n} \tag{10.88}$$

where s_0 and s_1 are the standard deviations of X_0 and X_1, respectively;

$R^2_{0\cdot123\cdots m}$ is the coefficient of multiple determination of X_0 by the m independent variables; $\bar{S}^2_{0\cdot123\cdots m}$ is the variance (or squared standard error of estimate) of the residuals from equation (10.86); and $R^2_{0\cdot23\cdots m}$ is the coefficient of multiple determination of X_0 by the other $m-1$ independent variables exclusive of X_1.

In the present example, $n = 196{,}428$ and each F-ratio in Table 10.12 has in its numerator the variance attributed to a specified independent variable (or variables) and in its denominator the quantity

$$\frac{\bar{S}^2_{0\cdot123\cdots m}}{196{,}428}$$

Since this denominator is relatively small, the F-ratios are extremely large, ranging from 555 to 9260! Any F-ratio of about 2 or more (given the number of degrees of freedom involved here) would be significant at the $\alpha = 0.01$ level.

The basic observations in Melichar's study consisted of the answers to questionnaires filled out by the reporting scientists. (No information which would have permitted the identification of individual scientists was included in the data made available for the salary analysis.) Some of the variables were inherently of the either-or type, such as sex, highest academic degree, type of employer, profession, and primary work activity. For these, zero-one variables offer the simplest and most logical (or at least the most obvious) approach. For example, the first two observations might involve the following values of the sex and the highest academic degree variables:

Observation	Sex: Male = 1 Female = 0	Highest Academic Degree		
		Ph. D.	M.A. or M.S.	B.A. or B.S.
1	1	1	0	0
2	0	0	1	0

The zero-one variables partition the observations into distinct categories.

Other variables (specifically, age and years of professional experience) were inherently continuous. Here, various alternatives might have been chosen. For example, data from many other sources have indicated that earnings increase with age up to a certain point and then level off and may even decline a bit (on a partial regression basis) during the last few years of full-time employment. So, it would have been possible to assume that the partial regression of salary on age was (say) a second-degree parabola

$$(10.89)\quad E(Y \mid \text{specified values of all other independent variables}) = f(A, A^2)$$

where Y is salary and A is age in years. In this form, the age variable would have used up only two degrees of freedom and would have permitted us to estimate the expected salary associated with being (say) 29.33 years old rather than 29.75 years old.

Actually, Melichar divided the age variable into seven age groups (plus nonreported); this used up seven degrees of freedom rather than two, but this was of no consequence given nearly 200,000 observations. Melichar was obligated, given the purpose of his study, to design his analysis in such a way that the results could be presented in a few small tables and summarized in a brief, expository text accessible to thousands of readers rather than to a limited number of specialists. Thus, two observations would be handled as follows with respect to age:

	Age						
Observation	*Under 30*	*30–34*	*35–39*	*40–44*	*45–54*	*55–64*	*65 and over*
1	0	1	0	0	0	0	0
2	0	0	1	0	0	0	0

The two scientists involved could be as close together in age as 34 and 35 or as far apart as 30 and 39. With perhaps 20,000 scientists in each of the age groups 30–34 and 35–39, the net difference between average salaries of the two groups (attributed to age) should be estimated within a very small range of error. This net difference might reasonably be regarded as applying to scientists of ages 32 and 37, respectively; it would be smaller than the actual net difference between average salaries of scientists aged 30 and 39 and larger than the actual net difference for scientists aged 34 and 35. With the relatively precise determination of each salary-on-age-group partial regression coefficient, it would not be unreasonable to fit a curve (such as a second-degree parabola) to the 7 data points provided by the 7 age-group coefficients and use this curve for estimating the expected salary associated with any age we wish to specify.

Partial regression coefficients with respect to zero-one variables. At first glance, it may seem surprising that the usual regression formulas apply to zero-one variables, but they do. For example, Table 10.13 indicates that 93.4 percent of the scientists included in the analysis were male and 6.6 percent were female. The "percentage difference" coefficients indicate that (other factors held constant) the expected salary of male scientists would be 1.1 percent higher, and the expected salary of female scientists 15.0 percent lower, than the expected salary of all scientists.

Similarly, Table 10.13 indicates that (other variables held constant) the expected salary of scientists with Ph.D. degrees is 15.5 percent above, and

Table 10.13 Net Relationships Between Professional Salaries and Specified Characteristics: All Professions, 1966 National Register

Characteristic	Percentage Difference from National Geometric Mean, 1966	Percentage of Respondents in Class
Highest academic degree		
Professional medical	38.9	2.5
Ph.D.	15.5	41.5
M.S.	−8.5	25.1
B.S.	−13.6	28.8
Other or not reported	−18.3	2.1
Primary work activity		
Management	14.9	23.0
Research and development	−1.5	35.4
Production and inspection	−5.5	7.9
Teaching	−9.4	19.9
Other or not reported	−2.3	13.8
Age		
Under 30	−12.8	13.8
30–34	−4.2	17.6
35–39	1.9	19.2
40–44	4.7	17.0
45–54	5.9	22.0
55–64	2.7	8.8
65 and over	−2.8	1.4
Not reported	−1.5	0.2
Sex		
Male	1.1	93.4
Female	−15.0	6.6
Field of Science		
Mathematics	13.3	9.6
Economics	11.6	5.7
Statistics	8.5	1.3
Physics	7.9	11.0
Anthropology	1.0	0.4
Meteorology	−0.4	1.7
Sociology	−0.6	1.5
Psychology	−1.4	8.2
Earth sciences	−2.7	7.9
Chemistry	−4.7	27.2
Biological sciences	−4.9	12.4
Linguistics	−6.7	0.5
Agricultural sciences	−15.5	4.8
Other	3.7	7.7

Source. Melichar, *op. cit.*, page 66.

that of scientists with Master's degrees 8.5 percent below, the expected salary of all scientists. Other coefficients indicate the average net salary advantages or disadvantages attributable to primary work activity and to the field of science in which a scientist's highest competence lies. On inspection, the partial regression coefficients for age groups in Table 10.13 appear to show a smooth pattern which could be approximated by a continuous curve.

Table 10.14 shows the partial regression coefficients (translated, like those in Table 10.13, from logarithmic to percentage form) of salary upon each of a number of "type of employer and years of experience" categories. Melichar had some reasons for believing that the partial regressions of salary upon years of professional experience might differ among the three main types of employers. After the fact, however, it appears that the percentage or logarithmic rates of salary increase with years of experience are quite similar for the different types of employers. Hence, years of experience could have been

Table 10.14 Net Relationship Between Professional Salaries and Employer-Experience Characteristics: All Professions, 1966 National Register

Type of Employer	Years of Experience						
	1 or less	2 to 4	5 to 9	10 to 14	15 to 19	20 to 29	30 and over
	Percentage Difference from National Geometric Mean						
Educational institutions							
Academic year base	−36.6	−31.9	−25.7	−17.6	−11.5	−4.3	0.0
Calendar year base	−28.8	−23.4	−16.5	−7.3	−0.5	7.6	12.4
Government							
Federal	−26.0	−15.8	−5.7	0.9	7.2	16.0	24.7
Other	−36.5	−27.8	−19.1	−13.5	−8.0	−0.5	7.0
Other employer							
Nonprofit organization	−20.9	−16.7	−9.0	−1.2	6.5	15.6	24.8
Industry or business	−10.8	−6.0	2.7	11.5	20.2	30.5	40.8
Self-employed	0.7	6.1	15.9	25.9	35.7	47.4	59.0

Source. Melichar, *op. cit.*, page 62.

handled with 6 degrees of freedom and type of employer (plus academic year versus calendar year salary bases in educational institutions) could have been handled with another 6, compared with the 29 required for the pattern in Table 10.14. But the apparent "waste" of 17 degrees of freedom is of no consequence with 196,428 observations, and Table 10.14 answers questions explicitly for each type of employer, whereas the other approach, using only 12 degrees of freedom instead of 29, might have left lingering doubts as to whether the shape of the salary-and-experience relationship really was the same in industry, government, and educational institutions.

Each row in Table 10.14 suggests a relatively smooth nonlinear partial regression relationship of salary to years of professional experience. The different widths of the experience categories and the open-ended class "30 years and over" leave some ambiguities as to how a curve of best fit should actually be derived but, for rough work, a free-hand curve might be sufficient for interpolating to estimate (say) the expected salaries associated with five years and nine years of experience, even though both are represented by the same partial regression coefficient in Table 10.14.

EXERCISES

Section 10.1

1. Derive the least-squares estimators for β_1 and β_2 when the regression equation is $Y = \beta_1 X_1 + \beta_2 X_2 + u$.
2. Write down the normal equations for estimating the coefficients of the regression equation:

$$Y = \alpha + \beta_1 X_1 + \beta_2 X_2 + \beta_3 X_3 + \beta_4 X_4 + u$$

3. Using the following data on weight (Y), height (X_1), and age (X_2), find the least-squares regression equation of Y on X_1 and X_2.

Weight Y	62	52	68	72	78	58	58	74
Height X_1	51	44	52	57	62	48	53	61
Age X_2	7	6	8	8	12	7	9	11

4. Compute the beta coefficients for the regression function computed in Exercise 3.

Section 10.2

5. Compute r_{Y1}, r_{Y2}, r_{12}, and $r_{Y1 \cdot 2}$ using the data presented in Exercise 3.
6. What does the coefficient of multiple determination, R^2, measure and why is it usually presented along with the estimated regression line?

7. Correlation measures only linear association and does not imply the existence or absence of causation. Proof of causation must depend on a priori considerations. Discuss.
8. If two independent variables in a regression function are closely correlated, they are redundant. Which variable should be eliminated should depend primarily on theoretical rather than statistical considerations. Discuss.

Section 10.3

The following data on the production and price of sugar is to be used for Exercises 9–11.

Year	Sugar Production (1000 MT) Q_t	Price ($/$MT$) P_t	Trend Value T
1950	290	102	−7
1951	271	127	−6
1952	313	106	−5
1953	424	82	−4
1954	458	73	−3
1955	497	74	−2
1956	439	75	−1
1957	516	97	0
1958	438	78	1
1959	504	72	2
1960	553	86	3
1961	615	104	4
1962	568	96	5
1963	638	102	6
1964	487	131	7

9. Estimate the following regression equations using the data on sugar production and prices.
   (a) $Q_t = \alpha + \beta P_t + u_t$.
   (b) $Q_t = \alpha + \beta P_t + \gamma T + u_t$.
   (c) $Q_t = \alpha + \beta P_{t-1} + u_t$.
10. Do the equations estimated in Exercise 9 appear to be demand or supply functions? Explain.
11. Which of the equations in Exercise 9 would you use for predicting sugar production? Explain.
12. Using the data presented in Exercise 3, test the hypothesis that the coefficient β_1 associated with height (X_1) is significantly different from zero.

13. Using the results obtained in part (b) of Exerci se 9, test the hypothesis that $(\beta + \gamma) = 0$ against the alternative $(\beta + \gamma) \neq 0$. Use the 5 percent level of significance.

Section 10.4

14. Suppose that for the regression equations, $Y_i = \beta_1 + \beta_2 X_i + u_i$, it is assumed that $E(u_i) = 0$, $E(u_i^2) = \sigma^2 X_i$, and $E(u_i u_j) = 0$ for $i \neq j$.
    (a) Derive the generalized least-squares estimators b_1^* and b_2^* of β_1 and β_2.
    (b) Derive a computing formula for $\text{Var}(b_2^*)$.

Section 10.5

15. Suppose we have a model of the form:

$$Y_t = \alpha + \beta X_t + u_t \qquad (\text{for } t = 1, 2, \ldots, n)$$

where there is autocorrelation caused by a linear trend in the Y_t and the X_t. For example, for the variable X, $X_t = \gamma + \delta t$ and similarly for the variable Y. Suggest a possible transformation of the data to eliminate the autocorrelation; specifiy the new transformed regression equation, and derive the least-squares estimators of the transformed regression.

16. Presence of serial correlation suggests that one or more pertinent variables may have been left out of the regression equation. Absence of serial correlation, on the other hand, indicates that no important variables have been omitted. Discuss.
17. Using the results obtained in part (a) of Exercise 9, test for the presence of autocorrelated error terms.

Section 10.6

18. Use the data on sugar production and prices presented for Exercise 9 to compute the generalized least-squares estimates for the regression model

$$Q_t = \alpha + \beta P_t + u_t$$

where:
(a) $E(u_t) = 0$.
(b) $E(u_t^2) = \sigma^2 X_t^2$ for all t.
(c) $E(u_t u_j) = 0$ for $t \neq j$.
(d) The P_t are assumed to be "fixed" on successive observations.

Section 10.7

19. Using the following data:
    (a) Estimate the regression equation

$$Y = \alpha + \beta_1 X_1 + \beta_2 X_2 + u$$

(b) Compute r_{12}.
(c) Explain how you would proceed to predict the value of Y associated with $X_1 = 75$ and $X_2 = 90$.

Y	190	340	350	400	300	450	370	180	280	300
X_1	20	45	30	57	40	62	50	22	60	34
X_2	40	65	70	80	50	80	70	35	40	60

Section 10.8

20. Suppose we have a multiplicative demand function of the form

$$Q_x = aP_x^b P_y^c$$

where Q_x is the quantity of good X demanded, P_x is the price of X, P_y is the price of a closely related good Y, and a, b, and c are the parameters to be estimated.

(a) How would you proceed to estimate this demand function?
(b) What is the coefficient of price elasticity for X equal to?
(c) What is the coefficient of cross elasticity equal to?

21. The demand function for rice in Peru was found to be

$$\hat{Q} = 1371.59 + \underset{(0.264)}{2.6999Y} - \underset{(0.019)}{0.167N} - \underset{(0.019)}{0.086P} \qquad R^2 = 0.87$$

where
$\hat{Q}$ is rice consumed measured in hundreds of metric tons
Y is an index of real gross national product with 1960 = 1000
N is the population measured in thousands of persons
P is the retail price of rice measured in soles per metric ton

(a) Does the retail price of rice have a significant effect on the quantity of rice demanded?
(b) The coefficient for population is negative; the opposite of that expected. What might account for this result?
(c) What factors would you consider and what additional information would be useful in determining if the demand function provides an accurate estimate of the effect of price on the quantity of rice consumed?

22. Are the signs of the regression coefficients in equation (10.82) in Section 10.8.3 what you would expect on the basis of economic theory? Explain your answer.

23. Could equation (10.82) in Section 10.8.3 be used for predicting the net change in inventories (ΔI_t) in period $t + 1$. What information or assumptions would be needed? Would the estimated relationship be more useful for testing economic theories or predicting changes in inventories? Explain.

24. Using the multiple regression functions fitted by Houthakker presented in Section 10.8.2, compute the coefficient of total expenditure elasticity for food.

SELECTED REFERENCES

Christ, Carl F., *Econometric Models and Methods*, John Wiley & Sons, Inc., New York, 1966.

Fox, Karl A., *Intermediate Economic Statistics*, John Wiley & Sons, Inc., New York, 1968.

Goldberger, Arthur S., *Topics in Regression Analysis*, The Macmillan Company, New York, 1968.

Johnston, J., *Econometric Methods*, McGraw-Hill Book Company, Inc., New York 1963.

Kane, Edward J., *Economic Statistics and Econometrics*, Harper & Row, Publishers, New York, 1968.

Yamane, Taro, *Statistics—An Introductory Analysis*, Harper & Row, Publishers, New York, 1964.

CHAPTER 11

Univariate Analysis of Time Series

Businessmen and economists frequently deal with statistical data which are collected and recorded at successive intervals of time and which are usually referred to as time series. There are a multiplicity of topics and techniques that might reasonably be included in an introductory treatment of the analysis of economic time series. We shall not be able to discuss elaborate methods of time series analysis such as those used by the United States Bureau of the Census. Our discussion will serve only to illustrate some of the basic techniques along with a few of the problems involved.

Almost every science gives considerable attention to phenomena that show quantitative variations over time. On the average, economists have probably been somewhat less successful in solving their time series problems than have scientists dealing with physical phenomena. It may be that other scientists have been more specific than economists as to why they expect particular variables to move in particular ways; the reasons for expecting cycles of certain lengths; and the nature of the other variables, if any, that determine or influence the time path of the variable of primary concern. Most of the time series that economists deal with are affected more by human nature than by mother nature. The complexities and variability of human nature will undoubtedly continue to challenge the ability of economists to predict future values of time series.

11.1 INTRODUCTION

Some of the early advances in automation and assembly-line production methods were stimulated by time and motion studies. A worker packing apples in wooden boxes, if asked what he was doing, would most likely say "packing apples." A time and motion expert would answer the question differently in terms of basic units of motion, perception, and decision. Through visual perception and some kind of thought process, the worker selects an apple that seems to him an appropriate candidate; moves his hand to the vicinity of the next vacant spot in the box; places it in the desired position; spreads his fingers and releases the apple. One of the most imaginative of the early scientific managers, Frank Gilbreth, coined the word "therbligs" (an approximation to Gilbreth spelled backwards) to describe these basic units of human motion and action.

Once an activity such as packing apples had been reduced to its basic "therbligs," two sorts of response were possible. The first would be to devise an alternative sequence of therbligs which would enable an average worker to pack more boxes per hour with the same effort. The other would be to consider whether a machine could be devised to execute an appropriate sequence of therbligs which would accomplish the desired result at a still lower cost. Even if the machine did a less tasteful job so that boxes packed by machine brought a slightly lower price than those packed by hand, construction of the machine would still be warranted from a firm's standpoint if the reduction in costs were expected to be greater than the reduction in market price. We shall begin at virtually the "therblig" level of concreteness in order to illustrate a few of the complexities involved in the analysis of aggregative economic time series.

Suppose we were to take an inventory of the electrical equipment in a private home at a particular moment. We could measure the current required to operate each light bulb, the electric stove, the air conditioner, the washing machine, the refrigerator, and the other electrical appliances in the house. Given accurate measuring and recording devices, we should be able to add up the flows of current through each electrical appliance, hour by hour, and arrive at an aggregative time series for total use of electricity by the household.

Next, let us suppose that the particular household is one of a thousand households that collectively constitute the household demand for electricity from a municipal power plant. The aggregative time series for the households could be added to obtain an aggregate time series of the total amount of electricity entering all household circuits served by the power plant. Since there are power losses between the plant's generating equipment and the household meter boxes, the quantity of current demanded at the power plant will be somewhat larger than the simple aggregate of demand at the household meter boxes. These losses are assumed to be small relative to total household demand and are ignored, therefore, in the following discussion.

Let us suppose that we have accurate information on the technological characteristics of each microelement in the system just described and a complete inventory of the electrical appliances in each household. A model of household consumption of electric power in this system might run somewhat as follows. If each appliance has only one setting while turned on, we may write the total power demand on the system at time t as

$$D_t = \sum_{i=1}^{m} d_{it} = \sum_{i=1}^{m} n_i k_i p_{it} \tag{11.1}$$

where n_i is the total number of appliances of the ith type, k_i is the number of kilowatts required by each unit of the ith type, and p_{it} is the proportion of the total number of units of the ith type turned on at time t.

At any given date, the n_i and k_i are constants, presumed known. If the plant manager could predict the annual time path of each of the p_i's, he could then estimate the total power requirement at the plant at different times of the day for each day of the year. This would allow him to do a better job of planning the work force and number of generators needed for tomorrow or next month.

Accurate prediction of the daily, weekly, and monthly fluctuations in the proportions p_i would be very difficult, however, because of the tremendous amount of data required. The proportion of air conditioners turned on would depend on things such as the outside temperature, the humidity, and the number of family members likely to be in the house rather than elsewhere at particular times of day. The proportion of all electric lights turned on would be a function of the time of day and a seasonal factor reflecting times of sunrise and sunset. Different within-day patterns would be found for different days of the week. The proportion of electric stoves turned on might be a function of the hour of the day and a seasonal factor representing differences between (say) summer and winter cooking habits.

An alternative approach would be to ignore the individual components that make up the aggregate time series of the total demand at the power plant and attempt to predict the future values of D_t on the basis of its past values. This approach will not provide any information as to how different factors affect D_t, but such information may not be necessary for short-run operating decisions. Ignoring the factors that affect D_t may reduce the accuracy of our predictions, but the cost of obtaining more accurate predictions may greatly exceed their value.

We shall discuss methods for analyzing individual time series in more detail later in this chapter. For the moment, we want only to emphasize that an investigator should always be aware of the "therbligs" that make up the economic time series with which he is working and the factors affecting the "therbligs." In the final analysis, he may decide to work only with the aggregate time series data but knowledge of the "therbligs" will often help him to decide how to analyze the aggregate data.

If the plant manager is responsible for long-run investment and expansion plans, he must also try to project the *trend* of demand for electrical power. Logically, this trend may be broken down into (1) a trend in the number of households and (2) a trend in the average consumption of electric power per household. The latter trend, in turn, may be broken down into the effects of changes in each of the variables n_i, k_i, and p_i. The trend in the total number of units of a particular kind of appliance installed (that is, in n_i) might follow a logistic curve and approach an upper limit or saturation point relative to any given total number of households. An increase in the efficiency of the ith type of appliance would reduce the number of kilowatts, k_i, required by each unit of that type of appliance. Even if the efficiency of the ith type of appliance increases, k_i could still trend upward if standards of appliance capacity or intensity of illumination increased. The trend in the p_i would be related to trends in consumer behavior.

Trends in the local household demand for electricity will also depend upon trends in household incomes and perhaps upon the stability of employment in the community. We might postulate that the total number of units of each kind of appliance owned in the community per 1000 households will depend upon (1) the average income per household, (2) the proportions of households in each of five or ten income classes, (3) the stability of employment, and (4) the expectations of members of the community concerning its economic prospects over the next several years. For example, households with a high degree of confidence in their economic prospects in the locality would be more likely to invest in central air conditioning than households that expected to leave the community within two or three years.

Still other explanatory factors must lie behind trend projections of the total number of households in the community. Will new plants, government offices, or other sources of employment locate in or near the community? Or is the community likely to lose jobs (and probably population) because certain components of its economic base are declining? Is the local factory modern and apparently competitive—or does it appear to be on its last legs?

So far, we have considered only *what* consumers do and not *why* they do it. Household members turn on lights because they want to do things that require or are facilitated by light and turn on stoves because they want hot beverages and cooked food. Other household wants are clean clothes, a cool house, well-refrigerated (or simply well-preserved) food, and perhaps certain types of television and radio programs. If some of these wants can be satisfied in alternative ways, through commercial laundries, laundromats, gas stoves, and irradiated food—the prices of (a) electrical appliances and (b) electric power relative to the prices of commercial services and the operating costs of gas stoves will be relevant to the demand for electric power.

Hence, the plant management may wish to influence the behavior of its household customers by means of its rate schedules and other policy instruments. Price schedules might be altered in order to reduce the peak-load and achieve a somewhat less variable pattern of current use within a day. In principle, it would be possible to install household meters which measured not only the total amount of electricity consumed but the amount consumed in particular time periods. If the peak use of air conditioners occurs in the early afternoon and present peak-load on the generators occurs in the late afternoon, favorable rates for households having central air conditioning units could be used to encourage their purchase. On the other hand, if air conditioners became sufficiently numerous to shift the peak-load on the system from late afternoon to early afternoon, the plant management might decide to withdraw the favorable rates for households with central air conditioners or (theoretically) even charge higher than average rates to those households that install central air conditioners after a specified date.

To the extent that the number of appliances owned (n_i), the average power or rate of current consumption required to operate a unit of each appliance (k_i), and the proportion of the total number of appliances in use at a given moment (p_i) can be influenced by the average price or price schedule charged for electricity, a private power company would be justified in estimating the probable effects on household demand for power of a wide range of possible variations in its prices. If there were only two kinds of appliances and if adjustments of n_i, k_i, and p_i to changes in r_i (the rate charged for current used in the ith type of appliance) were instantaneous, the economic framework in which the firm operated could be characterized as follows:

$$\hat{n}_1 = \alpha_1 + \beta_1 r_1 + \gamma_1 N + \delta_1 Y \tag{11.2}$$

$$\hat{p}_{1t} = \phi_{1t} + \theta_{1t} r_1 \qquad \text{for all } t \tag{11.3}$$

$$\hat{n}_2 = \alpha_2 + \beta_2 r_2 + \gamma_2 N + \delta_2 Y \tag{11.4}$$

$$\hat{p}_{2t} = \phi_{2t} + \theta_{2t} r_2 \qquad \text{for all } t \tag{11.5}$$

$$Q = \sum_t D_t = \sum_t (d_{1t} + d_{2t}) = \sum_t (\hat{n}_1 k_1 \hat{p}_{1t} + \hat{n}_2 k_2 \hat{p}_{2t}) \tag{11.6}$$

$$TR = \sum_t (r_1 d_{1t} + r_2 d_{2t}) \tag{11.7}$$

$$TC = TFC + f(Q) \tag{11.8}$$

$$TNR = TR - TC \tag{11.9}$$

where N is the total number of households in the community, Y is the average income per household, TR is the firm's total revenue, TC its total cost, TFC its total fixed cost, $f(Q)$ its total variable cost, TNR the firm's total net revenue (profit), and α_i, β_i, γ_i, δ_i, ϕ_{it}, and θ_{it} are constants. The summations are over all of the time units in the accounting period being used.

It is assumed that the plant manager will try to select those rates r_1 and r_2 that will maximize the firm's profit for the accounting period. The profit-maximizing values of r_1 and r_2 depend on the values of N and Y in this model. We assume the power company has no control over N and Y. Hence, its interest in them is limited to a desire for accurate *forecasts* of their future values. If the power company takes N and Y as given, and *if* it can make satisfactory estimates of the coefficients of the four equations (11.2) through (11.5), it can then determine the values of r_1 and r_2 that will maximize profits.

The mathematical model defined by equations (11.2) to (11.9) is unquestionably an extremely simplified description of the relationships between a few of the variables affecting the firm's profits. Most household appliances can be expected to last several years. Thus the number of units of the ith appliance, n_i, may not adjust instantaneously to changes in r_i or Y. Hence, equations (11.2) and (11.4) should probably include lagged values of r_i and Y, in which case they become ownership adjustment equations rather than short-run demand equations. Equations (11.3) and (11.5), relating the proportions of existing appliances actually turned on at a given moment to r_1 and r_2, would probably contain most of the short-run or instantaneous response of power consumption to rate changes. A utility commission may limit the firm's ability to change its rates. If so, the limits on r_1 and r_2 should be specified in the model.

It is seldom a simple task to develop models that will adequately characterize the relationships between economic variables. There are, however, some cases in which changes in an economic variable over time cannot logically be analyzed without specifying a mathematical model which relates the changes in a time series to changes in other economic variables. In such cases, the traditional approach to time series analysis, which attempts to *describe* the changes in a time series rather than explain them, is not likely to be very useful. Even so, the traditional or univariate approach to time series analysis still is used frequently in some areas of economic analysis particularly in adjusting data collected by state and federal agencies. We shall therefore present in this chapter a few of the standard techniques used in the univariate analysis of time series. The structural approach to time series analysis will be considered in more detail in Appendix A of this chapter.

11.2 THE UNIVARIATE APPROACH

The univariate approach to time series analysis is to attempt to partition a time series into four distinct components called the trend (T), cyclical fluctuation (C), seasonal variation (S), and irregular movements (I). The actual value (Y) of the time series is then assumed to behave according to a specified model. The most frequently used models are the additive model and the multiplicative model. In its abstract form, the basic equation of the additive model is:

$$Y = T + C + S + I \tag{11.10}$$

The standard multiplicative model, on the other hand, would be stated as

$$Y = TCSI \tag{11.11}$$

There are numerous variations of these basic models. The models $Y = TCS + I$ and $Y = TC + SI$ are two such variations.

Relatively little empirical work has been carried out to test the validity of the assumed models. The work that has been done suggests that the multiplicative model is probably reasonable for partitioning many of the commonly used aggregate time series in the United States.

Strictly speaking, the trend, seasonal, cyclical, and irregular components of a time series are defined by the mathematical calculations used to determine them. The mathematical processes chosen may depend, to some extent, on theoretical ideas about the nature of the components, but the *exact* definitions of the components are still the result of the calculation. The general definitions that follow provide starting points for developing specific mathematical definitions.

In broad terms, the *trend* component may be thought of as that part of the time series which displays a smooth or regular movement over a fairly long period of time. In a sense, the trend characterizes the gradual changes that occur in a time series. In some cases, it may be possible to describe such changes by a straight line or other mathematical function. At other times, it may be desirable to define the trend by a mathematical process which requires very weak a priori assumptions about the form of the trend.

The *seasonal* component of a time series can be thought of as that component which displays a regularly recurring pattern during subperiods of any specified period of time. For example:

(a) The sales of a supermarket may display a fairly regular *daily* pattern from *week* to *week*.

(b) Farm employment may display a fairly regular *monthly* pattern from *year* to *year*.

(c) The number of people traveling by bus in Chicago may have a regular *hourly* pattern on week *days*.

The regularly-recurring pattern in each case would be referred to as a *seasonal pattern*, even though there may be little, if any, relationship between the pattern and the seasons of the year.

Strictly speaking, the words "regularly recurring" might be thought to rule out the possibility of changes in the seasonal pattern. A precise mathematical definition of the seasonal component could easily specify the way in which the seasonal pattern changes over time. Thus the words "regularly recurring" should be interpreted in such a way as to allow for some changes in the form of the seasonal pattern.

There are some types of regular variation which it may be both possible and desirable to measure and exclude before attempting to determine seasonal variations. This is frequently true for what is usually referred to as calendar variation. Such variations should be defined in a manner consistent with the basic definition of seasonal variation being used. The type of information needed and the techniques used to measure these types of variation will depend on the cause of the variation. The technique used to measure regular variations in car sales due to periodic model changes may be very different from the technique used to measure regular variations due to holidays falling on different days of the week in different years.

During the 1920's and 1930's, economists spent a great deal of time attempting to throw some light on that formidable and complex phenomenon known as the "business cycle." The period covered by a business cycle was usually considered to be somewhat longer than the period covered by seasonal variations but not as long as that covered by the trend. The cyclical movement in a time series is not necessarily a business cycle, but the concept of the business cycle is sometimes a useful one in visualizing the meaning of a cyclical movement.

The *cyclical* component of a time series can be thought of as that part of the time series that displays a fairly long-term, but not necessarily regular, movement around the trend. It is usually difficult to justify assumptions about the length and form of the cyclical movement of a time series. As a result, it is usually necessary to treat the cyclical component of a time series as a residual that is estimated after the trend and seasonal components have been identified and irregular movements averaged out. An alternative approach is to treat the cyclical and trend components as a single component and not attempt to estimate them separately. This approach is especially common when the investigator is interested primarily in adjusting data for seasonal variations.

The *irregular* component of a time series can be thought of as that part of the time series which varies sporadically from period to period. Irregular movements may be due to chance events such as earthquakes, strikes, and floods. In a sense, the irregular component is a residual that the investigator is not able to "explain" in terms of the trend, seasonal, and cyclical variations. The irregular movements are usually smoothed out and "hidden" by averaging techniques.

It is usually assumed that the components of a time series can be regarded as independent of one another and therefore can be estimated successively rather than simultaneously. The assumption of independence is not always justified in the real world. For example, an "irregular" event such as a strike may lead to a wage settlement that will alter the trend of employment in an industry. Even if not justified, the assumption of independence is not likely to be dropped until workable procedures for simultaneous estimation of components are available. Meanwhile successive estimation techniques that allow the trend to change over time and use iterative procedures to filter the effect of extreme values in a time series will probably provide reasonable results even when the components of the time series are interdependent.

11.3 THE TREND

The projection of trends is usually an important part of any study involving long-term planning. The businessman considering the construction of a new plant wants to know what the demand for his product will be five, ten, or 15 years from now. A government official involved in an urban renewal program would like to know how the demand for low-cost housing is likely to change in different parts of the country during the next ten years. The engineer planning a new highway has to estimate the increase in traffic on the highway during the next 20 years or so. The required long-run projections are frequently based in part on projections of past trends.

The statistical procedure used to estimate the trend will depend partly on the distribution of the data and partly on the reason for estimating the trend. If the data appear to follow a linear trend and the investigator is primarily interested in long-run projections, then simple least squares may be a satisfactory technique for estimating the trend. If the form of the trend in the data appears to change over time or the researcher wants to estimate the trend in order to determine the seasonal pattern of the time series, then the method of moving averages may provide the best estimate of the trend.

11.3.1 Semi-Averages

The method of *semi-averages* is a very quick way to estimate a linear trend line. The data are first divided into two parts. Means of the values in each part are then computed and centered at the midpoints of the time intervals covered by the respective parts. The straight line joining the two means (or semi-averages) is the estimated trend line.

The data on the United States gold stock shown in Table 11.1 will be used to illustrate the method of semi-averages. Since the period 1957 to 1967

Table 11.1 United States Gold Stock

Year	X	Amount of Gold, Y (Billions of Dollars)	Estimated Trend Value Y^*
1957	−2	22.8	21.36
1958	−1	20.6	20.44
1959	0	19.5	$19.52 = \bar{Y}_1$
1960	1	17.8	18.60
1961	2	16.9	17.68
1962	3	16.0	16.76
1963	4	15.6	15.84
1964	5	15.4	14.92
1965	6	13.8	$14.00 = Y_6^*$
1966	7	13.2	13.08
1967	8	12.0	12.16

Source. Federal Reserve Bulletin, March 1968.

covers an odd number of years, the middle year, 1962, is omitted. The means of the two parts can then be based on the same number of years and are easily centered. The average gold stock was

$$\bar{Y}_1 = \frac{22.8 + 20.6 + 19.5 + 17.8 + 16.9}{5} = 19.52$$

during the five year period from 1957 to 1961 and

$$\bar{Y}_2 = \frac{15.6 + 15.4 + 13.8 + 13.2 + 12.0}{5} = 14.00$$

during the 1963 to 1967 period. The semi-average $\bar{Y}_1$ is centered on 1959 and $\bar{Y}_2$ is centered on 1965. The trend line is obtained directly from these semi-averages.

The equation for the linear trend line is $Y^* = a + bX$ where Y^* is the estimated *trend value* of the gold stock measured in billions of dollars at time X. The constants a and b depend on the basic data and the year for which X is equal to zero. If we let $X = 0$ at 1959, we have

$$a = Y_0^* = \bar{Y}_1 = 19.52$$

The slope coefficient, or *annual trend increment*, b, is the average annual change in the gold stock. Solving $Y^* = a + bX$ for b, we obtain

$$b = \frac{Y^* - a}{X}$$

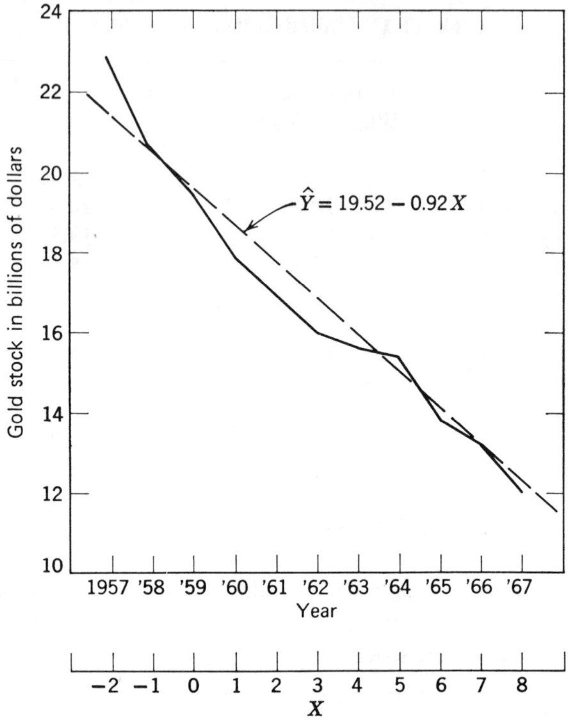

Figure 11.1 United States gold stock.

Since $a = Y_0{}^* = 19.52$ and $Y_6{}^* = \overline{Y}_2 = 14.00$, we have

$$b = \frac{14.00 - 19.52}{6} = \frac{-5.52}{6} = -0.92$$

and therefore

$$Y^* = 19.52 - 0.92X \tag{11.12}$$

This trend line is presented in Figure 11.1 along with the original data. The estimated trend values are obtained from (11.12) by substituting the values of X corresponding to the years for which trend values are needed. For 1962, we have $X = 3$ and $Y^* = 19.52 - 0.92(3) = 16.76$ billion dollars. The trend values for other years are presented in the last column of Table 11.1.

11.3.2 Least Squares Linear Trend Lines

The method of least squares is the most frequently used method for fitting trends. The least squares procedure for estimating a linear trend line is identical to the procedure discussed in Chapter 9. The normal equations for the trend line $Y^* = a + bX$ are

$$\sum Y = an + b\sum X$$
$$\sum XY = a\sum X + b\sum X^2$$

If the time scale is selected so that $\sum X = 0$, the normal equations reduce to

$$\sum Y = an \qquad \text{and} \qquad \sum XY = b\sum X^2$$

which are readily solved to obtain

$$a = \frac{\sum Y}{n} = \bar{Y} \qquad \text{and} \qquad b = \frac{\sum XY}{\sum X^2}$$

The data on policy loans by United States life insurance companies presented in Table 11.2 will be studied to illustrate the least squares method of

Table 11.2 Policy Loans by United States Life Insurance Companies (Figures in Billions of Dollars)

Year	X	Amount of Policy Loans Y	Trend Value of Policy Loans Y^*
1958	−4	4.2	4.035
1959	−3	4.6	4.590
1960	−2	5.2	5.145
1961	−1	5.7	5.700
1962	0	6.2	6.255
1963	1	6.6	6.810
1964	2	7.1	7.365
1965	3	7.6	7.920
1966	4	9.1	8.475

Source. 1967 Life Insurance Fact Book.

fitting trend lines. We find

$$a = \frac{\sum Y}{n} = \frac{56.3}{9} = 6.255$$

and

$$b = \frac{\sum XY}{\sum X^2} = \frac{33.3}{60} = 0.555$$

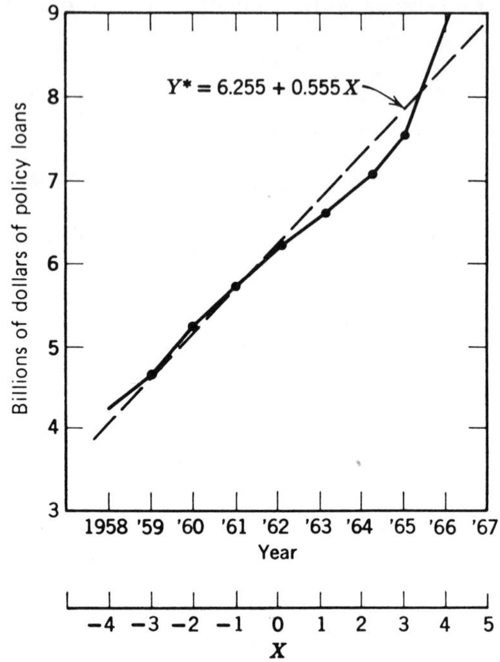

Figure 11.2 Policy loans of United States insurance companies.

The trend line is therefore equal to

$$Y^* = 6.255 + 0.555X \tag{11.13}$$

(X in units of 1 year with origin at 1962)

where Y is the total policy loans per year in billions of dollars.

The trend values in the last column of Table 11.2 were obtained by substituting the X's corresponding to the different years into the trend equation (11.13). The trend value of policy loans in 1959, for example, is

$$Y^* = 6.255 + 0.555(-3) = 4.590 \text{ billion dollars}$$

The actual values of Y and the trend line are shown in Figure 11.2.

When a time series has an even number of years, the midpoint of the time scale will fall between the two middle years. In this case, it is convenient to use a time scale with X measured in units of six months. The slope coefficient, b, in this case, is the semi-annual trend increment. The years are numbered $\ldots, -5, -3, -1, 1, 3, 5, \ldots$ and the sum of the X's is again equal to zero. Let us use the first six years of data on policy loans in Table 11.2 to illustrate this situation. The basic data and calculations are:

Year	X	Y	XY	X^2	Y^*
1958	−5	4.2	−21.0	25	4.182
1959	−3	4.6	−13.8	9	4.676
1960	−1	5.2	−5.2	1	5.170
1961	1	5.7	5.7	1	5.664
1962	3	6.2	18.6	9	6.158
1963	5	6.6	33.0	25	6.652
Totals		32.5	17.3	70	

To obtain the estimated trend line, we compute

$$a = \frac{32.5}{6} = 5.417 \qquad \text{and} \qquad b = \frac{17.3}{70} = 0.247$$

The trend line is then written as

$$Y^* = 5.417 + 0.247X$$

where X is in units of 6 months with origin at 1960–1961 and Y is annual loans measured in billions of dollars. (Verify that the trend value for 1958 in equal to 4.182.)

It is sometimes convenient to modify a trend equation by changing the origin of the time scale or by changing the units of measurement of X or Y. These modifications can be made without reference to the original data from which the estimated trend line was obtained. The following examples will serve to illustrate the procedures used. In all of the examples we shall use the hypothetical trend line

$$Y^* = 12.48 + 0.24X \tag{11.14}$$

where X is measured in annual units with origin at 1959 and Y is annual profits measured in millions of dollars.

Example 1: Suppose the origin of X is to be shifted from 1959 to 1965. Changing the origin of X does not change the annual trend increment b, but it does change the value of the intercept coefficient a. The new value of the intercept coefficient is easily obtained by noting that a is always equal to the trend value in the year for which X is equal to zero. For the original trend line, X is equal to 6 in 1965. When the origin of X is shifted to 1965, we have

$$a = 12.48 + 0.24(6) = 13.92$$

and

$$Y^* = 13.92 + 0.24X$$

where X is measured in annual units with origin at 1965 and Y is still annual profits in millions of dollars.

Example 2: Suppose that we want to change the units of Y^* in (11.14) from annual to monthly profits. This is easily accomplished by dividing both sides of (11.14) by 12, the number of months in a year. Making the divisions, we obtain

$$Y' = \frac{Y^*}{12} = \frac{12.48}{12} + \frac{0.24}{12}X = 1.04 + 0.02X$$

where X is measured in annual units with origin at 1959 and Y' is the estimated trend value of monthly profits measured in millions of dollars.

The trend value of monthly profits was 1.04 million dollars in 1959. The trend of monthly profits increased at the rate of 0.02 million dollars *per year*. Thus in 1963 (when $X = 4$), the trend value of monthly profits was $Y' = 1.04 + 0.02(4) = 1.12$ million dollars. To check this result, we note that

$$Y^* = 12.48 + 0.24(4) = 13.44 = 12(1.12)$$

That is, the estimated trend value of annual profits in 1963 is equal to 12 times the estimated trend value of monthly profits.

Example 3: Changing the units of measurement of X results in a change in the slope coefficient, b, but does not require a change in the intercept coefficient, a. Suppose the trend line is to be modified so that X is measured in units of six months. The slope coefficient is redefined as the semi-annual trend increment which is equal to one-half of the annual trend increment. The new trend line is therefore equal to $Y^* = 12.48 + 0.12X$ where X is measured in units of six months with origin at 1959 and Y is annual profits in millions of dollars. To compute the annual trend value of profits in 1956, we use $X = -6$ to obtain

$$Y^* = 12.48 + 0.12(-6) = 11.76 \text{ million dollars}$$

The trend value of monthly profits in 1956 would be $(11.76/12) = 0.98$ million dollars.

11.3.3 Parabolic Trends

If a trend cannot be described adequately with a straight line, the investigator may decide to fit the data with a parabola. The second degree polynomial equation

$$Y^* = a + bX + cX^2 \tag{11.15}$$

is frequently used in this case.

The coefficients a, b, and c may be found by the method of least squares. The problem is to find the numerical values of a, b, and c that will minimize the sum of squares

$$\phi = \sum (Y - Y^*)^2 = \sum (Y - a - bX - cX^2)^2 \tag{11.16}$$

Using calculus, we differentiate (11.16) partially with respect to a, b, and c to obtain a set of three normal equations:

$$\sum Y = an + b\sum X + c\sum X^2$$
$$\sum XY = a\sum X + b\sum X^2 + c\sum X^3$$
$$\sum X^2Y = a\sum X^2 + b\sum X^3 + c\sum X^4$$

If the time scale of X is selected to that $\sum X = 0$, then $\sum X^3$ is also equal to zero. The normal equations are thereby reduced to

$$\sum Y = an + c\sum X^2$$
$$\sum XY = b\sum X^2$$
$$\sum X^2Y = a\sum X^2 + c\sum X^4$$

Solving for a, b, and c, we obtain

$$a = \frac{\sum Y \sum X^4 - \sum X^2Y \sum X^2}{n\sum X^4 - (\sum X^2)^2}$$

$$b = \frac{\sum XY}{\sum X^2}$$

$$c = \frac{n\sum X^2Y - \sum Y \sum X^2}{n\sum X^4 - (\sum X^2)^2}$$

The data in Table 11.3 on public utility bonds owned by United States life insurance companies will be used to illustrate the procedure for computing a second degree parabolic trend line. The basic computations are presented in Table 11.4.

Using the formulas for a, b, and c; we obtain

$$a = \frac{161{,}180.6}{9{,}438} = 17.0777, \qquad b = \frac{24.5}{110} = 0.2227$$

and

$$c = \frac{-551.1}{9{,}438} = -0.05839$$

The trend line is therefore equal to

$$Y^* = 17.0777 + 0.2227X - 0.05839X^2 \tag{11.17}$$

Table 11.3 Public Utility Bonds Owned by United States Life Insurance Companies (in Billions of Dollars)

Year	X	Amounts of Bonds Held Y	Parabolic Trend Value Y^*	Linear Trend Value Y'
1956	−5	14.5	14.504	15.358
1957	−4	15.2	15.252	15.851
1958	−3	15.9	15.884	15.804
1959	−2	16.4	16.398	16.026
1960	−1	16.7	16.796	16.249
1961	0	17.0	17.077	16.472
1962	1	17.3	17.242	16.694
1963	2	17.3	17.289	16.917
1964	3	17.2	17.220	17.140
1965	4	17.0	17.034	17.362
1966	5	16.7	16.731	17.585

Source. 1967 Life Insurance Fact Book, p. 75.

Table 11.4 Computations Based on Public Utility Bonds Owned by United States Life Insurance Companies

X	Y	XY	X^2	X^2Y	X^4
−5	14.5	−72.5	25	362.5	625
−4	15.2	−60.8	16	243.2	256
−3	15.9	−47.7	9	143.1	81
−2	16.4	−32.8	4	65.6	16
−1	16.7	−16.7	1	16.7	1
0	17.0	0.0	0	0.0	0
1	17.3	17.3	1	17.3	1
2	17.3	34.6	4	69.2	16
3	17.2	51.6	9	154.8	81
4	17.0	68.0	16	272.0	256
5	16.7	83.5	25	417.5	625
Totals	181.2	24.5	110	1761.9	1958

$n\sum X^4$	$=21{,}538$	$\sum Y \sum X^4$	$=354{,}789.6$	$n\sum X^2Y$	$=19{,}380.9$
$(\sum X^2)^2$	$=12{,}100$	$\sum X^2Y \sum X^2$	$=193{,}809.0$	$\sum Y \sum X^2$	$=19{,}932.0$
difference	9,438	difference	161,180.6	difference	−551.1

where X is measured in units of one year with origin at 1961 and Y is equal to billions of dollars of utility bonds held in year X.

The trend line (11.17) fits the data extremely well. For example, for 1959 the actual value of Y is 16.4 while the estimated trend value is equal to

$$Y^* = 17.0777 + 0.2227(-2) - 0.05839(-2)^2 = 16.3987$$

The linear trend line fit by the method of least squares is equal to

$$Y' = 16.472 + 0.2227X$$

where X is measured in units of one year with origin at 1961. This equation does not adequately describe the trend of the data. Using the linear trend equation, the estimated trend value of Y for 1959 is equal to 16.0266, over 373 million dollars less than the actual value of bonds held in 1959. Both the linear and parabolic trend lines are shown in Figure 11.3. There is no doubt as to which trend line fits the data best in this particular case.

11.3.4 Exponential Trends

If a time series tends to increase geometrically rather than linearly, the trend of the series can often be described by the exponential equation

$$Y^* = ab^X \tag{11.18}$$

where a and b are constants and the time variable X appears as an exponent. If b is greater than 1, the trend value Y^* will grow by increasingly larger

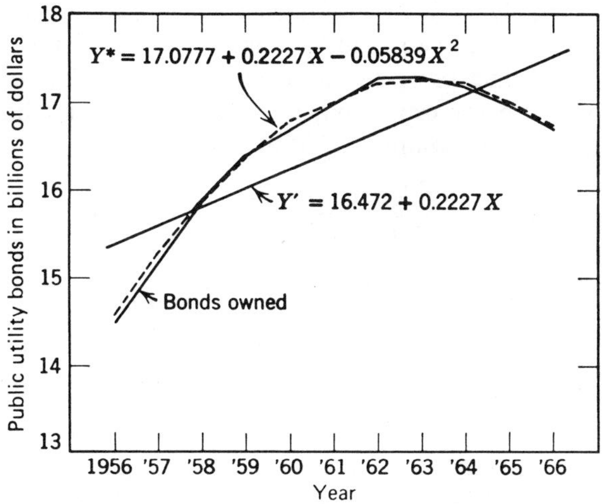

Figure 11.3 Public utility bonds owned by United States life insurance companies.

amounts each time period. If, on the other hand, b is between zero and 1, the trend value Y^* will approach zero.

Written in its logarithmic form, the exponential trend line becomes

$$\log Y^* = \log a + X \log b = A + BX \tag{11.19}$$

which is seen to be linear in $\log Y^*$ and X. Thus a time series that has an exponential trend will appear linear when plotted on semi-logarithmic paper. Conversely, plotting the original data on semi-logarithmic paper can be helpful when trying to decide if an exponential trend line should be used.

The exponential equation (11.18) is used frequently in economic theory and business finance. The compound interest formula, for example, is written as

$$A_n = P(1 + r)^n \tag{11.20}$$

where A_n is the amount due at the end of n periods on an original principal of P dollars invested at an interest rate r and compounded each period. To obtain (11.18) from (11.20), let $A_n = Y^*$, $P = a$, $(1 + r) = b$, and $n = X$.

The method of least squares can be used to estimate the constants a and b. The normal equations for (11.19) are

$$\sum \log Y = n \log a + (\log b) \sum X$$
$$\sum X \log Y = (\log a) \sum X + (\log b) \sum X^2$$

By selecting the origin so that $\sum X = 0$, the normal equations are reduced to

$$\sum \log Y = n \log a \qquad \text{and} \qquad \sum X \log Y = (\log b) \sum X^2$$

which are solved easily for $\log a$ and $\log b$. Once $\log a$ and $\log b$ are known, the constants a and b can be found by using antilogarithms.

The hypothetical data presented in Table 11.5 will serve to illustrate the calculation procedure for estimating an exponential trend line. Using the totals of the appropriate columns of Table 11.5, we obtain

$$\log a = \frac{\sum \log Y}{n} = \frac{14.3512}{9} = 1.59458$$

$$\log b = \frac{\sum X \log Y}{\sum X^2} = \frac{8.5710}{60} = 0.14285$$

By substituting $\log a$ and $\log b$ into (11.19), we find

$$\log Y^* = 1.59458 + 0.14285X \tag{11.21}$$

where X is in units of one year with origin at 1964 and Y is annual sales in thousands of dollars. This equation is shown along with the original data in the semi-log graph, Figure 11.4.

Table 11.5 Annual Sales of Big-B Company

Year	Time Variable X	Annual Sales[a] Y	Log Y	X^2	$X \log Y$	Trend Value Y^*
1960	−4	10	1.0000	16	−4.0000	10.5
1961	−3	17	1.2304	9	−3.6912	14.6
1962	−2	19	1.2787	4	−2.5574	20.3
1963	−1	27	1.4314	1	−1.4314	28.2
1964	0	39	1.5911	0	0.0	39.3
1965	1	55	1.7404	1	1.7404	54.6
1966	2	80	1.9031	4	3.8062	75.9
1967	3	100	2.0000	9	6.0000	105.5
1968	4	150	2.1761	16	8.7044	146.7
Totals	0	497	14.3512	60	8.5710	—

[a] Figures in thousands of dollars.

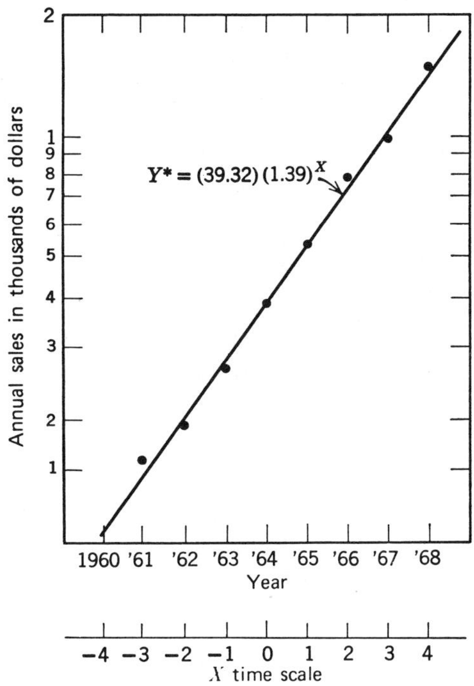

Figure 11.4 Annual sales of Big-B Company.

Using antilogarithms, the values of a and b are found to be 39.32 and 1.39, respectively. Thus, the trend line can also be written as

$$Y^* = (39.32)(1.39)^X \tag{11.22}$$

where the units of X and Y are the same as before.

To calculate the trend value for, say, 1967, we substitute $X = 3$ into (11.21) to obtain

$$\log Y^* = 1.59458 + 0.14285(3) = 2.02313$$

and then use antilogarithms to find $Y^* = 105.5$. The trend values can also be computed directly from the trend equation (11.22). For example. the trend value in 1961 is

$$Y^* = (39.32)(1.39)^{-3} = 14.6$$

Using (11.22), the average annual rate of growth of sales during the period from 1960 to 1968 is found to be $g = b - 1 = 0.39$, or 39 percent. It seems doubtful that sales would continue to increase at this rate for an extended period of time. Thus, the estimated trend line is likely to produce exaggerated forecasts of *future* sales even though it fits the data on *past* sales extremely well.

Time series that increase exponentially often appear to increase arithmetically if the rate of growth is small and data are available for only a few time periods. Consider the population figures for a hypothetical country in Table 11.6.

Suppose the true trend line for the population series is

$$Y^* = 1.16(1.03)^X \tag{11.23}$$

where Y^* is the trend value of population and X is the time variable measure

Table 11.6

Year	Time Variable X	Population (in Millions) Y	Linear Trend Value Y^*
1960	−4	1.03	1.022
1961	−3	1.06	1.057
1962	−2	1.09	1.091
1963	−1	1.12	1.125
1964	0	1.16	1.160
1965	1	1.19	1.194
1966	2	1.23	1.228
1967	3	1.27	1.262
1968	4	1.30	1.297

in units of 1 year. According to this trend equation, population is growing at the rate of 3 percent per year. The data for the 1960 to 1968 period appear to follow a linear rather than an exponential trend, however.

Fitting a linear trend line to the data by the method of least squares, we obtain

$$Y^* = 1.16 + 0.0343X$$

where X is in annual units with origin at 1964 and Y is population in millions of persons. The linear trend line fits the observed data very closely and will serve nicely for making population projections for the next five or six years. Suppose, however, that a population projection for the year 2000 is needed. The projection based on the linear trend line would be

$$Y^* = 1.16 + 0.0343(36) = 2.39 \text{ million persons}$$

If population continues to grow according to the exponential equation (11.23), there will, in fact, be over 3.33 million people in the country by year 2000.

For both examples presented in this section, the trend lines selected fit the observed data extremely well. Yet, it appeared doubtful that the estimated relationships were the appropriate ones to use for forecasting future values of the time series from which they were obtained. Obviously, the "goodness of fit" criteria is not adequate in itself for deciding on whether or not to use an estimated trend equation for making forecasts. There may be technical or theoretical reasons for not using the closest fitting trend line as a forecasting tool. The investigator may be aware of changes in other variables which, in his judgment, may affect the future values of the time series he is trying to forecast. Such information should not be ignored even if it cannot be incorporated into a mathematical function.

11.3.5 The Method of Moving Averages

Mathematical functions such as those discussed previously will not always provide an adequate description of the trend of a time series. An alternative approach is to use the method of moving averages to isolate the trend. The computations required to obtain a moving average are very easy, and yet the method is extremely flexible in the sense that the trend is not forced to conform to any particular mathematical function.

Given a set of numbers $Y_1, Y_2, \ldots, Y_n$; a k-period arithmetic moving average is defined by a sequence of arithmetic means:

$$\frac{Y_1 + Y_2 + \cdots + Y_k}{k}, \frac{Y_2 + Y_3 + \cdots + Y_{k+1}}{k}, \frac{Y_3 + Y_4 + \cdots + Y_{k+2}}{k},$$
$$\ldots, \frac{Y_{n+1-k} + Y_{n+2-k} + \cdots + Y_n}{k}$$

In a five-period moving average of a set of 16 numbers, for example, the mean corresponding to the original observation Y_6 is the average of Y_4, Y_5, Y_6, Y_7, and Y_8. The mean corresponding to Y_7 is the average of Y_5, Y_6, Y_7, Y_8, and Y_9. The moving average tends to "average out" the variation in the original time series. The longer the period of the moving average, the greater the "averaging out" effect.

The data in Table 11.7 will serve to illustrate the computational procedure for a 3-year moving average. We first find the 3-year moving totals. For 1947, 1948, and 1949, for example, we have $73.2 + 81.9 + 68.5 = 223.6$. This total is located to correspond with the middle year 1948. The next moving total is obtained by dropping the 1947 value and adding the 1950 value to obtain $81.9 + 68.5 + 82.8 = 233.2$. The 3-year moving averages are obtained by dividing the moving totals by 3.

Table 11.7 Employment by General Building Contractors in California (Figures in Thousands)

Year	Number of Employees	3-Year Moving Total	3-Year Moving Average	5-Year Moving Total	5-Year Moving Average
1947	73.2	—	—	—	—
1948	81.9 (peak)	223.6	74.53	—	—
1949	68.5	233.2	77.73	394.7	78.95
1950	82.8	239.6	79.86	407.1	81.42
1951	88.3 (peak)	256.7	85.56	412.6	82.52
1952	85.6	261.3	87.09	424.9	84.98
1953	87.4 (peak)	253.8	84.60	425.5	85.10
1954	80.8	251.6	83.86	418.7	83.74
1955	83.4 (peak)	245.7	81.90	414.8	82.96
1956	81.5	246.6	82.20	413.0	82.60
1957	81.7	248.8	82.93	424.2	84.84
1958	85.6	259.3	86.43	423.7	84.74
1959	92.0 (peak)	260.5	86.83	422.9	84.58
1960	82.9	255.6	85.20	425.9	85.18
1961	80.7	248.3	82.76	431.4	86.28
1962	84.7	256.5	85.50	431.5	86.30
1963	91.1	267.9	89.30	436.3	87.26
1964	92.1 (peak)	270.9	90.30	441.4	88.28
1965	87.7	265.6	88.53	—	—
1966	85.8	—	—	—	—

Source. Employment and Earning Statistics for States and Areas, 1939–1966, Bulletin No. 1370-4, U.S. Department of Labor, July 1967.

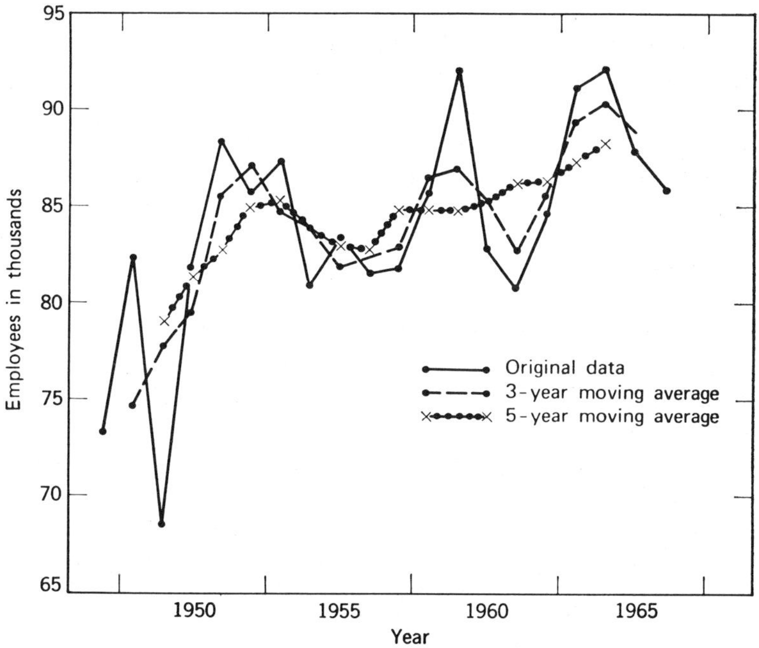

Figure 11.5 Employment by general building contractors in California.

The number of observations that should be included in a moving average designed to isolate the trend will depend on the length of the cyclical movements in the time series. Regular, periodic cycles can be eliminated by making the period of the moving average equal to, or a multiple of, that of the cycle. Thus, if a time series exhibits a regular 3-year cycle, a 3-year moving average should be used to isolate the trend. If a cycle has a 7-year duration, a 7-year moving average would be appropriate. In this way, the values at the peak of the cycle tend to be "cancelled-out" by the values at the trough provided, of course, that a regular, periodic cycle exists. The cycle is thereby "eliminated," leaving a "smoothed" time series of trend values.

Unfortunately, cycles in economic time series are usually not very regular. By graphing the original data, the investigator can usually determine if there is enough regularity to justify the use of moving averages and, at the same time, determine the appropriate period to use for the moving average. If the length of the cycle changes during the period for which data are available, the investigator may decide to use several moving averages of different periods to depict the trend in different parts of the time series. Inspection of the employment data in Figure 11.5, for example, suggests that a 3-year moving average may be appropriate for the 1947–1955 period. The length of the employment cycle increases after 1955, however. Thus, a 3-year moving average cannot be expected to eliminate all of the cyclical component of the series after 1955. Both 3-year and 5-year moving averages were computed for this time series. The 3-year moving average provides a reasonably good description of the upward trend from 1947 to 1951 and the downward trend from 1952 to 1956. A cyclical movement is evident in the 3-year moving average after 1956, however. The 5-year moving average provides an adequate description of the trend from 1956 to 1966 although it tends to underestimate the upward trend from 1947 to 1951 and the downward trend from 1952 to 1956.

An investigator interested in predicting future trend values is likely to find the mathematical trend lines discussed previously to be more helpful than the method of moving averages. To compute future values of a moving average requires knowledge of the future values of the time series. By assuming that the trend line has a certain mathematical form, it is possible to predict future trend values without future data.

Moving averages often provide a realistic description of past trends in a time series and, therefore, may be especially useful when studying deviations from the trend. An example of the use of moving averages to analyze seasonal variations will be presented in the following section.

11.4 SEASONAL VARIATIONS

The use of seasonally-adjusted data often facilitates the interpretation of economic developments. A businessman who finds his sales down by 30 percent in June may want to know if such a drop is part of an annually-recurring pattern before making decisions about a new sales campaign or the hiring of new salesmen. The Board of Governors of the Federal Reserve may want to know if the change in the money supply last month was typical of previous years before making decisions about this month's open-market operations. A grain dealer may want to know if grain prices last week were higher or lower than usual before he decides to sell grain. Part of the information needed in each of these cases could be obtained by comparing the observed data with seasonally adjusted data.

An understanding of seasonal variations can also be helpful in short-term forecasting. A businessman who knows that his sales of television sets usually increase by 20 percent in December can plan his inventory accordingly. An economist attempting to estimate Gross National Product (GNP) for the second quarter of next year will need to take seasonal variations into account as well as the trend of GNP. A bank may need to be able to forecast monthly changes in the demand for money when planning its short-term investments.

There are many different ways to measure seasonal variation. The basic goal of most approaches is to obtain a *seasonal index* that can then be used to adjust the original data for seasonal variations. For quarterly data, a seasonal index consists of 4 numbers, one for each quarter, which represent percentages of their average. If a seasonal index of first quarter sales is 90, for example, this means that first quarter sales have usually been 10 percent below sales of the average quarter.

The methods discussed in this section are useful for measuring *constant seasonal patterns*, that is, seasonal patterns that do not change over time. Seasonal indexes computed on the assumption of constant seasonal patterns must be used with considerable care if there is evidence that the seasonal pattern of the data may have changed over time.

11.4.1 Prior Adjustments

It is frequently advisable to adjust the basic data for such events as bad weather, holidays, or work stoppages before estimating the trend, seasonal, or cyclical components of the time series. Such adjustments are often especially important when estimating seasonal indexes because the effects of such "special events" are not always "averaged out" within the month or quarter in which they occur. There is no unique procedure for making prior adjustments. The basic purpose of such adjustments is to estimate what the values of the time series would have been in the absence of the special event or factor affecting the series.

It is frequently necessary to adjust the original data for variation in the length of the month. This can be done by estimating the number of "effective" days in each month and then multiplying the original data by the ratio

$$\frac{\text{number of effective days in an average month}}{\text{number of effective days in the given month}}$$

The "adjusted monthly" values can then be adjusted proportionately to insure that the annual totals are equal to those of the original data. The exact definition of "effective days" will depend on the time series of interest and the reason for making the adjustment. A short dock strike, for example, may affect imports and exports of some items. In this case, "effective days" may be appropriately defined as the number of days worked each month by dock workers. In other cases, "effective days" may be defined as (1) the number of days in the calendar month, (2) the number of working days in the month, (3) the number of calendar days less any holidays, (4) the number of days when the temperature was above 70° F., or (5) the number of days when there was no rainfall.

Some time series are affected by holidays that do not occur on a fixed date each year. Retail sales of hats, for example, may be relatively high in March if Easter falls early in April and relatively high in April if Easter falls in late April. In such cases, the assumption of a constant, monthly seasonal pattern is not justified and the techniques discussed in this section for estimating seasonal indexes would not provide useful seasonal index numbers for some months. There are several ways that this type of problem can be handled. In the case of Easter hat sales, it may be possible to avoid the effects of changing dates for Easter by aggregating the data and computing seasonal indexes on a bimonthly or quarterly basis rather than a monthly basis. If this is not satisfactory, the years in the time series could be divided into two sets—one for those years in which Easter falls in late March or early April and another for years in which Easter falls in late April. Two seasonal indexes would then be estimated. The seasonal index used for seasonal adjustment purposes in any particular year would depend on whether Easter fell early or late in April.

Occasionally, time series are characterized by abrupt changes in their seasonal patterns. The date for submitting individual income tax returns, for example, changed in 1955 from March 15 to April 15 and thus caused an abrupt change in the seasonal pattern of tax receipts. If feasible, such series should be partitioned into segments at the points where the abrupt changes occur and each segment analyzed separately.

Some series occasionally undergo substantial changes in the definition of the data. Such changes will often affect the trend but may or may not affect the seasonal pattern of the series. In the case of government statistical series, the appropriate procedure for adjusting an old time series to correspond with a new one is frequently outlined by the agency that compiles the series. In other cases, it may be necessary to drop part of the data before estimating the various components.

The seasonal patterns of many series are affected by the dates when new models are introduced. This is particularly true, for example, of time series which are affected by automobile production and sales. The problems of adjusting such series are similar to the problems of adjusting series affected by holidays with shifting dates. It may be possible in some cases to partition the data and estimate several seasonal indexes. In other cases, it may be useful to estimate "seasonal patterns" on a model year basis rather than a calendar year basis.

11.4.2 The Average Percentage Method

The average percentage method provides a quick and simple procedure for constructing a seasonal index. The first step in the average percentage

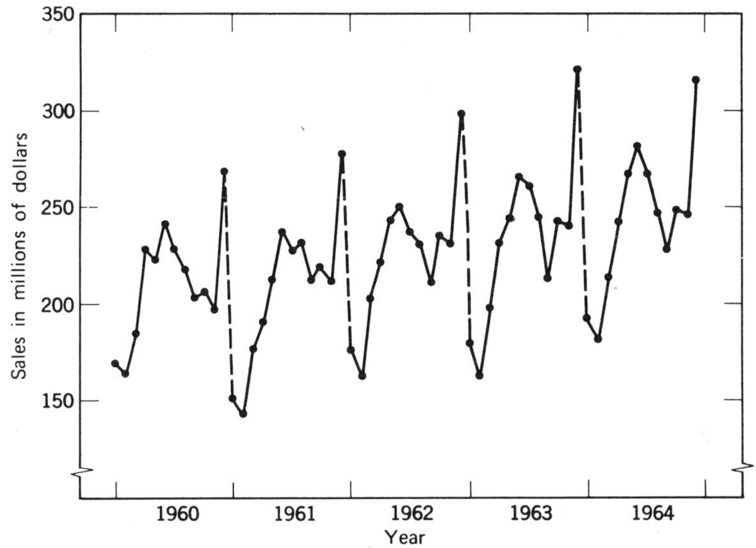

Figure 11.6 Sales of automobile tires, batteries, and accessories.

method is to express the data for each month (or subperiod) as an average for the year. The percentages for corresponding months (or subperiods) are then averaged to obtain the set of numbers constituting the seasonal index.

The data on sales of automobile tires, batteries, and accessories found in Table 11.8 and presented in Figure 11.6 will be used to illustrate this approach. The average monthly sales in 1960 were

$$\frac{170 + 165 + 185 + \cdots + 269}{12} = \frac{2541}{12} = 211.75$$

Each of the monthly figures for 1960 is then divided by 211.75 and the result multiplied by 100 to obtain a set of percentages for 1960. In this way we obtain

$$\frac{170(100)}{211.75} = 80 \qquad \text{for January 1960}$$

$$\frac{165(100)}{211.75} = 78 \qquad \text{for February}$$

$$\vdots$$

$$\frac{269(100)}{211.75} = 127 \qquad \text{for December}$$

The monthly percentages for the years 1961 to 1964 are obtained in the same way.

Table 11.8 Sales of Automobile Tires, Batteries, and Accessories

PART A: ORIGINAL DATA IN MILLIONS OF DOLLARS[a]

Month	Year 1960	1961	1962	1963	1964
January	170	151	177	180	192
February	165	143	164	163	182
March	185	176	203	197	214
April	229	190	221	232	242
May	224	213	243	244	266
June	242	238	250	265	282
July	229	229	236	261	267
August	219	232	230	245	246
September	204	214	212	214	229
October	206	218	235	243	248
November	199	211	231	241	246
December	269	277	299	318	317
Column totals	2541	2492	2701	2803	2931
Monthly average	211.75	207.67	225.08	233.58	244.25

PART B: PERCENTAGES OF MONTHLY AVERAGES

Month	Year 1960	1961	1962	1963	1964	Row Totals	Seasonal Index
January	80	73	79	77	79	388	77.6
February	78	69	73	70	74	364	72.8
March	88	85	90	84	88	435	87.0
April	108	91	98	99	99	495	99.0
May	106	102	108	104	109	529	105.8
June	114	115	111	114	115	569	113.8
July	108	110	105	112	109	544	108.8
August	104	112	102	105	101	524	104.8
September	96	103	94	92	94	479	95.8
October	97	105	104	104	101	511	102.2
November	94	102	103	103	101	503	100.6
December	127	133	133	136	130	659	131.8
Total							1200.0

[a] *Source. Survey of Current Business*, April 1966.

The percentages for corresponding months of the five-year period 1960 to 1964 are then averaged using either a mean or median to obtain the seasonal index. The arithmetic mean for January, for example, is

$$\frac{80 + 73 + 79 + 77 + 79}{5} = \frac{388}{5} = 77.6$$

In other words, sales in January have usually been $(100.0 - 77.6) = 22.4$ percent below average monthly sales for the year.

The mean of the numbers making up a seasonal index should be equal to 100. In this case, the sum of 12 monthly index numbers should be equal to 1200, which in fact it is for the example presented in Table 11.8. If the median is used instead of the mean or if rounding "errors" fail to average out, the seasonal index numbers should be adjusted by multiplying them by a suitable constant. For example, suppose the initial seasonal index obtained by the average percentage method for quarterly data is:

Quarter	*Initial Seasonal Index*
1st	80
2nd	115
3rd	75
4th	120
Sum	390

Each of the numbers in the index should then be multiplied by

$$\frac{400}{390} = 1.02464$$

to obtain

Quarter	*Final Seasonal Index*
1st	82.1
2nd	117.9
3rd	76.9
4th	123.1
Sum	400.0

Returning to the example presented in Table 11.8, let us adjust the 1964 data for seasonal variation. This is accomplished by dividing sales each month by the corresponding seasonal index number and multiplying by 100. Thus, January 1964 sales adjusted for seasonal variation are equal to

$$\frac{192.0(100)}{77.6} = 247.4 \text{ million dollars}$$

The seasonally adjusted sales for February 1964 are equal to

$$\frac{182.0(100)}{72.8} = 250.0 \text{ million dollars}$$

These figures indicate what sales would have been if there were no seasonal variations. The actual percentage decrease in sales from January to February was

$$\frac{(182 - 192)(100)}{192} = -5.2$$

Inspection of the seasonal index indicates that February sales are usually lower than January sales, and thus some decline in sales is to be expected. The seasonally adjusted (or deseasonalized) sales increased by 2.6 million dollars or

$$\frac{(250.0 - 247.4)(100)}{247.4} = 1.0 \text{ percent}$$

Thus, the decline in actual sales from January to February was not as large as would be expected normally. The difference between the actual and the normal change in sales is not very large, however, and would not appear to justify any special study as to why sales did not decline as much as expected.

The seasonally adjusted sales for the other months are as follows:

Month	*Seasonally-Adjusted Sales (in Millions of Dollars)*
January	247.4
February	250.0
March	246.0
April	244.4
May	251.4
June	247.8
July	245.4
August	234.7
September	239.0
October	242.7
November	244.5
December	240.5

Month-to-month changes in seasonally adjusted sales were usually relatively small in 1964. These sales declined slightly during the year but, in general, 1964 appears to have been a fairly typical year.

11.4.3 The Ratio-to-Trend Method

The *ratio-to-trend* or *percentage-of-trend* method of calculating a seasonal index allows the user to take the trend of the data into account with somewhat more precision than is possible with the average percentage method. The basic procedure is, first, to eliminate the trend from the original data by division and then to average out the irregular and cyclical variations to obtain a seasonal index.

The quarterly data on imports of building materials presented in Table 11.9 will be used to illustrate this method. The first step was to compute the quarterly average for each year and then to fit a least squares trend line to these averages. The linear trend line obtained is

$$\hat{Y} = 147.95 + 8.61X \tag{11.24}$$

where X is measured in annual units with origin at June 31, 1961, and $\hat{Y}$ represents the average quarterly imports of building materials.

In order to estimate the quarterly trend values, it is necessary to change from annual to quarterly the units of X and to shift the origin of the trend line to the middle of the second quarter of 1961. The quarterly increase in imports is equal to the annual increase divided by 4 or $(8.61/4) = 2.15$. The origin is shifted to the middle of the second quarter of 1961 by subtracting one-half of the quarterly increase from 147.95. The quarterly trend line is therefore equal to

$$\hat{Y} = 146.88 + 2.15X \tag{11.25}$$

where X is in quarterly units with origin at May 15, 1961.

The quarterly trend values presented in Part B of Table 11.9 are obtained from the trend line (11.25). The original data are then divided by the quarterly trend values for corresponding quarters and multiplied by 100 to obtain the ratio-to-trend values presented in Part C of Table 11.9. The ratio-to-trend value for the first quarter of 1956, for example, is equal to

$$\frac{117(100)}{101.73} = 115.0$$

In other words, imports in the first quarter of 1956 were 15 percent above the trend value for that quarter. This was somewhat unusual in the sense that imports in the first quarter of all other years in the series were always below the quarterly trend values.

The cyclical and irregular variations in the ratio-to-trend numbers is averaged out by summing the numbers corresponding to each quarter and dividing by the number of years in the series. Thus, the initial mean of the ratio-to-trend numbers for the first quarter is

$$\frac{115.0 + 81.6 + \cdots + 94.3}{11} = \frac{976.5}{11} = 88.8$$

Table 11.9 Imports of Building Materials

PART A: ORIGINAL DATA IN MILLIONS OF DOLLARS[a]

	Quarter				Quarterly Average		
Year	I	II	III	IV	Y	X	XY
1956	117	126	132	112	121.75	−5	−608.75
1957	90	104	113	100	101.75	−4	−407.00
1958	83	99	126	127	108.75	−3	−326.25
1959	125	170	165	144	151.00	−2	−302.00
1960	133	151	141	116	135.25	−1	−135.25
1961	106	142	152	137	134.25	0	0.0
1962	135	165	169	148	154.25	1	154.25
1963	136	170	190	165	165.25	2	330.50
1964	155	189	197	166	176.75	3	530.25
1965	155	186	201	183	181.25	4	725.00
1966	177	223	215	174	197.25	5	986.25
Totals					1627.50		947.00

PART B: TREND VALUES

	Quarter			
Year	I	II	III	IV
1956	101.73	103.88	106.03	108.18
1957	110.33	112.48	114.63	116.78
1958	118.93	121.08	123.23	125.38
1959	127.53	129.68	131.83	133.98
1960	136.13	138.28	140.43	142.58
1961	144.73	146.88	149.03	151.18
1962	153.33	155.48	157.63	159.78
1963	161.93	164.08	166.23	168.38
1964	170.53	172.68	174.83	176.98
1965	179.13	181.28	183.43	185.58
1966	187.73	189.88	192.03	194.18

PART C: RATIO-TO-TREND VALUES

Year	Quarter I	II	III	IV	
1956	115.0	121.3	124.4	103.5	
1957	81.6	92.5	98.6	85.6	
1958	69.8	81.8	102.2	101.3	
1959	98.0	131.1	125.2	107.5	
1960	97.7	109.2	100.4	81.3	
1961	73.2	96.7	102.0	90.6	
1962	88.0	106.1	107.2	92.6	
1963	84.0	103.6	114.3	98.0	
1964	88.4	109.4	112.7	93.8	
1965	86.5	102.6	109.6	98.6	
1966	94.3	117.4	111.9	89.6	
Totals	976.5	1171.7	1208.5	1042.4	
					Totals
Seasonal index based on the mean	88.8	106.5	109.9	94.8	400.0
Median	88.4	106.1	109.6	93.8	397.9
Seasonal index based on the median	88.8	106.7	110.2	94.3	400.0

[a] *Source. Survey of Current Business*, September 1967.

The sum of the initial means (or initial seasonal index numbers) should be equal to 400 when working with quarterly data. If, due to rounding errors, the means do not add to 400, the seasonal index numbers can be obtained by multiplying the initial means by an appropriate constant. If there are several extreme values in any particular quarter, they can be dropped before computing the mean of the ratio-to-trend numbers for that quarter or the quarterly medians may be used in place of means to obtain the seasonal index numbers.

There are several variations of the ratio-to-trend method depending on the procedure used to calculate the trend. One alternative would be to fit the trend line directly to the original quarterly data rather than the annual averages. In some cases, a nonlinear trend line may provide a better description of the trend and, if so, should be used to obtain the trend values.

We shall use the data on imports of building materials to illustrate the use of seasonal indexes in short-term forecasting. Suppose, for example, that a forecast of quarterly imports of building materials is needed for 1967. Preliminary estimates can be obtained from the trend equation (11.25). The trend value for the first quarter of 1967 is

$$\hat{Y} = 146.88 + 2.15(23) = 196.33 \text{ million dollars}$$

This trend value could serve as a forecast of first quarter imports if there were no seasonal variation. Inspection of the seasonal index indicates that first quarter imports are usually about 11 percent below the trend value. The final estimate of first quarter imports is therefore obtained by multiplying the trend value by the first quarter seasonal index number and dividing the result by 100. Carrying out these operations, we obtain

$$\frac{(196.33)(88.8)}{100} = 174.34 \text{ million dollars}$$

The quarterly projections for 1967 are summarized below.

Quarter	Trend Value	Seasonal Index (as a Proportion)	Estimated Quarterly Imports for 1967
1st	196.33	0.888	174.34
2nd	198.48	1.065	211.38
3rd	200.63	1.099	220.49
4th	202.78	0.948	192.24

It should be remembered that these forecasts are based entirely on past data on imports. Changes in tariffs, changes in interest rates affecting new construction, or an expanded urban renewal program are just a few of the things that could affect future imports. Such information could be formally taken into account in a detailed forecasting model, but the development of such a model would increase the cost of the forecast. The forecasts presented above may be sufficiently accurate for many purposes but should be periodically revised as new information becomes available.

11.4.4 The Ratio-to-Moving Average Method

The ratio-to-moving average or percentage of moving average method is probably the most widely used for measuring seasonal variation. Complex versions are used to seasonally adjust many of the time series compiled by various agencies of the United States government. The ratio-to-moving average method is relatively simple and yet is a substantial improvement over the ratio-to-trend method which does not provide a satisfactory means of taking the cyclical variation into account. This point will be explained in more detail after the ratio-to-moving average method has been presented.

The first step of the ratio-to-moving average method is to compute a 12-month moving average from the original data. (Or a 4-quarter moving average if quarterly data are being used.) A 2-month moving average is then used to center the 12-month moving average in the middle of the months. The result is called a centered 12-month moving average.

The original data are then expressed as a percentage of the centered 12-month moving average. The percentages for corresponding months are then averaged to obtain the seasonal index.

The data used earlier on quarterly imports of building materials will be used to illustrate the computations involved. The preliminary steps are presented in Table 11.10. Beginning with the moving average, the first and second numbers of the uncentered 4-quarter moving average are

$$\frac{(117 + 126 + 132 + 112)}{4} = \frac{487}{4} = 121.75$$

and

$$\frac{(126 + 132 + 112 + 90)}{4} = \frac{460}{4} = 115.00$$

The first number of the centered 4-quarter moving average is therefore equal to

$$\frac{(121.75 + 115.00)}{2} = 118.4$$

and is located at the third quarter of 1956.

The centered 4-quarter moving average is obtained more directly by adding consecutive 4-quarter moving totals and dividing the result by 8. Thus, for the third quarter of 1956, we have

$$\frac{(117 + 126 + 132 + 112) + (126 + 132 + 112 + 90)}{8} = \frac{947}{8} = 118.4$$

which is identical to the previous result.

We next compute the ratio-to-moving average percentages by dividing the original data by the centered moving average figures for corresponding quarters and multiplying by 100. For the third quarter of 1956, we have

$$\frac{132.0(100)}{118.4} = 111.5$$

Table 11.10 Imports of Building Materials

Year and Quarter	Data	4-Quarter Moving Total	2-Quarter Moving Total	4-Quarter Centered Moving Average	Ratio-to-Moving Average
1956					
I	117	—	—	—	—
II	126	487	—	—	—
III	132	460	947	118.4	111.5
IV	112	438	898	112.3	99.7
1957					
I	90	419	857	107.1	84.0
II	104	407	826	103.3	100.7
III	113	400	807	100.9	112.0
IV	100	395	795	99.4	100.6
1958					
I	83	408	803	100.4	82.7
II	99	435	843	105.4	93.9
III	126	477	912	114.0	110.5
IV	127	548	1025	128.1	99.1
1959					
I	125	587	1135	141.9	88.1
II	170	604	1191	148.9	114.2
III	165	612	1216	152.0	108.5
IV	144	593	1205	150.6	95.6
1960					
I	133	569	1162	145.3	91.5
II	151	541	1110	138.8	108.8
III	141	514	1055	131.9	106.9
IV	116	505	1019	127.4	91.0
1961					
I	106	516	1021	127.6	83.1
II	142	537	1053	131.6	107.9
III	152	566	1103	137.8	110.3
IV	137	589	1155	144.4	94.9
1962					
I	135	606	1195	149.4	90.4
II	165	617	1223	152.4	107.9
III	169	618	1235	154.7	109.2
IV	148	623	1241	155.1	95.4
1963					
I	136	644	1267	158.4	85.8
II	170	661	1305	163.1	104.2
III	190	680	1341	167.6	113.3
IV	165	699	1379	172.4	95.7

Table 11.10—*Continued*

Year and Quarter	Data	4-Quarter Moving Total	2-Quarter Moving Total	4-Quarter Centered Moving Average	Ratio-to-Moving Average
1964					
I	155	706	1405	175.6	88.3
II	189	707	1413	176.6	107.0
III	197	707	1414	176.7	111.5
IV	166	704	1411	176.4	94.1
1965					
I	155	708	1412	176.5	87.8
II	186	725	1433	179.1	103.8
III	201	747	1472	184.0	109.2
IV	183	784	1531	191.4	95.6
1966					
I	177	798	1582	197.7	89.5
II	223	789	1587	198.4	112.4
III	215	—	—	—	—
IV	174	—	—	—	—

Proceeding in this way, we obtain ratio-to-moving average percentages for all except the first and last two quarters of the series.

The ratio-to-moving average percentages for corresponding quarters are next averaged using either the mean or median to obtain the seasonal index. The four numbers in the index should add to 400. If not, they should be adjusted by multiplying each number by an appropriate constant.

The ratio-to-moving average figures for imports of building materials are presented in Table 11.11. Using means, the seasonal index is found to be:

Quarter	*Seasonal Index*
1st	87.1
2nd	106.1
3rd	110.3
4th	96.2

The sum of the means, 399.7, is so close to the required 400 that no adjustment is necessary.

We shall now compare the ratio-to-trend and ratio-to-moving average methods in more detail. Using the multiplicative model, the original data, Y, are equal to $TCSI$ where T is the trend, C the cyclical, S the seasonal, and I the irregular component. In the ratio-to-trend method, the trend component is first isolated and then each monthly (or quarterly) value is divided by the

Table 11.11 Ratio-to-Moving Average for Imports of Building Materials

Year	Quarter I	II	III	IV
1956	—	—	111.5	99.7
1957	84.0	100.7	112.0	100.6
1958	82.7	93.9	110.5	99.1
1959	88.1	114.2	108.5	95.6
1960	91.5	108.8	106.9	91.0
1961	83.1	107.9	110.3	94.9
1962	90.4	107.9	109.2	95.4
1963	85.8	104.2	113.3	95.7
1964	88.3	107.0	111.5	94.1
1965	87.8	103.8	109.2	95.6
1966	89.5	112.4	—	—
Totals	871.2	1060.8	1102.9	961.7
Seasonal index based on mean	87.1	106.1	110.3	96.2
Median	87.9	107.4	110.4	95.6
Seasonal index based on median	87.6	107.1	110.0	95.3

corresponding trend value T to obtain $Y/T = CSI$. The CSI values are averaged next to eliminate the cyclical and irregular components. Some of the cyclical variation may remain in the resulting seasonal index, however, if the number of years in the time series is not a multiple of the number of years in the cycle.

In the ratio-to-moving average method, the centered moving average serves to eliminate the seasonal and irregular component and is thus equivalent to the values given by TC. The monthly (or quarterly) values Y are then divided by the corresponding moving average values to obtain $Y/CT = SI$. The SI values for corresponding months (or quarters) are then averaged to eliminate the irregular component I. Very little, if any, cyclical variation should remain in the resulting seasonal index.

The difference in the two methods is readily seen by inspection of Figure 11.7. Notice that, during the period 1956 to 1961, most of the values for imports lie either above or below the trend line. As a result, the ratio-to-trend values tend to be considerably larger than the corresponding seasonal index numbers in some years and considerably less in others. This is readily seen by comparing the 1956 and 1957 ratio-to-trend values in Part C of Table 11.9 with the seasonal index figures at the bottom of that table.

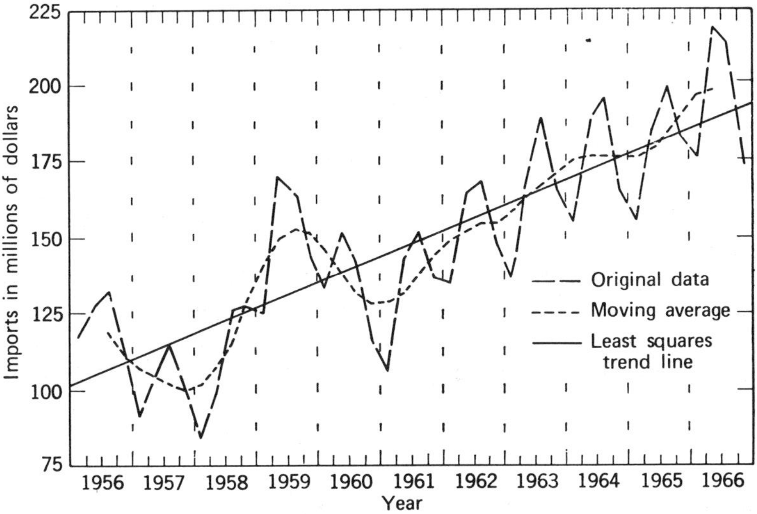

Figure 11.7 Imports of building materials.

The ratio-to-moving average values reflect both the trend and strong cyclical movements of the 1956 to 1961 period. With two minor exceptions, the values for imports in any particular year during this period are consistently below the moving average values for the first and fourth quarters and above the moving average values for the second and third. Of course, this is as expected once the seasonal index is known.

11.5 CYCLICAL VARIATIONS

The cyclical component of a time series is frequently obtained as a residual by dividing the original data by the estimates of the trend and seasonal factors and taking out irregular movements by using a moving average. We shall use the first part of the data on imports of building materials presented in Part A of Table 11.9 to illustrate this procedure. The trend values and seasonal index have already been computed and are found in Parts B and C of Table 11.9. The trend-seasonal values are obtained by multiplying the trend values by the seasonal index numbers for corresponding quarters. Thus the trend-seasonal for the first quarter of 1956 is equal to

$$TS = \frac{(101.7)(88.8)}{100} = 90.3$$

The original data are then divided by the trend-seasonal values and multiplied by 100 to obtain the cyclical-irregular components. For the first quarter of 1956, we have

$$CI = \frac{117(100)}{90.3} = 129.5$$

The cyclical-irregular components are, in a sense, "percentages of normal." That is, they indicate how much higher or lower imports are relative to what they would have been if there were no cyclical or irregular movements in the time series.

A short moving average is commonly used to eliminate, insofar as possible, the irregular component. The cyclical relatives presented in the last column of Table 11.12 were obtained by using a 3-quarter moving average. The cyclical relative of the second quarter of 1956, for example, is equal to

$$\frac{129.5 + 113.9 + 113.3}{3} = 118.9$$

The moving average used must be kept short in order to avoid smoothing out the cyclical component as well as the irregular variation.

The cyclical relatives for the 1956 to 1960 period are shown in Figure 11.8. The cycle obtained appears to be about three and one-half years in length. A more complete analysis would indicate that the character of the cycle changes considerably after 1960. This is evident from the moving average presented in Figure 11.7.

The procedure that we have used to find the cyclical relatives does not require the assumption that the cyclical pattern is consistent in amplitude and period. The cyclical relatives obtained merely describe the past cyclical movements of the time series. They cannot be used to project future cyclical

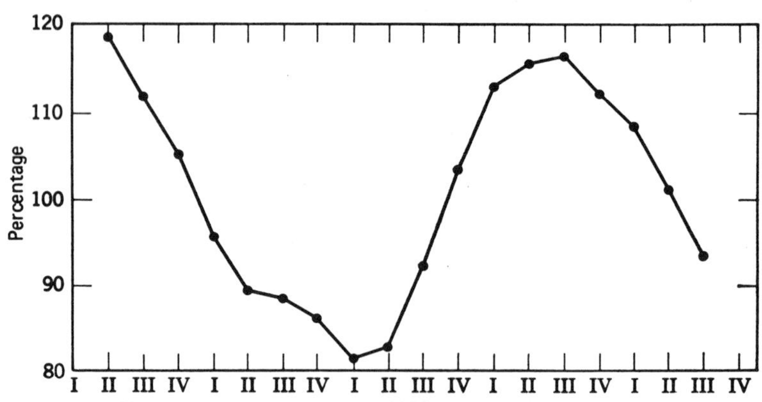

Figure 11.8 Cyclical relatives.

Table 11.12 Cyclical Relatives

Year and Quarter	Imports *TSCI*	Trend *T*	Seasonal *S*	Trend-Seasonal *TS*	Cyclical-Irregular *CI*	Cyclical Relatives *C*
1956						
I	117	101.7	88.8	90.3	129.5	—
II	126	103.9	106.5	110.6	113.9	118.9
III	132	106.0	109.9	116.5	113.3	112.1
IV	112	108.2	94.8	102.6	109.2	104.8
1957						
I	90	110.3	88.8	97.9	91.9	96.0
II	104	112.5	106.5	119.8	86.8	89.5
III	113	114.6	109.9	125.9	89.7	88.9
IV	100	116.8	94.8	110.7	90.3	86.2
1958						
I	83	118.9	88.8	105.6	78.6	81.9
II	99	121.1	106.5	128.9	76.8	82.8
III	126	123.2	109.9	135.4	93.1	92.2
IV	127	125.4	94.8	118.9	106.8	103.4
1959						
I	125	127.5	88.8	113.2	110.4	113.4
II	170	129.7	106.5	138.1	123.1	115.8
III	165	131.8	109.9	144.8	114.0	116.8
IV	144	134.0	94.8	127.0	113.4	112.5
1960						
I	133	136.1	88.8	120.9	110.0	108.6
II	151	138.3	106.5	147.3	102.5	101.3
III	141	140.4	109.9	154.3	91.4	93.2
IV	116	142.6	94.8	135.2	85.8	—

movements unless the cyclical pattern is consistent or changes in a consistent manner. Unfortunately, this is almost never the case when dealing with economic data. Thus it is seldom possible to predict the economic cycle of a time series from a historical study of the series itself.

APPENDIX A

THE STRUCTURAL ANALYSIS OF ECONOMIC TIME SERIES

In this appendix, we shall concentrate on interpreting the systematic components of economic time series. The key word in such interpretations is *why*. Why does a particular variable trace out a certain time path? What causes the seasonal pattern? What causes the trend? In what ways, if any, can knowledge of these causes improve the decisions of a business firm, a public agency, or a maker of government policy?

A.1 STRUCTURAL ANALYSIS OF SEASONAL, TREND, AND CYCLICAL COMPONENTS OF ECONOMIC TIME SERIES

This section presents a structural approach to the analysis of various components of economic time series. Seasonal, trend, and cyclical movements will be discussed in turn. In each case, it appears that the movements in one time series can hardly be explained without pointing to movements in other time series and a model—sometimes a rather loose and intuitive one—relating these series to those of primary concern.

A.1.1 Structural Analysis of Seasonal Patterns

Figure A.1 shows a hypothetical consumer demand function for eggs which has an elasticity of -0.5 at the point (100,100) which we will take as

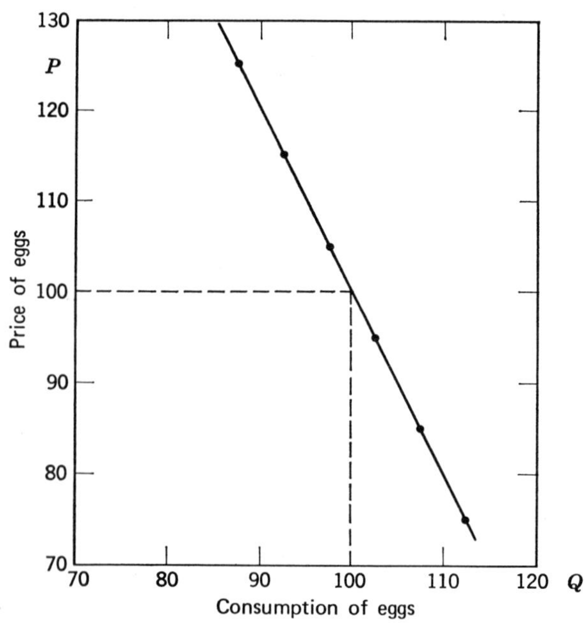

Figure A.1 Consumer demand for eggs (schematic).

the point of annual averages or means. According to this relationship, a 10 percent decrease in the retail price of eggs is associated with a 5 percent increase in consumption. It is assumed that this consumer demand function applies equally to all 12 months of the year. Suppose that there is no storage activity for eggs, so that the quantity of eggs offered to consumers by retailers in a given month is precisely equal to the farm production of eggs for that month. Let us further assume the monthly pattern of egg production which is shown in Table A.1.

Table A.1

Month	Production (Units)	Price in Absence of Storage (Cents per Unit)
January	97.5	105
February	102.5	95
March	107.5	85
April	112.5	75
May	112.5	75
June	107.5	85
July	102.5	95
August	97.5	105
September	92.5	115
October	87.5	125
November	87.5	125
December	92.5	115
Average	100.0	100.0

The specific formula for the demand function in Figure A.1 is $P = 300 - 2Q$. If consumption is equal to production retail prices must move in the manner shown in the third column of Table A.1. The ratio of the extreme ranges, $(112.5 - 87.5)/(75 - 125) = -25/50 = -0.5$, which in the present symmetrical and straight-line pattern of price-quantity points is identical with the elasticity of demand at the point of means. These particular sets of monthly consumption (equals production) and monthly prices are shown as the solid lines in Figure A.2.

Now, suppose that there are some firms engaged in the seasonal storage of eggs. For simplicity, we assume that the eggs do not deteriorate in quality while in storage but will sell at the same price as the current month's production. If the cost of storing a dozen eggs is two cents a month, and if storage operators are working under pure competition and have a clear knowledge of the seasonal pattern of egg production and of the consumer

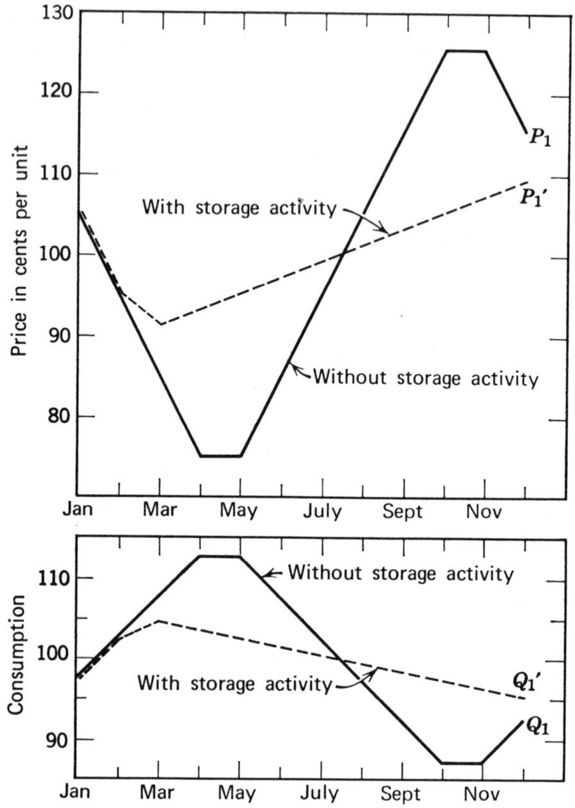

Figure A.2 Seasonal patterns of price and consumption of eggs with and without a storage activity (schematic).

demand curve, the resulting price and consumption patterns should be as indicated in the dashed line in Figure A.2.

Obviously, storage operators would be buying eggs in the months in which production is large and prices are low and they would be selling eggs in months of below-average production and above-average price. In equilibrium, as the marginal cost of storage is assumed to be two cents a month, the expected selling price of eggs must increase by two cents a month to justify a storage operation. The price pattern resulting under the conditions of our particular problem is one in which prices rise by two cents a month from a low of 91 cents in March to a high of 109 cents in December. Conversely, consumption (as distinct from production) would decline from a high of 104.5 units in March to a low of 95.5 units in December.

The paired observations of price and *consumption* fall on the demand function shown in Figure A.1; the ratio of consumption range to price range

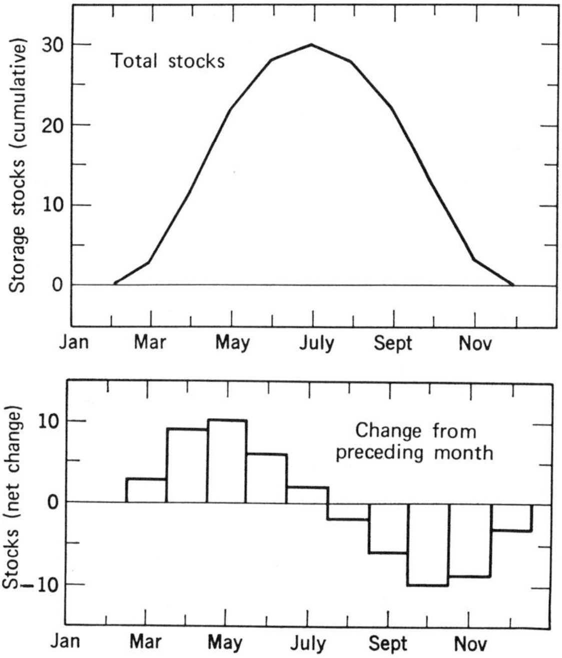

Figure A.3 Seasonal variations in storage holdings of eggs (schematic).

is $(104.5 - 95.5)/(91 - 109) = -9/18 = -0.5$, as before. However, the ratio of the seasonal range in *production* to the new seasonal range in price becomes $(112.5 - 87.5)/(91 - 109) = -25/18 = -1.4$. This ratio or "pseudo-elasticity" is about three times as large as the true elasticity of consumer demand.

Intuitively, it appears that (except for discontinuities and special circumstances) a storage cost of one cent per month would reduce the amplitude of seasonal price variation from 18 cents to about nine cents; with the amplitude of seasonal variation in production unchanged, the "pseudo-elasticity" would rise to -2.8. An increase in storage costs to three or four cents a month would clearly increase the amplitude of price variation and reduce the "pseudo-elasticity."

A complete explanation of Figures A.2 and A.3 would involve some simple operations research considerations. Figure A.3 indicates that the consumer demand function for eggs, the seasonal pattern of egg production, and the level of storage costs collectively result in a strong and regular seasonal pattern of storage stocks and of net movements of eggs into and out of storage if all disturbances and uncertainty are ruled out.

The traditional approach to seasonal analysis would have applied the same routine methods of determining seasonal indexes to each of the five time series of (1) egg production, (2) egg consumption, (3) retail price, (4) total storage holdings, and (5) net movements of eggs into and out of storage. In effect, the process of data adjustment would have treated each of these series as if it were completely independent of the others.

Our analysis implies that the seasonal patterns in the five series are closely bound together by means of a consumer demand function and a storage cost function, operating in conjunction with a predetermined pattern of egg production.

Given a structural analysis of seasonal variation, we should be able to estimate the effects of changes in individual elements of the structure. For example, if monthly storage costs increased successively from two to three cents and from three to four cents, the amplitude of seasonal price variation would increase from 18 to 27 to 30 cents; peak storage stocks would fall from 30.00 to 22.75 to 19.00 units; the total number of "unit-months" of storage would fall from 160.00 to 117.75 to 91.00; and the gross income to all storage operators would rise from 320.00 "cents" to 353.25 to 364.00.

This level of precision is, of course, not found in the real world. As a matter of fact, seasonal patterns of egg storage, like seasonal patterns of many other economic time series, have changed considerably in recent years. A structural analysis of these changes can largely explain them while an empirical analysis can only decribe or extrapolate them on the basis of unanalyzed assumptions that certain trends will continue.

Obviously, a reduction in the seasonal amplitude of egg production would also reduce the seasonal amplitude of price variations and of storage holdings. If an egg producer takes any particular seasonal pattern of egg prices as given, he is faced with the problem of choosing a seasonal pattern that maximizes his net profits. It would be an oversimplification to say that the cost of egg production varied from month to month. It might be more realistic to regard the egg producer as being confronted with a number of alternative activities or processes for producing eggs, each requiring a different seasonal pattern of labor inputs, feed purchases, short-term credit costs, and costs of buildings and equipment. From the complete range of activities available and the limitations upon his labor supply, his line of credit, and his aversion to debt (or preference for liquidity), the "rational" egg producer would be viewed as selecting that level and seasonal pattern of production which maximized his expected profits given the expected seasonal pattern of egg prices.

Changes in the array of production techniques available set forces in motion which tend to rearrange the seasonal pattern of egg production. If the rest of the demand and storage cost mechanism remains as indicated in our example, these changes in the seasonal pattern of production will be associated with changes in the seasonal patterns of prices, consumption, and storage activity.

A.1.2 *Structural Analysis of Trends*

The trend in a particular variable Y may often be separable into the net effects of trends in certain other variables $X_1, X_2, \ldots X_m$. If none of the X_i's are subject to influence by the firm or agency interested in explaining the trend in Y, the problem of projecting or predicting Y in the future becomes one of projecting or predicting the time paths of each of the relevant X_i's. Estimation of trends might take the form either of statistical regression analysis or of a synthesis of engineering relationships and/or institutional relationships *with* regression analysis.

A. *Input-Output Relationships and Trend Analysis*

The input-output mechanism often helps to explain trends in aggregative production variables which are too highly intercorrelated to be separated by regression analysis. Consider the following model in which the gross outputs X_1 and X_2 are expressed as linear functions of the deliveries to final demand, F_1 and F_2:

$$X_1 = 1.8F_1 + 0.1F_2 \tag{A.1}$$

$$X_2 = 0.4F_1 + 1.6F_2 \tag{A.2}$$

In an expanding economy, both categories of final demand, F_1 and F_2, tend to increase strongly. Through the input-output mechanism, a unit increase in F_1 tends to raise X_1 by 1.8 units and X_2 by 0.4 units. Similarly, a unit increase in F_2 tends to raise X_1 by 0.1 units and X_2 by 1.6 units.

In the following tabulation, F_1 and F_2 both follow exactly linear trends and are therefore perfectly correlated:

Year	F_1	F_2	$X_{1.1} + X_{1.2} = X_1$	$X_{2.1} + X_{2.2} = X_2$
0	10	20	$(18.0 + 2.0) = 20.0$	$(4.0 + 32.0) = 36.0$
1	11	21	$(19.8 + 2.1) = 21.9$	$(4.4 + 33.6) = 38.0$
2	12	22	$(21.6 + 2.2) = 23.8$	$(4.8 + 35.2) = 40.0$
3	13	23	$(23.4 + 2.3) = 25.7$	$(5.2 + 36.8) = 42.0$
4	14	24	$(25.2 + 2.4) = 27.6$	$(5.6 + 38.4) = 44.0$
5	15	25	$(27.0 + 2.5) = 29.5$	$(6.0 + 40.0) = 46.0$

The input-output mechanism causes X_1 to move according to the sum of two linear trends, $X_1 = X_{1.1} + X_{1.2}$, where $X_{1.1} = 1.8F_1$ and $X_{1.2} = 0.1F_2$. Also, X_2 moves as the sum of the two linear trends, $X_{2.1} = 0.4F_1$ and $X_{2.2} = 1.6F_2$. The sum of the two linear trend components of X_1 is a linear trend; the same is true in the case of X_2. The result is that X_1, X_2, F_1, and F_2 are *all* perfectly correlated; it is impossible to separate out the net effects of F_1 and F_2 upon X_1 and X_2, respectively, by any statistical technique!

If F_1 and F_2 have short-run deviations, f_1 and f_2, respectively, from their trends and f_1 and f_2 are not perfectly correlated, these deviations will be propagated through the input-output mechanism of the economy into X_1 and X_2. The expected deviation in X_1 will be $x_1 = 1.8f_1 + 0.1f_2$, and the expected deviation in x_2 will be $x_2 = 0.4f_1 + 1.6f_2$. The deviations in X_1 and X_2 will not be perfectly correlated with each other or with either f_1 or f_2. If there are errors of measurement or aggregation in all variables, the regression of x_1 on f_1 and f_2 may give a good approximation to the input-output coefficient connecting x_1 to f_1, but the net regression coefficient of x_1 upon f_1 might appear to be of borderline significance.

In the conventional input-output model, the deliveries to final demand, F_1 and F_2, are independent or exogenous variables and X_1 and X_2 are dependent or endogenous. Equations (A.1) and (A.2) permit us to make conditional forecasts of X_1 and X_2 *given* F_1 and F_2. If we were free to choose sets of values of F_1 and F_2 which were completely uncorrelated with one

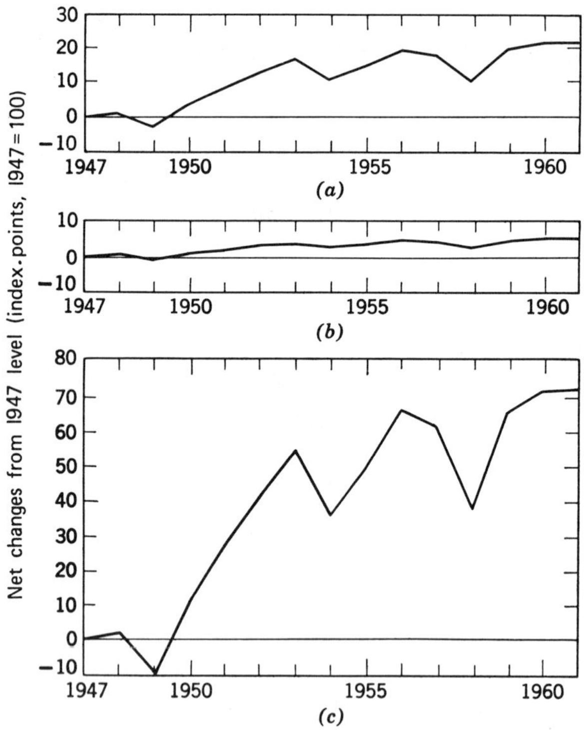

Figure A.4 Net interindustry effects of changes in production of machinery upon required production of steel and electric power, 1947–1961. (*a*) Effects on steel. (*b*) Effects on electric power. (*c*) Machinery production.

another and which covered wide ranges, we could regress the resulting values of X_1 and X_2 on the assigned values of F_1 and F_2 and obtain quite good estimates of all four coefficients along with their standard errors.

Figure A.4 illustrates the net input-output effects of changes in production in the machinery and equipment industry in the United States upon required production of steel and electric power over the 1947–1961 period, *assuming* (1) that the value of purchases of steel and electric power per dollar of machinery sales remain constant at the 1947 level and (2) that the complete structure of technological relationships connecting other industries in the economy either remains as in 1947 or that changes tend to be offsetting and neutral with respect to the interactions between machinery production, steel, and electric power.

In 1947, the machinery industry purchased about 0.12 dollars worth of steel and 0.0047 dollars worth of electric power per dollar of machinery output or sales. The machinery industry used about 1 percent of total electric power production in 1947.

If, for the sake of illustration, we regard the production of machinery as an exogenous or independent variable, the direct and indirect effects of an increase of one index point (1947 = 100) in machinery output would require an increase of about 0.3 index points in steel production and perhaps 0.07 index points in the production of electric power. (These figures are not precise, but the orders of magnitude as of 1947 are approximated from the 1947 interindustry relations study for the United States.)

On these assumptions, the time path of machinery production in Figure A.4*c* would have accounted for an increase of 20 points in the index of steel production and five points in the index of electric power production between 1947 and 1961. If each of the two series in Figure A.4*a* and A.4*b* were regressed on the index of machinery production in A.4*c*, all observations would fall on the regression lines, the squared correlation coefficients would equal 1, and the slopes of the regression lines would be 0.3 for steel and 0.07 for electric power; the constant terms would be 0 in each instance.

B. *Trends in Population*

One of the most important variables in economic projections and trend analysis is population. For a large country such as the United States with a limited amount of immigration and emigration, the general inclination of economists is to treat population growth as an exogenous, noncontrollable factor. In a broad sense, however, human population is often an endogenous or dependent variable influenced by political and economic conditions.

Three "population pyramids" or percentage distributions by age and sex are shown in Figure A.5. A striking feature of the United States population pyramid in 1960 was the fact that the numbers in the age groups 20–24 and

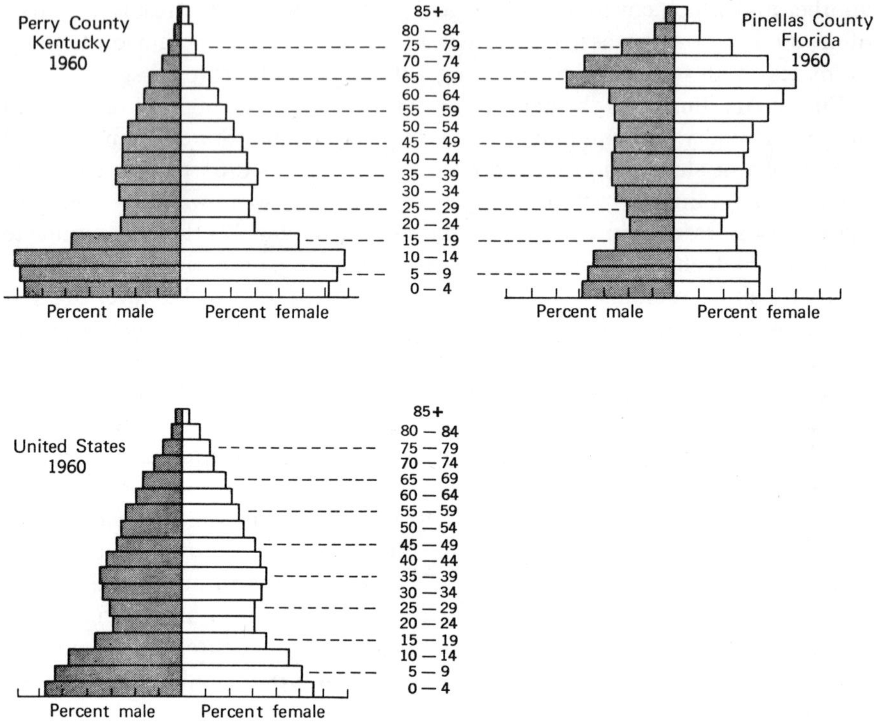

Figure A.5 Population pyramids for the United States and selected counties in 1960. (*Source*. Bureau of the Census, U.S. Census of Population: 1960, *General Populations Characteristics*. Washington: U.S. Government Printing Office, 1961).

25–29 were smaller than the numbers in the age groups 30–34, 35–39, and 40–44. Working backward, the proximate cause is found in the fact that the number of babies born in the 1930's was smaller than the number born in each of the two preceding decades. This was largely a result of the depressed economic circumstances of the 1930's during which many adults were unemployed and many young people delayed marrying and starting families because of the difficulties and uncertainties of finding jobs.

Population changes in small subdivisions of a country are highly endogenous. Projections for a particular county must take into account the array of economic and other forces which are tending to redistribute jobs and people among areas. Perry County, Kentucky, shows evidence of a very high birth rate and a very high rate of out-migration. Apparently, job opportunities in Perry County have not expanded rapidly enough to accommodate (on a net basis) more than half of their young people as they reached working age. Pinellas County, Florida, evidently serves in part as a retirement area for elderly people. The number of persons aged 65–69 is larger than the number in any other five-year group.

Thus, in forecasting population, routine and empirical methods may prove to be exceedingly unreliable. Only when we understand the economic and other variables that influence people in their decisions do we begin to find satisfying explanations for past population trends or convincing support for future projections.

C. *Trends in the Stock of Capital and Production Capacity*

Changes in the stock of capital (plant and equipment) result from the difference between gross investment (new construction and installation of new machines and equipment) and depreciation plus other forms of capital consumption.

The technological characteristics and efficiency of machinery produced in 1960 are different and generally better than those of machinery produced in 1950. Therefore, the age distribution of the nation's stock of machinery and equipment is relevant to projections of (1) total productive capacity and (2) the time path of investment demand. For example, a high rate of investment in new machinery during one five-year period might be followed by a low rate in the succeeding five years and an increased rate in the third five-year period owing to a bunching of replacement demand.

Analogously, young people entering the labor force in the early 1970's have different kinds of educations (at least in matters of emphasis and detail) than workers educated in previous decades and have different implications for the potential gross national product at full employment.

D. *Growth Models and Decision Models—Trends or Targets?*

Since World War II, governments in most democratic nations have assumed increased responsibilities for economic growth as well as for economic stability. Most modern economies are "mixed economies" in the sense that government agencies and enterprises (schools, post offices, the defense establishment, and so on) account for a significant fraction of production and final demand. Moreover, many of the activities of private firms and individuals operate within constraints (tax policies, antitrust laws, pure food and drug laws, and others) which are subject to change by legislative or administrative bodies. As awareness of the importance of government actions for the behavior of the private economy grows and as measurement of structural relationships in the economy becomes more dependable, there will be a greater tendency for public bodies to consider the effects of alternative policies upon the probable future course of the private economy.

A knowledge of the structure of the private economy would enable us to project economic trends in it under the assumption that government policies are unchanged or are changed in specified ways. In contrast, *unconditional forecasts* of the values of certain economic variables in 1975 would also imply unconditional forecasts of the actions that will be undertaken by government and the degree of success of these actions in achieving their intended objectives.

In particular, many countries and groups of countries are concerned with increasing their rates of economic growth and are evaluating their economic policies in relation to this objective. Logically, we may think of a regression equation connecting the national income or the gross national product in future years with (1) each variable which might be used directly or indirectly as a policy instrument and (2) each of the more important noncontrollable factors.

The implementation of realistic means for accelerating economic growth in any country entails some economic and/or social costs in addition to benefits. This implies difficult problems of measurement to which economists should nevertheless address themselves, and even more difficult problems of obtaining political and social consensus as to the values of different targets and the full costs of the proposed means of implementation.

A.1.3 Structural Analysis of Cyclical Fluctuations

The input-output or interindustry relations effects underlying Figure A.4 help to explain cyclical fluctuations as well as trends in economic time series of production. During the 1950's and 1960's, economists have given increasing attention to the analysis of inventory fluctuations in individual firms and industries, and for the national economy as a whole. If there are random or irregular variations in the final demand for a product, retailers, wholesalers, and manufacturers of the product respond to these changes partly through offsetting fluctuations in inventories. During a business recession, however, partly as a result of technical input-output relationships, the final sales of many firms in many industries level off or decline simultaneously.

If many firms have as one of their policy targets the maintenance of a desired ratio of inventories to sales, a decline in final sales will be followed by a larger absolute decline in factory production. In effect, the "market" for inventories becomes supersaturated. When sales increase again, the desired level of inventories also rises and production is increased sufficiently to meet the increased final demand and also the desired increase in inventories.

The following tables illustrate a few important aspects of the structure of the United States economy which influence the cyclical patterns of major macroeconomic variables.

Table A.2 indicates that production of durable manufactures declined by an average of 15 percent in the three recessions while nondurable manufactures declined only 5 percent. If we make a special grouping of the industrial

Table A.2 Industrial Production: Changes in Major Groups of Industries from Peak to Trough and Trough to Recovery During Three Cyclical Episodes, United States, 1948–1958

Industry Group[b]	Average Percent Changes: Peak to Trough (3 Cycles) in Percent[a]
Industrial production total	−11.0
Durable manufactures	−15.2
Nondurable manufactures	−5.0
Minerals	−12.1
Manufactures: special groupings	
Metal manufactures	−16.8
Consumer nondurables	−3.8
Miscellaneous manufactures	−10.4

Sources. Computed from basic data in *Business Statistics*, 1957, Biennial Edition (a supplement to the *Survey of Current Business*); also various issues of *Survey of Current Business* and *Federal Reserve Bulletin.*

[a] Averages for the three special groups of manufacturing industries are based on simple arithmetic averages of published index numbers for 5, 6, and 9 industries, respectively; figures for the "total," "durable manufactures," "nondurable manufactures," and "minerals" indexes are as officially published.

[a] Relative importance of the components may be estimated from 1956 figures on "production and related workers" employed: all manufacturing industries, 13.2 million; metal manufactures, 5.6 million; consumer nondurables, 4.2 million; and miscellaneous manufactures, 3.4 million. (About 0.8 million wage and salary workers were engaged in minerals production.)

production components into metal manufactures, miscellaneous manufactures, and manufactures of consumer nondurable goods, we find that the average decline in the production of metal manufactures was more than four times as large percentagewise as the average decline in the output of consumer nondurables. This latter fact tallies with the relative stability of personal consumption expenditures during these recessions primarily as a result of various built-in stabilizers and patterns of corporate behavior as well as some special counter-cyclical actions by the government.

Table A.3 shows the absolute and percentage changes in nonagricultural employment from peak to trough in each of the three recessions, by broad industry groups. On a net basis, the declines were concentrated in manufacturing and predominantly in the durable goods industries.

Table A.4 reflects the interaction of cyclical variations and trends in employment in major industry groups. Thus, employment in manufacturing,

Table A.3 Nonagricultural Employment: Absolute and Percentage Changes from Peak to Trough in Three Recessions, by Industry Group, United States, 1948–1958[a]

Industry Group	Actual Employment August 1957	Absolute Changes (Thousands)				Percent Changes			
		Average Three Recessions	November 1948 to July 1949	July 1953 to July 1954	August 1957 to April 1958	Average Three Recessions	November 1948 to July 1949	July 1953 to July 1954	August 1957 to April 1958
Total[b]	52,457	−1,948	−1,692	−1,749	−2,403	−4.0	−3.8	−3.5	−4.6
Manufacturing	16,826	−1,655	−1,629	−1,753	−1,583	−10.0	−10.6	−10.0	−9.4
Durable-goods industries	9,863	−1,248	−1,070	−1,376	−1,297	−13.2	−12.9	−13.4	−13.2
Nondurable-goods industries	6,963	−392	−514	−377	−286	−5.5	−7.3	−5.2	−4.1
Mining	820	−78	−69	−79	−97	−9.0	−5.9	−9.3	−11.8
Contract construction	2,805	−97	−114	3	−181	−3.8	−5.1	0.1	−6.5
Transportation and public utilities	4,184	−249	−202	−250	−294	−5.9	−4.8	−5.9	−7.0
Wholesale and retail trade	11,402	−173	−132	−35	−352	−1.6	−1.4	−0.3	−3.1
Finance, insurance, and real estate	2,354	32	17	78	2	1.6	1.0	3.8	0.1
Service and miscellaneous	6,477	0	−5	130	−125	0.1	−0.1	2.3	−1.9
Government	7,694	123	90	157	122	1.6	1.2	2.0	1.6

Sources. Survey of Current Business, selected monthly issues, 1949 to 1958: *Business Statistics*, 1957 Biennial Edition (a supplement to the *Survey of Current Business*), p. 59.

[a] Seasonally adjusted. Includes all full- and part-time wage and salary workers in nonagricultural establishments. Excludes proprietors, self-employed persons, domestic servants, and personnel of the armed forces.

[b] Annual figures for this series on "nonagricultural employment" have borne the following relationships to "total civilian employment" as published in *Monthly Reports on the Labor Force*, USDC: 1948, 74.9 percent; 1953, 80.2 percent; 1957, 80.2 percent; average of the three peak years, 78.4 percent. If the "total" percentage changes were adjusted to a "total civilian employment basis," the figures in this

Table A.4 Nonagricultural Employment: Actual Employment, Average Percentage Change from Peak to Trough, and Number of Recessions in Which Employment Decreased, Remained Stable, or Increased, by Industry Group, United States, 1948–1958

Industry Group	Actual Employment, August 1957 (Millions)	Average Change from Peak to Trough in Three Recessions (Percent)	Number of Recessions during which Employment: Decreased	Remained Stable[a]	Increased
Total	52.5	−4.0[b]	3	—	—
Manufacturing	16.8	−10.0	3	—	—
Mining	0.8	−9.0	3	—	—
Transportation and public utilities	4.2	−5.9	3	—	—
Contract construction	2.8	−3.8	2	1	—
Wholesale and retail trade	11.4	−1.6	2	1	—
Service and miscellaneous	6.5	0.1	1	1	1
Finance, insurance, and real estate	2.4	1.6	—	1	2
Government	7.7	1.6	—	—	3

[a] Changed less than 0.5 percent.

[b] Equivalent to −3.1 percent in terms of total civilian employment, including self-employed persons, persons engaged in agriculture, and certain other categories not included in "nonagricultural employment." Note, however, that total civilian employment had a long-run growth trend of slightly more than 1 percent a year, or approximately 0.8 percent during the average recession period of nine months from peak to trough.

mining, transportation, and public utilities decreased during each of the three recessions. Employment in contract construction and wholesale and retail trade decreased slightly in two of the recessions and remained stable in one. Employment in service and miscellaneous industries showed extremely small changes in the three recessions; employment in finance, insurance, and real estate actually increased during two of them and government employment (state, local, and federal combined) increased during each of the three recessions.

Table A.5 shows for the 1953–1954 recession the relationship between regional variations in employment change and the degree of regional concentration of employment in the durable goods manufacturing industries. In this recession, the decline in employment in durable goods manufactures

Table A.5 Nonagricultural Employment: Regional Variations in Employment Change Related to Degree of Regional Concentration in Durable-Goods Industries, United States, July 1953 to July 1954

Geographic Division	(1) Percent of Nonagricultural Workers Employed in Durable-Goods Industries, July 1953	(2) Percent Change in Total Nonagricultural Employment, July 1953 to July 1954	(3) Percent Change in Total Nonagricultural Employment, March–April 1940 to March 1953
United States	16.8[a]	−3.7[ab]	69.2[a]
Geographic Divisions			
East North Central	32.8	−7.0	69.4
New England	23.1	−4.6	41.7
Middle Atlantic	20.3	−4.4	48.3
Pacific	18.6	−3.1	98.9
East South Central	14.5	−3.7	59.7
West North Central	13.4	−2.5	61.0
South Atlantic	10.8	−3.1	66.7
West South Central	9.8	−1.9	85.8
Mountain	7.8	−3.0	90.9

Source. Based on data prepared by Edmond L. Kanwit in connection with his article, "Patterns of Recent Employment Changes—Area and National," in *Survey of Current Business*, June 1955, pp. 15–20.

[a] Simple arithmetic averages of regional figures.

[b] Simple regression of Column (2) upon Column (1) is as follows:

$$X_1 = -0.763 - \underset{(0.029)}{0.175} X_2 \; ; \quad r^2 = .84$$

accounted for nearly 80 percent of the decline in total nonagricultural employment. As might be expected, regions having a large proportion of their labor forces engaged in durable goods manufactures showed much larger than average percentage declines in total nonagricultural employment.

Although there have been some approximate regularities in the business cycles in the United States since 1945, the lengths of the cycles have varied considerably. Moreover, the Federal government is obligated under the Employment Act of 1946 to apply whatever instruments it reasonably can to maintain high levels of employment and production, to prolong the upswings of business cycles, to avert or postpone the next downturn, and if a recession begins, to moderate its amplitude and shorten its duration. In this context of an active and self-conscious economic stabilization policy, it seems unwise to place great emphasis on techniques of time series analysis which assume cycles of constant amplitude and duration or which deal with individual time series one at a time.

A.2 THE IMPLICATIONS OF A STRUCTURAL ANALYSIS OF TIME SERIES FOR MACROECONOMIC POLICY

The rapid advance of knowledge about the structure and functioning of a national economy has made it impossible for a modern government to ignore the effects of its own policies upon output, employment, and income in the private as well as the public sectors of the economy. A Dutch economist, Jan Tinbergen, formulated the nature of the relation of a modern government to the national economy in what appears to be a particularly fruitful way.

Tinbergen published his conceptualization of the nature of national economic policy in 1952 and more elaborately in 1956. Figure A.6 is an effort to summarize Tinbergen's approach in graphic form.

Tinbergen classified macroeconomic variables into four groups from the standpoint of an official responsible for national economic policy. In the first place, there are some variables which are beyond his power to influence. These must be taken simply as facts of life and as neither instruments nor targets of economic policy. It is important, however, for the policy maker to obtain the best forecasts he can of the future values of these *noncontrollable variables*. Prices of imports and the level of demand for exports from a given country would be examples of noncontrollable factors.

The policy maker also sees one set of variables as being the primary *targets* or objectives of economic policy. These variables enter into the policy maker's preference function, W, which is analogous to the indifference surface of an individual consumer. The policy maker is assumed to be capable of ranking different combinations of values of target variables such as the level of

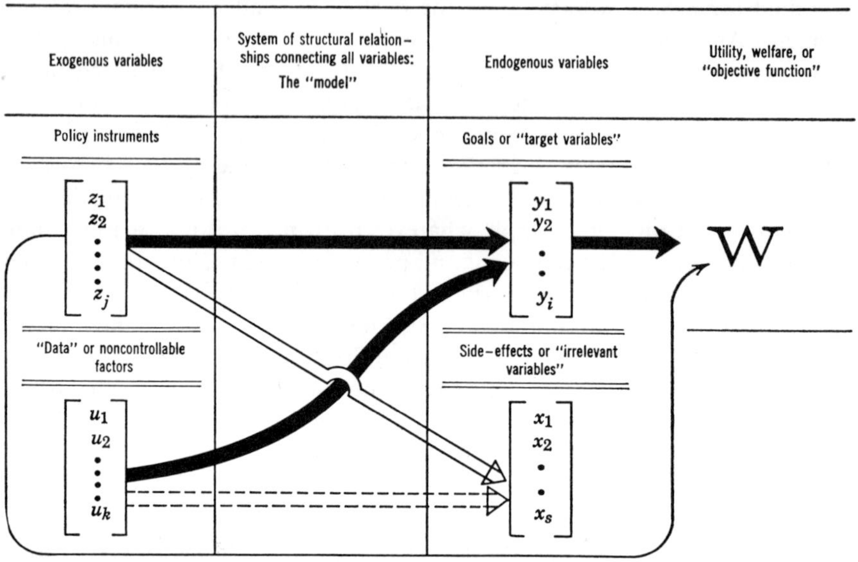

Figure A.6 The theory of economic policy. Classification of variables based on J. Tinbergen. "Data" are variables not subject to control by the policy maker or level of government that sets the goals and uses the policy instruments in question.

employment, the real gross national product, the rate of change in the consumer price level, and others and stating that, in his judgment, one combination of values is better than, as good as, or worse than other specified alternative combinations.

The third group of variables in Tinbergen's conceptualization consists of the *instruments* available to the policy maker for influencing the economy in the direction of the desired targets. These instruments may include the level of government purchases of goods and services, changes in tax rates, changes in the instruments of monetary policy (excess reserve requirements, rediscount rates, and the like), and many others. (The usability or practical availability of different instruments varies considerably from country to country and from one decade to another in a given country.)

There is a fourth group of economic variables which, for a particular policy maker at a particular time, are regarded as *irrelevant* or neutral in their implications for policy. Some of these may be regarded simply as byproducts of achieving the primary target variables—for example, if target levels are achieved for (1) the gross national product in constant dollars and (2) the price level or implicit price deflator for the gross national product, the *money value* of the gross national product is automatically determined as the product of the two target variables.

The essence of the problem of macroeconomic policy is this: to use the available instruments of economic policy in such a way as to achieve the desired levels of the target variables as closely as possible despite the fact that the time paths of the target variables are strongly influenced by factors which are, for the policy maker, noncontrollable. Formally, this problem has the same structure as does statistical decision theory.

This conceptualization of the policy maker as a decision maker and steersman for the economy by no means implies rigid or sweeping government controls over the economy and the individuals and firms in it. In selecting instruments such as a reduction in income tax rates or an acceleration of investment in a national highway network, one of the considerations would typically be the avoidance of needlessly detailed controls over individual firms or citizens.

The active role played by modern governments in promoting economic growth and reducing cyclical fluctuations inevitably affects the approach of economic statisticians to the analysis of the cyclical and trend components of time series. A lagging growth rate brings forth efforts to stimulate growth; forecasts of a recession bring forth efforts to avert it or at least to reduce its depth and shorten its duration. Univariate approaches to trend and cycle analysis are becoming increasingly vulnerable to the more and more sophisticated use of national income and product accounts and macroeconomic models as bases for government efforts to change trends and cycles.

EXERCISES

Section 11.1

1. The weekly sales of the Little Giant department store for the past two weeks are reported to be:

Week	*Gross Sales in Dollars*
1	\$20,000
2	\$28,000

Suppose that daily sales are an important variable in an economic model which you have developed but that the Little Giant department store is unable to supply the needed information on its daily sales. Further research reveals a recent study that summarizes the daily sales patterns of department stores open seven days a week as follows:

Day	*Index of Sales* (*Wednesday* = 100)
Sunday	120
Monday	80
Tuesday	90
Wednesday	100
Thursday	110
Friday	140
Saturday	160

(a) Assuming that no other information is available, how could the index of daily sales be used to estimate the daily sales of the Little Giant department store?

(b) Would the above index be of any use if the Little Giant department store were not open on Sunday? Explain.

2. A highway department survey records the number of automobiles passing a certain point each hour of the day during a one-month period. Let N_{it} represent the number of cars passing the point on day i during hour t where $i = 1, 2, \ldots, 30$ and $t = 1, 2, \ldots, 24$. Explain how the numbers N_{it} could be broken down into smaller parts and what factors may affect the parts.

Section 11.2

3. With which component of a time series would you associate each of the following?

(a) A strike by automobile workers.

(b) An increase in rice yields due to the development of new rice varieties.

(c) An increase in automobile sales in November.

(d) A decline in ice cream sales in June due to cold weather.

(e) A two-year decline in the construction of private homes.

4. Suppose that data were available on monthly sales of textbooks by the university book store during the past 10 years.

(a) Would you expect these data to show a definite trend? Explain why or why not.

(b) Would you expect these data to show a seasonal pattern? If so, what type of pattern and why?

(c) What factors might cause a cyclical variation in the monthly sales?

(d) What factors would tend to cause irregular variations in the data?

Section 11.3

5. A trend line fitted to data on production of iron (in tons) by a large steel company for the years 1920–1960 is

$$\hat{Y} = 15{,}000 + 125X \quad \text{(origin 1940, } X \text{ in years)}$$

(a) During the period 1920 to 1960, the company's production of iron increased on the average by (?) tons per year.
(b) The trend value for 1965 is (?).
(c) Shift the origin of the trend line to 1950.

6. Using the following data on total payroll of the aluminum castings industry:
(a) Construct a 3-year moving average.
(b) Construct a 4-year centered moving average.
(c) Graph the original data and the 3-year moving average on the same graph.

Year	Total Payroll of the Aluminum Castings Industry in Millions of Dollars
1958	159
1959	197
1960	198
1961	195
1962	229
1963	223
1964	241
1965	265
1966	301

Source. BDSA *Industrial Profiles*, 1958–1966 (Washington: U S. Department of Commerce, 1968).

7. Prove that, if every number in a time series is increased by a constant, say, k, the moving average is also increased by the amount k.

8. A sequence has 24 numbers.
(a) How many numbers will there be in a moving average of order 4?
(b) Of order 3?
(c) Of order N?

9. Using the following data on exports by ferrous foundries:
(a) Construct a graph showing exports.
(b) Draw a free-hand trend line.
(c) Compute a trend line using the method of semi-averages.
(d) Compute the least squares trend line $\hat{Y} = a + bX$ where Y is exports and X is years.
(e) Change the units of X in the trend line found in part (d) from annual units to quarterly units.
(f) Fit a parabolic trend line of the form $Y = a + bX + cX^2$ to the data.
(g) Which of the trend lines computed in (c), (d), and (f) above appear to fit the data best?

Year	Exports of Products by Ferrous Foundries (in Millions of Dollars)
1958	19
1959	19
1960	17
1961	16
1962	18
1963	30
1964	28
1965	46
1966	56
1967	53

Source. BDSA, *U.S. Industrial Outlook, 1968* (Washington: U.S. Department of Commerce, 1967.)

10. Show that a 2-year centered moving average is equivalent to a 3-year weighted moving average with weights 1, 2, 1, respectively.

Section 11.4

11. Suppose that, for some time series of quarterly values, the multiplicative model is correct, except that the pattern of seasonal fluctuation is changing over time.
   (a) Would a 4-quarter moving average still be a good indicator of the *TCI* component?
   (b) How would seasonal fluctuations be measured in this case?

12. In obtaining a seasonal index from a time series of monthly values by the ratio-to-moving average method, why do we require that the sum of the 12 index values equal 1200?

13. An ice cream manufacturer finds that the seasonal index of sales in January is 40 and 160 in June. How would you interpret these results?

14. Given the following time series:
   (a) Find the least squares trend line for these data.
   (b) Predict the trend value for this series for 1975.
   (c) Discuss the probable accuracy of prediction obtained in (b) above.

Year	*Y*
1964	9
1965	7
1966	10
1967	6
1968	5
1969	7

15. Two series showing adjusted and unadjusted sales of a department store for 1968 are shown below. The increase in sales from the third to the fourth quarter was:

(a) More than seasonally expected.
(b) Less than seasonally expected.
(c) About what you would expect.

Indexes of Store Sales

Quarter	Unadjusted	Seasonally Adjusted
1st	232	279
2nd	284	290
3rd	298	339
4th	399	302

Explain your conclusion.

16. The seasonal index of sales for the ABC Sales Company for January is 40. If actual sales in January 1969 were $120,000, what are seasonally adjusted sales equal to?

17. Employment in the agricultural chemicals industry increased from 10,000 in May to 15,000 in June. The seasonally adjusted data showed a decline from 10,000 to 7500.
(a) How do you explain the decline in this seasonally adjusted data?
(b) What is the seasonal index for June?

18. Suppose that you are working with monthly data for the period 1950 to 1968.
(a) Explain how you would adjust the original data to make allowances for the leap years during this period.
(b) Would the adjustments recommended in part (a) have a significant effect on a seasonal index computed from the basic data? Explain.

19. In constructing a typical seasonal index, it is better to compute the specific seasonals by dividing the original monthly figures by a moving average rather than by the value of a least-squares line for each month. Do you agree or disagree? Explain.

20. Why go to all the trouble of computing a seasonal index and using it to adjust the original data when a moving average of the original data also tends to be free of seasonal variation? Discuss.

21. After monthly sales have been expressed as a percentage of a centered 12-month moving average, the following results are obtained for a small potato chip factory.

Year	*May*	*June*	*July*
1965	122	105	109
1966	124	105	119
1967	121	95	101
1968	125	90	105
1969	120	100	91

What is the seasonal index for these three months?

Section 11.5

22. A 12-month moving average contains which of the following types of variation?
   (a) Seasonal variability.
   (b) Trend and irregular variations.
   (c) Cycle and irregular variations.
   (d) Cycle, trend, and seasonal variations.
   (e) Cycle, trend, and irregular variations.
   Explain you conclusion.

23. Which component of the following time series would probably be of most interest and how would you recommend estimating the component?
   (a) Annual population figures for the United States between 1900 and 1960.
   (b) Monthly sales of television sets in New Jersey between 1960 and 1970.
   (c) Quarterly figures on the number of persons employed by the steel industry between 1945 and 1965.
   (d) Average rainfall by month in Iowa between 1955 and 1968.
   (e) Annual sales of automobiles in the United States between 1945 and 1970.

SELECTED REFERENCES

Croxton, F. E. and D. J. Cowden, *Practical Business Statistics*, Prentice-Hall, Inc., Englewood Cliffs, N.J., 1948.

Davis, H. T., *The Analysis of Economic Time Series*, The Principia Press, Bloomington, Indiana, 1941.

Ferber, R. and P. J. Verdoorn, *Research Methods in Economics and Business*, The Macmillan Company, New York, 1962.

Fox, K. A., *Intermediate Economic Statistics*, John Wiley and Sons, Inc., New York, 1968.

Granger, C. W. J. and M. Hatanaka, *Spectral Analysis of Economic Time Series*, Princeton University Press, Princeton, 1964.

Organization for Economic Co-operation and Development, *Seasonal Adjustment on Electronic Computers*, Proceedings of an International Conference held in Paris; O.E.C.D., 1960.

CHAPTER 12

Identification, Causal Ordering, and Estimation in Economic Models

There are several clusters of problems which are frequently encountered in the construction of large-scale economic models. We shall discuss several of these problems in the following sections under the headings *identification*, *causal ordering*, and appropriate methods of *estimation* in multi-equation economic models. Although professional discussion on these problems has tended to focus on large-scale models of national economies, the problems are also common to the simpler models which have been discussed in this book. We shall discuss the identification problem with the aid of examples from demand, supply, and price structures for particular commodities. The basic principles can, of course, be generalized to other economic contexts.

12.1 THE IDENTIFICATION PROBLEM IN ECONOMIC MODELS

In Chapter 10 we indicated that multiple regression analysis can, in principle, be extended to include any number of *independent* variables. And, so far as the computations are concerned, any one of the variables may be placed in the dependent position. For example, we may express the retail price of beef as a function of the quantity of beef available for consumption, along with other variables. But we can also express the quantity of beef available for consumption (and purchased by consumers) as a function of the retail price of beef, along with the other variables in the preceding equation.

It may be argued that there is a logical inconsistency between these two equations. In the first, we use quantity consumed to "explain" retail price and, in the second, we use retail price to "explain" quantity consumed. If we are indeed interested in cause and effect relationships, one of these equations would seem to be preferable to the other.

We might also argue that there is a "structural" relationship between price and consumption—that they are in fact connected by a consumer demand curve. A demand curve is an exact mathematical construct. It is immaterial whether we write it as $P = a + bQ$ or as $Q = -a/b + (1/b)P$. We know, however, that if P and Q are statistical time series subject to the disturbances and measurement errors of the real world, our least-squares estimate of b obtained by regressing P on Q will, in general, not equal the reciprocal of the coefficient $(1/b)$ when the latter is estimated as the least-squares regression of Q on P.

If for reasons of logic or exposition we need to manipulate a "reversible" demand curve, we could choose either the regression of P on Q or the regression of Q on P and treat it as if it were in fact reversible. Or, we may choose as our reversible demand curve one with a slope somewhere between the two regression lines. For example, in the 1920's Henry Schultz used a reproducible mathematical procedure to arrive at reversible "weighted regressions" in demand analysis.[1]

It should be noted that any procedure by which we arrive at a reversible demand curve somewhere in between the two least-squares regressions implies that both variables are on the same logical footing—that is, both partake of the characteristics of dependent variables. If only the relation between price and consumption is logically required to be reversible, the remaining variables in the multiple regression equation may be treated as independent variables in the usual sense.

But if price and consumption are both logically dependent upon the remaining variables in the equation, we are faced with a contradiction. We have only one equation with which to determine the values of two variables! Logically, the values of retail price and consumer purchases in any time period represent the point of intersection of a consumer demand curve and a producer's (or dealer's) supply curve. If the independent variables in the equation we have considered thus far clearly belong to a consumer demand curve but not in a producer's supply curve, it appears that we should specify a supply curve with equal care, including in it variables (raw material costs, unit labor costs, and the like) which would influence the quantity that producers would offer for sale at a given retail price.

Now we are clearly beyond the domain of conventional least-squares methods. We need to estimate two different relationships between two jointly dependent variables, one relationship being a reversible demand curve and the other being a reversible supply curve (by "reversible" here, we simply

[1] Henry Schultz, *Statistical Laws of Supply and Demand*, (Chicago: The University of Chicago Press, 1928).

mean that in the demand curve the "slope" of consumption on price is the reciprocal of the "slope" of price on consumption; a similarly reciprocal relationship should exist between the price-quantity slope and the quantity-price slope in the supply curve).

The requirement that we estimate *two* relationships between price and consumption for the same time period raises a certain uneasiness. How will we be able to identify one regression coefficient (or similar measure) with the slope of the demand curve and another regression coefficient (or related measure) with the slope of the simultaneous supply curve? The answer is not self-evident. It will be elaborated in the following section.

12.1.1 The Identification Problem: (1) The Same Two Endogenous Variables in Each of Two Equations; No Exogenous Variables

Suppose we are simply given two time series on retail price, P, and consumption, Q, for a specified commodity. The P and Q values for each year are points of intersection of a demand curve and a supply curve. If we consider only three observations, we have three points on a supply curve; we also have three points on a demand curve. Two points are sufficient to determine a straight line; with three points we can also detect deviations from a simple (or at least a stable) straight-line relationship.

Figures 12.1*a* and 12.1*b* will clarify our estimation problem. If both the supply curve and the demand curve have remained absolutely fixed for three years, our three observations should be identical. In the diagrams, the "average" positions of both the supply curve and the demand curve pass through the point at which $P = 100$ and $Q = 100$.

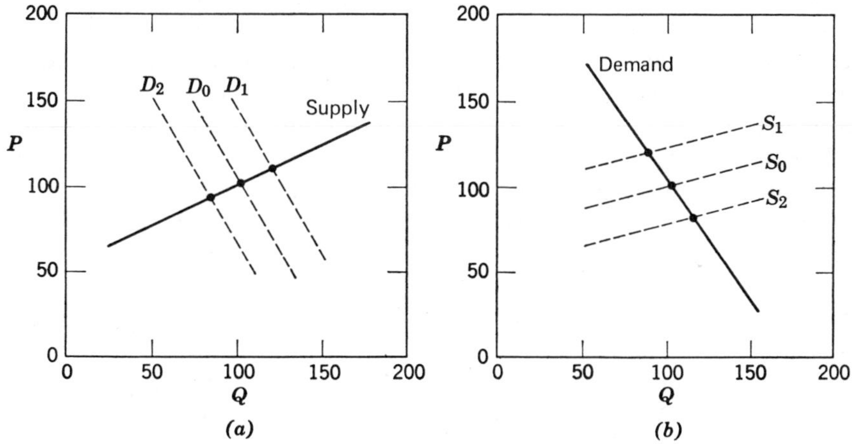

Figure 12.1 The identification problem: one function fixed, the other shifting. (*a*) Supply curve fixed, demand curve shifting. (*b*) Demand curve fixed, supply curve shifting.

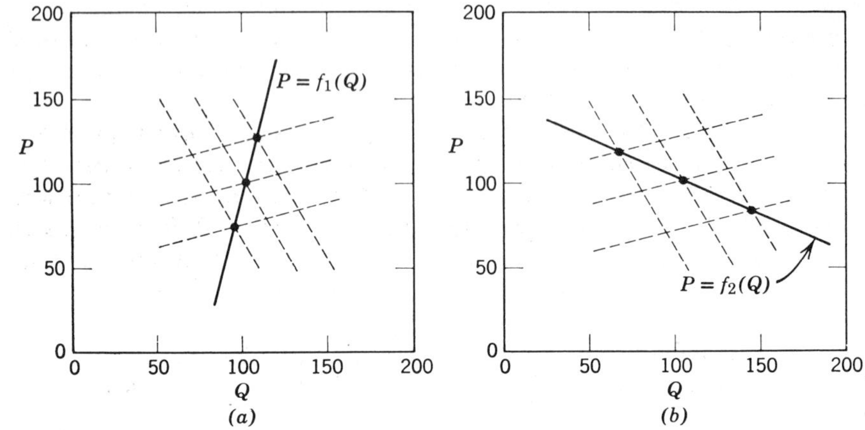

Figure 12.2 The identification problem: both functions shifting, with shifts in the two functions highly correlated. (*a*) Shifts positively correlated. (*b*) Shifts negatively correlated.

If the supply curve has remained fixed but the demand curve has shifted each year (perhaps because of changes in consumer income), our three observations will lie on the supply curve, as in Figure 12.1*a*. If, on the other hand, the demand curve has remained fixed for three years while the supply curve has shifted (perhaps as a result of fluctuations in the price of a major raw material), the three points will lie on the demand curve as in Figure 12.1*b*.

Clearly, we cannot obtain this information from the observations of P and Q alone. If the three observations fall along a positively sloping line, we may be encouraged to believe that we are approximating the supply curve. Conversely, if the three points indicate a negatively sloping relationship, we are encouraged to believe we are approximating the demand curve. Note, however, that if supply has remained stable, our observations will give us no information about the slope of the demand curve. And, if the demand curve has remained fixed, our observations give us no information about the slope of the supply curve.

In general, both the demand curve and the supply curve will have shifted from year to year. In Figure 12.2*a* we assume that the shifts are positively associated, with the highest position of the supply curve occurring in the same year as the highest position of the demand curve. The line joining these observations, $P = f_1(Q)$, has a steep positive slope. The positive sign suggests that we have an approximation to the supply curve, and the steepness of the slope suggests that supply must be highly inelastic. But it is not enough simply to have the correct sign. If a producer thought that other producers in the aggregate would make only negligible increases in output in response to a substantial increase in retail price, his own production and marketing strategy might go sorely awry.

Figure 12.2*b* shows the results of negatively correlated shifts in the two functions—that is, the highest position of the supply curve occurs in the same year as the lowest position of the demand curve and vice versa. In this case, the three observations lie on a negatively sloping line. The negative slope is characteristic of demand functions, so we may be led to believe that the line $P = f_2(Q)$ approximates the demand curve, and that demand is quite elastic (a 25 percent drop in price is associated with a 50 percent increase in consumption). But, in fact, the demand curve is moderately *inelastic* (a 25 percent decrease in retail price will result in only a 15 or 16 percent increase in consumption). If interpreted as *the* demand curve, $P = f_2(Q)$ suggests that total revenue could be increased considerably by reducing price when actually total revenue would be reduced by a price cut.[2]

The three observations in Figure 12.2*a* fall on a straight line because we have assumed a perfect positive linear association between disturbances in the demand and the supply functions. Similarly, the three points in Figure 12.2*b* lie on a straight line because we have assumed a perfect negative linear association between the disturbances. If the actual correlation between the disturbances is somewhere between 1 and -1, and if we assume that the demand curve and the supply curve can take only the three distinct positions indicated, any three of the nine points of intersection might turn up as the P and Q observations for the three years of our sample.

Three points could be selected from the total of nine points in 84 different combinations. Of these, three combinations would lie along a demand curve, three would lie along a supply curve, and 78 would yield regression lines which did not coincide with either a demand or a supply curve.

Two of these "nonstructural" regressions might be particularly attractive, since the three observations would lie on straight lines, implying perfect positive or negative correlation between the observed values of P and Q. These are the cases shown in Figure 12.2*a* and 12.2*b*. In the remaining 76 possible cases, the three observations would not lie on a straight line. If it were known that P and Q were estimated without error or with only negligible errors, the relatively low correlations between P and Q in many of the 76 cases would perhaps serve to warn us that all was not well.

The simple two-equation economic model represented in Figures 12.1 and 12.2 can be stated in algebraic form as follows:

(12.1) Demand: $$P = a_1 + b_1 Q + u$$

(12.2) Supply: $$P = a_2 + b_2 Q + v$$

[2] This would be true of an individual producer who had essentially a monopoly, within a limited price range, in a given trade area. Otherwise, if the demand function applies to the total sales of an industry including several or many firms, its inelasticity implies that an industry-wide price cut would reduce industry-wide gross revenue.

Assume we have a specific number of time series observations on P and Q from which we hope to derive a supply curve and a demand curve.[3] We can then express P and Q as deviations from their means as follows:

$$p = b_1 q + u \tag{12.3}$$

$$p = b_2 q + v \tag{12.4}$$

where $p = (P - \bar{P})$ and $q = (Q - \bar{Q})$.

As the two functions are shifted by the disturbances u and v, we may solve for p and q as functions of u and v, using Cramer's rule.

Write

$$p - b_1 q = u$$

$$p - b_2 q = v$$

Then

$$p = \frac{\begin{vmatrix} u & -b_1 \\ v & -b_2 \end{vmatrix}}{\begin{vmatrix} 1 & -b_1 \\ 1 & -b_2 \end{vmatrix}} = \frac{-b_2 u + b_1 v}{-b_2 + b_1} = \frac{b_1 v - b_2 u}{b_1 - b_2}$$

$$q = \frac{\begin{vmatrix} 1 & u \\ 1 & v \end{vmatrix}}{\begin{vmatrix} 1 & -b_1 \\ 1 & -b_2 \end{vmatrix}} = \frac{v - u}{-b_2 + b_1} = \frac{v - u}{b_1 - b_2}$$

Now, suppose we regress p on q by least squares. The formula for the least-squares regression coefficient of p on q is

$$B_{pq} = \frac{\sum pq}{\sum q^2} = \frac{\sum [(b_1 v - b_2 u)/(b_1 - b_2)][(v - u)/(b_1 - b_2)]}{\sum [(v - u)/(b_1 - b_2)]^2} \tag{12.5}$$

Hence, canceling out $1/(b_1 - b_2)^2$ in numerator and denominator, we obtain

$$B_{pq} = \frac{\sum (b_1 v^2 - b_1 uv - b_2 uv + b_2 u^2)}{\sum (v^2 - 2uv + u^2)}$$

The summations are taken over the number of observations to which the regression is fitted. Dividing both numerator and denominator by the number of observations, we obtain

$$B_{pq} = \frac{b_1 s_v^2 - (b_1 + b_2) r_{uv} s_u s_v + b_2 s_u^2}{s_v^2 - 2 r_{uv} s_u s_v + s_u^2} \tag{12.6}$$

[3] We shall use small letters frequently in this chapter to denote parameters. This will tend to simplify the notation and is consistent with much of the existing literature on the material to be covered.

From equation (12.6) we can see a number of things.[4] First, if the supply curve has not shifted, $s_v = 0$; two terms in both numerator and denominator vanish, and $B_{pq} = b_2 s_u^2/s_u^2$ or b_2. In other words, the least-squares regression coefficient is the slope of the supply curve as in Figure 12.1*a*.

Conversely, if the demand curve has not shifted, $s_u = 0$; two terms in both numerator and denominator vanish, and $B_{pq} = b_1 s_v^2/s_v^2$ or b_1. The least-squares regression coefficient in this case is the slope of the demand curve as in Figure 12.1*b*.

In brief, if we set out to estimate a demand curve and a supply curve with no more information than a series of paired observations on P and Q, the most likely outcome is that we do not obtain a good estimate of either curve; if we are lucky, we may get a good estimate of one, but in no case can we obtain good estimates of both curves.[5]

12.1.2 The Identification Problem: (2) Two Endogenous Variables in Each of Two Equations and a Different Exogenous Variable in Each Equation

There is a way forward, however, if the disturbances u and v are thought to result from fluctuations in certain other specific and quantifiable variables. For example, we might have a good basis in experience and in economic theory for thinking that a considerable portion of the disturbances in the demand function result from changes in the disposable income of consumers, Y. Also, we may know that a large proportion of the shifts in the supply function result from fluctuations in the cost of some major raw material, say, Z.

If we have time series on Y and Z, an obvious procedure is to rewrite Equations (12.1) and (12.2) as

$$\text{Demand:} \qquad P = a_1 + b_1 Q + c_1 Y + e_1$$

$$\text{Supply:} \qquad P = a_2 + b_2 Q + c_2 Z + e_2$$

We might then run the least-squares regression of P on Q and Y (or Q on P and Y) and take the resulting equations as approximations to the demand function. Similarly, we might run the least-squares regressions of P on Q and Z (or Q on P and Z) and take these equations as approximations to the supply function.

[4] The value of B_{pq} would not be changed if we divided both numerator and denominator by the number of observations *minus one*, so that s_u and s_v would be properly adjusted for degrees of freedom.

[5] We rule out the "cobweb" model in this statement and insist that the P and Q observations to be correlated refer to the same time interval—their values emerge simultaneously and not in sequence.

At this point, it will advance our discussion if we unveil the model by means of which the actual observations in Figures 12.1 and 12.2 were generated. The demand and supply equations are as follows:

$$\text{(12.7)} \qquad \text{Demand:} \qquad P = 260 - 1.6Q + 0.8(Y - 100)$$

$$\text{(12.8)} \qquad \text{Supply:} \qquad P = 70 + 0.3Q + 0.6(Z - 100)$$

Thus, the disturbances in Equations (12.1) and (12.2) are

$$\text{(12.9)} \qquad u = 0.8(Y - 100)$$

and

$$\text{(12.10)} \qquad v = 0.6(Z - 100)$$

Each unit change in Y above or below 100 changes the constant term a_1 of the demand curve $P = a_1 + b_1 Q$ by 0.8 units. Similarly, each change of one unit in Z changes the constant term a_2 in the supply curve $P = a_2 + b_2 Q$ by 0.6 units.

Knowing the complete model of equations (12.7) and (12.8), we can generate the observations in Figure 12.2*a* in the following manner:

Y	*Demand Curve*	*Z*	*Supply Curve*
140	$P = 292 - 1.6Q$	140	$P = 94 + 0.3Q$
100	$P = 260 - 1.6Q$	100	$P = 70 + 0.3Q$
60	$P = 228 - 1.6Q$	60	$P = 46 + 0.3Q$

Using Cramer's rule, we solve for P and Q for each pair of values of Y and Z (we first transpose $-1.6Q$ and $0.3Q$ to the left sides of the equations and change their signs):

$Y = Z = 140$

$$P = \frac{\begin{vmatrix} 292 & 1.6 \\ 94 & -0.3 \end{vmatrix}}{\begin{vmatrix} 1 & 1.6 \\ 1 & -0.3 \end{vmatrix}} = \frac{-87.6 - 150.4}{-0.3 - 1.6} = \frac{238.0}{1.9} = \underline{125.3}$$

$$Q = \frac{\begin{vmatrix} 1 & 292 \\ 1 & 94 \end{vmatrix}}{\begin{vmatrix} 1 & 1.6 \\ 1 & -0.3 \end{vmatrix}} = \frac{94 - 292}{-1.9} = \frac{198}{1.9} = \underline{104.2}$$

$Y = Z = 60$

$$P = \frac{\begin{vmatrix} 228 & 1.6 \\ 46 & -0.3 \end{vmatrix}}{\begin{vmatrix} 1 & 1.6 \\ 1 & -0.3 \end{vmatrix}} = \frac{-68.4 - 73.6}{-1.9} = \frac{142}{1.9} = \underline{74.7}$$

$$Q = \frac{\begin{vmatrix} 1 & 228 \\ 1 & 46 \end{vmatrix}}{\begin{vmatrix} 1 & 1.6 \\ 1 & -0.3 \end{vmatrix}} = \frac{46 - 228}{-1.9} = \frac{182}{1.9} = \underline{95.8}$$

When $Y = Z = 100$, it is clear by inspection of the equations $P = 260 - 1.6Q$ and $P = 70 + 0.3Q$ that they are satisfied by $P = 100$, $Q = 100$.

The slope of the line connecting the three points just derived is

$$b_{f1} = \frac{125.3 - 100}{104.2 - 100} = \frac{25.3}{4.2} = 6$$

and the full equation is $P - 100 = 6(Q - 100) = 6Q - 600$, or $P = -500 + 6Q$, which bears little resemblance to either the true supply curve or the true demand curve.

The *range* of u is 64 and the *range* of v is 48. We could divide both the numerator and the denominator of Equation (12.6) by s_v^2 without affecting B_{pq}; i.e., the ratio of s_u/s_v is the important factor, not their absolute magnitudes. In the present case, $s_u^2 = [(32)^2 + 0 + (-32)^2]/(3 - 1)$ and $s_v^2 = [(24)^2 + 0 + (-24)^2]/(3 - 1)$, so $s_u = 32$ and $s_v = 24$. The ratiois 4/3, so we will take $s_u = 4$ and $s_v = 3$ when substituting into equation (12.6):

$$B_{pq} = \frac{-1.6(9) - (-1.6 + 0.3)(1)(12) + 0.3(16)}{9 - 2(1)(12) + 16} \tag{12.11}$$

$$= \frac{-14.4 + 15.6 + 4.8}{25 - 24} = \frac{1.2 + 4.8}{1} = 6$$

This checks with the slope already obtained for $P = f_1(Q)$ and confirms the algebra underlying equation (12.9).

If $r = -1$ but b_1, b_2, s_u, and s_v are unchanged, equation (12.11) becomes

$$B_{pq} = \frac{-14.4 - 15.6 + 4.8}{25 + 24} = \frac{-25.2}{49} = -0.5143 \tag{12.12}$$

Algebraic generation of the three observations in Figure 12.2*b* gives us approximately $B_{pq} = (115.2 - 100)/(70.5 - 100) = 15.2/-29.5 = -0.515$, the slight discrepancy from -0.5143 resulting from rounding errors. This is only about one-third as steep as the slope of the true demand curve.

12.1.3 Identification Through Experimental Control of One or More Variables

If we were free to experiment, we could readily solve the identification problem. On the one hand, we could vary the price, P, in a number of stores according to a carefully planned experimental design (we would have to "partial out" the effects of income too), and obtain estimates of the consumer demand function as a regression (simple or partial) upon the controlled retail price. The supply curve of producers would simply not be relevant for this experiment. As a practical matter, it would be hard to carry out a corresponding experiment with producers, as their sophistication would quickly tell them that something peculiar was going on. *Possibly* interviews could be conducted with a number of producers who would enter into the spirit of the inquiry and tell us quite accurately how they would change their outputs in response to specified changes in retail price and the cost of their principal raw material (P and Z, respectively). There would be no way for consumers to influence the results of this hypothetical experiment in production response.

Wold and Strotz (1960) have argued that, in principle, all identification problems involving the behaviors of two groups of economic agents can be eliminated if we measure our prices, quantities, and other relevant variables in terms of the appropriate time unit. For example, suppose that retailers change the price of a perishable commodity to consumers once a week. If consumers generally use up any particular purchase of the commodity within the same week, accurate weekly observations on the prices charged by producers and the quantities purchased by consumers should enable us to estimate the consumer demand curve.

Similarly, dealers may look at their raw material costs and sales for the preceding week in determining the price they will charge during the current week. In each case, that of consumers and that of producers, if we choose the proper time unit and make observations on the proper variables (including time lags, where they exist), we should be able to make a clean-cut identification of the consumer demand functions and the producer supply functions. Wold and Strotz have argued that the kind of identification problem represented in Figures 12.1 and 12.2 is the result of "aggregation over time," so that several supply response cycles (with coefficient b_2) and several demand response cycles (with coefficient b_1) are averaged together in the annual or quarterly observations for P and Q.

12.1.4 Estimation by the Method of Reduced Forms

The identification problem was formulated quite clearly by E. J. Working in 1927. Ezekiel (1928) demonstrated that this problem could be solved for most agricultural products because this year's supply was determined by *previous* prices and not by the current year's price.

However, it remained for Haavelmo (1943) to generalize the identification problem and to develop a method based on modern probability theory for solving it in a good many situations. In the situation which Haavelmo characterized as the "just-identified" case, the solution is simple and precise but at the same time most ingenious. The procedure is called the method of *reduced forms.*

Equations (12.7) and (12.8) can be rewritten, incorporating 0.8 (−100) and 0.6 (−100) into their respective constant terms:

(12.13) Demand: $P = 180 - 1.6Q + 0.8Y$

(12.14) Supply: $P = 10 + 0.3Q + 0.6Z$

We will arbitrarily transpose Q into the "dependent" position for the ensuing discussion:

(12.15) Demand: $Q = 112.5 - 0.625P + 0.5Y$

(12.16) Supply: $Q = -33.3 + 3.33P - 2Z$

For notational convenience, we will write the coefficients of (12.15) as $a_1 = 112.5$, $b_1 = -0.625$, and $c_1 = 0.5$, and those of (12.16) as $a_2 = -33.3$, $b_2 = 3.33$, and $c_2 = -2$.

Note that both equations are exact functions, with no disturbances or errors, because the mysterious disturbances of the two-variable equations (12.1) and (12.2) have been assumed to be exact functions of Y and Z.

There is a logical difference between Y and Z on the one hand and P and Q on the other. The values of Y and Z are determined independently of or prior to P and Q in a causal sense. If we *could* directly control the values of P and Q, we would find that they had no discernible effects upon the current period values of Y and Z. We may call Y and Z *exogenous* variables in the present model, meaning that their values are determined *outside of* the model.

In contrast, we may call P and Q *endogenous* variables, as their values are determined *inside of* the model; P and Q are *jointly dependent* upon the values of the exogenous variables Y and Z.

From the mathematical standpoint, the values of Y and Z in any time period can be taken as given numbers; the two equations are then logically sufficient to determine the values of the two jointly dependent variables, P and Q. It is quite possible to solve equations (12.15) and (12.16) for P and Q as functions of Y and Z; we shall do this on the intuition that this procedure will throw some light on the identification problem.

For greater generality, we will deal with the algebraic forms of (12.15) and (12.16), namely,

$$\text{(12.17)} \qquad \text{Demand:} \qquad Q = a_1 + b_1 P + c_1 Y$$

$$\text{(12.18)} \qquad \text{Supply:} \qquad Q = a_2 + b_2 P + c_2 Z$$

We bring $b_1 P$ and $b_2 P$ to the left sides of the equations and apply Cramer's rule to

$$Q - b_1 P = a_1 + c_1 Y + 0.Z$$
$$Q - b_2 P = a_2 + 0.Y + c_2 Z$$

the zeros being inserted for formal reasons. Then

$$Q = \frac{\begin{vmatrix} a_1 + c_1 Y + 0.Z & -b_1 \\ a_2 + 0.Y + c_2 Z & -b_2 \end{vmatrix}}{\begin{vmatrix} 1 & -b_1 \\ 1 & -b_2 \end{vmatrix}} = \frac{(-a_1 b_2 - c_1 b_2 Y) + a_2 b_1 + c_2 b_1 Z}{-b_2 + b_1}$$

or

$$\text{(12.19)} \qquad Q = \left(\frac{-a_1 b_2 + a_2 b_1}{b_1 - b_2}\right) - \left(\frac{b_2 c_1}{b_1 - b_2}\right) Y + \left(\frac{b_1 c_2}{b_1 - b_2}\right) Z$$

Also,

$$P = \frac{\begin{vmatrix} 1 & a_1 + c_1 Y + 0.Z \\ 1 & a_2 + 0.Y + c_2 Z \end{vmatrix}}{b_1 - b_2} = \frac{a_2 + c_2 Z - a_1 - c_1 Y}{b_1 - b_2}$$

or

$$\text{(12.20)} \qquad P = \left(\frac{a_2 - a_1}{b_1 - b_2}\right) - \left(\frac{c_1}{b_1 - b_2}\right) Y + \left(\frac{c_2}{b_1 - b_2}\right) Z$$

Equations (12.19) and (12.20) are known as *reduced form equations* or, taken together, as the *reduced form* of the model represented by (12.17) and (12.18). Each reduced form equation contains a single endogenous variable, so there is no logical reason why we should not estimate the coefficients of each equation separately by least squares. Since (in the present illustration) P and Q are exact linear functions of Y and Z, each equation, when fitted to time series observations, will yield an R^2 of one and the standard errors of the regression coefficients will be zero.

For convenience, let us rewrite Equation (12.20) as

$$\text{(12.21)} \qquad P = A_1 + B_1 Y + C_1 Z$$

and equation (12.19) as

$$\text{(12.22)} \qquad Q = A_2 + B_2 Y + C_2 Z$$

where the coefficients of (12.21) and (12.22) are assumed to have been estimated by least squares. Statistically, these equations give us minimum variance estimates of P and Q. In a certain sense, they contain the maximum amount of information concerning the determination of the jointly dependent variables that can be extracted from the data.

The real payoff to our reduced form equations becomes apparent when we look more closely at the coefficients of equations (12.19) and (12.20). First, we note that $b_1 = C_2/C_1$, since

$$\frac{C_2}{C_1} = \frac{b_1 c_2/(b_1 - b_2)}{c_2/(b_1 - b_2)} = \frac{b_1 c_2}{c_2} = b_1$$

Next, we see that $b_2 = B_2/B_1$, since

$$\frac{B_2}{B_1} = \frac{b_2 c_1/(b_1 - b_2)}{c_1/(b_1 - b_2)} = \frac{b_2 c_1}{c_1} = b_2$$

These two coefficients give us the value of $(b_1 - b_2)$, which appears in the denominators of all the terms.

Now, $c_1 = B_1[-(b_1 - b_2)]$, and $c_2 = C_1(b_1 - b_2)$. Only a_1 and a_2 remain to be determined. From (12.19) and (12.20), respectively, we obtain

$$-a_1 b_2 + a_2 b_1 = A_2(b_1 - b_2) \tag{12.23}$$

and

$$-a_1 + a_2 = A_1(b_1 - b_2) \tag{12.24}$$

Multiplying the second equation by $-b_2$ and adding, we have

$$\begin{array}{rl} -a_1 b_2 + a_2 b_1 &= A_2(b_1 - b_2) \\ a_1 b_2 - a_2 b_2 &= A_1(b_1 - b_2)(-b_2) \\ \hline a_2(b_1 - b_2) &= (b_1 - b_2)(A_2 - b_2 A_1) \end{array} \tag{12.25}$$

so $a_2 = A_2 - b_2 A_1$. Finally, from equation (12.24), $a_1 = a_2 - A_1(b_1 - b_2)$. This completes the solution. From the numerical values of the least-squares regression coefficients A_1, B_1, C_1 and A_2, B_2, C_2, we have succeeded in deriving the numerical values of the structural coefficients of our demand curve (a_1, b_1, c_1) and of our supply curve (a_2, b_2, c_2)!

The demonstration that reduced form equations could be used to derive consistent and reversible estimates of structural relations between jointly dependent or endogenous variables was perhaps Haavelmo's major contribution. The procedure is by no means obvious to an economist who has grown up in the tradition of least-squares regression analysis. Although a knowledge of matrix algebra is not necessary in order to understand the relation of structural coefficients to reduced form coefficients in a two-equation model, matrix algebra is almost essential in generalizing the identification problem and the reduced forms method of estimation for models of three equations or more.

Furthermore, we have taken the simplest case, that in which the model represented by equations (12.7) and (12.8) is mathematically exact. In real situations, there would be some variation in P and Q which could not be explained by Y and Z. Also, suppose that Y and Z, though logically distinct entities, happened to be highly correlated during the sample period. It is clear from equations (12.21) and (12.22) that, in the usual case, the R^2's will be less than 1; the regression coefficients will have standard errors, and, if Y and Z are highly intercorrelated, these standard errors may be rather large. In other words, we do not escape the problems of intercorrelation or (in the extreme) multicollinearity by using the method of reduced forms.

Two other situations are possible with respect to identification in our two-equation model. In one of these, the model is said to be *underidentified.* This would be the case if, for example, the variable Z did not appear in the supply equation. In that case, Z would not appear in either of the reduced form equations. The coefficient b_2 could be estimated as the ratio of the coefficients of Y in equations (12.19) and (12.20) (the terms in Z being omitted). However, it would be impossible to determine the remaining coefficients, b_1, c_1, a_1, and a_2, from information provided by this revised model.

The *overidentified* case may be illustrated by adding another variable, W, to the demand equation (12.7) and leaving the supply equation (12.8) as it is, with P expressed as a function of Q and Z. In this case, a term in W (perhaps an index number of supplies or prices of competing commodities) would be added to both equations (12.19) and (12.20). Thus, if we add a term $D_1 W$ to equation (12.21) and $D_2 W$ to equation (12.22),

$$D_1 = \frac{d_1}{(b_1 - b_2)}$$

and

$$D_2 = \frac{b_2 d_1}{(b_1 - b_2)}$$

We can obtain an estimate of b_2 as follows:

$$b_2 = \frac{D_2}{D_1}$$

But we can also obtain an estimate of b_2 as

$$b_2 = \frac{C_2}{C_1}$$

Since the reduced form coefficients D_2, D_1, C_2, and C_1 are all subject to sampling errors and the two reduced form equations are estimated independently of one another, the two estimates of b_2 will usually differ. This ambivalence cannot be confined to the coefficient b_2 alone because the expression $(b_1 - b_2)$ constitutes the denominator of every other reduced form coefficient. Hence, some reproducible method is required for reconciling the two divergent sets of estimates and arriving at some rational compromise.

Intuitively, if the standard errors of the reduced form coefficients are fairly small, one would expect some kind of weighted average of the two sets of estimates for the coefficients to yield a reasonable solution. In models of three equations or more, overidentification may be of a multiple character; three or more alternative means of estimating the coefficients may be available. A reproducible method of arriving at a unique set of estimates is virtually indispensable, and such a procedure has in fact been devised.

Before going further, we should perhaps indicate the numerical values of the reduced form coefficients derived from our equations (12.15) and (12.16); they are

$$P = 36.864 + 0.12642Y + 0.50568Z \tag{12.26}$$

$$Q = 89.55 + 0.42225Y - 0.31605Z \tag{12.27}$$

These coefficients have been derived by means of logarithms and longhand arithmetic so there may be at least rounding errors. Referring to the detailed structure of the reduced form coefficients in equations (12.19) and (12.20), we calculate

$$b_1 = \frac{-0.31605}{0.50568} = -0.625$$

$$b_2 = \frac{0.42225}{0.12642} = 3.34$$

$$c_1 = 3.955(0.12642) = 0.5$$

$$c_2 = -3.955(0.50568) = -2$$

$$a_2 = 89.55 - 36.864(3.33) = -33.3$$

and

$$a_1 = -33.3 + 145.8 = 112.5$$

These values reproduce coefficient by coefficient the original values of the structural coefficients in equations (12.15) and (12.16).

In the overidentified case, the computations originally developed by Haavelmo and his colleagues in the 1940's are extremely laborious when carried out with nothing more than desk calculators. More recently, standard routines have been developed for carrying out these calculations on electronic computers. The computations in the overidentified case are more expensive than certain alternative methods, but not sufficiently so as to warrant avoiding the complex method if it seems to be indicated by the nature of an applied problem.

One of the simpler methods that has been proposed is the method of *two-stage least squares*. Assume that, for some logical reason, it seems best to us to write P in the dependent position in the demand function, $P = f(Q, Y)$. We regress Q upon Y and Z, which are all of the exogenous variables in the present model; this gives us the reduced form equation (12.19). The values of Q estimated from this equation, $\hat{Q}$, are exact functions of the exogenous variables Y and X. Therefore, the *statistical* properties of $\hat{Q}$ must be the statistical properties of the exogenous variables; the disturbances in the two equations (if we recognize the existence of disturbances in addition to the exogenous variables) are by hypothesis uncorrelated with Y and Z. At any rate, the time series of $\hat{Q}$ is substituted for the time series of the original Q's; we then calculate the least-squares regression of P on $\hat{Q}$ and Y as an estimate of our structural demand function.

Similarly, if it seems logical to put Q in the dependent position in our structural supply function, we may calculate the least-squares regression of P upon Y and Z, obtaining the estimate $\hat{P}$. We then calculate the least-squares regression of Q on $\hat{P}$ and Z, taking the result as an estimate of the structural supply curve.

Advocates of this method feel that it is equally applicable both to the just-identified and the over-identified equations of multiple equation models. A numerical example of two-stage least-squares estimation will be presented in a later section.

12.2 SOME MULTIPLE EQUATION MODELS IN WHICH LEAST-SQUARES ESTIMATION PERFORMS WELL

The results obtained using different methods to estimate multi-equation models suggest that true simultaneity is less prevalent or at least less damaging to least-squares estimation in macroeconomic models than one would expect, given the theoretical interrelationships among all commodities and decision makers in an economic system. They also suggest that an attempt to explain *why* least-squares estimation appears to work well in particular fields of economic analysis may be enlightening. We shall once again draw some examples from demand analysis.

Examples from the food and agricultural sectors are of interest because many of them show close approximations to the cobweb or recursive model, even in annual data. Until the 1950's, very little econometric work had been done with quarterly data. Many sectors of the United States economy appear to provide examples of recursiveness in terms of quarterly data, which have become available since 1950. The heavy emphasis on interdependent models maintained by leading econometricians in the 1940's and 1950's may have been based partly on an implicit assumption that the most important economic time series were, and would continue to be, available only on an annual basis. Monthly data should disclose an even larger proportion of recursive models than do quarterly data. Hence the methodological implications of the following pages are applicable to situations in any sector of the economy (and for any selections of time units) which are characterized by recursive structures or close approximations of them.

Figure 9.21 expresses a hypothesis about the economic structure or model responsible for generating time series observations on a number of major variables which logically belong in consumer demand, producers' supply, and other relationships in the hog and pork economy. The directions of the arrows express our hypothesis about the direction of causal relations *given* the time unit (one year) and the market (the United States as a whole) to which our time series data refer. The widths of the arrows reflect our expectations that fluctuations in specified variables will cause large changes in the variables to which arrows lead but that fluctuations in other variables will have only moderate effects. Similar diagrams could be drawn for any other food product.

Let us assume that our primary object is to obtain a structural estimate of the consumer demand function for a food product. Equation (12.28) is typical of the simpler demand functions for food products which have been estimated by economists in the United States:

$$X_1(t) = b_0 X_2(t) + c_0 X_3(t) + d_1(t) \tag{12.28}$$

where X_1 is the retail price and X_2 the per capita consumption of the product; X_3 is the per capita disposable income of consumers; d_1 is a random disturbance; and b_0 and c_0 are "true" values of structural coefficients. In order to show (within the framework of simultaneous-equations theory) that this demand function can be approximated by a single least-squares equation, we must show on logical grounds (or assume) that the disturbances are distributed independently of the explanatory variables, consumption and disposable income. If this is so, we may regard consumption and disposable income as predetermined variables. Since the disturbances are then reflected only in retail price, the demand function must logically be fitted with price in the dependent position. A demand function with the attributes just described may be called a *uniequational complete model* to indicate that it contains only one genuinely endogenous variable and can properly be estimated by ordinary least squares.

The argument in the next few pages is directed toward showing that demand functions for a number of farm and food products probably meet the statistical specifications of the uniequational complete model. It is impossible to show affirmatively that the disturbances in a given case are distributed independently of the explanatory variables because the disturbances are (by definition) not directly observable. Thus, it is *possible* that a noneconomic variable such as summer temperature, which affects consumer demand for lemons, will be correlated with disturbances arising from minor economic factors. Similarly, it is *possible* that apple production, though causally determined by weather and by economic influences prior to harvest, will somehow be correlated with nonmeasurable disturbances in the demand function for apples during the subsequent marketing season. But there is certainly no a priori reason to expect that the disturbances will be dependent (in a probability sense) upon the variables in question.

In general, we shall argue that relevant variables whose values are determined prior to the current time period or outside a given model may be used as explanatory variables in the estimation of demand functions by the method of least squares. Such variables are independently determined or predetermined in the usual or logical sense in which this word is used. The nonanswerable question of whether the disturbances are distributed independently of the explanatory variables will be disregarded. Instead, we shall concentrate on the answerable question of whether certain variables entering into demand functions for foods (and farm products) are predetermined in a logical sense, or nearly enough, to be used as explanatory variables without leading to seriously biased estimates of demand elasticities.

12.2.1 Is Consumption of the Given Commodity Predetermined?

For perishable food products, production in the United States is very nearly equal to domestic consumption. It will facilitate the exposition to consider first the logical status of production as a variable determined independently of the current price of the commodity, assuming for the moment that production and consumption are identical.

If production is not predetermined, its current value must be influenced by the current values of other endogenous variables, particularly price. This at once implies the existence of a second structural equation in which production is expressed as a function of current price in addition to other relevant variables. As production is now dependent on price, it can no longer be regarded as distributed independently of the disturbances affecting price. A least-squares regression of price upon production (equals consumption) and consumer income would give biased estimates of the structural demand coefficients. Thus, the alternative to treating production as a predetermined variable is the simultaneous fitting of two equations, a supply curve and a demand curve.

For many farm products, it is clear on logical grounds that production is a predetermined variable. The production of a crop depends on acreage planted, which is influenced by economic and other considerations before planting time, and yield, which in any given year is strongly influenced by weather. For livestock products, the question might be approached on a partly statistical basis. What part of the observed variance in production can be explained by (1) variables whose values were actually known prior to the beginning of the current period, (2) variables whose values, though not *known* in advance, must clearly have been *determined* prior to the current period,[6] (3) exogenous or noneconomic variables, such as weather and disease, and (4) errors of measurement? If by such a procedure we can explain 95 percent or so of the observed variation in production, we may conclude that, for practical purposes, production is a predetermined variable. The residual variation sets an upper limit to the possible endogenous or jointly determined element in production and to the bias which might be involved in the least-squares regression of price on production.

If production of a perishable commodity is predetermined and if consumption is identically equal to production, consumption itself can obviously be treated as a predetermined variable. But can consumption be so treated under more general conditions?

United States consumption of many farm products differs from (predetermined) production due to variations in exports, imports, or stocks, but it is still highly correlated with production. For example, during 1922–1941 the correlation coefficient between year-to-year changes in production and consumption of meat was 0.98 ($r^2 = 0.95$). If the disturbance or unexplained residual in the relationship between consumption and production is random, it seems intuitively clear that the degree of bias in the least-squares estimate of b_0 in equation (12.28) will not exceed the percent of total variance in consumption which is uncorrelated with variations in production.

Thus, it appears that consumption may be treated as an explanatory variable in this case also though perhaps with certain minor adjustments. One of these would be to regard not actual consumption but consumption as estimated from its regression upon production as a predetermined variable. The remaining variation in actual consumption would be regarded as a random disturbance-and-error component attaching to the consumption variable itself. The least-squares regression of price upon consumption would

[6] For example, the number of milk cows on farms January 1 is causally determined as of that date but is not reported as a statistic until mid-February.

tend to be biased toward zero by the random measurement error component.

This may be shown as follows: Let

$$Y = a + \beta X + u \tag{12.29}$$

where X is a fixed or "true" value and u is a random disturbance, $E(u) = 0$. If we estimate β by fitting the least-squares regression of Y on X, we obtain

$$b = \frac{\sum xy}{\sum x^2} = \frac{\beta \sum x^2 + \sum xu}{\sum x^2} \tag{12.30}$$

where $x = (X - \overline{X})$ and $y = (Y - \overline{Y})$. Since $E(\sum xu) = 0$, we have

$$E(b) = \beta \tag{12.31}$$

Suppose, however, that in place of the "true" value, X, we have an estimated value, $X' = X + \Delta$, in which Δ is an error considered fixed. Then the least-squares regression of Y on X gives us

$$b' = \frac{\sum yx'}{\sum x'^2} = \frac{\beta(\sum x^2 + \sum x\delta) + \sum xu + \sum u\delta}{\sum x^2 + \sum \delta^2 + 2\sum x\delta} \tag{12.32}$$

in which the small letters x and y again denote deviations from means. Since $E(\sum xu)$ and $E(\sum u\delta)$ are both zero,

$$E(b') = \beta \frac{\sum x^2 + \sum x\delta}{\sum x^2 + \sum \delta^2 + 2\sum x\delta} \tag{12.33}$$

In particular, if

$$\sum \delta^2 > -\sum x\delta \tag{12.34}$$

it follows that

$$|E(b')| < |\beta| \tag{12.35}$$

Condition (12.34) does not appear to be particularly restrictive.

Under the special assumption that the δ's are random and independent of the true x's, $E(\sum x\delta) = 0$, and we have

$$E(b') = \beta\left(\frac{\sum x^2}{\sum x^2 + \sum \delta^2}\right) \tag{12.36}$$

In this case,

$$|E(b')| < |\beta| \text{ for all possible values of } \sum \delta^2 \neq 0 \tag{12.37}$$

The bias in this coefficient arising from the disturbances is less obvious and could be either positive or negative depending upon their correlation with the disturbances attaching to the "true" demand function.

The possible "disturbance bias" in the situation under discussion may be shown as follows.

Assume a "true" demand function,

$$p = \beta q + u \tag{12.38}$$

and a consumption-production relationship,

$$q = \gamma z + v \tag{12.39}$$

where p, q, and z are, respectively, price, consumption, and production, all in deviation form: u and v are random disturbances distributed independently of z, which is a predetermined variable; $E(u)$ and $E(v) = 0$.

If we attempt to estimate β by fitting the least-squares regression of p on q, we obtain

$$b = \frac{\sum pq}{\sum q^2} = \frac{\sum (\beta q + u)q}{\sum q^2} = \frac{\beta \sum q^2 + \sum qu}{\sum q^2} = \beta + \frac{\sum qu}{\sum q^2} \tag{12.40}$$

Substituting $(\gamma z + v)$ for q in the numerator, we have

$$b = \beta + \frac{\sum (\gamma z + v)u}{\sum q^2} = \beta + \frac{\sum \gamma zu + \sum vu}{\sum q^2} \tag{12.41}$$

Under our assumptions, z and u are not correlated, so $E(\sum zu) = 0$ and the expected value of the term $\sum \gamma zu / \sum q^2$ vanishes. If we define $b_{uv} = E[uv/\sum v^2]$, we may write

$$E\left[\frac{\sum vu}{\sum q^2}\right] = E\left[\frac{\sum vu}{\sum q^2}\frac{\sum v^2}{\sum v^2}\right] = E\left[\frac{\sum vu}{\sum v^2}\frac{\sum v^2}{\sum q^2}\right] \tag{12.42}$$

or

$$E\left[\frac{\sum uv}{\sum q^2}\right] = b_{uv} E\left[\frac{\sum v^2}{\sum q^2}\right]$$

Hence,

$$E(b) = \beta + b_{uv}\left[\frac{\sigma_v^2}{\sigma_q^2}\right] \tag{12.43}$$

where σ_v^2 and σ_q^2 are the population variances of v and q, respectively.

If the disturbances u and v are not correlated, $b_{uv} = 0$, and b is an unbiased estimate of β. Alternatively, if σ_v^2 is small relative to σ_q^2, the bias in b will generally be small, since it seems unlikely that the regression of *disturbances* in (12.38) upon disturbances in (12.39), b_{uv}, will be as large in absolute value as the structural coefficient, β, relating p and q.

In the present model, β might also be estimated as follows. By least squares, fit the reduced-form equations

$$p = dz + u' \tag{12.44}$$

$$q = cz + v \tag{12.45}$$

in which u' is the random disturbance $(u + \beta v)$. Since d is an estimate of $(\beta\gamma)$ and c is an estimate of γ, the estimate of the structural parameter, β, is given by d/c. This estimate should be asymptotically unbiased even if q contains random measurement errors as well as the effects of disturbances.

12.2.2 Is Consumer Income Predetermined?

If this question is answered in the negative, we are implying that the applicable model includes an equation "explaining" the disposable personal income of consumers as a function of other variables. In an article by Girshick and Haavelmo,[7] disposable income was treated as a function only of predetermined variables, so that the income equation could be fitted independently by the method of least squares. This did not preclude the possibility that disturbances in this equation were correlated with disturbances in other equations.

A more interesting possibility is that disposable income may also be a function of one or more other endogenous variables, such as price and possibly consumption of a given commodity. If this is so, we are immediately confronted with the problem of deriving this equation simultaneously with the demand function (and also with a supply function if the consumption variable is not wholly or approximately predetermined). This alternative should be kept in mind during the following discussion.

First, if the commodity in question is relatively unimportant—that is, if total expenditures for it are very small relative to the national income—even relatively large (percentage) gyrations in the value of the commodity can have little effect on the level of aggregate consumer income.

Pork, beef, and fluid milk are the three most important farm food products in terms of retail value. Expenditures for each of these products are equivalent to 2 or 3 percent of disposable personal income. It is difficult to see how variations in the supply of one of these commodities, operating chiefly through the coefficients of consumer demand functions, could account for more than 2 or 3 percent of the total variation in disposable income. In fact, since the bulk of the variation in disposable income is usually attributed to changes in

[7] M. A. Girshick, and T. Haavelmo, "Statistical Analysis of the Demand for Food: Examples of Simultaneous Estimation of Structural Equations," *Econometrica*, **15**, April 1947, 79–111.

investment and government expenditures, the 2 or 3 percent figure is almost certainly too high.

Various factors may be cited in support of this proposition. Among these are the relative stability of agricultural production, especially that of livestock products, and the apparently limited degree of competition (in demand) between major farm products on a short-run or year-to-year basis. The fact that consumer demand elasticities for pork and beef are between -0.5 and -1.0 tends to restrict the income effects which might otherwise flow from these commodities. It seems doubtful, therefore, that "feedback effects" on disposable income resulting from variations in the price of pork, beef, or fluid milk would stand out above the errors of measurement in the published disposable income series, even if the latter were assessed at not more than 1 percent of the observed variance. If this conclusion is justified with respect to the major farm products, it applies even more strongly to the minor ones.

In the preceding paragraphs, we have rejected on intuitive grounds the hypothesis that an equation expressing the generation of consumer income must be fitted simultaneously with the consumer demand function. This hypothesis can also be treated formally.

If disposable income y is determined jointly with the price of the given commodity p_1, we may write our structural equations as follows:

(12.46) Demand: $$p_1 = cy + bq_1 + u$$

(12.47) Income: $$y = \gamma p_1 + \delta z + v$$

where q_1 is (predetermined) consumption of the given commodity and z is an aggregate of all the predetermined elements in disposable income. We wish to test the hypothesis that y is an endogenous variable which must be estimated simultaneously with p_1. If this is the case, y is not independent of u, and equation (12.46) is not a uniequational complete model.

The reduced-form equations are

(12.48) $$p_1 = \left(\frac{b}{1 - \gamma c}\right)q_1 + \left(\frac{c\delta}{1 - \gamma c}\right)z + \left(\frac{u + cv}{1 - \gamma c}\right)$$

(12.49) $$y = \left(\frac{\gamma b}{1 - \gamma c}\right)q_1 + \left(\frac{\delta}{1 - \gamma c}\right)z + \left(\frac{\gamma u + v}{1 - \gamma c}\right)$$

Each of these equations may be fitted separately by the method of least squares. An estimate of the structural coefficient, γ, is then derived by dividing the coefficient of q_1 in equation (12.49) by the corresponding coefficient in equation (12.48).

The latter regression coefficient (of p_1 upon q_1) will be statistically significant in most analyses for farm and food products. But it seems quite unlikely that the regression of y upon q_1 will be significantly different from zero for any commodity which absorbs (say) 3 percent or less of disposable income. If this is so, the estimate of γ will also be nonsignificant, and we may assume $\gamma = 0$ without loss of statistical accuracy.

If $\gamma = 0$, the reduced-form equations become

$$p_1 = bq_1 + c\delta z + (cv + u) \tag{12.50}$$

$$y = \delta z + v \tag{12.51}$$

The structural coefficient b is estimated without significant bias in the least-squares fitting of equation (12.50). The coefficient c might still be estimated as the ratio $c\delta/\delta$. However, if y is very highly correlated with z, the regression of p_1 upon y in a least-squares fitting of equation (12.46) will give a value of c not materially different from the above ratio.

For the above reasons, it appears safe to ignore equation (12.47) in demand analyses for individual foods and to take a least-squares fitting of equation (12.46) as an estimate of the structural demand function (assuming that q_1 is predetermined).

12.2.3 Applications to Pork and Beef

The complication that seems most likely in estimating demand functions for food livestock products is the existence of a simultaneous supply function.

Suppose that consumption *is* determined simultaneously with price. Then the two structural equations are identical with (12.17) and (12.18) except for the addition of disturbance terms, u and v:

$$\text{Demand:} \quad Q = a_1 + b_1 P + c_1 Y + u \tag{12.52}$$

$$\text{Supply:} \quad Q = a_2 + b_2 P + c_2 Z + v \tag{12.53}$$

where P is retail price, Q is consumption, Y is disposable consumer income, and Z is an estimate of production based wholly on exogenous and predetermined variables. If the variables are in logarithmic form, b_1 is the elasticity of consumer demand and b_2 is the elasticity of producer supply.

We have noted before that these equations form a just-identified model. The corresponding reduced form equations may be written precisely as in equations (12.21) and (12.22). The only difference is that Z stands for another variable now, but with the same statistical properties as the Z of our hypothetical example.

Table 12.1 provides data for four time series relating to pork which meet the requirements of the above model.

As no new *statistical* calculations are involved in the just-identified case, we shall present only the final equations obtained in an attempt to derive simultaneous demand and supply equations for pork. The same symbols are

Table 12.1 Basic Data for "Just-Identified" Model of Supply and Demand for Pork, United States, 1922–1941

Year	Original Arithmetic Values				Logarithms				First Differences of Logarithms			
	Retail Price of Pork	Consumption of Pork	Disposable Personal Income	"Predetermined Elements in Pork Production"								
T	P_t	Q_t	Y_t	Z_t	P_t	Q_t	Y_t	Z_t	P_t	Q_t	Y_t	Z_t
	Cents per Pound	*Pounds per Capita*	*Dollars per Capita*	*Pounds per Capita*								
1922	26.8	65.7	541	74.0	1.428	1.818	2.733	1.869	—	—	—	—
1923	25.3	74.2	616	84.7	1.403	1.870	2.790	1.928	−0.025	0.052	0.057	0.059
1924	25.3	74.0	610	80.2	1.403	1.869	2.785	1.904	0.000	−0.001	−0.005	−0.024
1925	31.1	66.8	636	69.9	1.493	1.825	2.803	1.844	0.090	−0.044	0.018	−0.060
1926	33.3	64.1	651	66.8	1.522	1.807	2.814	1.825	0.029	−0.018	0.011	−0.019
1927	31.2	67.7	645	71.6	1.494	1.831	2.810	1.855	−0.028	0.024	−0.004	0.030
1928	29.5	70.9	653	73.6	1.470	1.851	2.815	1.867	−0.024	0.020	0.005	0.012
1929	30.3	69.6	682	71.2	1.481	1.843	2.834	1.852	0.011	−0.008	0.019	−0.015
1930	29.1	67.0	604	69.6	1.464	1.826	2.781	1.843	−0.017	−0.017	−0.053	−0.009
1931	23.7	68.4	515	68.0	1.375	1.835	2.712	1.833	−0.089	0.009	−0.069	−0.010
1932	15.6	70.7	390	74.8	1.193	1.849	2.591	1.874	−0.182	0.014	−0.121	0.041
1933	13.9	69.6[a]	364	73.6[b]	1.143	1.843	2.561	1.867	−0.050	−0.006	−0.030	−0.007
1934	18.8	63.1[a]	411	70.2[b]	1.274	1.800	2.614	1.846	0.131	−0.043	0.053	−0.021
1935	27.4	48.4	459	46.5	1.438	1.685	2.662	1.667	0.164	−0.115	0.048	−0.179
1936	26.9	55.1	517	57.6	1.430	1.741	2.713	1.760	−0.008	0.056	0.051	0.093
1937	27.7	55.8	551	58.7	1.442	1.747	2.741	1.769	0.012	0.006	0.028	0.009
1938	24.5	58.2	506	58.0	1.389	1.765	2.704	1.763	−0.053	0.018	−0.037	−0.006
1939	22.2	64.7	538	67.2	1.346	1.811	2.731	1.827	−0.043	0.046	0.027	0.064
1940	19.3	73.5	576	73.7	1.286	1.866	2.760	1.867	−0.060	0.055	0.029	0.040
1941	24.7	68.4	697	66.5	1.393	1.835	2.843	1.823	0.107	−0.031	0.083	−0.044

[a] Excludes quantities purchased and distributed under government emergency programs. Per capita consumption including these quantities was 70.7 pounds in 1933 and 64.4 pounds in 1934.

[b] Adjusted by subtracting quantities purchased for government emergency programs.

used as in equations (12.52) and (12.53); the basic data were first differences of logarithms of annual observations for the period 1922–1941 (see Table 12.1).

The reduced-form equations are as follows:

$$\hat{P} = -0.0101 + \underset{(0.1339)}{1.0813Y} - \underset{(0.1159)}{0.8320Z} \qquad R^2 = 0.893$$

$$\hat{Q} = 0.0026 - \underset{(0.0673)}{0.0018Y} + \underset{(0.0582)}{0.6839Z} \qquad R^2 = 0.898$$

The numbers in parentheses are standard errors of the net regression coefficients. The coefficients of the structural equations are calculated in this way:

$$b_2 = \frac{-0.0018}{1.0813} = -0.0017$$

$$b_1 = \frac{0.6839}{-0.8320} = -0.8220$$

$$c_1 = 1.0813\{-[-0.8220 - (-0.0017)]\} = 1.0813(0.8203) = 0.8870$$

$$c_2 = -0.8320(-0.8203) = 0.6825$$

To compute the constant terms a_1 and a_2, we solve the following two simultaneous equations:

$$0.0017a_2 - 0.0017a_1 = (-0.0108)(-0.8203)(0.0017) = 0.0000$$

$$-0.8220a_2 - (-0.0017)a_1 = (0.0026)(-0.8203) = -0.0021$$

Adding the two equations, we obtain

$$-0.8203a_2 = -0.0021 + (0.0017)(0.0089) = -0.0021$$

$$a_2 = \frac{-0.0021}{-0.8203} = 0.0026$$

Then, substituting $a_2 = 0.0026$ into the first equation divided by 0.0017, we have $0.0026 - a_1 = 0.0089$, or $a_1 = -0.0063$. This completes the set of coefficients for the two structural equations. Consequently, the structural equations can be written:

Demand function: $\quad Q = -0.0063 - 0.8220P + 0.8870Y + u$

Supply function: $\quad Q = 0.0026 - 0.0017P + 0.6825Z + v$

Because the data were in logarithmic form, the coefficients represent *approximately* the percentage changes in Q associated with changes of 1 percent in each of the other variables. In particular, the coefficients of P are estimates of the elasticities of demand and (simultaneous or concurrent) supply for pork. Although standard errors of the structural coefficients are not presented here, such errors (appropriate for large samples) can be computed.

For each observation, values of u and v are computed from the following forms of the last two equations:

$$u_t = Q_t + 0.0063 + 0.8220P_t - 0.8870Y_t$$
$$v_t = Q_t - 0.0026 + 0.0017P_t - 0.6825Z_t$$

The subscript t simply refers to the particular year for which the disturbances are being calculated. The disturbances for each year, and their successive differences, are shown in Table 12.2. The von Neumann ratios prove to be:

Demand function: $$K = \left(\frac{0.008017}{0.003553}\right)\left(\frac{19}{18}\right) = 2.256403(1.055556) = 2.381760$$

Supply function: $$K = \left(\frac{0.007122}{0.003181}\right)\left(\frac{19}{18}\right) = 2.238919(1.055556) = 2.363304$$

Table 12.2 Disturbances Computed from Structural Equations

	Disturbances in Demand Function		Disturbances in Supply Function	
Year	u_t	$u_{t+1}-u_t$	v_t	$v_{t+1}-v_t$
1923	−0.013	—	0.009	—
1924	0.010	0.023	0.013	0.004
1925	0.020	0.010	−0.005	−0.018
1926	0.002	−0.018	−0.008	−0.003
1927	0.011	0.009	0.001	0.009
1928	0.002	−0.009	0.009	0.008
1929	−0.010	−0.012	0.000	−0.009
1930	0.022	0.032	−0.014	−0.014
1931	0.003	−0.019	0.013	0.027
1932	−0.022	−0.025	−0.017	−0.030
1933	−0.014	0.008	−0.004	0.013
1934	0.024	0.038	−0.031	−0.027
1935	−0.016	−0.040	0.005	0.036
1936	0.010	0.026	−0.010	−0.015
1937	−0.003	−0.013	−0.003	0.007
1938	0.014	0.017	0.019	0.022
1939	−0.007	−0.021	0.000	−0.019
1940	−0.014	−0.007	0.025	0.025
1941	−0.010	0.004	−0.003	−0.028

We find no evidence of significant autocorrelation in either set of disturbances.

It is interesting to compare the equations just derived with their least-squares counterparts. The demand function for pork has been fitted by least squares in two different ways. When Q is treated as the dependent variable, we obtain

$$\text{Least squares:} \quad \hat{Q} = -0.0049 - \underset{(0.0594)}{0.7205P} + \underset{(0.0967)}{0.7646Y} \qquad R^2 = 0.903$$

$$\text{Structural:} \quad Q = -\,0.0063 - 0.8220P + 0.8870Y + u$$

In each case, the structural coefficient is within less than two standard errors of the least-squares coefficient; the differences are not statistically significant according to usual criteria.

When P is treated as the independent variable, we obtain

$$\text{Least squares:} \quad \hat{P} = -0.0070 - \underset{(0.1032)}{1.2518Q} + \underset{(0.0861)}{1.0754Y} \qquad R^2 = 0.956$$

For easier comparison, we divide all terms in the structural demand equation by the coefficient of P (using the positive sign) and transpose P and Q to opposite sides of the equality sign, obtaining

$$\text{Structural:} \quad P = -0.0077 - 1.2165Q + 1.0791Y + \left(\frac{u}{0.8220}\right)$$

The coefficients of the two equations are almost identical—they differ by small fractions of one standard error. Evidently the least-squares equation with price dependent gives an excellent approximation to the structural demand function. And the logic of the simultaneous-equations approach supports the choice of price as the dependent variable in our least-squares demand function *if* there is no simultaneous response of supply to price.

The corresponding comparison for the supply function (with Q dependent in the least-squares equation) is:

$$\text{Least squares:} \quad Q = 0.0022 - \underset{(0.0522)}{0.0788P} + \underset{(0.0734)}{0.6090Z} \qquad R^2 = 0.910$$

$$\text{Structural:} \quad Q = 0.0026 - 0.0017P + 0.6825Z + v$$

The coefficient of P in the least-squares supply function is nonsignificant according to the usual criteria and has a negative sign, whereas normally we would expect a true supply function to show a positive response of quantity to price.

If we discard the price variable from the least-squares supply function as nonsignificant, we obtain simply

$$\text{Least squares:} \qquad \underset{(0.0857)}{Q = 0.0025 + 0.6841Z} \qquad R^2 = 0.898$$

This is almost identical with the structural equation, recognizing that the coefficient of P in the latter is negligibly small and would prove to be statistically nonsignificant if its standard error was calculated.

In the simple two-equation model for pork, the similarity of the least-squares demand function to the corresponding structural equation has both theoretical and statistical explanations. In the first place, 90 percent or more of the variation in pork *production* was attributable either to variables which were predetermined as of January 1 or to exogenous factors, such as the effects of weather and disease upon the number of pigs saved per litter. Under 1922–1941 conditions, most hogs were marketed at the age of eight or nine months; the gestation period for pigs is about four months. With this built-in lag of 12 or 13 months between sow breeding and hog slaughter (equals pork production), hog producers had very little latitude to change the *current* year's production in response to the current year's price of pork or hogs. Hence, there was little reason to expect a significant net regression coefficient between the current price of pork and the current quantity of pork in the supply equation. As a matter of fact, this structural coefficient was nonsignificant, being based on a nonsignificant coefficient in the reduced form of the model. Moreover, as was already noted, the structural coefficient was negative, whereas one would ordinarily expect an increase in price to induce an increase in supply.

If we discard P from the supply equation as nonsignificant, we are left with the least-squares equation

$$Q = 0.0025 + \underset{(0.0857)}{0.6841}Z$$

as a basis for estimating current pork consumption from a composite of predetermined factors affecting the production of pork. This new estimating equation also helps to explain why the least-squares demand function is so much like its structural counterpart. If 90 percent of the variation in Q is associated with a predetermined variable, its statistical properties will be very similar to those of a predetermined variable. If any bias is involved in treating Q as a predetermined variable in the demand equation, it would seem intuitively that the bias would not exceed 10 percent.

On this argument, our least-squares demand equation fitted with price as the dependent variable is roughly consistent with the theory underlying the simultaneous-equations method.

Simultaneous-equations methods for "overidentified" models. The basic problem of *overidentification* can be illustrated in a two-equation demand and supply model for beef. The structural equations are assumed to be

$$\text{(12.54)} \qquad \text{Demand:} \qquad Q = a_1 + b_1 P + c_1 Y + d_1 W + u$$

$$\text{(12.55)} \qquad \text{Supply:} \qquad Q = a_2 + b_2 P + c_2 Z + v$$

where Z is an estimate of beef production based on wholly predetermined variables and W is consumption of other meats, assumed predetermined for present purposes. The reduced form of this model is

$$\text{(12.56)} \qquad P = \left(\frac{a_2 - a_1}{b_1 - b_2}\right) - \left(\frac{c_1}{b_1 - b_2}\right)Y + \left(\frac{c_2}{b_1 + b_2}\right)Z - \left(\frac{d_1}{b_1 - b_2}\right)W + \left(\frac{v - u}{b_1 - b_2}\right)$$

$$\text{(12.57)} \qquad Q = \left(\frac{a_2 t - a_1 b_2}{b_1 - b_2}\right) - \left(\frac{c_1 b_2}{b_1 - b_2}\right)Y + \left(\frac{c_2 b_1}{b_1 - b_2}\right)Z - \left(\frac{d_1 b_2}{b_1 - b_2}\right)W + \left(\frac{b_1 v - b_2 u}{b_1 + b_2}\right)$$

We can estimate b_1 as a ratio of the coefficients of Z in the two equations; but b_2 can be estimated in two ways, one as a ratio of the coefficients of Y and the other as a ratio of the coefficients of W. If the two estimates of b_2 differ, we can also obtain different estimates for a_1, a_2, c_1, c_2, and d_1.

The reduced-form equations fitted to logarithms of the variables during 1922–1941 by least squares are as follows, neglecting the constant terms:

$$\underset{}{P = A_1 + \underset{(0.1061)}{0.8185Y} - \underset{(0.1877)}{0.8521Z} - \underset{(0.1631)}{0.4346W} + f_1(u, v)} \qquad R^2 = 0.87$$

$$Q = A_2 + \underset{(0.0545)}{0.0509Y} + \underset{(0.0964)}{0.8801Z} - \underset{(0.0838)}{0.0899W} + f_2(u, v) \qquad R^2 = 0.87$$

Then $b_1 = 0.8801/-0.8521 = -1.0329$; this is our structural estimate of the elasticity of demand for beef. The two alternative values of b_2 are

$$b_{2(y)} = \frac{0.0509}{0.8185} = 0.0622$$

and

$$b_{2(w)} = \frac{-0.0899}{-0.4346} = 0.2068$$

Some pragmatic comments are also warranted in the present case. First, neither of the reduced-form coefficients in the numerators of the ratios from which b_2 is calculated differ significantly from zero; one is slightly smaller than its standard error and the other is only slightly larger. Starting from such unpromising materials, it seems likely that any compromise estimate of b_2, which must be analogous to a weighted average of $b_{2(y)}$ and $b_{2(w)}$, will also be nonsignificant. If we discard Y and W from the second reduced-form equation, we obtain a least-squares estimate of Q as a function of the composite predetermined variable Z:

$$Q = A_2' + \underset{(0.0900)}{0.8878Z} + f_2'(u, v) \qquad r^2 = 0.85$$

If we estimate c_1 and d_1 on the assumption that $b_2 = 0$, we obtain

$$c_1 = -(0.8185)(-1.0329) = 0.8454$$

and

$$d_1 = -(0.4346)(-1.0329) = -0.4489$$

Our structural demand function would then be

$$Q = a_1' - 1.0329P + 0.8454Y - 0.4489W + u$$

Transposing P and Q and dividing through by the coefficient of P, the structural equation becomes

$$P = a_1'' - 0.9682Q + 0.8185Y - 0.4346W + u'$$

The same variables fitted by least squares with P in the dependent position give the following result:

$$\hat{P} = a_1''' - \underset{(0.1179)}{1.0645Q} + \underset{(0.0620)}{0.8815Y} - \underset{(0.0887)}{0.5247W} \qquad R^2 = 0.95$$

The respective coefficients differ by about one standard error in each case, or about 0.7 standard error of their differences *if* the regressions had been fitted to two random samples from the same universe. The similarities between the two equations can be rationalized on the same basis as in the case of pork.[8]

In concluding this section, we should like to make one point which would suggest to experienced price analysts some of the difficulties to be expected in the measurement of simultaneous supply functions. Our ability to establish statistically significant coefficients in the subordinate or less influential members of a simultaneous-equation model will be limited both by specification

[8] The structural equations for the beef model have been estimated by Joan Friedman, using the simultaneous-equations method theoretically appropriate in the overidentified case. The equations, with standard errors of the respective coefficients, are as follows:

Demand: $$P = a_1'' - \underset{(0.24)}{0.97Q} + \underset{(0.13)}{0.89Y} - \underset{(0.18)}{0.47W}$$

Supply: $$Q = a_2'' + \underset{(0.06)}{0.09P} + \underset{(0.11)}{0.94Q}$$

As expected, the coefficient of P in the supply function was nonsignificant. All other coefficients of the least-squares counterpart equations differed from corresponding coefficients of the structural equations by much less than one standard error of the latter coefficients.

and data problems. For example, the reduced-form equation upon which depends our ability to estimate the simultaneous elasticity of supply will frequently take the following form:

$$q = k_0 + k_1 z + k_2 y + v \tag{12.58}$$

where z is a combination of the predetermined variables underlying commodity production and y is the disposable income of consumers. The statistical significance of b_2, the elasticity of supply, depends on that of k_2—that is, we must show a statistically significant association between the production of a specific commodity and total consumer income *after* the influence of factors directly and obviously affecting commodity production has been eliminated. As the influence of consumer income upon production operates through the retail price of the commodity and then, via the marketing system, upon and through its farm price, data, disturbance, and aggregation problems at each of these levels may obscure the relationship between production and income.

The foregoing does not deny the possibility of establishing significant simultaneous supply elasticities for some farm products. It does suggest that the data requirements for the simultaneous-equations approach are more exacting than has been commonly realized, and probably more exacting than many of our existing agricultural time series can support.

12.3 CAUSAL ORDERING, RECURSIVENESS, AND INTERDEPENDENCY: A FRAMEWORK FOR CHOOSING METHODS OF ESTIMATION

The preceding section has pointed out a number of reasons why a least-squares regression equation gives us good estimates of a consumer demand function in a multiple equation structure *if the structure as a whole possesses certain properties*. One justification advanced for treating consumption as an independent variable was the near-identity of current consumption and current production and the fact that current production was almost wholly determined by a combination of *past* values of economic variables (and current values of exogenous variables such as weather).

A simple time lag would be sufficient to produce identification in the model of equations (12.59) and (12.60). The original model was, in deviation form,

$$\text{Demand:} \quad p = b_1 q + u \tag{12.59}$$

$$\text{Supply:} \quad p = b_2 q + v \tag{12.60}$$

where p and q refer to the same time unit. When we introduce time lags, we must be careful to designate the time unit to which each price or quantity applies, so we rewrite the model in which p and q are simultaneously determined as

$$\text{Demand:} \quad p_{(t)} = b_1 q_{(t)} + u_{(t)} \tag{12.61}$$

$$\text{Supply:} \quad p_{(t)} = b_2 q_{(t)} + v_{(t)} \tag{12.62}$$

Now, suppose that the current supply, $q_{(t)}$, is actually determined by last year's price, $p_{(t-1)}$. We rewrite the supply equation as

(12.63) Supply: $$q_{(t)} = \beta_2 \, p_{(t-1)} + w_{(t)}$$

The following diagram may be helpful in visualizing the operation of the model consisting of equations (12.61) and (12.62):

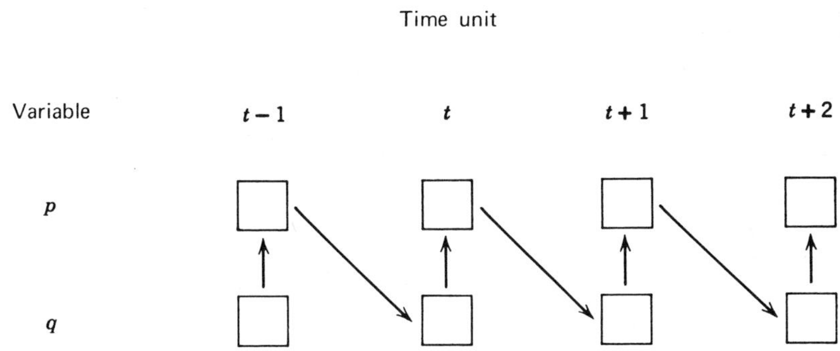

The supply equation is represented by the arrows slanting downward from $p_{(t-1)}$ to $q_{(t)}$, $p_{(t)}$ to $q_{(t+1)}$, and $p_{(t+1)}$ to $q_{(t+2)}$. The demand equation is represented by the vertical arrows leading from $q_{(t)}$ to $p_{(t)}$, $q_{(t+1)}$ to $p_{(t+1)}$, and $q_{(t+2)}$ to $p_{(t+2)}$.

By a slight extension of this diagram we can include every element in the model:

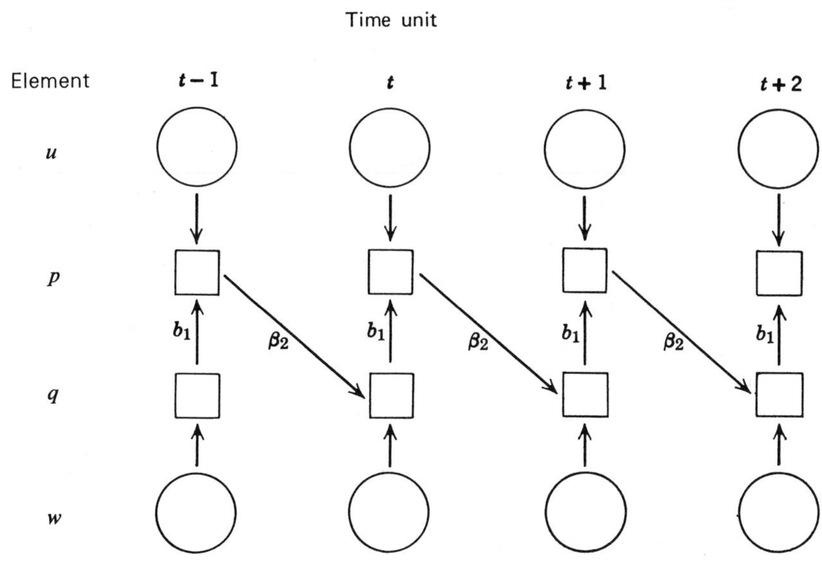

If $w_{(t)}$ is not correlated with $u_{(t)}$, the supply equation can be fitted by least squares with $q_{(t)}$ as the dependent variable and the demand equation can be fitted separately by least squares with $p_{(t)}$ as the dependent variable.

In one sense, our model still includes only two variables, price and quantity. But in a statistical sense, it includes *three* variables, since $p_{(t-1)}$ is a fixed number, fully determined in Period $(t-1)$ before the opening day of Period (t). When used as an independent variable in estimating $q_{(t)}$, $p_{(t-1)}$ is logically equivalent to one of the controlled variables in a regression model based on experimental data. Causality moves forward in time, and the regression of $p_{(t-1)}$ on $q_{(t)}$ would have no explanatory value.

It will be useful to rearrange terms in equations (12.61) and (12.62) as follows:

$$\begin{aligned} &\text{Supply:} \qquad -\beta_2 p_{(t-1)} + q_{(t)} = w_{(t)} \\ &\text{Demand:} \qquad -b_1 q_{(t)} + p_{(t)} \qquad = u_{(t)} \end{aligned}$$

We can also display the elements of these equations in the following form:

$$\begin{array}{rc} & \begin{array}{ccccc} (p_{(t-1)} & & q_{(t)} & p_{(t)}) & \qquad (w_{(t)} \quad u_{(t)}) \end{array} \\ \begin{array}{r} \text{Supply:} \\ \text{Demand:} \end{array} & \left[\begin{array}{c:cc} -\beta_2 & 1 & \\ & -b_1 & 1 \end{array}\right] = \left[\begin{array}{cc} 1 & \\ & 1 \end{array}\right] \end{array}$$

We can reconstitute the equations simply by multiplying each variable or disturbance by the coefficient or coefflcients in the column immediately below it.

If this pattern of coefficients represents the true economic (and statistical) structure of the model, it is clear that neither $p_{(t)}$ nor $u_{(t)}$ enter into the supply function at all. The estimation of the supply function coefficient β_2 is a "private matter" involving only $q_{(t)}$, $p_{(t-1)}$, and $w_{(t)}$.

The value of $q_{(t)}$, then, has also been fully determined by $p_{(t-1)}$ and $w_{(t)}$. We may now regard $q_{(t)}$ as a fixed number and a logically independent variable as we enter it into the demand function. The estimation of the demand function coefficient b_1 becomes a "private matter" involving only $q_{(t)}$, $p_{(t)}$, and $u_{(t)}$.

From the arrow diagrams of this model, it is clear that the impacts of $p_{(t-1)}$ on $q_{(t)}$ and of $q_{(t)}$ on $p_{(t)}$ are linked together in a repetitive or recursive causal chain. This kind of model is frequently referred to as a "recursive model" to distinguish it from an "interdependent" or simultaneous model such as that formed by equations (12.61) and (12.62). The elements of that model may be displayed as

$$\begin{array}{rc} & \begin{array}{cccc} (q_{(t)} & p_{(t)}) & \qquad (v_{(t)} & u_{(t)}) \end{array} \\ \begin{array}{r} \text{Supply:} \\ \text{Demand:} \end{array} & \begin{bmatrix} -b_1 & 1 \\ -b_2 & 1 \end{bmatrix} = \begin{bmatrix} 1 & \\ & 1 \end{bmatrix} \end{array}$$

The arrays formed by the coefficients of the endogenous variables $q_{(t)}$ and $p_{(t)}$ are characteristic of recursive and simultaneous models, respectively. In the recursive model, this array or matrix is *triangular*. In a purely recursive model, no coefficients of endogenous variables will appear above the major diagonal of this matrix if the equations are arranged in the order of their causal relationship in the model. To the extent that one or more coefficients of endogenous variables appear above the major diagonal, we have a simultaneous or interdependent model.

A slightly larger model will help to illustrate the relationship of triangularity to causal ordering, namely, the model of Figure 9.20; the supply and demand structure for potatoes. The variables in the model are

X_1 = weather

X_2 = disposable personal income

X_3 = charge for marketing services

X_4 = production

X_5 = consumption (quantity offered for sale by retailers)

X_6 = retail price

X_7 = farm price (current year)

and $X_8 = X_{7(t-1)}$ = farm price (previous year)

We regard four of these variables as exogenous (weather, disposable personal income, the charge for marketing services, and farm price in the previous year) and four as endogenous (production, consumption, retail price, and farm price in the current year).

These variables can be grouped into the following equations:

Production:	$X_4 = f(X_8, X_1) + u_4$
Supply available to consumers:	$X_5 = f(X_4) + u_5$
Consumer demand:	$X_6 = f(X_5, X_2) + u_6$
Relation between farm price and retail price:	$X_7 = f(X_6, X_3) + u_7$

Assuming that (1) the functions are linear, (2) the variables are expressed as deviations from their means, and (3) the disturbances u_4, u_5, u_6, and u_7 are not correlated with one another, we may write the model as follows:

$$
\begin{aligned}
x_4 - b_{48}x_8 - b_{41}x_1 &= u_4 \\
x_5 - b_{54}x_4 \qquad\qquad &= u_5 \\
x_6 - b_{65}x_5 - b_{62}x_2 &= u_6 \\
x_7 - b_{76}x_6 - b_{73}x_3 &= u_7
\end{aligned}
$$

The elements may be arranged as:

$$
\begin{array}{cccc:cccc}
\multicolumn{4}{c}{\textit{Exogenous variables}} & \multicolumn{4}{c}{\textit{Endogenous variables}} \\
(x_8 & x_1 & x_2 & x_3 & x_4 & x_5 & x_6 & x_7) \\
\hline
-b_{48} & -b_{41} & & & 1 & & & \\
 & & & & -b_{54} & 1 & & \\
 & & -b_{62} & & & -b_{65} & 1 & \\
 & & & -b_{73} & & & -b_{76} & 1
\end{array}
=
\begin{array}{cccc}
\multicolumn{4}{c}{\textit{Disturbances}} \\
(u_4 & u_5 & u_6 & u_7) \\
\hline
1 & & & \\
 & 1 & & \\
 & & 1 & \\
 & & & 1
\end{array}
$$

The coefficient matrix linking the endogenous variables to one another has no coefficients above the major diagonal; it is therefore triangular and the model is fully recursive. Although we have assumed no correlation among u_4, u_5, u_6, and u_7, it is well to remember (1) that they *could* be correlated in the population and (2) if we could actually measure the disturbances in particular samples, we would find that the correlations among them were nonzero and varied from sample to sample. In the remainder of this section, we shall ignore the disturbance terms and concentrate on the patterns of coefficients between variables.

It is clear that the current values of the four endogenous variables are determined in a certain logical sequence or *causal order* which is reflected in the triangularity of the coefficient matrix. We can diagram this causal ordering as is shown in Figure 12.3. We assign the four exogenous variables to Causal Order 0, since they are not determined or caused within the model. They come as predetermined given numbers, ready for use in explaining the movements of the endogenous variables.

If any endogenous variable is a function only of exogenous variables, we assign it to Causal Order 1. This is true of X_4, production, in the present model, which is determined by X_8, the previous year's farm price, and X_1, weather.

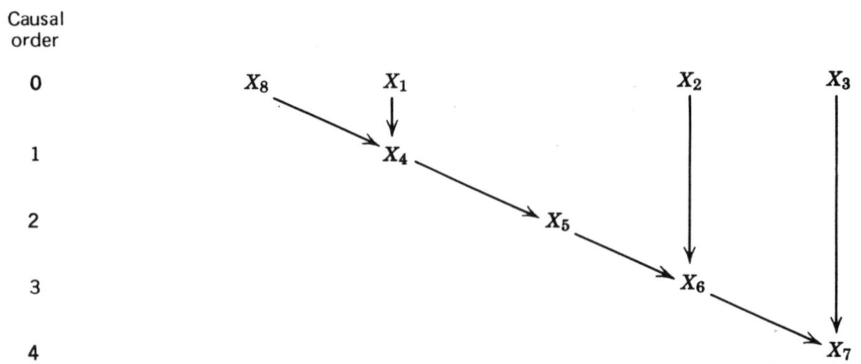

Figure 12.3

At this point, the values of five variables—the four exogenous variables and X_4—are fully determined. No endogenous variable other than X_4 could be determined at Causal Order 1 because no other endogenous variables could be explained exclusively by exogenous variables. However, given X_4, we can now determine the value of X_5; we assign X_5 to Causal Order 2. No other endogenous variable can be determined at Causal Order 2 because the value of X_5 is a prerequisite for the estimation of X_6 and X_7.

The value of X_6 is determined, then, at Causal Order 3; it is a function of X_5 and X_2. Finally, the value of X_7 is determined at Causal Order 4 as a function of X_6 and X_3; X_4 could not have been determined logically at any earlier stage.

If the error terms in the four equations are independent of (uncorrelated with) one another, each equation can be estimated optimally by least squares. If the error terms are not independent of one another, triangularity would justify the following procedure. Regress X_4 on X_1; then, regress X_5 on $\hat{X}_4$, where $\hat{X}_4$ is estimated from the first regression equation; next, regress X_6 upon X_2 and upon $\hat{X}_5$ as estimated from the second sequential regression; and finally, regress X_7 upon X_3 and upon $\hat{X}_6$ as estimated from the third sequential regression.

As $\hat{X}_4$, $\hat{X}_5$, and $\hat{X}_6$ are exact functions of the exogenous variables, they are independent of the error terms and are suitable regressors for the successive least-squares equations. This sequential procedure based on causal ordering is logically very close to two-stage least squares which will be discussed later in the chapter.

The concepts of a "basic model" and a "derived model" also throw some light on the nature of interdependence in economic models. We may conceive of a complete model incorporating explicitly all the actions of each and every individual economic agent—thus, a model pushing the disaggregation process over subjects and goods to its ultimate microlimit—and in which time is introduced in such a way that periods are sufficiently short to correspond to the actions of the subjects; let us call this a "basic model." This type is always recursive. Each subject acts on the basis of information acquired in or before time t − 1, prior to the moment of his action at time t. His action as known to others or as it affects the economic environments of others becomes part of the information which other subjects use in deciding on their own actions in time $t + 1$ or later.

A derived model, in contrast to a basic model, is an aggregated and gross approximation of reality. The availability of data often determines the unit of time selected as well as the degree of aggregation, introducing essentially arbitrary elements into the model. These elements may be responsible for the interdependent nature of many derived models.

Two examples may suffice to clarify this last point. First, many time series are available only on a quarterly or even an annual basis. Time lags between the plans and the actions of economic subjects and between the actions and their effects on other variables in the system do not correspond to the temporal breakdown of the time series. In most cases, the basic lags are of a shorter duration than the minimum time span between successive observations in the statistical time series. If this is the case, interdependency may appear in the derived model, simply as a consequence of data limitations. The longer the unit of time in the derived model, the more "simultaneous" relationships will be found in it which, in fact, were not simultaneously determined, as they took place at different points in time within the unit specified by the model.

For example, suppose the basic model of interaction between consumers and retail food stores is

$$\text{Consumer Demand:} \qquad q_{(t)} = b_1 p_{(t)} + u_{(t)}$$

$$\text{Retailers' Supply:} \qquad q_{(t)} = b_2\, p_{(t-1)} + w_{(t)}$$

We assume that, each week, retailers offer consumers a definite supply $q_{(t)}$, and the average price for the week $p_{(t)}$ seeks the level necessary to move the quantity $q_{(t)}$ into consumption. Toward the end of Week (t), retailers decide on the basis of $p_{(t)}$ how large a supply, $q_{(t+1)}$ they will order from wholesalers and offer for sale to consumers in Week ($t + 1$). This quantity moves into consumption at the price $p_{(t+1)}$.

If this weekly response mechanism operated throughout the year, then from Week 52 in Year 0 to Week 52 in Year 1 there would have been 52 supply responses, each relating $q_{(t)}$ to $p_{(t-1)}$ according to the coefficient b_2, and 52 price adjustments along the consumer demand curve, each relating $p_{(t)}$ to $q_{(t)}$ according to the coefficient b_1. The annual or even quarterly observations on $p_{(t)}$ and $q_{(t)}$ would present us with a serious problem of interdependence and identification. However, using weekly data, we could estimate quite readily b_1 and b_2, the structural coefficients of the basic behavioral model.

A second kind of interdependency in the derived model when none existed in the basic model could follow from some types of aggregation procedures. If four variables are causally related so that A causes B which, in turn, is the cause of C which causes D, grouping together, respectively, A and C in Variable I and B and D in Variable II introduces interdependency between I and II where the original (basic) system was recursive.

Although the logical arguments for recursiveness are very strong so far as basic models are concerned, we are nevertheless often faced with the practical problem of constructing important economic models from annual or quarterly data. Whatever the cause of interdependency in these relatively gross time units, we must nevertheless take account of it in our estimation procedures and/or by drawing on sources of information other than the annual or quarterly time series.

Section 12.4 will deal with one of the several techniques which have been devised for estimating the structural coefficients of interdependent equations directly from the data. The straightforward logic and computational simplicity of the method of two-stage least squares makes it a useful introduction to more complicated methods which are usually discussed in more advanced courses in econometrics.

12.4 TWO-STAGE LEAST SQUARES AND THE ESTIMATION OF INTERDEPENDENT EQUATIONS

Given the same demand and supply model for pork as before, including random disturbances u and v,

(12.64) Demand: $$Q = a_1 + b_1 P + c_1 Y + u$$

(12.65) Supply: $$Q = a_2 + b_2 P + c_2 Z + v$$

we can demonstrate the method of two-stage least squares.

We assume that P and Q are jointly dependent endogenous variables and that Y and Z are exogenous variables. By hypothesis, the disturbances in the two structural equations are not correlated with Y and Z.

Now, consider the corresponding reduced-form equations, including disturbance terms which are combinations of the structural disturbances, u and v:

(12.66) $$P = A_1 + B_1 Y + C_1 Z + \frac{v - u}{b_1 - b_2}$$

(12.67) $$Q = A_2 + B_2 Y + C_2 Z + \frac{b_1 v - b_2 u}{b_1 - b_2}$$

If we fit (12.66) by least squares, we obtain a series of estimated values of P, namely, $\hat{P}$, which are exact linear functions of Y and Z. If u and v are uncorrelated with Y and Z, exact linear combinations of u and v should be uncorrelated with exact linear combinations of Y and Z. The disturbance term in (12.66) is just such a function of u and v; it should therefore be uncorrelated with $\hat{P}$.

The method of least squares applied to (12.66) produces this result in the particular set of data which consitutes our sample. The hypotheses about zero correlations in the preceding two paragraphs relate to the parent population—hence, to the *expected values* of correlation coefficients rather than to their values in individual samples. Just as the standard error of estimate in a single-equation least-squares model is only an estimate of the standard error in the population, so the residual variance about $\hat{P} = A_1 + B_1 Y + C_1 Z$

is only an estimate of the variance of $(v - u)/(b_1 - b_2)$ in the population underlying our two-equation model.

The same reasoning applies in equation (12.67) to the least-squares residuals $Q - \hat{Q}$. Let us call the two series of reduced-form residuals or disturbances D_1 and D_2. Knowing that $D_1 = (v - u)/(b_1 - b_2)$ and $D_2 = (b_1 v - b_2 u)/(b_1 - b_2)$, we can derive estimates of the structural disturbances u and v as follows.

From the equations defining D_1 and D_2, we write

$$v - u = (b_1 - b_2)D_1 \tag{12.68}$$

$$b_1 v - b_2 u = (b_1 - b_2)D_2 \tag{12.69}$$

We then multiply both sides of the first equation by $(-b_2)$, obtaining

$$-b_2 v + b_2 u = -b_2(b_1 - b_2)D_1 \tag{12.70}$$

Adding equations (12.70) and (12.69), we obtain

$$(b_1 - b_2)v = (b_1 - b_2)(D_2 - b_2 D_1)$$

so

$$v = D_2 - b_2 D_1$$

Then, since $(v - u) = (b_1 - b_2)D_1$, we have $u = v - (b_1 - b_2)D_1$. For each observation on the variables P, Q, Y, and Z, D_1 and D_2 will be known numbers; b_1 and b_2 will also be known numbers, so numerical estimates of u and v can be calculated readily.

The first "stage" of two-stage least squares consists of fitting each reduced-form equation by least squares. In the pork example, then, we obtain the reduced-form equations previously cited:

$$\hat{P} = -0.0101 + \underset{(0.1339)}{1.0813}Y - \underset{(0.1159)}{0.8320}Z \qquad R^2 = 0.893 \tag{12.71}$$

$$\hat{Q} = 0.0026 - \underset{(0.0673)}{0.0018}Y + \underset{(0.0582)}{0.6839}Z \qquad R^2 = 0.898 \tag{12.72}$$

The second "stage" of two-stage least squares is equally simple. In the structural demand function, we replace P by $\hat{P}$ and fit the resulting equation $Q = f(\hat{P}, Y)$, by least squares. In the structural supply function, we replace P by $\hat{P}$ and fit the resulting equation, $Q = f(\hat{P}, Z)$, by least squares.

There is a certain ambiguity in the second stage, for we could also transpose P to the left side of equations (12.64) and (12.65), obtaining

$$\text{Demand:} \qquad P = -\frac{a_1}{b_1} + \frac{1}{b_1}Q - \frac{c_1}{b_1}Y + \frac{1}{b_1}u \tag{12.73}$$

$$\text{Supply:} \qquad P = -\frac{a_2}{b_2} + \frac{1}{b_2}Q - \frac{c_2}{b_2}Z + \frac{1}{b_2}v \tag{12.74}$$

In each of these equations, we could replace Q by $\hat{Q}$ and fit each of the equations $P = f(\hat{Q}, Y)$ and $P = f(\hat{Q}, Z)$ by least squares. Or we could take P as dependent in the demand function and Q as dependent in the supply function —or Q dependent in the demand function and P dependent in the supply function.

It is a simple matter to run all these alternatives on an electronic computer. The cases are as follows:

Case I

Demand:

$$Q = \underset{(0.0033)}{-0.0057} - \underset{(0.0695)}{0.8221\hat{P}} + \underset{(0.1067)}{0.8854Y} + u; \quad R^2 = 0.8991; \quad \bar{S} = 0.0140$$

Supply:

$$Q = \underset{(0.0033)}{0.0025} - \underset{(0.0623)}{0.0022\hat{P}} + \underset{(0.0829)}{0.6820Z} + v; \quad R^2 = 0.8977; \quad \bar{S} = 0.0141$$

Case II

Demand:

$$P = \underset{(0.0065)}{-0.0069} - \underset{(0.1698)}{1.2164\hat{Q}} + \underset{(0.1342)}{1.0763Y} - \frac{u'}{b_1'}; \quad R^2 = 0.8930; \quad \bar{S} = 0.0282$$

Supply:

$$P = \underset{(0.0011)}{0.0025} - \underset{(39.7828)}{96.3080\hat{Q}} + \underset{(27.2116)}{64.9210Z} - \frac{v'}{b_2'}; \quad R^2 = 0.6033; \quad \bar{S} = 0.0543$$

Case III

Demand:

$$P = \underset{(0.0065)}{-0.0069} - \underset{(0.1698)}{1.2164\hat{Q}} + \underset{(0.1342)}{1.0763Y} - \frac{u'}{b_1'}; \quad R^2 = 0.8930; \quad \bar{S} = 0.0282$$

Supply:

$$Q = \underset{(0.0033)}{-0.0025} - \underset{(0.0623)}{0.0022\hat{P}} + \underset{(0.0829)}{0.6820Z} + v; \quad R^2 = 0.8977; \quad \bar{S} = 0.0141$$

Case IV

Demand:

$$Q = \underset{(0.0033)}{-0.0057} - \underset{(0.0695)}{0.8221\hat{P}} + \underset{(0.1067)}{0.8854Y} + u; \quad R^2 = 0.8991; \quad \bar{S} = 0.0140$$

Supply:

$$P = 0.0025 - 96.3080\hat{Q} + 64.9210Z - \frac{v'}{b_2'}; \quad R^2 = 0.6033; \quad \bar{S} = 0.0543$$

(0.0011) (39.7828) (27.2116)

If our objective is still to arrive at unique estimates of coefficients of reversible demand curves and supply curves, it is reasonable to divide through each of the equations which was fitted with P in the dependent position by the coefficient of $\hat{Q}$, obtaining equations with $\hat{Q}$ on the left-hand side. We know that the least-squares net regression of Q on P does not exactly equal the reciprocal of the least-squares net regression of P on Q. While we are not yet sure of the properties of $\hat{Q}$ as an "independent" variable, we write the disturbances of equations fitted with P in the dependent position as $(-u'/b_1')$ and $(-v'/b_2')$ to allow for the possibility that $b_1' \neq b_1$, $b_2' \neq b_2$, $u' \neq u$, and $v' \neq v$.

Table 12.3 Two-Stage Least Squares: Comparisons of Disturbances Computed in Alternative Manners (Pork Example)

	Disturbances in Demand Functions		Disturbances in Supply Functions	
Year	u	u'	v	v'
1923	0.013	0.013	0.009	0.009
1924	0.009	0.009	0.013	0.013
1925	0.020	0.020	−0.005	−0.005
1926	0.002	0.002	−0.007	−0.008
1927	0.010	0.010	0.001	0.001
1928	0.002	0.002	0.009	0.009
1929	−0.010	−0.010	0.000	0.000
1930	0.022	0.021	−0.013	−0.014
1931	0.003	0.002	0.013	0.012
1932	−0.023	−0.023	−0.017	−0.018
1933	0.015	0.015	−0.004	−0.004
1934	0.023	0.024	−0.031	−0.030
1935	−0.017	−0.017	0.005	0.005
1936	0.010	0.010	−0.010	−0.009
1937	−0.003	−0.003	−0.003	−0.003
1938	0.013	0.013	0.019	0.019
1939	−0.008	−0.007	0.000	0.000
1940	−0.014	−0.014	0.025	0.025
1941	−0.011	−0.011	−0.003	−0.003

For easy comparison, the two estimates of the demand curve are shown directly above one another:

$$\text{Demand:}\quad Q = -0.0057 - 0.8221\hat{P} + 0.8854Y + u$$
$$\hat{Q} = -0.0057 - 0.8221P + 0.8848Y + u'$$

$$\text{Supply:}\quad Q = 0.0025 - 0.0022\hat{P} + 0.6820Z + v$$
$$\hat{Q} = 0.0026 - 0.0104P + 0.6741Z + v'$$

The estimated structural disturbances u and u' are almost identical with each other and with those computed in Table 12.2; v and v' are also nearly identical with each other and with those computed in Table 12.2. The comparison of estimated structural disturbances, u and u', v and v' is presented in Table 12.3.

For the four cases outlined, the relevant correlation matrices of the disturbances would be as follows.

Case I

	u	v
u	1.0000	
v	−0.2889	1.0000

Case II

	u'	v'
u'	1.0000	
v'	−0.2896	1.0000

Case III

	u'	v'
u'	1.0000	
v	−0.2989	1.0000

Case IV

	u	v'
u	1.0000	
v'	−0.2805	1.0000

The correlation matrix of P, Q, $\hat{P}$, $\hat{Q}$, Y, and Z is also of interest:

	P	Q	$\hat{P}$	$\hat{Q}$	Y	Z
P	1.0000					
Q	−0.7247	1.0000				
$\hat{P}$	0.9458	−0.6816	1.0000			
$\hat{Q}$	−0.6803	0.9475	−0.7209	1.0000		
Y	0.7414	−0.1248	0.7837	−0.1346	1.0000	
Z	−0.6772	0.9474	−0.7173	0.9999	−0.1294	1.0000

Interpretation of the results. Some peculiarities of the present example are listed below.

1. As Q is highly correlated with Z ($r_{QZ} = 0.9474$; $r_{QZ}^2 = 0.8976$) and is almost negligibly correlated with Y ($r_{QY} = -0.1248$; $r_{QY}^2 = 0.0156$), $\hat{Q} = f(Y, Z)$ proves to be very nearly an exact linear function of Z ($r_{\hat{Q}Z} = 0.9999$); Y has almost no effect on $\hat{Q}$.

In the second-stage calculations, $\hat{Q}$ is virtually an alias for Z, except that the standard deviations are different: $s_{\hat{Q}} = 0.03948$ and $s_Z = 0.05772$. If we multiply the regression coefficient of P and Z in the reduced form by $(s_Z/s_{\hat{Q}}) = 1.46$, we obtain $-0.8320Z(1.46) = -1.2147\hat{Q}$, almost identical with the regression coefficient of P on $\hat{Q}$ in the second stage or structural demand function of Case II.

2. Also, there is very little intercorrelation between Y and Z ($r_{YZ} = -0.1294$; $r_{YZ}^2 = 0.0167$) so the effects of Y and Z on P are essentially independent and additive.

3. It appears that the pork demand and supply model is almost purely recursive. In Case III form, the model can be displayed as

$$\begin{array}{cccc} Z & Y & \hat{Q} & \hat{P} \\ \hline \end{array} \qquad a$$

$$\begin{bmatrix} -0.6820 & & : 1 & 0.0022 \\ & -1.0763 & : 1.2164 & 1 \end{bmatrix} + \begin{bmatrix} 0.0025 \\ 0.0069 \end{bmatrix} = \begin{bmatrix} 0 \\ 0 \end{bmatrix}$$

The coefficient above the diagonal (0.0022) is negligibly small and statistically nonsignificant.

Since our example is so nearly recursive, two-stage least squares has not improved on or significantly changed the results of ordinary least squares. On the positive side, however, two-stage least squares has given us sound results for the demand function—results almost identical with and, therefore, as good as ordinary least squares.

4. When the second-stage supply function was fitted with $\hat{Q}$ and Z as independent variables, we encountered an extreme case of multicollinearity. It is clear that two-stage least squares is as vulnerable to high intercorrelation among the exogenous variables as is single-stage least squares. We shall see shortly that the second stage estimation procedure is more vulnerable than the first, for the second stage used variables such as $\hat{Q}$ and $\hat{P}$, which are exact linear functions of the exogenous variables, as independent variables alongside of certain of the exogenous variables. Although extremes such as the $\hat{Q}$ and Z intercorrelation may be rare, the possibility cannot be ignored.

12.4.1 Two-Stage Least Squares in the Overidentified Case

Now, suppose we decide that the "true" model determining the demand for and the supply of pork includes an additional variable, W, the per capita supply of beef—which we regard as a wholly predetermined variable along with Y and Z. The new variable, in terms of original units (pounds), logarithms

Table 12.4 Beef Supply (Consumption) per Capita, United States, 1922–1941

Year	Beef Supply (Consumption) per Capita: Pounds	Logarithms[a] (3 Decimal Places)	First Differences of Logarithms W
1922	59.1	1.772	—
1923	59.6	1.775	0.003
1924	59.5	1.775	0.000
1925	59.5	1.775	0.000
1926	60.3	1.780	0.005
1927	54.5	1.736	−0.044
1928	48.7	1.688	−0.048
1929	49.7	1.696	0.008
1930	48.9	1.689	−0.007
1931	48.6	1.687	−0.002
1932	46.7	1.669	−0.018
1933	51.5	1.712	0.043
1934	55.9	1.747	0.035
1935	52.9	1.723	−0.024
1936	58.1	1.764	0.041
1937	55.2	1.742	−0.022
1938	54.4	1.736	−0.006
1939	54.7	1.738	0.002
1940	54.9	1.740	0.002
1941	60.9	1.785	0.045

[a] To base 10.

and first differences of logarithms, is shown in Table 12.4. The new variable appears in the structural demand function but not in the structural supply function. Hence, our revised model is

(12.75) Demand: $$Q = a_1 + b_1 P + c_1 Y + d_1 W + u$$

(12.76) Supply: $$Q = a_2 + b_2 P + c_2 Z + v$$

We noted earlier that the method of reduced forms applied to this model gives two alternative estimates of the structural coefficients; elaborate computations are required (in the limited information maximum likelihood approach) to achieve a unique set of estimates which is in some sense "best."

To focus on the distinctive features resulting from the addition of W, we shall decide in advance that the second-stage estimates will use $P = f(\hat{Q}, Y, W)$ in the demand function and $Q = f(\hat{P}, Z)$ in the supply function. This corresponds to Case III of the just-identified model.

The reduced-form equations now express P and Q, respectively, as functions of all three exogenous or predetermined variables:

$$P = f(Y, Z, W) \qquad \text{and} \qquad Q = f(Y, Z, W)$$

The fitted reduced-form equations are:

$$\hat{P} = \underset{(0.0066)}{-0.0101} + \underset{(0.1439)}{1.0375\,Y} - \underset{(0.1180)}{0.8471Z} + \underset{(0.2706)}{0.2368\,W}; \quad R^2 = 0.8986; \quad \bar{S} = 0.0283$$

$$\hat{Q} = \underset{(0.0032)}{0.0025} + \underset{(0.0695)}{0.0331\,Y} + \underset{(0.0570)}{0.6959Z} - \underset{(0.1307)}{0.1885\,W}; \quad R^2 = 0.9102; \quad \bar{S} = 0.0137$$

The second stage gives us the following estimates of the structural equations:

Demand:
$$P = \underset{(0.0065)}{-0.0069} - \underset{(0.1692)}{1.2169\hat{Q}_w} + \underset{(0.1428)}{1.0752\,Y} + \underset{(0.2674)}{0.0037\,W} - \frac{u''}{b_1''}$$
$$R^2 = 0.8989; \qquad \bar{S} = 0.0283$$

Supply:
$$Q = \underset{(0.0033)}{0.0025} - \underset{(0.0617)}{0.0108\hat{P}_w} + \underset{(0.0824)}{0.6738Z} + v'$$
$$R^2 = 0.8979; \qquad \bar{S} = 0.0141$$

In this example, the resulting equations are very close to the three-variable demand function of Case II and the supply function of Case I, the same two functions that are paired as Case III.

Where computations are concerned, two-stage least squares can be applied just as easily to overidentified as to just-identified models. From the standpoint of convenience, then, there is a good deal to say for two-stage least squares in the overidentified case.

In the underidentified case, the method of reduced forms enabled us to estimate one coefficient of the model

$$\text{Demand:} \quad Q = f(P, Y, u)$$
$$\text{Supply:} \quad Q = f(P, v)$$

but not the others. If we try to apply two-stage least squares to this model, we obtain:

Reduced forms

(1) $\hat{Q} = f(Y)$ $\quad \hat{Q} = \underset{(0.0098)}{0.0015} - \underset{(0.2008)}{0.1041\,Y}; \qquad r^2 = 0.0156; \qquad \bar{S} = 0.0425$

(2) $\hat{P} = f(Y)$ $\quad \hat{P} = -0.0088 + 1.2058\,Y; \quad r^2 = 0.5498; \quad \bar{S} = 0.0561$

$$\qquad\qquad (0.0130) \quad (0.2646)$$

Choosing $P = f(\hat{Q}, Y, u)$ as our demand function and $Q = f(\hat{P}, v)$ as our supply function, these are our second stage equations:

Demand: $$P = 0.0133 + 13.8174\hat{Q} + 0.2260\,Y - \frac{u'}{b_1'};$$

$$(0.0501) \quad (30.1843) \quad (3.1396)$$

$$R^2 = 0.5556; \qquad \bar{S} = 0.0574$$

Supply: $$Q = 0.00074 - 0.0858\hat{P} + v; \qquad r^2 = 0.0154; \qquad \bar{S} = 0.0426$$

$$(0.00977) \quad (0.1664)$$

Unfortunately, $\hat{Q}$ is an exact linear function of Y, so an attempt to fit the above demand function involves multicollinearity in the absolute sense.

The correlation matrix of P, $\hat{Q}$, and Y is

	(1) P	(2) $\hat{Q}$	(3) Y
(1) P	1.0000	r_{12}	$r_{13} = r_{12}$
(2) $\hat{Q}$	r_{12}	1.0000	1.0000
(3) Y	$r_{13} = r_{12}$	1.0000	1.0000

As $\hat{Q}$ correlates perfectly with Y, the correlation coefficients of any third variable (such as P) with Q and Y, respectively, will be identical. If P is the dependent variable in a least-squares multiple regression equation, the regression coefficients and their standard arrors will be calculated as ratios with $1 - r_{QY}{}^2$ in the denominators. Since $r_{\hat{Q}Y}{}^2 = 1$, this denominator is zero so the coefficients and standard errors become infinite.

The values actually obtained for the above correlation matrix are

	P	$\hat{Q}$	Y
P	1.0000	−0.7453	0.7414
$\hat{Q}$	−0.7453	1.0000	−0.9962
Y	0.7414	−0.9962	1.0000

Two points should be noted. First, the sign of the regression coefficient of $\hat{Q}$ on Y in the reduced form was negative, so we have perfect *negative* correlation between $\hat{Q}$ and Y. Second, because of rounding errors, the correlation coefficient between $\hat{Q}$ and Y is computed as -0.9962 instead of -1.0000. By carrying more decimal places, we could have caused the coefficient to approach -1.0000 as closely as we liked.

As a further consequence of these small rounding errors, the correlation coefficients obtained between P and $\hat{Q}$ and between P and Y are not identical in absolute value but differ by a fraction of 1 percent, that is, -0.7453 as against 0.7414. Finally, if the rounding errors had been reduced, the standard errors of the regression coefficients of P on $\hat{Q}$ and P on Y in the structural demand function would have become even larger than the very large figures shown.

It should be clear that perfect negative correlation produces the same effects as perfect positive correlation, for the determinant of the correlation submatrix of $\hat{Q}$ and Y is

$$1 - r_{\hat{Q}Y}{}^2 = \begin{vmatrix} 1 & -1 \\ -1 & 1 \end{vmatrix} = 1^2 - (-1)^2 = 1 - 1 = 0$$

With rounding errors, the determinant is

$$1 - r_{\hat{Q}Y}{}^2 = \begin{vmatrix} 1.0000 & -0.9962 \\ -0.9962 & 1.0000 \end{vmatrix} = (1.0000)^2 - (-0.9962)^2 = 1.0000 - 0.9924 = 0.0076$$

The standard errors of the net regression coefficients may be calculated from the formulas

$$s_{b_{P\hat{Q}\cdot Y}} = \sqrt{\frac{1 - R^2_{P\cdot\hat{Q}Y}}{(1 - r_{\hat{Q}Y}{}^2)(n-3)}} \cdot \frac{s_P}{s_{\hat{Q}}}$$

and

$$s_{b_{PY\cdot\hat{Q}}} = \sqrt{\frac{1 - R^2_{P\cdot\hat{Q}Y}}{(1 - r_{\hat{Q}Y}{}^2)(n-3)}} \cdot \frac{s_P}{s_Y}$$

Therefore, if $r_{\hat{Q}Y}{}^2$ is identically zero, each of the standard errors becomes infinite.

It is less clear what is accomplished in the second-stage supply function. From the reduced-form equation, $\hat{P} = f(Y)$, P is an exact linear function of Y. If we now regress Q on $\hat{P}$, the regression estimates, $\hat{Q}$, will also be an exact linear function of Y. From a common-sense standpoint, a supply function which makes the current output of pork a linear function of consumer income (and nothing else) leaves much to be desired.

We may also consider the implications of two-stage least squares in the model,

$$\text{Demand:} \quad P = f(Q, Y, W, u)$$
$$\text{Supply:} \quad Q = f(P, v)$$

The reduced-form equations are $\hat{Q} = f(Y, W)$ and $\hat{P} = f(Y, W)$.

What happens when we try to estimate the demand function as $P = f(\hat{Q}, Y, W)$? We again have strict multicollinearity, but in a less obvious form. The correlation matrix of the four variables is

	(1) P	(2) $\hat{Q}$	(3) Y	(4) W
(1) P	1.0000	r_{12}	r_{13}	r_{14}
(2) $\hat{Q}$	r_{12}	1.0000	r_{23}	r_{24}
(3) Y	r_{13}	r_{23}	1.0000	r_{34}
(4) W	r_{14}	r_{24}	r_{34}	1.0000

The reduced-form equation $\hat{Q} = f(Y, W)$ implies that $R^2_{2.34} = 1$, and that the determinant of the 3 by 3 correlation matrix involving r_{23}, r_{24}, and r_{34} equals zero. For, in order that $R^2_{2.34}$ may equal 1, the numerator of the ratio in

$$R^2_{2.34} = 1 - \frac{\begin{vmatrix} 1 & r_{23} & r_{24} \\ r_{23} & 1 & r_{34} \\ r_{24} & r_{34} & 1 \end{vmatrix}}{\begin{vmatrix} 1 & r_{34} \\ r_{34} & 1 \end{vmatrix}}$$

must equal zero. Hence, as this numerator appears in the denominators of the formulas for regression coefficients and their standard errors when P is the dependent variable, standard errors become infinite, as in the three-variable instance.

A supply function in which $Q = f(\hat{P}) = f(Y, W)$ makes the estimates of production an exact linear function of Y and W, two variables which appear only in the structural demand function, does not appear very attractive.

12.5 FINAL COMMENTS ON ESTIMATION IN MULTIPLE EQUATION MODELS

During the late 1950's and early 1960's, the journal literature on statistical estimation of simultaneous economic relations expanded rapidly. It is not feasible for an introductory book such as this one to survey the many ramifications of this literature.

This chapter has tried to establish some perspective on the relative importance of simultaneity and the so-called identification problem in economic models. As in the earlier chapters of this book, our perspective has been that of economists who are interested in appropriate statistical techniques as an aid to the measurement and clarification of economic relationships.

The spirit of our attempt is closely paralleled in a statement by L. R. Klein (1960).

"If econometric results are today more useful than in the past, this is only partly a result of the particular method of estimation but much more significantly a product of painstaking research of a more pedestrian nature. The building of institutional reality into *a priori* formulations of economic relationships and the refinement of basic data collection have contributed much more to the improvement of empirical econometric results than have more elaborate methods of statistical inference. I look towards improvements in precision of econometric judgments of the order of magnitude of 50 percent as a result of a better knowledge of the functioning of economic institutions, through the use of new measurements on variables, and through the use of more accurate data. In contrast, I would expect marginal improvements of 5 or 10 percent through the use of more powerful methods of statistical inference. All routes to improvement must be followed since any gains, no matter how small, are precious, yet different contributions should be kept in proper perspective. The adoption of more powerful methods of mathematical statistics is no panacea."[9]

Our comparisons between different methods of estimation applied to the same data have shown more similarities than differences. In the area of demand and supply analysis, many cases could be found which, if we limited ourselves to official data, would be more favorable to simultaneous methods of estimation than the pork and beef examples we have presented. On the other hand, if an economic problem involving demand and supply relationships is really important either to a business firm or to a public agency, more direct and intensive methods of data collection, experimental design, surveys and interviews can resolve many identification problems that seem bewildering or inconclusive on the basis of published data which are highly aggregated over firms and over time.

For example, if we refer to Figures 12.1 and 12.2, we may find situations in which a particular firm knows the coefficients of its own supply function quite accurately. It may have a near monopoly of a certain line of goods or services in a limited trade area. Even if one or two competitors are involved in addition to his own firm, the proprietor may know enough through long experience to specify the coefficients of the supply function that is relevant to his "identification problem."

Consider equation (12.6) which we used to express the identification

[9] L. R. Klein, "Single Equation vs. Equation System Methods of Estimation in Econometrics," *Econometrica*, **28** (4), October 1960, p. 867.

problem:

$$B_{pq} = \frac{b_1 s_v^2 - (b_1 + b_2) r_{uv} s_u s_v + b_2 s_u^2}{s_v^2 - 2r_{uv} s_u s_v + s_u^2}$$

If the supplier knew the values of b_2 and s_v in his own supply function, he could substitute their numerical values in the equation. He could estimate B_{pq} as the least-squares regression coefficient of P on Q. If this numerical value also were substituted into equation (12.6), the equation would express the slope of the demand function, b_1, as a function of s_u and r_{uv}. On our assumptions, the supplier could specify the values of v in each time period; if he knew a quantifiable variable (weather, consumer income) that accounted for most of the shifts in consumer demand, he could correlate this variable with his own private time series on v and estimate r_{uv}. He could then express b_1 as a function only of s_u. From conversations with customers, or from family budget data, or from other types of investigation and introspection, he might set reasonable upper and lower bounds for s_u (which, in our example, is expressed in units of the price variable).

Although this procedure is inelegant, it may be quite practical, particularly if knowledge of the consumer demand function will contribute several thousand dollars a year to the supplier's net income and/or to his peace of mind.

EXERCISES

Section 12.1

1. What is the "identification problem?"
2. Suppose we have a model of the form

$$Y = a_1 + b_1 X + c_1 Z$$
$$Y = a_2 + b_2 X + c_2 W$$

where X and Y are endogenous variables and W and Z are exogenous. The reduced forms are estimated to be

$$Y = 50 + 4Z + 2W$$
$$X = 100 - 2Z + 6W$$

What are the estimates of b_1, b_2, c_1, and c_2?

Section 12.2

3. Suppose we have a supply and demand model of the form

Demand: $Q_t = a_1 + b_1 P_t + c_1 Y_t + u_t$

Supply: $Q_t = a_2 + b_2 P_t + c_2 Z_t + v_t$

Estimate the supply and demand equations using the following data and the method of reduced forms.

t	Q_t	P_t	Y_t	Z_t
1	64	67	54	19
2	64	73	62	16
3	56	79	64	21
4	62	76	65	23
5	77	60	66	22
6	80	59	64	20
7	82	58	68	24
8	85	57	60	25
9	84	59	64	20
10	80	66	68	7
11	81	65	70	8
12	86	64	72	12

Section 12.3

4. What are the main advantages derived from having a recursive model rather than a strictly interdependent model?

5. How would you change the supply and demand model presented in Exercise 3 into a recursive model?

Section 12.4

6. Estimate the supply and demand model presented in Exercise 3 using the method of two-stage least squares.

SELECTED REFERENCES

Fox, K. A., "Structural Analysis and the Measurement of Demand for Farm Products," *Review of Economics and Statistics*, **37**, (1), pp. 57–66, 1954.

Fox, K. A., J. K. Sengupta, and E. Thorbecke, *The Theory of Quantitative Economic Policy*, North-Holland Publishing Co., Amsterdam; and Rand McNally and Co., Chicago, pp. 51–72, 1966.

Girshick, M. A. and T. A. Haavelmo, "Statistical Analysis of the Demand for Food: Examples of Simultaneous Estimation of Structural Equations," *Econometrica*, **15**, April 1947, pp. 79–111.

Konijn, H. A., "Identification and Estimation in a Simultaneous Equation Model with Errors in the Variables," *Econometrica*, **30**, January 1962 pp. 79–87.

Koopmans, T. C., "Identification Problems in Economic Model Construction," in *Studies in Econometric Method*, W. Hood and T. C. Koopmans (eds.), John Wiley and Sons, Inc., New York, 1953.

Schultz, Henry, *Statistical Laws of Supply and Demand*, The University of Chicago Press, Chicago, 1928.

Simon, H. A., "Causal Ordering and Identifiability," in *Studies in Econometric Method*, W. Hood and T. C. Koopmans (eds.), John Wiley and Sons, Inc., New York, 1953.

Strotz, R. and H. Wold, "Recursive versus Nonrecursive Systems: An Attempt at Synthesis," *Econometrica*, **28**, April 1960, pp. 417–427.

APPENDIX I

Statistical Tables

Table I Areas of a Standard Normal Distribution

An entry in the table is the proportion under the entire curve which is between $z = 0$ and a positive value of z. Areas for negative values of z are obtained by symmetry.

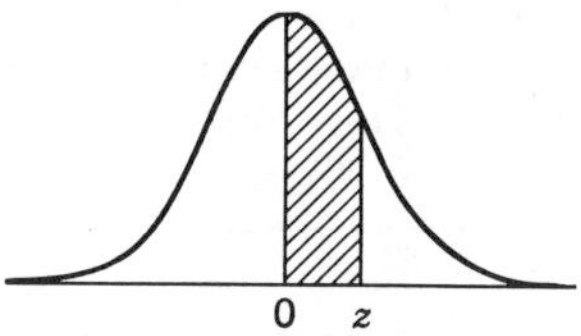

z	.00	.01	.02	.03	.04	.05	.06	.07	.08	.09
0.0	.0000	.0040	.0080	.0120	.0160	.0199	.0239	.0279	.0319	.0359
0.1	.0398	.0438	.0478	.0517	.0557	.0596	.0636	.0675	.0714	.0753
0.2	.0793	.0832	.0871	.0910	.0948	.0987	.1026	.1064	.1103	.1141
0.3	.1179	.1217	.1255	.1293	.1331	.1368	.1406	.1443	.1480	.1517
0.4	.1554	.1591	.1628	.1664	.1700	.1736	.1772	.1808	.1844	.1879
0.5	.1915	.1950	.1985	.2019	.2054	.2088	.2123	.2157	.2190	.2224
0.6	.2257	.2291	.2324	.2357	.2389	.2422	.2454	.2486	.2517	.2549
0.7	.2580	.2611	.2642	.2673	.2703	.2734	.2764	.2794	.2823	.2852
0.8	.2881	.2910	.2939	.2967	.2995	.3023	.3051	.3078	.3106	.3133
0.9	.3159	.3186	.3212	.3238	.3264	.3289	.3315	.3340	.3365	.3389
1.0	.3413	.3438	.3461	.3485	.3508	.3531	.3554	.3577	.3599	.3621
1.1	.3643	.3665	.3686	.3708	.3729	.3749	.3770	.3790	.3810	.3830
1.2	.3849	.3869	.3888	.3907	.3925	.3944	.3962	.3980	.3997	.4015
1.3	.4032	.4049	.4066	.4082	.4099	.4115	.4131	.4147	.4162	.4177
1.4	.4192	.4207	.4222	.4236	.4251	.4265	.4279	.4292	.4306	.4319
1.5	.4332	.4345	.4357	.4370	.4382	.4394	.4406	.4418	.4429	.4441
1.6	.4452	.4463	.4474	.4484	.4495	.4505	.4515	.4525	.4535	.4545
1.7	.4554	.4564	.4573	.4582	.4591	.4599	.4608	.4616	.4625	.4633
1.8	.4641	.4649	.4656	.4664	.4671	.4678	.4686	.4693	.4699	.4706
1.9	.4713	.4719	.4726	.4732	.4738	.4744	.4750	.4756	.4761	.4767
2.0	.4772	.4778	.4783	.4788	.4793	.4798	.4803	.4808	.4812	.4817
2.1	.4821	.4826	.4830	.4834	.4838	.4842	.4846	.4850	.4854	.4857
2.2	.4861	.4864	.4868	.4871	.4875	.4878	.4881	.4884	.4887	.4890
2.3	.4893	.4896	.4898	.4901	.4904	.4906	.4909	.4911	.4913	.4916
2.4	.4918	.4920	.4922	.4925	.4927	.4929	.4931	.4932	.4934	.4936
2.5	.4938	.4940	.4941	.4943	.4945	.4946	.4948	.4949	.4951	.4952
2.6	.4953	.4955	.4956	.4957	.4959	.4960	.4961	.4962	.4963	.4964
2.7	.4965	.4966	.4967	.4968	.4969	.4970	.4971	.4972	.4973	.4974
2.8	.4974	.4975	.4976	.4977	.4977	.4978	.4979	.4979	.4980	.4981
2.9	.4981	.4982	.4982	.4983	.4984	.4984	.4985	.4985	.4986	.4986
3.0	.4987	.4987	.4987	.4988	.4988	.4989	.4989	.4989	.4990	.4990

Table II Student's t Distribution

The first column lists the number of degrees of freedom (ν). The headings of the other columns give probabilities (α) for t to exceed numerically the entry value. Thus for $\alpha = 0.10$ and $v = 5$, we have $P(t \geq 2.0150) = P(t \leq -2.0150) = 0.05 = \alpha/2$.

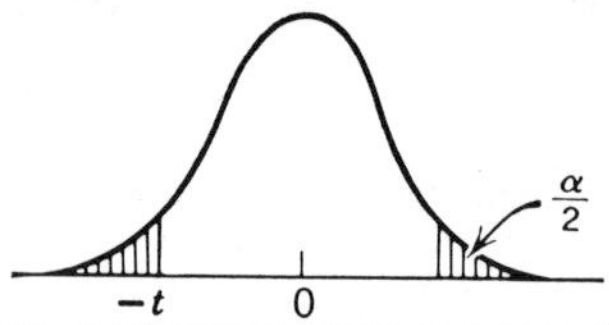

ν \ α	0.50	0.25	0.10	0.05	0.025	0.01	0.005
1	1.00000	2.4142	6.3138	12.7060	25.4520	63.6570	127.3200
2	0.81650	1.6036	2.9200	4.3027	6.2053	9.9248	14.0890
3	0.76489	1.4226	2.3534	3.1825	4.1765	5.8409	7.4533
4	0.74070	1.3444	2.1318	2.7764	3.4954	4.6041	5.5976
5	0.72669	1.3009	2.0150	2.5706	3.1634	4.0321	4.7733
6	0.71756	1.2733	1.9432	2.4469	2.9687	3.7074	4.3168
7	0.71114	1.2543	1.8946	2.3646	2.8412	3.4995	4.0293
8	0.70639	1.2403	1.8595	2.3060	2.7515	3.3554	3.8325
9	0.70272	1.2297	1.8331	2.2622	2.6850	3.2498	3.6897
10	0.69981	1.2213	1.8125	2.2281	2.6338	3.1693	3.5814
11	0.69745	1.2145	1.7959	2.2010	2.5931	3.1058	3.4966
12	0.69548	1.2089	1.7823	2.1788	2.5600	3.0545	3.4284
13	0.69384	1.2041	1.7709	2.1604	2.5326	3.0123	3.3725
14	0.69242	1.2001	1.7613	2.1448	2.5096	2.9768	3.3257
15	0.69120	1.1967	1.7530	2.1315	2.4899	2.9467	3.2860
16	0.69013	1.1937	1.7459	2.1199	2.4729	2.9208	3.2520
17	0.68919	1.1910	1.7396	2.1098	2.4581	2.8982	3.2225
18	0.68837	1.1887	1.7341	2.1009	2.4450	2.8784	3.1966
19	0.68763	1.1866	1.7291	2.0930	2.4334	2.8609	3.1737
20	0.68696	1.1848	1.7247	2.0860	2.4231	2.8453	3.1534
21	0.68635	1.1831	1.7207	2.0796	2.4138	2.8314	3.1352
22	0.68580	1.1816	1.7171	2.0739	2.4055	2.8188	3.1188
23	0.68531	1.1802	1.7139	2.0687	2.3979	2.8073	3.1040
24	0.68485	1.1789	1.7109	2.0639	2.3910	2.7969	3.0905
25	0.68443	1.1777	1.7081	2.0595	2.3846	2.7874	3.0782
26	0.68405	1.1766	1.7056	2.0555	2.3788	2.7787	3.0669
27	0.68370	1.1757	1.7033	2.0518	2.3734	2.7707	3.0565
28	0.68335	1.1748	1.7011	2.0484	2.3685	2.7633	3.0469
29	0.68304	1.1739	1.6991	2.0452	2.3638	2.7564	3.0380
30	0.68276	1.1731	1.6973	2.0423	2.3596	2.7500	3.0298
40	0.68066	1.1673	1.6839	2.0211	2.3289	2.7045	2.9712
60	0.67862	1.1616	1.6707	2.0003	2.2991	2.6603	2.9146
120	0.67656	1.1559	1.6577	1.9799	2.2699	2.6174	2.8599
∞	0.67449	1.1503	1.6449	1.9600	2.2414	2.5758	2.8070

Source: This table is abridged from Table III of Fisher and Yates: *Statistical Tables for Biological, Agricultural and Medical Research*, published by Oliver Boyd Ltd., Edinburgh, and by permission of the authors and publishers.

Table III *F* Distribution

5% (Roman Type) and 1% (Boldface Type) Points for the Distribution of *F*

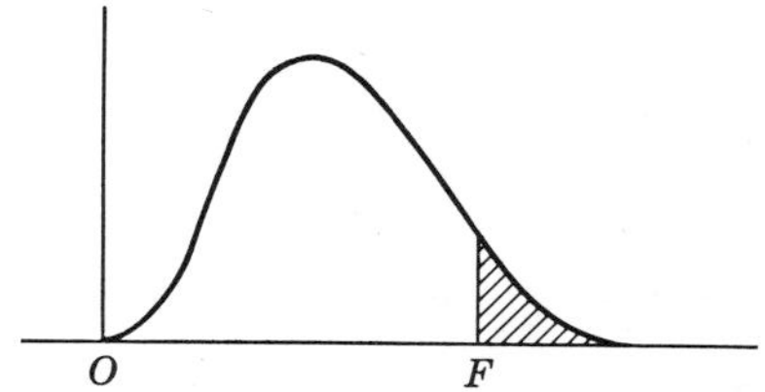

Degrees of Freedom for Denominator (v_2)	Degrees of Freedom for Numerator (v_1)																						
	1	2	3	4	5	6	7	8	9	10	11	12	14	16	20	24	30	40	50	75	100	200	500	∞
1	161 **4052**	200 **4999**	216 **5403**	225 **5625**	230 **5764**	234 **5859**	237 **5928**	239 **5981**	241 **6022**	242 **6056**	243 **6082**	244 **6106**	245 **6142**	246 **6169**	248 **6208**	249 **6234**	250 **6258**	251 **6286**	252 **6302**	253 **6323**	253 **6334**	254 **6352**	254 **6361**	254 **6366**
2	18.51 **98.49**	19.00 **99.01**	19.16 **99.17**	19.25 **99.25**	19.30 **99.30**	19.33 **99.33**	19.36 **99.34**	19.37 **99.36**	19.38 **99.38**	19.39 **99.40**	19.40 **99.41**	19.41 **99.42**	19.42 **99.43**	19.43 **99.44**	19.44 **99.45**	19.45 **99.46**	19.46 **99.47**	19.47 **99.48**	19.47 **99.48**	19.48 **99.49**	19.49 **99.49**	19.49 **99.49**	19.50 **99.50**	19.50 **99.50**
3	10.13 **34.12**	9.55 **30.81**	9.28 **29.46**	9.12 **28.71**	9.01 **28.24**	8.94 **27.91**	8.88 **27.67**	8.84 **27.49**	8.81 **27.34**	8.78 **27.23**	8.76 **27.13**	8.74 **27.05**	8.71 **26.92**	8.69 **26.83**	8.66 **26.69**	8.64 **26.60**	8.62 **26.50**	8.60 **26.41**	8.58 **26.30**	8.57 **26.27**	8.56 **26.23**	8.54 **26.18**	8.54 **26.14**	8.53 **26.12**
4	7.71 **21.20**	6.94 **18.00**	6.59 **16.69**	6.39 **15.98**	6.26 **15.52**	6.16 **15.21**	6.09 **14.98**	6.04 **14.80**	6.00 **14.66**	5.96 **14.54**	5.93 **14.45**	5.91 **14.37**	5.87 **14.24**	5.84 **14.15**	5.80 **14.02**	5.77 **13.93**	5.74 **13.83**	5.71 **13.74**	5.70 **13.69**	5.68 **13.61**	5.66 **13.57**	5.65 **13.52**	5.64 **13.48**	5.63 **13.46**
5	6.61 **16.26**	5.79 **13.27**	5.41 **12.06**	5.19 **11.39**	5.05 **10.97**	4.95 **10.67**	4.88 **10.45**	4.82 **10.27**	4.78 **10.15**	4.74 **10.05**	4.70 **9.96**	4.68 **9.89**	4.64 **9.77**	4.60 **9.63**	4.56 **9.65**	4.53 **9.47**	4.50 **9.38**	4.46 **9.29**	4.44 **9.24**	4.42 **9.17**	4.40 **9.13**	4.38 **9.07**	4.37 **9.04**	4.36 **9.02**
6	5.99 **13.74**	5.14 **10.92**	4.76 **9.73**	4.53 **9.16**	4.39 **8.75**	4.28 **8.47**	4.21 **8.26**	4.15 **8.10**	4.10 **7.98**	4.06 **7.87**	4.03 **7.79**	4.00 **7.72**	3.96 **7.60**	3.92 **7.52**	3.87 **7.39**	3.84 **7.31**	3.81 **7.23**	3.77 **7.14**	3.75 **7.09**	3.72 **7.02**	3.71 **6.99**	3.69 **6.94**	3.68 **6.90**	3.67 **6.88**
7	5.59 **12.25**	4.74 **9.55**	4.35 **8.45**	4.12 **7.85**	3.97 **7.46**	3.87 **7.19**	3.79 **7.00**	3.73 **6.84**	3.68 **6.71**	3.63 **6.62**	3.60 **6.54**	3.57 **6.47**	3.52 **6.35**	3.49 **6.27**	3.44 **6.15**	3.41 **6.07**	3.38 **5.98**	3.34 **5.90**	3.32 **5.85**	3.29 **5.78**	3.28 **5.75**	3.25 **5.70**	3.24 **5.67**	3.23 **5.65**
8	5.32 **11.26**	4.46 **8.65**	4.07 **7.59**	3.84 **7.01**	3.69 **6.63**	3.58 **6.37**	3.50 **6.19**	3.44 **6.03**	3.39 **5.91**	3.34 **5.82**	3.31 **5.74**	3.28 **5.67**	3.23 **5.56**	3.20 **5.48**	3.15 **5.36**	3.12 **5.28**	3.08 **5.20**	3.05 **5.11**	3.03 **5.06**	3.00 **5.00**	2.98 **4.96**	2.96 **4.91**	2.94 **4.88**	2.93 **4.86**
9	5.12 **10.56**	4.26 **8.02**	3.86 **6.99**	3.63 **6.42**	3.48 **6.06**	3.37 **5.80**	3.29 **5.62**	3.23 **5.47**	3.18 **5.35**	3.13 **5.26**	3.10 **5.18**	3.07 **5.11**	3.02 **5.00**	2.98 **4.92**	2.93 **4.80**	2.90 **4.73**	2.86 **4.64**	2.82 **4.56**	2.80 **4.51**	2.77 **4.45**	2.76 **4.41**	2.73 **4.36**	2.72 **4.33**	2.71 **4.31**
10	4.96 **10.04**	4.10 **7.56**	3.71 **6.55**	3.48 **5.99**	3.33 **5.64**	3.22 **5.39**	3.14 **5.21**	3.07 **5.06**	3.02 **4.95**	2.97 **4.85**	2.91 **4.78**	2.91 **4.71**	2.86 **4.60**	2.82 **4.52**	2.77 **4.41**	2.74 **4.33**	2.70 **4.25**	2.67 **4.17**	2.64 **4.12**	2.61 **4.05**	2.59 **4.01**	2.56 **3.96**	2.55 **3.93**	2.54 **3.91**

Table III *F* (*continued*)

Degrees of Freedom for Denominator (v_2)	Degrees of Freedom for Numerator (v_1)																						
	1	2	3	4	5	6	7	8	9	10	11	12	14	16	20	24	30	40	50	75	100	200	500	∞
11	4.84 **9.65**	3.98 **7.20**	3.59 **6.22**	3.36 **5.67**	3.20 **5.32**	3.09 **5.07**	3.01 **4.88**	2.95 **4.74**	2.90 **4.63**	2.86 **4.54**	2.82 **4.46**	2.79 **4.40**	2.74 **4.29**	2.70 **4.21**	2.65 **4.10**	2.61 **4.02**	2.57 **3.94**	2.53 **3.86**	2.50 **3.80**	2.47 **3.74**	2.45 **3.70**	2.42 **3.66**	2.41 **3.62**	2.40 **3.60**
12	4.75 **9.33**	3.88 **6.93**	3.49 **5.95**	3.26 **5.41**	3.11 **5.06**	3.00 **4.82**	2.92 **4.65**	2.85 **4.50**	2.80 **4.39**	2.76 **4.30**	2.72 **4.22**	2.69 **4.16**	2.64 **4.05**	2.60 **3.98**	2.54 **3.86**	2.50 **3.78**	2.46 **3.70**	2.42 **3.61**	2.40 **3.56**	2.36 **3.49**	2.35 **3.46**	2.32 **3.41**	2.31 **3.38**	2.30 **3.36**
13	4.67 **9.07**	3.80 **6.70**	3.41 **5.74**	3.18 **5.20**	3.02 **4.86**	2.92 **4.62**	2.84 **4.44**	2.77 **4.30**	2.72 **4.19**	2.67 **4.10**	2.63 **4.02**	2.60 **3.96**	2.55 **3.85**	2.51 **3.78**	2.46 **3.67**	2.42 **3.59**	2.38 **3.51**	2.34 **3.42**	2.32 **3.37**	2.38 **3.30**	2.26 **3.27**	2.24 **3.21**	2.22 **3.18**	2.21 **3.16**
14	4.60 **8.86**	3.74 **6.51**	3.34 **5.56**	3.11 **5.03**	2.96 **4.69**	2.85 **4.46**	2.77 **4.28**	2.70 **4.14**	2.65 **4.03**	2.60 **3.94**	2.56 **3.86**	2.53 **3.80**	2.48 **3.70**	2.44 **3.62**	2.39 **3.51**	2.35 **3.43**	2.31 **3.34**	2.27 **3.26**	2.24 **3.21**	2.21 **3.14**	2.19 **3.11**	2.16 **3.06**	2.14 **3.02**	2.13 **3.00**
15	4.54 **8.68**	3.68 **6.36**	3.29 **5.42**	3.06 **4.89**	2.90 **4.56**	2.79 **4.32**	2.70 **4.14**	2.64 **4.00**	2.59 **3.89**	2.55 **3.80**	2.51 **3.73**	2.48 **3.67**	2.43 **3.56**	2.39 **3.48**	2.33 **3.36**	2.29 **3.29**	2.25 **3.20**	2.21 **3.12**	2.18 **3.07**	2.15 **3.00**	2.12 **2.97**	2.10 **2.92**	2.08 **2.89**	2.07 **2.87**
16	4.49 **8.53**	3.63 **6.23**	3.24 **5.29**	3.01 **4.77**	2.85 **4.44**	2.74 **4.20**	2.66 **4.03**	2.59 **3.89**	2.54 **3.78**	2.49 **3.69**	2.45 **3.61**	2.42 **3.55**	2.37 **3.45**	2.33 **3.37**	2.28 **3.25**	2.24 **3.18**	2.20 **3.10**	2.16 **3.01**	2.13 **2.96**	2.09 **2.89**	2.07 **2.86**	2.04 **2.80**	2.02 **2.77**	2.01 **2.75**
17	4.45 **8.40**	3.59 **6.11**	3.20 **5.18**	2.96 **4.67**	2.81 **4.34**	2.70 **4.10**	2.62 **3.93**	2.55 **3.79**	2.50 **3.68**	2.45 **3.59**	2.41 **3.52**	2.38 **3.45**	2.33 **3.35**	2.29 **3.27**	2.23 **3.16**	2.19 **3.08**	2.15 **3.00**	2.11 **2.92**	2.08 **2.86**	2.04 **2.79**	2.02 **2.76**	1.99 **2.70**	1.97 **2.67**	1.96 **2.65**
18	4.41 **8.28**	3.55 **6.01**	3.16 **5.09**	2.93 **4.53**	2.77 **4.25**	2.66 **4.01**	2.58 **3.85**	2.51 **3.71**	2.46 **3.60**	2.41 **3.51**	2.37 **3.44**	2.34 **3.37**	2.29 **3.27**	2.25 **3.19**	2.19 **3.07**	2.15 **3.00**	2.11 **2.91**	2.07 **2.83**	2.04 **2.78**	2.00 **2.71**	1.98 **2.68**	1.95 **2.62**	1.93 **2.59**	1.92 **2.57**
19	4.38 **8.18**	3.52 **5.93**	3.13 **5.01**	2.90 **4.60**	2.74 **4.17**	2.63 **3.94**	2.55 **3.77**	2.48 **3.63**	2.43 **3.62**	2.38 **3.43**	2.34 **3.36**	2.31 **3.30**	2.26 **3.19**	2.21 **3.12**	2.15 **3.00**	2.11 **2.92**	2.07 **2.84**	2.02 **2.76**	2.00 **2.70**	1.96 **2.63**	1.94 **2.60**	1.91 **2.54**	1.90 **2.51**	1.88 **2.49**
20	4.35 **8.10**	3.49 **5.85**	3.10 **4.94**	2.87 **4.43**	2.71 **4.10**	2.60 **3.87**	2.52 **3.71**	2.45 **3.56**	2.40 **3.45**	2.35 **3.37**	2.31 **3.30**	2.28 **3.23**	2.23 **3.13**	2.18 **3.05**	2.12 **2.94**	2.08 **2.86**	2.04 **2.77**	1.99 **2.69**	1.96 **2.63**	1.92 **2.56**	1.90 **2.53**	1.87 **2.47**	1.85 **2.44**	1.84 **2.42**
21	4.32 **8.02**	3.47 **5.78**	3.07 **4.87**	2.84 **4.37**	2.68 **4.04**	2.57 **3.81**	2.49 **3.65**	2.42 **3.51**	2.37 **3.40**	2.32 **3.31**	2.28 **3.24**	2.25 **3.17**	2.20 **3.07**	2.15 **2.99**	2.09 **2.88**	2.05 **2.80**	2.00 **2.72**	1.96 **2.63**	1.93 **2.58**	1.89 **2.51**	1.87 **2.47**	1.84 **2.42**	1.82 **2.38**	1.81 **2.36**
22	4.30 **7.94**	3.44 **5.72**	3.05 **4.82**	2.82 **4.31**	2.66 **3.99**	2.55 **3.76**	2.47 **3.59**	2.40 **3.45**	2.35 **3.35**	2.30 **3.26**	2.26 **3.18**	2.23 **3.12**	2.18 **3.02**	2.13 **2.94**	2.07 **2.83**	2.03 **2.75**	1.98 **2.67**	1.93 **2.58**	1.91 **2.53**	1.87 **2.46**	1.84 **2.42**	1.81 **2.37**	1.80 **2.33**	1.78 **2.31**
23	4.28 **7.88**	3.42 **5.66**	3.03 **4.76**	2.80 **4.26**	2.64 **3.94**	2.53 **3.71**	2.45 **3.54**	2.38 **3.41**	2.32 **3.30**	2.28 **3.21**	2.24 **3.14**	2.20 **3.07**	2.14 **2.97**	2.10 **2.89**	2.04 **2.78**	2.00 **2.70**	1.96 **2.62**	1.91 **2.53**	1.88 **2.48**	1.84 **2.41**	1.82 **2.37**	1.79 **2.32**	1.77 **2.28**	1.76 **2.26**
24	4.26 **7.82**	3.40 **5.61**	3.01 **4.72**	2.78 **4.22**	2.62 **3.90**	2.51 **3.67**	2.43 **3.50**	2.36 **3.36**	2.30 **3.25**	2.26 **3.17**	2.22 **3.09**	2.18 **3.03**	2.13 **2.93**	2.09 **2.85**	2.02 **2.74**	1.98 **2.66**	1.94 **2.58**	1.89 **2.49**	1.86 **2.44**	1.82 **2.36**	1.80 **2.33**	1.76 **2.27**	1.74 **2.23**	1.73 **2.21**

Table III F (*continued*)

Degrees of Freedom for Denominator (v_2)	Degrees of Freedom for Numerator (v_1)																						
	1	2	3	4	5	6	7	8	9	10	11	12	14	16	20	24	30	40	50	75	100	200	500	∞
25	4.24 **7.77**	3.38 **5.57**	2.99 **4.68**	2.76 **4.18**	2.60 **3.86**	2.49 **3.63**	2.41 **3.46**	2.34 **3.32**	2.28 **3.21**	2.24 **3.13**	2.20 **3.05**	2.16 **2.99**	2.11 **2.89**	2.06 **2.81**	2.00 **2.70**	1.96 **2.62**	1.92 **2.54**	1.87 **2.45**	1.84 **2.40**	1.80 **2.32**	1.77 **2.29**	1.74 **2.23**	1.72 **2.19**	1.71 **2.17**
26	4.22 **7.72**	3.37 **5.53**	2.89 **4.64**	2.74 **4.14**	2.59 **3.82**	2.47 **3.59**	2.39 **3.42**	2.32 **3.29**	2.27 **3.17**	2.22 **3.09**	2.18 **3.02**	2.15 **2.96**	2.10 **2.86**	2.05 **2.77**	1.99 **2.66**	1.95 **2.58**	1.90 **2.50**	1.85 **2.41**	1.82 **2.36**	1.78 **2.28**	1.76 **2.25**	1.72 **2.19**	1.70 **2.15**	1.69 **2.13**
27	4.21 **7.68**	3.35 **5.49**	2.96 **4.60**	2.73 **4.11**	2.57 **3.79**	2.46 **3.56**	2.37 **3.39**	2.30 **3.26**	2.25 **3.14**	2.20 **3.06**	2.16 **2.98**	2.13 **2.93**	2.08 **2.83**	2.03 **2.74**	1.97 **2.63**	1.93 **2.55**	1.88 **2.47**	1.84 **2.38**	1.80 **2.33**	1.76 **2.25**	1.74 **2.21**	1.71 **2.16**	1.68 **2.12**	1.67 **2.10**
28	4.20 **7.64**	3.34 **5.54**	2.95 **4.57**	2.71 **4.07**	2.56 **3.76**	2.44 **3.53**	2.36 **3.36**	2.29 **3.23**	3.24 **3.11**	2.19 **3.03**	2.15 **2.95**	2.12 **2.90**	2.06 **2.80**	2.02 **2.71**	1.96 **2.60**	1.91 **2.52**	1.87 **2.44**	1.81 **2.35**	1.78 **2.30**	1.75 **2.22**	1.72 **2.18**	1.69 **2.13**	1.67 **2.09**	1.65 **2.06**
29	4.18 **7.60**	3.33 **5.52**	2.93 **4.54**	2.70 **4.04**	2.54 **3.73**	2.43 **3.50**	2.35 **3.33**	2.28 **3.20**	2.22 **3.08**	2.18 **3.00**	2.14 **2.92**	2.10 **2.87**	2.05 **2.77**	2.00 **2.68**	1.94 **2.57**	1.90 **2.49**	1.85 **2.41**	1.80 **2.32**	1.77 **2.27**	1.73 **2.19**	1.71 **2.15**	1.68 **2.10**	1.65 **2.06**	1.64 **2.03**
30	4.17 **7.56**	3.32 **5.39**	2.92 **4.51**	2.69 **4.02**	2.53 **3.70**	2.42 **3.47**	2.34 **3.30**	2.27 **3.17**	2.21 **3.06**	2.16 **2.98**	2.12 **2.90**	2.09 **2.84**	2.04 **2.74**	1.99 **2.66**	1.93 **2.55**	1.89 **2.47**	1.84 **2.38**	1.79 **2.29**	1.76 **2.24**	1.72 **2.16**	1.69 **2.13**	1.66 **2.07**	1.64 **2.03**	1.62 **2.01**
32	4.15 **7.50**	3.30 **5.34**	2.90 **4.46**	2.67 **3.97**	2.51 **3.66**	2.40 **3.42**	2.32 **3.25**	2.25 **3.12**	2.19 **3.01**	2.14 **2.94**	2.10 **2.86**	2.07 **2.80**	2.02 **2.70**	1.97 **2.62**	1.91 **2.51**	1.86 **2.42**	1.82 **2.34**	1.76 **2.25**	1.74 **2.20**	1.69 **2.12**	1.67 **2.08**	1.64 **2.02**	1.61 **1.98**	1.59 **1.96**
34	4.13 **7.44**	3.28 **5.29**	2.88 **4.42**	2.65 **3.93**	2.49 **3.61**	2.38 **3.38**	2.30 **3.21**	2.23 **3.08**	2.17 **2.97**	2.12 **2.89**	2.08 **2.82**	2.05 **2.76**	2.00 **2.66**	1.95 **2.58**	1.89 **2.47**	1.84 **2.38**	1.80 **2.30**	1.74 **2.21**	1.71 **2.15**	1.67 **2.08**	1.64 **2.04**	1.61 **1.98**	1.59 **1.94**	1.57 **1.91**
36	4.11 **7.39**	3.26 **5.25**	2.86 **4.38**	2.63 **3.89**	2.48 **3.58**	2.36 **3.35**	2.28 **3.18**	2.21 **3.04**	2.15 **2.94**	2.10 **2.86**	2.06 **2.78**	2.03 **2.72**	1.89 **2.62**	1.93 **2.54**	1.87 **2.43**	1.82 **2.35**	1.78 **2.26**	1.72 **2.17**	1.69 **2.12**	1.65 **2.04**	1.62 **2.00**	1.59 **1.94**	1.56 **1.90**	1.55 **1.87**
38	4.10 **7.35**	3.25 **5.21**	2.85 **4.34**	2.62 **3.86**	2.46 **3.54**	2.35 **3.32**	2.26 **3.15**	2.19 **3.02**	2.14 **2.91**	2.09 **2.82**	2.05 **2.75**	2.02 **2.69**	1.96 **2.59**	1.92 **2.51**	1.85 **2.40**	1.80 **2.32**	1.76 **2.22**	1.71 **2.14**	1.67 **2.08**	1.63 **2.00**	1.60 **1.97**	1.57 **1.90**	1.54 **1.86**	1.53 **1.84**
40	4.08 **7.31**	3.23 **5.18**	2.84 **4.31**	2.61 **3.83**	2.45 **3.51**	2.34 **3.29**	2.25 **3.12**	2.18 **2.99**	2.12 **2.88**	2.07 **2.80**	2.04 **2.73**	2.00 **2.66**	1.95 **2.56**	1.90 **2.49**	1.84 **2.37**	1.79 **2.29**	1.74 **2.20**	1.69 **2.11**	1.66 **2.05**	1.61 **1.97**	1.59 **1.94**	1.55 **1.88**	1.53 **1.84**	1.51 **1.81**
42	4.07 **7.27**	3.22 **5.15**	2.83 **4.29**	2.59 **3.80**	2.44 **3.49**	2.32 **3.26**	2.24 **3.10**	2.17 **2.96**	2.11 **2.86**	2.06 **2.77**	2.02 **2.70**	1.99 **2.64**	1.94 **2.54**	1.89 **2.46**	1.82 **2.35**	1.78 **2.26**	1.73 **2.17**	1.68 **2.08**	1.64 **2.02**	1.60 **1.94**	1.57 **1.91**	1.54 **1.85**	1.51 **1.80**	1.49 **1.78**
44	4.06 **7.24**	3.21 **5.12**	2.82 **4.26**	2.58 **3.78**	2.43 **3.46**	2.31 **3.24**	2.23 **3.07**	2.16 **2.94**	2.10 **2.84**	2.05 **2.75**	2.01 **2.68**	1.98 **2.62**	1.92 **2.52**	1.88 **2.44**	1.81 **2.32**	1.76 **2.24**	1.72 **2.15**	1.66 **2.06**	1.63 **2.00**	1.58 **1.92**	1.56 **1.88**	1.52 **1.82**	1.50 **1.78**	1.48 **1.75**
46	4.05 **7.21**	3.20 **5.10**	2.81 **4.24**	2.57 **3.76**	2.42 **3.44**	2.30 **3.22**	2.22 **3.05**	2.14 **2.92**	2.09 **2.82**	2.04 **2.73**	2.00 **2.66**	1.97 **2.60**	1.91 **2.50**	1.87 **2.42**	1.80 **2.30**	1.75 **2.22**	1.71 **2.13**	1.65 **2.04**	1.62 **1.98**	1.57 **1.90**	1.54 **1.86**	1.51 **1.80**	1.48 **1.76**	1.46 **1.72**
∞	3.84 **6.64**	2.99 **4.60**	2.60 **3.78**	2.37 **3.32**	2.21 **3.02**	2.09 **2.80**	2.01 **2.64**	1.94 **2.51**	1.88 **2.41**	1.83 **2.32**	1.79 **2.24**	1.75 **2.18**	1.69 **2.07**	1.64 **1.99**	1.57 **1.87**	1.52 **1.79**	1.46 **1.69**	1.40 **1.59**	1.35 **1.52**	1.28 **1.41**	1.24 **1.36**	1.17 **1.25**	1.11 **1.15**	1.00 **1.00**

Source: Reproduced by permission from *Statistical Methods*, 5th edition, by George W. Snedecor, © 1956 by Iowa State University Press.

Table IV The χ^2 Distribution

The first column lists the number of degrees of freedom (ν). The headings of the other columns give probabilities α for χ^2 to exceed the entry value. For $\nu > 100$, treat $\sqrt{2\chi^2} - \sqrt{2\nu - 1}$ as a standard normal variable.

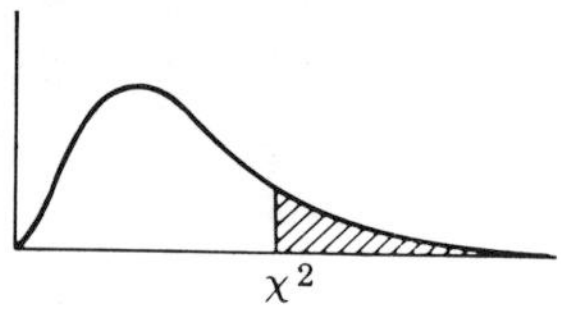

ν \ α	0.995	0.975	0.050	0.025	0.010	0.005
1	$0.0^4 3927$	$0.0^3 9821$	3.84146	5.02389	6.63490	7.87944
2	0.010025	0.050636	5.99147	7.37776	9.21034	10.5966
3	0.071721	0.215795	7.81473	9.34840	11.3449	12.8381
4	0.206990	0.484419	9.48773	11.1433	13.2767	14.8602
5	0.411740	0.831211	11.0705	12.8325	15.0863	16.7496
6	0.675727	1.237347	12.5916	14.4494	16.8119	18.5476
7	0.989265	1.68987	14.0671	16.0128	18.4753	20.2777
8	1.344419	2.17973	15.5073	17.5346	20.0902	21.9550
9	1.734926	2.70039	16.9190	19.0228	12.6660	23.5893
10	2.15585	3.24697	18.3070	20.4831	23.2093	25.1882
11	2.60321	3.81575	19.6751	21.9200	24.7250	26.7569
12	3.07382	4.40379	21.0261	23.3367	26.2170	28.2995
13	3.56503	5.00874	22.3621	24.7356	27.6883	29.8194
14	4.07468	5.62872	23.6848	26.1190	29.1413	31.3193
15	4.60094	6.26214	24.9958	27.4884	30.5779	32.8013
16	5.14224	6.90766	26.2962	28.8454	31.9999	34.2672
17	5.69724	7.56418	27.5871	30.1910	33.4087	35.7185
18	6.26481	8.23075	28.8693	31.5264	34.8053	37.1564
19	6.84398	8.90655	30.1435	32.8523	36.1908	38.5822
20	7.43386	9.59083	31.4104	34.1696	37.5662	39.9968
21	8.03366	10.28293	32.6705	35.4789	38.9321	41.4010
22	8.64272	10.9823	33.9244	36.7807	40.2894	42.7956
23	9.26042	11.6885	35.1725	38.0757	41.6384	44.1813
24	9.88623	12.4001	36.4151	39.3641	42.9798	45.5585
25	10.5197	13.1197	37.6525	40.6465	44.3141	46.9278
26	11.1603	13.8439	38.8852	41.9232	45.6417	48.2899
27	11.8076	14.5733	40.1133	43.1944	46.9630	49.6449
28	12.4613	15.3079	41.3372	44.4607	48.2782	50.9933
29	13.1211	16.0471	42.5569	45.7222	49.5879	52.3356
30	13.7867	16.7908	43.7729	46.9792	50.8922	53.6720
40	20.7065	24.4331	55.7585	59.3417	63.6907	66.7659
50	27.9907	32.3574	67.5048	71.4202	76.1539	79.4900
60	35.5346	40.4817	79.0819	83.2976	88.3794	91.9517
70	43.2752	48.7576	90.5312	95.0231	100.425	104.215
80	51.1720	57.1532	101.879	106.629	112.329	116.321
90	59.1963	65.6466	113.145	118.136	124.116	128.299
100	67.3276	74.2219	124.342	129.561	135.807	140.169

Source: This table is abridged from Table 8 of *Biometrika Tables for Statisticians*, Vol. 1, 3rd edition, 1966, and is published with permission of the editor of *Biometrika*.

Table V Binomial Distribution Function $_nC_xp^x(1-p)^{n-x}$

Entries in the table are values of $_nC_xp^x(1-p)^{n-x}$ for the indicated values of n, x and p. When $p > 0.5$, the value of $_nC_xp^x(1-p)^{n-x}$ for a given n, x and p is obtained by finding the tabular entry for the given n, with $n-x$ in place of the given x, and $1-p$ in place of the given p.

		p									
n	x	.05	.10	.15	.20	.25	.30	.35	.40	.45	.50
1	0	.9500	.9000	.8500	.8000	.7500	.7000	.6500	.6000	.5500	.5000
	1	.0500	.1000	.1500	.2000	.2500	.3000	.3500	.4000	.4500	.5000
2	0	.9025	.8100	.7225	.6400	.5625	.4900	.4225	.3600	.3025	.2500
	1	.0950	.1800	.2550	.3200	.3750	.4200	.4550	.4800	.4950	.5000
	2	.0025	.0100	.0225	.0400	.0625	.0900	.1225	.1600	.2025	.2500
3	0	.8574	.7290	.6141	.5120	.4219	.3430	.2746	.2160	.1664	.1250
	1	.1354	.2430	.3251	.3840	.4219	.4410	.4436	.4320	.4084	.3750
	2	.0071	.0270	.0574	.0960	.1406	.1890	.2389	.2880	.3341	.3750
	3	.0001	.0010	.0034	.0080	.0156	.0270	.0429	.0640	.0911	.1250
4	0	.8145	.6561	.5220	.4096	.3164	.2401	.1785	.1296	.0915	.0625
	1	.1715	.2916	.3685	.4096	.4219	.4116	.3845	.3456	.2995	.2500
	2	.0135	.0486	.0975	.1536	.2109	.2646	.3105	.3456	.3675	.3750
	3	.0005	.0036	.0115	.0256	.0469	.0756	.1115	.1536	.2005	.2500
	4	.0000	.0001	.0005	.0016	.0039	.0081	.0150	.0256	.0410	.0625
5	0	.7738	.5905	.4437	.3277	.2373	.1681	.1160	.0778	.0503	.0312
	1	.2036	.3280	.3915	.4096	.3955	.3602	.3124	.2592	.2059	.1562
	2	.0214	.0729	.1382	.2048	.2637	.3087	.3364	.3456	.3369	.3125
	3	.0011	.0081	.0244	.0512	.0879	.1323	.1811	.2304	.2757	.3125
	4	.0000	.0004	.0022	.0064	.0146	.0284	.0488	.0768	.1128	.1562
	5	.0000	.0000	.0001	.0003	.0010	.0024	.0053	.0102	.0185	.0312
6	0	.7351	.5314	.3771	.2621	.1780	.1176	.0754	.0467	.0277	.0156
	1	.2321	.3543	.3993	.3932	.3560	.3025	.2437	.1866	.1359	.0938
	2	.0305	.0984	.1762	.2458	.2966	.3241	.3280	.3110	.2780	.2344
	3	.0021	.0146	.0415	.0819	.1318	.1852	.2355	.2765	.3032	.3125
	4	.0001	.0012	.0055	.0154	.0330	.0595	.0951	.1382	.1861	.2344
	5	.0000	.0001	.0004	.0015	.0044	.0102	.0205	.0369	.0609	.0938
	6	.0000	.0000	.0000	.0001	.0002	.0007	.0018	.0041	.0083	.0156
7	0	.6983	.4783	.3206	.2097	.1335	.0824	.0490	.0280	.0152	.0078
	1	.2573	.3720	.3960	.3670	.3115	.2471	.1848	.1306	.0872	.0547
	2	.0406	.1240	.2097	.2753	.3115	.3177	.2985	.2613	.2140	.1641
	3	.0036	.0230	.0617	.1147	.1730	.2269	.2679	.2903	.2918	.2734
	4	.0002	.0026	.0109	.0287	.0577	.0972	.1442	.1935	.2388	.2734

Table V (*continued*) Binomial Distribution Function $_nC_xp^x(1-p)^{n-x}$

n	x	.05	.10	.15	.20	.25	.30	.35	.40	.45	.50
	5	.0000	.0002	.0012	.0043	.0115	.0250	.0466	.0774	.1172	.1641
	6	.0000	.0000	.0001	.0004	.0013	.0036	.0084	.0172	.0320	.0547
	7	.0000	.0000	.0000	.0000	.0001	.0002	.0006	.0016	.0037	.0078
8	0	.6634	.4305	.2725	.1678	.1001	.0576	.0319	.0168	.0084	.0039
	1	.2793	.3826	.3847	.3355	.2670	.1977	.1373	.0896	.0548	.0312
	2	.0515	.1488	.2376	.2936	.3115	.2965	.2587	.2090	.1569	.1094
	3	.0054	.0331	.0839	.1468	.2076	.2541	.2786	.2787	.2568	.2188
	4	.0004	.0046	.0185	.0459	.0865	.1361	.1875	.2322	.2627	.2734
	5	.0000	.0004	.0026	.0092	.0231	.0467	.0808	.1239	.1719	.2188
	6	.0000	.0000	.0002	.0011	.0038	.0100	.0217	.0413	.0703	.1094
	7	.0000	.0000	.0000	.0001	.0004	.0012	.0033	.0079	.0164	.0312
	8	.0000	.0000	.0000	.0000	.0000	.0001	.0002	.0007	.0017	.0039
9	0	.6302	.3874	.2316	.1342	.0751	.0404	.0207	.0101	.0046	.0020
	1	.2985	.3874	.3679	.3020	.2253	.1556	.1004	.0605	.0339	.0176
	2	.0629	.1722	.2597	.3020	.3003	.2668	.2162	.1612	.1110	.0703
	3	.0077	.0446	.1069	.1762	.2336	.2668	.2716	.2508	.2119	.1641
	4	.0006	.0074	.0283	.0661	.1168	.1715	.2194	.2508	.2600	.2461
	5	.0000	.0008	.0050	.0165	.0389	.0735	.1181	.1672	.2128	.2461
	6	.0000	.0001	.0006	.0028	.0087	.0210	.0424	.0743	.1160	.1641
	7	.0000	.0000	.0000	.0003	.0012	.0039	.0098	.0212	.0407	.0703
	8	.0000	.0000	.0000	.0000	.0001	.0004	.0013	.0035	.0083	.0176
	9	.0000	.0000	.0000	.0000	.0000	.0000	.0001	.0003	.0008	.0020
10	0	.5987	.3487	.1969	.1074	.0563	.0282	.0135	.0060	.0025	.0010
	1	.3151	.3874	.3474	.2684	.1877	.1211	.0725	.0403	.0207	.0098
	2	.0746	.1937	.2759	.3020	.2816	.2335	.1757	.1209	.0763	.0439
	3	.0105	.0574	.1298	.2013	.2503	.2668	.2522	.2150	.1665	.1172
	4	.0010	.0112	.0401	.0881	.1460	.2001	.2377	.2508	.2384	.2051
	5	.0001	.0015	.0085	.0264	.0584	.1029	.1536	.2007	.2340	.2461
	6	.0000	.0001	.0012	.0055	.0162	.0368	.0689	.1115	.1596	.2051
	7	.0000	.0000	.0001	.0008	.0031	.0090	.0212	.0425	.0746	.1172
	8	.0000	.0000	.0000	.0001	.0004	.0014	.0043	.0106	.0229	.0439
	9	.0000	.0000	.0000	.0000	.0000	.0001	.0005	.0016	.0042	.0098
	10	.0000	.0000	.0000	.0000	.0000	.0000	.0000	.0001	.0003	.0010

Table V (*continued*) Binomial Distribution Function ${}_nC_x p^x(1-p)^{n-x}$

n	x	p = .05	.10	.15	.20	.25	.30	.35	.40	.45	.50
11	0	.5688	.3138	.1673	.0859	.0422	.0198	.0088	.0036	.0014	.0005
	1	.3293	.3835	.3248	.2362	.1549	.0932	.0518	.0266	.0125	.0054
	2	.0867	.2131	.2866	.2953	.2581	.1998	.1395	.0887	.0513	.0269
	3	.0137	.0710	.1517	.2215	.2581	.2568	.2254	.1774	.1259	.0806
	4	.0014	.0158	.0536	.1107	.1721	.2201	.2428	.2365	.2060	.1611
	5	.0001	.0025	.0132	.0388	.0803	.1321	.1830	.2207	.2360	.2256
	6	.0000	.0003	.0023	.0097	.0268	.0566	.0985	.1471	.1931	.2256
	7	.0000	.0000	.0003	.0017	.0064	.0173	.0379	.0701	.1128	.1611
	8	.0000	.0000	.0000	.0002	.0011	.0037	.0102	.0234	.0462	.0806
	9	.0000	.0000	.0000	.0000	.0001	.0005	.0018	.0052	.0126	.0269
	10	.0000	.0000	.0000	.0000	.0000	.0000	.0002	.0007	.0021	.0054
	11	.0000	.0000	.0000	.0000	.0000	.0000	.0000	.0000	.0002	.0005
12	0	.5405	.2824	.1422	.0687	.0317	.1038	.0057	.0022	.0008	.0002
	1	.3413	.3766	.3012	.2062	.1267	.0712	.0368	.0174	.0075	.0029
	2	.0988	.2301	.2924	.2835	.2323	.1678	.1088	.0639	.0339	.0161
	3	.0173	.0852	.1720	.2362	.2581	.2397	.1954	.1419	.0923	.0537
	4	.0021	.0213	.0683	.1329	.1936	.2311	.2367	.2128	.1700	.1208
	5	.0002	.0038	.0193	.0532	.1032	.1585	.2039	.2270	.2225	.1934
	6	.0000	.0005	.0040	.0155	.0401	.0792	.1281	.1766	.2124	.2256
	7	.0000	.0000	.0006	.0033	.0115	.0291	.0591	.1009	.1489	.1934
	8	.0000	.0000	.0001	.0005	.0024	.0078	.0199	.0420	.0762	.1208
	9	.0000	.0000	.0000	.0001	.0004	.0015	.0048	.0125	.0277	.0537
	10	.0000	.0000	.0000	.0000	.0000	.0002	.0008	.0025	.0068	.0161
	11	.0000	.0000	.0000	.0000	.0000	.0000	.0001	.0003	.0010	.0029
	12	.0000	.0000	.0000	.0000	.0000	.0000	.0000	.0000	.0001	.0002
13	0	.5133	.2542	.1209	.0550	.0238	.0097	.0037	.0013	.0004	.0001
	1	.3512	.3672	.2774	.1787	.1029	.0540	.0259	.0113	.0045	.0016
	2	.1109	.2448	.2937	.2680	.2059	.1388	.0836	.0453	.0220	.0095
	3	.0214	.0997	.1900	.2457	.2517	.2181	.1651	.1107	.0660	.0349
	4	.0028	.0277	.0838	.1535	.2097	.2337	.2222	.1845	.1350	.0873
	5	.0003	.0055	.0266	.0691	.1258	.1803	.2154	.2214	.1989	.1571
	6	.0000	.0008	.0063	.0230	.0559	.1030	.1546	.1968	.2169	.2095
	7	.0000	.0001	.0011	.0058	.0186	.0442	.0833	.1312	.1775	.2095
	8	.0000	.0000	.0001	.0011	.0047	.0142	.0336	.0656	.1089	.1571
	9	.0000	.0000	.0000	.0001	.0009	.0034	.0101	.0243	.0495	.0873

Table V (*continued*) Binomial Distribution Function ${}_nC_x p^x(1-p)^{n-x}$

n	x	.05	.10	.15	.20	.25	.30	.35	.40	.45	.50
	10	.0000	.0000	.0000	.0000	.0001	.0006	.0022	.0065	.0162	.0349
	11	.0000	.0000	.0000	.0000	.0000	.0001	.0003	.0012	.0036	.0095
	12	.0000	.0000	.0000	.0000	.0000	.0000	.0000	.0001	.0005	.0016
	13	.0000	.0000	.0000	.0000	.0000	.0000	.0000	.0000	.0000	.0001
14	0	.4877	.2288	.1028	.0440	.0178	.0068	.0024	.0008	.0002	.0001
	1	.3593	.3559	.2539	.1539	.0832	.0407	.0181	.0073	.0027	.0009
	2	.1229	.2570	.2912	.2501	.1802	.1134	.0634	.0317	.0141	.0056
	3	.0259	.1142	.2056	.2501	.2402	.1943	.1366	.0845	.0462	.0222
	4	.0037	.0349	.0998	.1720	.2202	.2290	.2022	.1549	.1040	.0611
	5	.0004	.0078	.0352	.0860	.1468	.1963	.2178	.2066	.1701	.1222
	6	.0000	.0013	.0093	.0322	.0734	.1262	.1759	.2066	.2088	.1833
	7	.0000	.0002	.0019	.0092	.0280	.0618	.1082	.1574	.1952	.2095
	8	.0000	.0000	.0003	.0020	.0082	.0232	.0510	.0918	.1398	.1833
	9	.0000	.0000	.0000	.0003	.0018	.0066	.0183	.0408	.0762	.1222
	10	.0000	.0000	.0000	.0000	.0003	.0014	.0049	.0136	.0312	.0611
	11	.0000	.0000	.0000	.0000	.0000	.0002	.0010	.0033	.0093	.0222
	12	.0000	.0000	.0000	.0000	.0000	.0000	.0001	.0005	.0019	.0056
	13	.0000	.0000	.0000	.0000	.0000	.0000	.0000	.0001	.0002	.0009
	14	.0000	.0000	.0000	.0000	.0000	.0000	.0000	.0000	.0000	.0001
15	0	.4633	.2059	.0874	.0352	.0134	.0047	.0016	.0005	.0001	.0000
	1	.3658	.3432	.2312	.1319	.0668	.0305	.0126	.0047	.0016	.0005
	2	.1348	.2669	.2856	.2309	.1559	.0916	.0476	.0219	.0090	.0032
	3	.0307	.1285	.2184	.2501	.2252	.1700	.1110	.0634	.0318	.0139
	4	.0049	.0428	.1156	.1876	.2252	.2186	.1792	.1268	.0780	.0417
	5	.0006	.0105	.0449	.1032	.1651	.2061	.2123	.1859	.1404	.0916
	6	.0000	.0019	.0132	.0430	.0917	.1472	.1906	.2066	.1914	.1527
	7	.0000	.0003	.0030	.0138	.0393	.0811	.1319	.1771	.2013	.1964
	8	.0000	.0000	.0005	.0035	.0131	.0348	.0710	.1181	.1647	.1964
	9	.0000	.0000	.0001	.0007	.0034	.0116	.0298	.0612	.1048	.1527
	10	.0000	.0000	.0000	.0001	.0007	.0030	.0096	.0245	.0515	.0916
	11	.0000	.0000	.0000	.0000	.0001	.0006	.0024	.0074	.0191	.0417
	12	.0000	.0000	.0000	.0000	.0000	.0001	.0004	.0016	.0052	.0139
	13	.0000	.0000	.0000	.0000	.0000	.0000	.0001	.0003	.0010	.0032
	14	.0000	.0000	.0000	.0000	.0000	.0000	.0000	.0000	.0001	.0005
	15	.0000	.0000	.0000	.0000	.0000	.0000	.0000	.0000	.0000	.0000

Table V (*continued*) Binomial Distribution Function ${}_nC_x p(1-p)^{n-x}$

n	x	.05	.10	.15	.20	.25	.30	.30	.40	.45	.50
16	0	.4401	.1853	.0743	.0281	.0100	.0033	.0010	.0003	.0001	.0000
	1	.3706	.3294	.2097	.1126	.0535	.0228	.0087	.0030	.0009	.0002
	2	.1463	.2745	.2775	.2111	.1336	.0732	.0353	.0150	.0056	.0018
	3	.0359	.1423	.2285	.2463	.2079	.1465	.0888	.0468	.0215	.0085
	4	.0061	.0514	.1311	.2001	.2252	.2040	.1553	.1014	.0572	.0278
	5	.0008	.0137	.0555	.1201	.1802	.2099	.2008	.1623	.1123	.0667
	6	.0001	.0028	.0180	.0550	.1101	.1649	.1982	.1983	.1684	.1222
	7	.0000	.0004	.0045	.0197	.0524	.1010	.1524	.1889	.1969	.1746
	8	.0000	.0001	.0009	.0055	.0197	.0487	.0923	.1417	.1812	.1964
	9	.0000	.0000	.0001	.0012	.0058	.0185	.0442	.0840	.1318	.1746
	10	.0000	.0000	.0000	.0002	.0014	.0056	.0167	.0392	.0755	.1222
	11	.0000	.0000	.0000	.0000	.0002	.0013	.0049	.0142	.0337	.0667
	12	.0000	.0000	.0000	.0000	.0000	.0002	.0011	.0040	.0115	.0278
	13	.0000	.0000	.0000	.0000	.0000	.0000	.0002	.0008	.0029	.0085
	14	.0000	.0000	.0000	.0000	.0000	.0000	.0000	.0001	.0005	.0018
	15	.0000	.0000	.0000	.0000	.0000	.0000	.0000	.0000	.0001	.0002
	16	.0000	.0000	.0000	.0000	.0000	.0000	.0000	.0000	.0000	.0000

Linear interpolation with respect to p will generally not be accurate to more than two decimal places, and sometimes less.

For extensive tables of ${}_nC_x p^x(1-p)^{n-x}$ see *Tables of the Binomial Probability Distribution*, Applied Mathematics Series 6 (Washington, D.C.: National Bureau of Standards, 1950).

Table VI Poisson Distribution Function $e^{-m}m^x/x!$

Entries in the table are values of $e^{-m}m^x/x!$ for the indicated values of x and m.

x	m = 0.1	0.2	0.3	0.4	0.5	0.6	0.7	0.8	0.9	1.0
0	.9048	.8187	.7408	.6703	.6065	.5488	.4966	.4493	.4066	.3679
1	.0905	.1637	.2222	.2681	.3033	.3293	.3476	.3595	.3659	.3679
2	.0045	.0164	.0333	.0536	.0758	.0988	.1217	.1438	.1647	.1839
3	.0002	.0011	.0033	.0072	.0126	.0198	.0284	.0383	.0494	.0613
4	.0000	.0001	.0002	.0007	.0016	.0030	.0050	.0077	.0111	.0153
5	.0000	.0000	.0000	.0001	.0002	.0004	.0007	.0012	.0020	.0031
6	.0000	.0000	.0000	.0000	.0000	.0000	.0001	.0002	.0003	.0005
7	.0000	.0000	.0000	.0000	.0000	.0000	.0000	.0000	.0000	.0001

x	m = 1.1	1.2	1.3	1.4	1.5	1.6	1.7	1.8	1.9	2.0
0	.3329	.3012	.2725	.2466	.2231	.2019	.1827	.1653	.1496	.1353
1	.3662	.3614	.3543	.3452	.3347	.3230	.3106	.2975	.2842	.2707
2	.2014	.2169	.2303	.2417	.2510	.2584	.2640	.2678	.2700	.2707
3	.0738	.0867	.0998	.1128	.1255	.1378	.1496	.1607	.1710	.1804
4	.0203	.0260	.0324	.0395	.0471	.0551	.0636	.0723	.0812	.0902
5	.0045	.0062	.0084	.0111	.0141	.0176	.0216	.0260	.0309	.0361
6	.0008	.0012	.0018	.0026	.0035	.0047	.0061	.0078	.0098	.0120
7	.0001	.0002	.0003	.0005	.0008	.0011	.0015	.0020	.0027	.0034
8	.0000	.0000	.0001	.0001	.0001	.0002	.0003	.0005	.0006	.0009
9	.0000	.0000	.0000	.0000	.0000	.0000	.0001	.0001	.0001	.0002

x	m = 2.1	2.2	2.3	2.4	2.5	2.6	2.7	2.8	2.9	3.0
0	.1225	.1108	.1003	.0907	.0821	.0743	.0672	.0608	.0550	.0498
1	.2572	.2438	.2306	.2177	.2052	.1931	.1815	.1703	.1596	.1494
2	.2700	.2681	.2652	.2613	.2565	.2510	.2450	.2384	.2314	.2240
3	.1890	.1966	.2033	.2090	.2138	.2176	.2205	.2225	.2237	.2240
4	.0992	.1082	.1169	.1254	.1336	.1414	.1488	.1557	.1622	.1680
5	.0417	.0476	.0538	.0602	.0668	.0735	.0804	.0872	.0940	.1008
6	.0146	.0174	.0206	.0241	.0278	.0319	.0362	.0407	.0455	.0504
7	.0044	.0055	.0068	.0083	.0099	.0118	.0139	.0163	.0188	.0216
8	.0011	.0015	.0019	.0025	.0031	.0038	.0047	.0057	.0068	.0081
9	.0003	.0004	.0005	.0007	.0009	.0011	.0014	.0018	.0022	.0027
10	.0001	.0001	.0001	.0002	.0002	.0003	.0004	.0005	.0006	.0008
11	.0000	.0000	.0000	.0000	.0000	.0001	.0001	.0001	.0002	.0002
12	.0000	.0000	.0000	.0000	.0000	.0000	.0000	.0000	.0000	.0001

Table VI (*continued*) Poisson Distribution Function $e^{-m}m^x/x!$

x	m = 3.1	3.2	3.3	3.4	3.5	3.6	3.7	3.8	3.9	4.0
0	.0450	.0408	.0369	.0334	.0302	.0273	.0247	.0224	.0202	.0183
1	.1397	.1304	.1217	.1135	.1057	.0984	.0915	.0850	.0789	.0733
2	.2165	.2087	.2008	.1929	.1850	.1771	.1692	.1615	.1539	.1465
3	.2237	.2226	.2209	.2186	.2158	.2125	.2087	.2046	.2001	.1954
4	.1734	.1781	.1823	.1858	.1888	.1912	.1931	.1944	.1951	.1954
5	.1075	.1140	.1203	.1264	.1322	.1377	.1429	.1477	.1522	.1563
6	.0555	.0608	.0662	.0716	.0771	.0826	.0881	.0936	.0989	.1042
7	.0246	.0278	.0312	.0348	.0385	.0425	.0466	.0508	.0551	.0595
8	.0095	.0111	.0129	.0148	.0169	.0191	.0215	.0241	.0269	.0298
9	.0033	.0040	.0047	.0056	.0066	.0076	.0089	.0102	.0116	.0132
10	.0010	.0013	.0016	.0019	.0023	.0028	.0033	.0039	.0045	.0053
11	.0003	.0004	.0005	.0006	.0007	.0009	.0011	.0013	.0016	.0019
12	.0001	.0001	.0001	.0002	.0002	.0003	.0003	.0004	.0005	.0006
13	.0000	.0000	.0000	.0000	.0001	.0001	.0001	.0001	.0002	.0002
14	.0000	.0000	.0000	.0000	.0000	.0000	.0000	.0000	.0000	.0001

x	m = 4.1	4.2	4.3	4.4	4.5	4.6	4.7	4.8	4.9	5.0
0	.0166	.0150	.0136	.1023	.0111	.0101	.0091	.0082	.0074	.0067
1	.0679	.0630	.0583	.0540	.0500	.0462	.0427	.0395	.0365	.0337
2	.1393	.1323	.1254	.1188	.1125	.1063	.1005	.0948	.0894	.0842
3	.1904	.1852	.1798	.1743	.1687	.1631	.1574	.1517	.1460	.1404
4	.1951	.1944	.1933	.1917	.1898	.1875	.1849	.1820	.1789	.1755
5	.1600	.1633	.1662	.1687	.1708	.1725	.1738	.1747	.1753	.1755
6	.1093	.1143	.1191	.1237	.1281	.1323	.1362	.1398	.1432	.1462
7	.0640	.0686	.0732	.0778	.0824	.0869	.0914	.0959	.1002	.1044
8	.0328	.0360	.0393	.0428	.0463	.0500	.0537	.0575	.0614	.0653
9	.0150	.0168	.0188	.0209	.0232	.0255	.0280	.0307	.0334	.0363
10	.0061	.0071	.0081	.0092	.0104	.0118	.0132	.0147	.0164	.0181
11	.0023	.0027	.0032	.0037	.0043	.0049	.0056	.0064	.0073	.0082
12	.0008	.0009	.0011	.0014	.0016	.0019	.0022	.0026	.0030	.0034
13	.0002	.0003	.0004	.0005	.0006	.0007	.0008	.0009	.0011	.0013
14	.0001	.0001	.0001	.0001	.0002	.0002	.0003	.0003	.0004	.0005
15	.0000	.0000	.0000	.0000	.0001	.0001	.0001	.0001	.0001	.0002

Table VI (*continued*) Poisson Distribution Function $e^{-m} m^x/x!$

x	5.1	5.2	5.3	5.4	5.5	5.6	5.7	5.8	5.9	6.0
0	.0061	.0055	.0050	.0045	.0041	.0037	.0033	.0030	.0027	.0025
1	.0311	.0287	.0265	.0244	.0225	.0207	.0191	.0176	.0162	.0149
2	.0793	.0746	.0701	.0659	.0618	.0580	.0544	.0509	.0477	.0446
3	.1348	.1293	.1239	.1185	.1133	.1082	.1033	.0985	.0938	.0892
4	.1719	.1681	.1641	.1600	.1558	.1515	.1472	.1428	.1383	.1339
5	.1753	.1748	.1740	.1728	.1714	.1697	.1678	.1656	.1632	.1606
6	.1490	.1515	.1537	.1555	.1571	.1584	.1594	.1601	.1605	.1606
7	.1086	.1125	.1163	.1200	.1234	.1267	.1298	.1326	.1353	.1377
8	.0692	.0731	.0771	.0810	.0849	.0887	.0925	.0962	.0998	.1033
9	.0392	.0423	.0454	.0486	.0519	.0552	.0586	.0620	.0654	.0688
10	.0200	.0220	.0241	.0262	.0285	.0309	.0334	.0359	.0386	.0413
11	.0093	.0104	.0116	.0129	.0143	.0157	.0173	.0190	.0207	.0225
12	.0039	.0045	.0051	.0058	.0065	.0073	.0082	.0092	.0102	.0113
13	.0015	.0018	.0021	.0024	.0028	.0032	.0036	.0041	.0046	.0052
14	.0006	.0007	.0008	.0009	.0011	.0013	.0015	.0017	.0019	.0022
15	.0002	.0002	.0003	.0003	.0004	.0005	.0006	.0007	.0008	.0009
16	.0001	.0001	.0001	.0001	.0001	.0002	.0002	.0002	.0003	.0003
17	.0000	.0000	.0000	.0000	.0000	.0001	.0001	.0001	.0001	.0001

x	6.1	6.2	6.3	6.4	6.5	6.6	6.7	6.8	6.9	7.0
0	.0022	.0020	.0018	.0017	.0015	.0014	.0012	.0011	.0010	.0009
1	.0137	.0126	.0116	.0106	.0098	.0090	.0082	.0076	.0070	.0064
2	.0417	.0390	.0364	.0340	.0318	.0296	.0276	.0258	.0240	.0223
3	.0848	.0806	.0765	.0726	.0688	.0652	.0617	.0584	.0552	.0521
4	.1294	.1249	.1205	.1162	.1118	.1076	.1034	.0992	.0952	.0912
5	.1579	.1549	.1519	.1487	.1454	.1420	.1385	.1349	.1314	.1277
6	.1605	.1601	.1595	.1586	.1575	.1562	.1546	.1529	.1511	.1490
7	.1399	.1418	.1435	.1450	.1462	.1472	.1480	.1486	.1489	.1490
8	.1066	.1099	.1130	.1160	.1188	.1215	.1240	.1263	.1284	.1304
9	.0723	.0757	.0791	.0825	.0858	.0891	.0923	.0954	.0985	.1014

Table VI (*continued*) Poisson Distribution Function $e^{-m}m^x/x!$

x	6.1	6.2	6.3	6.4	6.5	6.6	6.7	6.8	6.9	7.0
10	.0441	.0469	.0498	.0528	.0558	.0588	.0618	.0649	.0679	.0710
11	.0245	.0265	.0285	.0307	.0330	.0353	.0377	.0401	.0426	.0452
12	.0124	.0137	.0150	.0164	.0179	.0194	.0210	.0227	.0245	.0264
13	.0058	.0065	.0073	.0081	.0089	.0098	.0108	.0119	.0130	.0142
14	.0025	.0029	.0033	.0037	.0041	.0046	.0052	.0058	.0064	.0071
15	.0010	.0012	.0014	.0016	.0018	.0020	.0023	.0026	.0029	.0033
16	.0004	.0005	.0005	.0006	.0007	.0008	.0010	.0011	.0013	.0014
17	.0001	.0002	.0002	.0002	.0003	.0003	.0004	.0004	.0005	.0006
18	.0000	.0001	.0001	.0001	.0001	.0001	.0001	.0002	.0002	.0002
19	.0000	.0000	.0000	.0000	.0000	.0000	.0000	.0001	.0001	.0001

x	7.1	7.2	7.3	7.4	7.5	7.6	7.7	7.8	7.9	8.0
0	.0008	.0007	.0007	.0006	.0006	.0005	.0005	.0004	.0004	.0003
1	.0059	.0054	.0049	.0045	.0041	.0038	.0035	.0032	.0029	.0027
2	.0208	.0194	.0180	.0167	.0156	.0145	.0134	.0125	.0116	.0107
3	.0492	.0464	.0438	.0413	.0389	.0366	.0345	.0324	.0305	.0286
4	.0874	.0836	.0799	.0764	.0729	.0696	.0663	.0632	.0602	.0573
5	.1241	.1204	.1167	.1130	.1094	.1057	.1021	.0986	.0951	.0916
6	.1468	.1445	.1420	.1394	.1367	.1339	.1311	.1282	.1252	.1221
7	.1489	.1486	.1481	.1474	.1465	.1454	.1442	.1428	.1413	.1396
8	.1321	.1337	.1351	.1363	.1373	.1382	.1388	.1392	.1395	.1396
9	.1042	.1070	.1096	.1121	.1144	.1167	.1187	.1207	.1224	.1241
10	.0740	.0770	.0800	.0829	.0858	.0887	.0914	.0941	.0967	.0993
11	.0478	.0504	.0531	.0558	.0585	.0613	.0640	.0667	.0695	.0722
12	.0283	.0303	.0323	.0344	.0366	.0388	.0411	.0434	.0457	.0481
13	.0154	.0168	.0181	.0196	.0211	.0227	.0243	.0260	.0278	.0296
14	.0078	.0086	.0095	.0104	.0113	.0123	.0134	.0145	.0157	.0169
15	.0037	.0041	.0046	.0051	.0057	.0062	.0069	.0075	.0083	.0090
16	.0016	.0019	.0021	.0024	.0026	.0030	.0033	.0037	.0041	.0045
17	.0007	.0008	.0009	.0010	.0012	.0013	.0015	.0017	.0019	.0021
18	.0003	.0003	.0004	.0004	.0005	.0006	.0006	.0007	.0008	.0009
19	.0001	.0001	.0001	.0002	.0002	.0002	.0003	.0003	.0003	.0004
20	.0000	.0000	.0001	.0001	.0001	.0001	.0001	.0001	.0001	.0002
21	.0000	.0000	.0000	.0000	.0000	.0000	.0000	.0000	.0001	.0001

Table VI (*continued*) Poisson Distribution Function $e^{-m}m^x/x!$

	m									
x	8.1	8.2	8.3	8.4	8.5	8.6	8.7	8.8	8.9	9.0
0	.0003	.0003	.0002	.0002	.0002	.0002	.0002	.0002	.0001	.0001
1	.0025	.0023	.0021	.0019	.0017	.0016	.0014	.0013	.0012	.0011
2	.0100	.0092	.0086	.0079	.0074	.0068	.0063	.0058	.0054	.0050
3	.0269	.0252	.0237	.0222	.0208	.0195	.0183	.0171	.0160	.0150
4	.0544	.0517	.0491	.0466	.0443	.0420	.0398	.0377	.0357	.0337
5	.0882	.0849	.0816	.0784	.0752	.0722	.0692	.0663	.0635	.0607
6	.1191	.1160	.1128	.1097	.1066	.1034	.1003	.0972	.0941	.0911
7	.1378	.1358	.1338	.1317	.1294	.1271	.1247	.1222	.1197	.1171
8	.1395	.1392	.1388	.1382	.1375	.1366	.1356	.1344	.1332	.1318
9	.1256	.1269	.1280	.1290	.1299	.1306	.1311	.1315	.1317	.1318
10	.1017	.1040	.1063	.1084	.1104	.1123	.1140	.1157	.1172	.1186
11	.0749	.0776	.0802	.0828	.0853	.0878	.0902	.0925	.0948	.0970
12	.0505	.0530	.0555	.0579	.0604	.0629	.0654	.0679	.0703	.0728
13	.0315	.0334	.0354	.0374	.0395	.0416	.0438	.0459	.0481	.0504
14	.0182	.0196	.0210	.0225	.0240	.0256	.0272	.0289	.0306	.0324
15	.0098	.0107	.0116	.0126	.0136	.0147	.0158	.0169	.0182	.0194
16	.0050	.0055	.0060	.0066	.0072	.0079	.0086	.0093	.0101	.0109
17	.0024	.0026	.0029	.0033	.0036	.0040	.0044	.0048	.0053	.0058
18	.0011	.0012	.0014	.0015	.0017	.0019	.0021	.0024	.0026	.0029
19	.0005	.0005	.0006	.0007	.0008	.0009	.0010	.0011	.0012	.0014
20	.0002	.0002	.0002	.0003	.0003	.0004	.0004	.0005	.0005	.0006
21	.0001	.0001	.0001	.0001	.0001	.0002	.0002	.0002	.0002	.0003
22	.0000	.0000	.0000	.0000	.0001	.0001	.0001	.0001	.0001	.0001

For extensive tables of $e^{-m}m^x/x!$ see Molina, E. C.: *Poisson's Exponential Binomial Limit* (New York: D. Van Nostrand, 1942).

Table VII 95 Percent Confidence Intervals for Proportions[a]

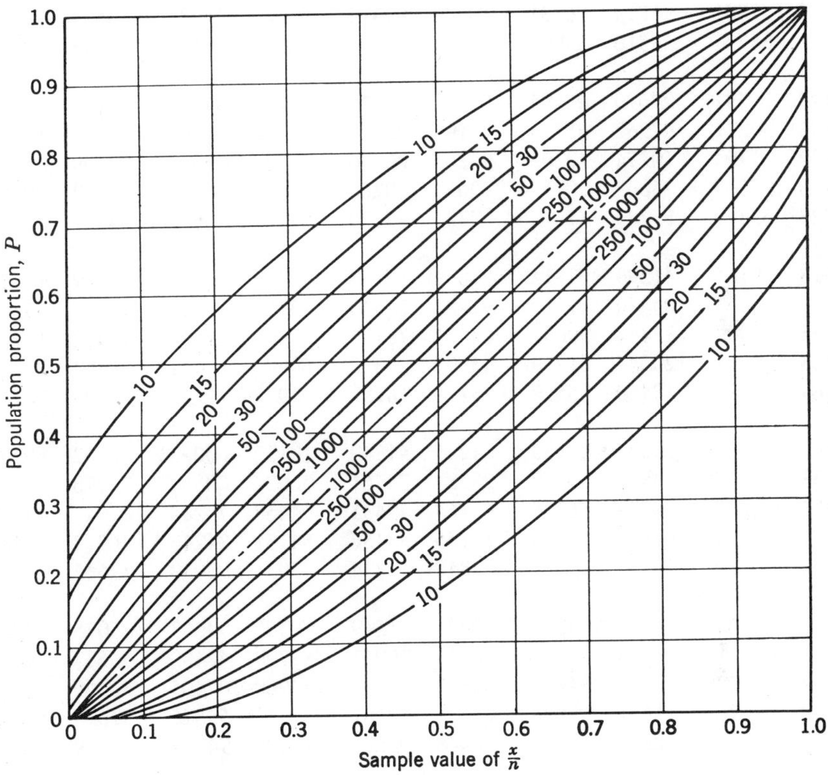

[a] This table is reproduced from C. J. Clopper and E. S. Pearson, "The use of confidence or fiducial limits illustrated in the case of the binomial," *Biometrika*, **26**, by permission of the *Biometrika* trustees.

Table VIII Random Digits

03991	10461	93716	16894	98953	73231	39528	72484	82474	25593
38555	95554	32886	59780	09958	18065	81616	18711	53342	44276
17546	73704	92052	46215	15917	06253	07586	16120	82641	22820
32643	52861	95819	06831	19640	99413	90767	04235	13574	17200
69572	68777	39510	35905	85244	35159	40188	28193	29593	88627
24122	66591	27699	06494	03152	19121	34414	82157	86887	55087
61196	30231	92962	61773	22109	78508	63439	75363	44989	16822
30532	21704	10274	12202	94205	20380	67049	09070	93399	45547
03788	97599	75867	20717	82037	10268	79495	04146	52162	90286
48228	63379	85783	47619	87481	37220	91704	30552	04737	21031
88618	19161	41290	67312	71857	15957	48545	35247	18619	13674
71299	23853	05870	01119	92784	26340	75122	11724	74627	73707
27954	58909	82444	99005	04921	73701	92904	13141	32392	19763
80863	00514	20247	81759	45197	25332	69902	63742	78464	22501
33564	60780	48460	85558	15191	18782	94972	11598	62095	36787
90899	75754	60833	25983	01291	41349	19152	00023	12302	80783
78038	70267	43529	06318	38384	74761	36024	00867	76378	41605
55986	66485	88722	56736	66164	49431	94458	74284	05041	49807
87539	08823	94813	31900	54155	83436	54158	34243	46978	35482
16818	60311	74457	90561	72848	11834	75051	93029	47665	64382
34677	58300	74910	64345	19325	81549	60365	94653	35075	33949
45305	07521	61318	31855	14413	70951	83799	42402	56623	34442
59747	67277	76503	34513	39663	77544	32960	07405	36409	83232
16520	69676	11654	99893	02181	68161	19322	53845	57620	52606
68652	27376	92852	55866	88448	03584	11220	94747	07399	37408
79375	95220	01159	63267	10622	48391	31751	57260	68980	05339
33521	26665	55823	47641	86225	31704	88492	99382	14454	04504
59589	49067	66821	41575	49767	04037	30934	47744	07481	83828
20554	91409	96277	48257	50816	97616	22888	48893	27499	98748
59404	72059	43947	51680	43852	59693	78212	16993	35902	91386
42614	29297	01918	28316	25163	01889	70014	15021	68971	11403
34994	41374	70071	14736	65251	07629	37239	33295	18477	65622
99385	41600	11133	07586	36815	43625	18637	37509	14707	93997
66497	68646	78138	66559	64397	11692	05327	82162	83745	22567
48509	23929	27482	45476	04515	25624	95096	67946	16930	33361
15470	48355	88651	22596	83761	60873	43253	84145	20368	07126
20094	98977	74843	93413	14387	06345	80854	09279	41196	37480

Table VIII (*continued*) Random Digits

73788	06533	28597	20405	51321	92246	80088	77074	66919	31678
60530	45128	74022	84617	72472	00008	80890	18002	35352	54131
44372	15486	65741	14014	05466	55306	93128	18464	79982	68416
18611	19241	66083	24653	84609	58232	41849	84547	46850	52326
58319	15997	08355	60860	29735	47762	46352	33049	69248	93460
61199	67940	55121	29281	59076	07936	11087	96294	14013	31792
18627	90872	00911	98936	76355	93779	52701	08337	56303	87315
00441	58997	14060	40619	29549	69616	57275	36898	81304	48585
32624	68691	14845	46672	61958	77100	20857	73156	70284	24326
65961	73488	41839	55382	17267	70943	15633	84924	90415	93614
20288	34060	39685	23309	10061	68829	92694	48297	39904	02115
59362	95938	74416	53166	35208	33374	77613	19019	88152	00080
99782	93478	53152	67433	35663	52972	38688	32486	45134	63545
27767	43584	85301	88977	29490	69714	94015	64874	32444	48277
13025	14338	54066	15243	47724	66733	74108	88222	88570	74015
80217	36292	98525	24335	24432	24896	62880	87873	95160	59221
10875	62004	90391	61105	57411	06368	11748	12102	80580	41867
54127	57326	26629	19087	24472	88779	17944	05600	60478	03343
60311	42824	37301	42678	45990	43242	66067	42792	95043	52680
49739	71484	92003	98086	76668	73209	54244	91030	45547	70818
78626	51594	16453	94614	39014	97066	30945	57589	31732	57260
66692	13986	99837	00582	81232	44987	69170	37403	86995	90307
44071	28091	07362	97703	76447	42537	08345	88975	35841	85771
59820	96163	78851	16499	87064	13075	73035	41207	74699	09310
25704	91035	26313	77463	55387	72681	47341	43905	31048	56699
22304	90314	78438	66276	18396	73538	43277	58874	11466	16082
17710	59621	15292	76139	59526	52113	53856	30743	08670	84741
25852	58905	55018	56374	35824	71708	30540	27886	61732	75454
46780	56487	75211	10271	36633	68424	17374	52003	70707	70214
59849	96169	87195	46092	26787	60939	59202	11973	02902	33250
47670	07654	30342	40277	11049	72049	83012	09832	25571	77628
94304	71803	73465	09819	58869	35220	09504	96412	90193	79568
08105	59987	21437	36786	49226	77837	98524	97831	65704	09514
64281	61826	18555	64937	64654	25843	41145	42820	14924	39650
66847	70495	32350	02985	01755	14750	48968	38603	70312	05682
72461	33230	21529	53424	72877	17334	39283	04149	90850	64618
21032	91050	13058	16218	06554	07850	73950	79552	24781	89683
95362	67011	06651	16136	57216	39618	49856	99326	40902	05069

Table VIII (*continued*) Random Digits

49712	97380	10404	55452	09971	59481	37006	22186	72682	07385
58275	61764	97586	54716	61459	21647	87417	17198	21443	41808
89514	11788	68224	23417	46376	25366	94746	49580	01176	28838
15472	50669	48139	36732	26825	05511	12459	91314	80582	71944
12120	86124	51247	44302	87112	21476	14713	71181	13177	55292
95294	00556	70481	06905	21785	41101	49386	54480	23604	23554
66986	34099	74474	20740	47458	64809	06312	88940	15995	69321
80620	51790	11436	38072	40405	68032	60942	00307	11897	92674
55411	85667	77535	99892	71209	92061	92329	98932	78284	46347
95083	06783	28102	57816	85561	29671	77936	63574	31384	51924
90726	57166	98884	08583	95889	57067	38101	77756	11657	13897
68984	83620	89747	98882	92613	89719	39641	69457	91339	22502
36421	16489	18059	51061	67667	60631	84054	40455	99396	63680
92638	40333	67054	16067	24700	71594	47468	03577	57649	63266
21036	82808	77501	97427	76479	68562	43321	31370	28977	23896
13173	33365	41468	85149	49554	17994	91178	10174	29420	90438
86716	38746	94559	37559	49678	53119	98189	81851	29651	84215
92581	02262	41615	70360	64114	58660	96717	54244	10701	41393
12470	56500	50273	93113	41794	86861	39448	93136	25722	08564
01016	00857	41396	80504	90670	08289	58137	17820	22751	36518
34030	60726	25807	24260	71529	78920	47648	13885	70669	93406
50259	46345	06170	97965	88302	98041	11947	56203	19324	20504
73959	76145	60808	54444	74412	81105	69181	96845	38525	11600
46874	37088	80940	44893	10408	36222	14004	23153	69249	05747
60883	52109	19516	90120	46759	71643	62342	07589	08899	05985

Table IX The Durbin-Watson d Statistic

Significance points of d_L and d_U: 5%

n	$k'=1$		$k'=2$		$k'=3$		$k'=4$		$k'=5$	
	d_L	d_U	d_L	d_U	d_L	d_U	d_L	d_U	d_L	d_U
15	1.08	1.36	0.95	1.54	0.82	1.75	0.69	1.97	0.56	2.21
16	1.10	1.37	0.98	1.54	0.86	1.73	0.74	1.93	0.62	2.15
17	1.13	1.38	1.02	1.54	0.90	1.71	0.78	1.90	0.67	2.10
18	1.16	1.39	1.05	1.53	0.93	1.69	0.82	1.87	0.71	2.06
19	1.18	1.40	1.08	1.53	0.97	1.68	0.86	1.85	0.75	2.02
20	1.20	1.41	1.10	1.54	1.00	1.68	0.90	1.83	0.79	1.99
21	1.22	1.42	1.13	1.54	1.03	1.67	0.93	1.81	0.83	1.96
22	1.24	1.43	1.15	1.54	1.05	1.66	0.96	1.80	0.86	1.94
23	1.26	1.44	1.17	1.54	1.08	1.66	0.99	1.79	0.90	1.92
24	1.27	1.45	1.19	1.55	1.10	1.66	1.01	1.78	0.93	1.90
25	1.29	1.45	1.21	1.55	1.12	1.66	1.04	1.77	0.95	1.89
26	1.30	1.46	1.22	1.55	1.14	1.65	1.06	1.76	0.98	1.88
27	1.32	1.47	1.24	1.56	1.16	1.65	1.08	1.76	1.01	1.86
28	1.33	1.48	1.26	1.56	1.18	1.65	1.10	1.75	1.03	1.85
29	1.34	1.48	1.27	1.56	1.20	1.65	1.12	1.74	1.05	1.84
30	1.35	1.49	1.28	1.57	1.21	1.65	1.14	1.74	1.07	1.83
31	1.36	1.50	1.30	1.57	1.23	1.65	1.16	1.74	1.09	1.83
32	1.37	1.50	1.31	1.57	1.24	1.65	1.18	1.73	1.11	1.82
33	1.38	1.51	1.32	1.58	1.26	1.65	1.19	1.73	1.13	1.81
34	1.39	1.51	1.33	1.58	1.27	1.65	1.21	1.73	1.15	1.81
35	1.40	1.52	1.34	1.58	1.28	1.65	1.22	1.73	1.16	1.80
36	1.41	1.52	1.35	1.59	1.29	1.65	1.24	1.73	1.18	1.80
37	1.42	1.53	1.36	1.59	1.31	1.66	1.25	1.72	1.19	1.80
38	1.43	1.54	1.37	1.59	1.32	1.66	1.26	1.72	1.21	1.79
39	1.43	1.54	1.38	1.60	1.33	1.66	1.27	1.72	1.22	1.79
40	1.44	1.54	1.39	1.60	1.34	1.66	1.29	1.72	1.23	1.79
45	1.48	1.57	1.43	1.62	1.38	1.67	1.34	1.72	1.29	1.78
50	1.50	1.59	1.46	1.63	1.42	1.67	1.38	1.72	1.34	1.77
55	1.53	1.60	1.49	1.64	1.45	1.68	1.41	1.72	1.38	1.77
60	1.55	1.62	1.51	1.65	1.48	1.69	1.44	1.73	1.41	1.77
65	1.57	1.63	1.54	1.66	1.50	1.70	1.47	1.73	1.44	1.77
70	1.58	1.64	1.55	1.67	1.52	1.70	1.49	1.74	1.46	1.77
75	1.60	1.65	1.57	1.68	1.54	1.71	1.51	1.74	1.49	1.77
80	1.61	1.66	1.59	1.69	1.56	1.72	1.53	1.74	1.51	1.77
85	1.62	1.67	1.60	1.70	1.57	1.72	1.55	1.75	1.52	1.77
90	1.63	1.68	1.61	1.70	1.59	1.73	1.57	1.75	1.54	1.78
95	1.64	1.69	1.62	1.71	1.60	1.73	1.58	1.75	1.56	1.78
100	1.65	1.69	1.63	1.72	1.61	1.74	1.59	1.76	1.57	1.78

Table IX (*continued*) *Significance points of d_L and d_U: 2.5%*

n	$k'=1$		$k'=2$		$k'=3$		$k'=4$		$k'=5$	
	d_L	d_U	d_L	d_U	d_L	d_U	d_L	d_U	d_L	d_U
15	0.95	1.23	0.83	1.40	0.71	1.61	0.59	1.84	0.48	2.09
16	0.98	1.24	0.86	1.40	0.75	1.59	0.64	1.80	0.53	2.03
17	1.01	1.25	0.90	1.40	0.79	1.58	0.68	1.77	0.57	1.98
18	1.03	1.26	0.93	1.40	0.82	1.56	0.72	1.74	0.62	1.93
19	1.06	1.28	0.96	1.41	0.86	1.55	0.76	1.72	0.66	1.90
20	1.08	1.28	0.99	1.41	0.89	1.55	0.79	1.70	0.70	1.87
21	1.10	1.30	1.01	1.41	0.92	1.54	0.83	1.69	0.73	1.84
22	1.12	1.31	1.04	1.42	0.95	1.54	0.86	1.68	0.77	1.82
23	1.14	1.32	1.06	1.42	0.97	1.54	0.89	1.67	0.80	1.80
24	1.16	1.33	1.08	1.43	1.00	1.54	0.91	1.66	0.83	1.79
25	1.18	1.34	1.10	1.43	1.02	1.54	0.94	1.65	0.86	1.77
26	1.19	1.35	1.12	1.44	1.04	1.54	0.96	1.65	0.88	1.76
27	1.21	1.36	1.13	1.44	1.06	1.54	0.99	1.64	0.91	1.75
28	1.22	1.37	1.15	1.45	1.08	1.54	1.01	1.64	0.93	1.74
29	1.24	1.38	1.17	1.45	1.10	1.54	1.03	1.63	0.96	1.73
30	1.25	1.38	1.18	1.46	1.12	1.54	1.05	1.63	0.98	1.73
31	1.26	1.39	1.20	1.47	1.13	1.55	1.07	1.63	1.00	1.72
32	1.27	1.40	1.21	1.47	1.15	1.55	1.08	1.63	1.02	1.71
33	1.28	1.41	1.22	1.48	1.16	1.55	1.10	1.63	1.04	1.71
34	1.29	1.41	1.24	1.48	1.17	1.55	1.12	1.63	1.06	1.70
35	1.30	1.42	1.25	1.48	1.19	1.55	1.13	1.63	1.07	1.70
36	1.31	1.43	1.26	1.49	1.20	1.56	1.15	1.63	1.09	1.70
37	1.32	1.43	1.27	1.49	1.21	1.56	1.16	1.62	1.10	1.70
38	1.33	1.44	1.28	1.50	1.23	1.56	1.17	1.62	1.12	1.70
39	1.34	1.44	1.29	1.50	1.24	1.56	1.19	1.63	1.13	1.69
40	1.35	1.45	1.30	1.51	1.25	1.57	1.20	1.63	1.15	1.69
45	1.39	1.48	1.34	1.53	1.30	1.58	1.25	1.63	1.21	1.69
50	1.42	1.50	1.38	1.54	1.34	1.59	1.30	1.64	1.26	1.69
55	1.45	1.52	1.41	1.56	1.37	1.60	1.33	1.64	1.30	1.69
60	1.47	1.54	1.44	1.57	1.40	1.61	1.37	1.65	1.33	1.69
65	1.49	1.55	1.46	1.59	1.43	1.62	1.40	1.66	1.36	1.69
70	1.51	1.57	1.48	1.60	1.45	1.63	1.42	1.66	1.39	1.70
75	1.53	1.58	1.50	1.61	1.47	1.64	1.45	1.67	1.42	1.70
80	1.54	1.59	1.52	1.62	1.49	1.65	1.47	1.67	1.44	1.70
85	1.56	1.60	1.53	1.63	1.51	1.65	1.49	1.68	1.46	1.71
90	1.57	1.61	1.55	1.64	1.53	1.66	1.50	1.69	1.48	1.71
95	1.58	1.62	1.56	1.65	1.54	1.67	1.52	1.69	1.50	1.71
100	1.59	1.63	1.57	1.65	1.55	1.67	1.53	1.70	1.51	1.72

Table IX (*continued*) *Significance points of d_L and d_U: 1%*

n	$k'=1$		$k'=2$		$k'=3$		$k'=4$		$k'=5$	
	d_L	d_U	d_L	d_U	d_L	d_U	d_L	d_U	d_L	d_U
15	0.81	1.07	0.70	1.25	0.59	1.46	0.49	1.70	0.39	1.96
16	0.84	1.09	0.74	1.25	0.63	1.44	0.53	1.66	0.44	1.90
17	0.87	1.10	0.77	1.25	0.67	1.43	0.57	1.63	0.48	1.85
18	0.90	1.12	0.80	1.26	0.71	1.42	0.61	1.60	0.52	1.80
19	0.93	1.13	0.83	1.26	0.74	1.41	0.65	1.58	0.56	1.77
20	0.95	1.15	0.86	1.27	0.77	1.41	0.68	1.57	0.60	1.74
21	0.97	1.16	0.89	1.27	0.80	1.41	0.72	1.55	0.63	1.71
22	1.00	1.17	0.91	1.28	0.83	1.40	0.75	1.54	0.66	1.69
23	1.02	1.19	0.94	1.29	0.86	1.40	0.77	1.53	0.70	1.67
24	1.04	1.20	0.96	1.30	0.88	1.41	0.80	1.53	0.72	1.66
25	1.05	1.21	0.98	1.30	0.90	1.41	0.83	1.52	0.75	1.65
26	1.07	1.22	1.00	1.31	0.93	1.41	0.85	1.52	0.78	1.64
27	1.09	1.23	1.02	1.32	0.95	1.41	0.88	1.51	0.81	1.63
28	1.10	1.24	1.04	1.32	0.97	1.41	0.90	1.51	0.83	1.62
29	1.12	1.25	1.05	1.33	0.99	1.42	0.92	1.51	0.85	1.61
30	1.13	1.26	1.07	1.34	1.01	1.42	0.94	1.51	0.88	1.61
31	1.15	1.27	1.08	1.34	1.02	1.42	0.96	1.51	0.90	1.60
32	1.16	1.28	1.10	1.35	1.04	1.43	0.98	1.51	0.92	1.60
33	1.17	1.29	1.11	1.36	1.05	1.43	1.00	1.51	0.94	1.59
34	1.18	1.30	1.13	1.36	1.07	1.43	1.01	1.51	0.95	1.59
35	1.19	1.31	1.14	1.37	1.08	1.44	1.03	1.51	0.97	1.59
36	1.21	1.32	1.15	1.38	1.10	1.44	1.04	1.51	0.99	1.59
37	1.22	1.32	1.16	1.38	1.11	1.45	1.06	1.51	1.00	1.59
38	1.23	1.33	1.18	1.39	1.12	1.45	1.07	1.52	1.02	1.58
39	1.24	1.34	1.19	1.39	1.14	1.45	1.09	1.52	1.03	1.58
40	1.25	1.34	1.20	1.40	1.15	1.46	1.10	1.52	1.05	1.58
45	1.29	1.38	1.24	1.42	1.20	1.48	1.16	1.53	1.11	1.58
50	1.32	1.40	1.28	1.45	1.24	1.49	1.20	1.54	1.16	1.59
55	1.36	1.43	1.32	1.47	1.28	1.51	1.25	1.55	1.21	1.59
60	1.38	1.45	1.35	1.48	1.32	1.52	1.28	1.56	1.25	1.60
65	1.41	1.47	1.38	1.50	1.35	1.53	1.31	1.57	1.28	1.61
70	1.43	1.49	1.40	1.52	1.37	1.55	1.34	1.58	1.31	1.61
75	1.45	1.50	1.42	1.53	1.39	1.56	1.37	1.59	1.34	1.62
80	1.47	1.52	1.44	1.54	1.42	1.57	1.39	1.60	1.36	1.62
85	1.48	1.53	1.46	1.55	1.43	1.58	1.41	1.60	1.39	1.63
90	1.50	1.54	1.47	1.56	1.45	1.59	1.43	1.61	1.41	1.64
95	1.51	1.55	1.49	1.57	1.47	1.60	1.45	1.62	1.42	1.64
100	1.52	1.56	1.50	1.58	1.48	1.60	1.46	1.63	1.44	1.65

Source. Reproduced by permission of the editor and authors, from J. Durbin and G. S. Watson, "Testing for serial correlation in least squares regression, (II)." *Biometrika*, **38**, 1951, pp. 159–178.

Table X

Corresponding Values of r (Correlation Coefficient) and z

Example

r:upper scale $\quad r = 0.7$

z:lower scale $\quad \rightarrow z = 0.867$

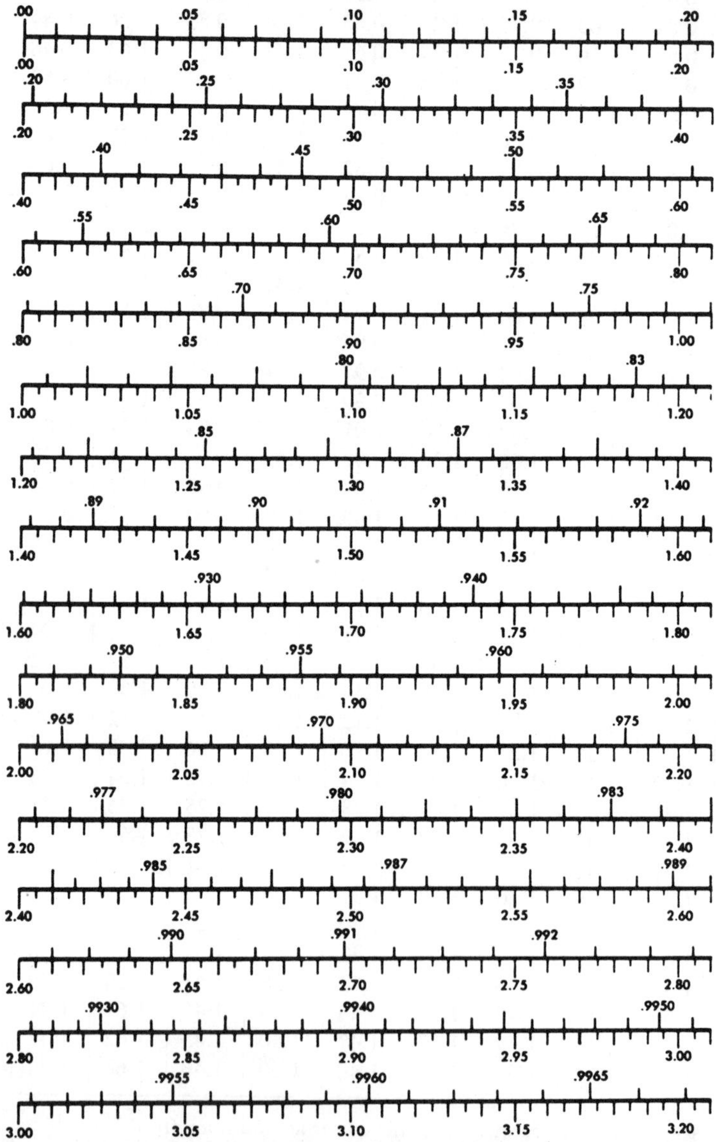

Source: Reproduced by permission from *Statistical Methods*, 5th edition, by George W. Snedecor, © 1956 by Iowa State State University Press.

APPENDIX II

Mathematical Appendix

I. INTRODUCTION

It seems likely that persons using this book will have had varying degrees of preparation in mathematics. The central purpose of the book is to provide an introduction to statistics to persons primarily interested in economics and business. Hence, we have built into the text from Chapter 2 onward those mathematical concepts and demonstrations which pertain directly to statistics and probability. Some rules of algebra, some operations involving summation signs, and some probability concepts (illustrated in part with Venn diagrams) have been explained as needed without explicit reference to set theory. Students who have some background in set theory will find valuable insights in Section II of this Appendix.

From Chapter 4 on, we have used some concepts of differential and integral calculus. In Chapter 10 (multiple regression analysis), we have used the notation and concepts of matrix algebra. It may be convenient for a good many readers to have a brief statement here of the concepts we have drawn upon in these two fields.

In addition, we shall list several references in which more extended presentations can be found. Some of these references, as will be noted, have been designed specifically for students of economics.

Some Preliminaries: Rules of Algebra and Summations

Rules of algebra. Three basic rules of algebra are used repeatedly in the text. These are:

1. The *commutative* laws:

The results of addition and multiplication are independent of the order in which the terms or factors are taken. That is,

$$A + B = B + A \qquad \text{and} \qquad AB = BA$$

2. The *associative* laws:

(a) The *sum* of three or more terms is independent of the manner in which the terms are grouped. That is,

$$A + (B + C) = (A + B) + C$$

(b) The *product* of three or more factors is independent of the manner in which the factors are grouped. That is,

$$(AB)\,C = A\,(BC)$$

3. The *distributive* law:

The product of a factor multiplied by the sum of two or more terms is equal to the sum of the products obtained by using the factor and each of the terms separately. That is,

$$A(B + C) = AB + AC$$

Rules of summations. The symbol $\sum$ is a mathematical shorthand notation which means to add. Therefore

$$\sum_{i=1}^{n} x_i = x_1 + x_2 + x_3 + \cdots + x_n$$

means to add all the x's for i going from 1 to n. The letter i is called a subscript. In some cases, a variable x may have more than one subscript. In such cases, it is possible to sum over any one of the subscripts without summing over the others. Thus, the quantity $\sum_{i=1}^{n} x_{ij}$ is not necessarily equal to the quantity $\sum_{j=1}^{m} x_{ij}$. The two subscripts imply a rectangular array of x's with n rows (that is, $i = 1, 2, 3, \ldots, n$) and m columns (that is, $j = 1, 2, 3, \ldots, m$):

$$\begin{bmatrix} x_{11} & x_{12} & x_{13} & \cdots & x_{1m} \\ x_{21} & x_{22} & x_{23} & \cdots & x_{2m} \\ x_{31} & x_{32} & x_{33} & \cdots & x_{3m} \\ \vdots & \vdots & \vdots & & \vdots \\ x_{n1} & x_{n2} & x_{n3} & \cdots & x_{nm} \end{bmatrix}$$

Then $\sum_{i=1}^{n} x_{i1} = x_{11} + x_{21} + x_{31} + \cdots + x_{n1}$ is the sum of the n items in the first column. Similarly, $\sum_{i=1}^{n} x_{i2} = x_{12} + x_{22} + x_{32} + \cdots + x_{n2}$ is the sum of the n items in the second column.

In contrast, $\sum_{j=1}^{m} x_{1j} = x_{11} + x_{12} + x_{13} + \cdots + x_{1m}$ is the sum of the m items in the first row; similarly, $\sum_{j=1}^{m} x_{2j} = x_{21} + x_{22} + x_{23} + \cdots + x_{2m}$ is the sum of the m items in the second row.

In general, where the x_{ij} notation is used, with $i = 1, 2, 3, \ldots n$ and $j = 1, 2, 3, \ldots m$, the number of rows is different from the number of columns (that is, $m \neq n$) and the x_{ij} can be any real numbers whatsoever. It is important, therefore, to specify very clearly just which row or column of numbers is to be summed when the $\sum$ notation is used for variables with two subscripts.

Four basic rules of summations have been used in the text. They are:

Rule I

If all of the items in a sum are multiplied by the same constant, then the constant can be placed before the summation. That is,

$$\sum_{i=1}^{n} cx_i = cx_1 + cx_2 + cx_3 + \cdots + cx_n = c(x_1 + x_2 + x_3 + \cdots + x_n)$$
$$= c\sum_{i=1}^{n} x_i$$

where c is a constant. This follows from the *distributive* law of algebra.

Rule II

The sum of two or more additive series of items can be obtained by summing each of the series separately and then adding their sums. That is,

$$\sum_{i=1}^{n} (x_i + y_i) = (x_1 + y_1) + (x_2 + y_2) + (x_3 + y_3) + \cdots + (x_n + y_n)$$
$$= (x_1 + x_2 + x_3 + \cdots + x_n) + (y_1 + y_2 + y_3 + \cdots + y_n)$$
$$= \sum_{i=1}^{n} x_i + \sum_{i=1}^{n} y_i$$

This follows from the *associative* law of algebra for sums.

Rule III

The sum of the differences of two or more series of items can be obtained by adding each series separately and performing the subtraction operation on these sums. That is,

$$\sum_{i=1}^{n} (x_i - y_i) = (x_1 - y_1) + (x_2 - y_2) + (x_3 - y_3) + \cdots + (x_n - y_n)$$
$$= (x_1 + x_2 + x_3 + \cdots + x_n) - (y_1 + y_2 + y_3 + \cdots + y_n)$$
$$= \sum_{i=1}^{n} x_i - \sum_{i=1}^{n} y_i$$

The *associative* law applies to subtractions as well as to additions.

Rule IV

Items which are multiplied by themselves or other items can be added up separately after the multiplication operation has been completed. That is,

$$\sum_{i=1}^{n} x_i(x_i + y_i) = x_1(x_1 + y_1) + x_2(x_2 + y_2) + x_3(x_3 + y_3) + \cdots + x_n(x_n + y_n)$$
$$= (x_1^2 + x_1y_1) + (x_2^2 + x_2y_2) + (x_3^2 + x_3y_3) + \cdots$$
$$+ (x_n^2 + x_ny_n)$$

$$= (x_1{}^2 + x_2{}^2 + x_3{}^2 + \cdots + x_n{}^2)$$

$$+ (x_1 y_1 + x_2 y_2 + x_3 y_3 + \cdots + x_n y_n)$$

$$= \sum_{i=1}^{n} x_i{}^2 + \sum_{i=1}^{n} x_i y_i$$

This follows from the distributive law and the associative law for sums.

II. SET THEORY IN RELATION TO PROBABILITY AND STATISTICS[1]

We shall denote by R a set of elements e which will be called *sample* points or event points, or simply *points*. The number of sample points may be finite or infinite. R is called the *sample space.* A sample point may be thought of as a possible outcome of a trial, experiment, or operation performed under a given set of conditions. The sample space R is simply the set of all possible outcomes which could be realized when an operation is performed under the given set of conditions.

A set E of sample points, which is a subset of the points in R, will be called an *event.* ("Set" and "event" will be used interchangeably.) A set E may consist of one point or of several or many points and will sometimes be referred to as an *e set.*

If e is an element of E, we write

(1) $$e \in E; \qquad \text{that is,} \qquad \text{"}e \text{ belongs to } E\text{"}$$

If E_1 and E_2 are two sets such that every point in E_1 is contained in E_2, then E_1 is called a subset of E_2 and we write

(2) $$E_1 \subset E_2 \qquad \text{or} \qquad E_2 \supset E_1$$

If E_1 consists of a single point, e, then $E_1 \subset E_2$ is equivalent to the statement $e \in E_2$.

Two sets E_1 and E_2 are *equal* if $E_1 \subset E_2$ and $E_1 \supset E_2$; in this case, we write

(3) $$E_1 = E_2$$

The *empty set* (or null set) is the set which contains no points. If E contains no points, we write

(4) $$E = \varnothing$$

The empty set $\varnothing$ is a subset of every set $E \subset R$.

[1] Based on S. S. Wilks, *Mathematical Statistics* (New York: John Wiley & Sons, Inc., 1962), pp. 1–7.

The set of all points contained in *both* E_1 and E_2 is called the *intersection* or *product* of E_1 and E_2 and is denoted by

(5) $$E_1 \cap E_2$$

sometimes read "E_1 cap E_2" or "E_1 intersection E_2."

The sets E_1 and E_2 are *disjoint*, that is, contain no common points, if

(6) $$E_1 \cap E_2 = \varnothing$$

The intersection of a sequence of sets $E_1, E_2, \ldots E_n$is denoted by

(7) $$\bigcap_{\alpha=1}^{n} E_\alpha$$

The set of all points contained in *at least one* of the sets E_1 and E_2 is called the *union* or *sum* of E_1 and E_2 and is denoted by

(8) $$E_1 \cup E_2$$

sometimes read "E_1 cup E_2." Note that if E_1 is any *e* set and $E_2 = \varnothing$, then $E_1 \cup E_2 = E_1 \cup \varnothing = E_1$.

The union of a finite sequence of sets $E_1, E_2, \ldots, E_n$ is denoted by

(9) $$\bigcap_{\alpha=1}^{n} E_\alpha$$

The set consisting of all points of E_1 not contained in E_2 is called the *difference* between E_1 and E_2 and is written as

(10) $$E_1 - E_2$$

If $E_1 \subset E_2$, then clearly $E_1 - E_2$ is the empty set. If $E_1 \cap E_2 = \varnothing$, then $E_1 - E_2 = E_1$. If $E_2 \subset E_1$, $E_1 - E_2$ is called the *proper difference* of E_1 and E_2. In this case, it is sometimes convenient to say that $E_1 - E_2$ is obtained by *subtracting* E_2 from E_1.

The proper difference $E_1 - E_2$ is sometimes called the *complement of* E_2 *relative to* E_1 and may be written as

(11) $$E_1 - E_2 = \bar{E}_2(E_1)$$

In the special case where E_1 is the entire sample space R, then $R - E_2$ is called the *complement* of E_2 and is written

(12) $$R - E_2 = \bar{E}_2$$

More generally, we have

(13) $$E_1 - E_2 = E_1 \cap \bar{E}_2$$

To illustrate schematically the notions of the union, intersection, and difference of sets, we use a *Venn diagram*. Let the sample space R be represented by the set of points inside the rectangle in Figure 1 and let E_1 and E_2

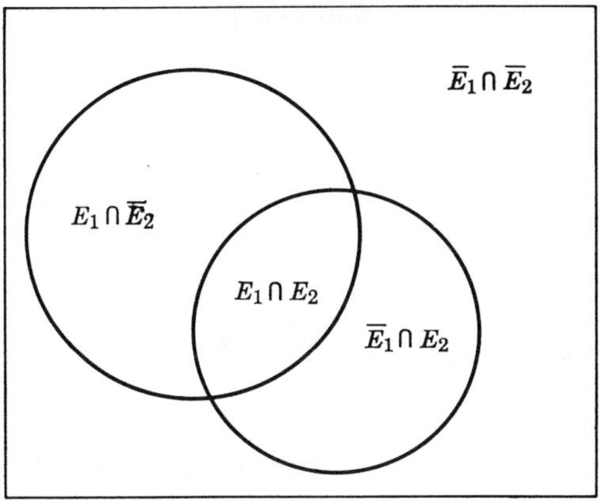

Figure 1. Venn diagram illustrating the four basic disjoint sets $E_1 \cap E_2$, $E_1 \cap \bar{E}_2$, $\bar{E}_1 \cap E_2$, and $\bar{E}_1 \cap \bar{E}_2$ generated by E_1 and E_2. (*Source*, Samuel S. Wilks, *Mathematical Statistics* (New York: John Wiley and Sons, Inc., 1962), p.5.)

be the set of points inside the large and small circles, respectively. Then the *union* $E_1 \cup E_2$ is the set of points enclosed by at least one of the circles; the *intersection* $E_1 \cap E_2$ is the set of points inside both circles; the *difference* $E_1 - E_2$, that is, $(E_1 \cap \bar{E}_2)$, is the set of points inside the larger circle but not the smaller; the *difference* $E_2 - E_1$ or $(\bar{E}_1 \cap E_2)$ is the set of points inside the smaller circle but not the larger; and $\bar{E}_1 \cap \bar{E}_2 = R - (E_1 \cup E_2)$ is the set of points inside the rectangle but not contained within either circle.

These set theory notions have the following interpretations in the language of events (outcomes of trials or experiments).

(i) $E_1 \subset E_2$ means that the occurrence of E_1 implies the occurrence of E_2; that is, a sample point e cannot occur in E_1 without occurring in E_2.

(ii) $E = \varnothing$, the empty set, means that event E cannot occur.

(iii) $E = R$, the entire sample space, means that event E must occur.

(iv) The intersection $E_1 \cap E_2$ is the event consisting of the joint occurrence of events E_1 and E_2.

(v) The union $E_1 \cup E_2$ is the event consisting of the occurrence of at least one of the events E_1 and E_2.

(vi) The difference $E_1 - E_2$ is the event denoting the occurrence of E_1 but not E_2.

It should be noted that all the definitions and operations (1) to (13) apply not only to sample points and sets of sample points (events) but also to elements and sets of elements of any kind. The elements may be point sets, people, automobiles, or anything else. However, in the present context we wish to emphasize the connections between set theory and probability theory.

The following statements apply to sets of any kind.

The operations of taking the union and product (intersection) of sets are (a) *commutative*; that is,

$$E_1 \cup E_2 = E_2 \cup E_1 \tag{14}$$

$$E_1 \cap E_2 = E_2 \cap E_1 \tag{15}$$

(b) *associative*; that is,

$$(E_1 \cup E_2) \cup E_3 = E_1 \cup (E_2 \cup E_3) \tag{16}$$

$$(E_1 \cap E_2) \cap E_3 = E_1 \cap (E_2 \cap E_3) \tag{17}$$

and (c) *distributive*; that is,

$$E_1 \cap (E_2 \cup E_3) = (E_1 \cap E_2) \cup (E_1 \cap E_3) \tag{18}$$

It is also evident that E_1 and the proper difference $E_2 - (E_1 \cap E_2)$ are disjoint and that

$$E_1 \cup E_2 = E_1 \cup [E_2 - (E_1 \cap E_2)] \tag{19}$$

Similarly, $E_1 \cap E_2$ and the proper differences $E_1 - (E_1 \cap E_2)$ and $E_2 - (E_1 \cap E_2)$ are disjoint so that

$$E_1 \cup E_2 = (E_1 \cap E_2) \cup [E_1 - (E_1 \cap E_2)] \cup [E_2 - (E_1 \cap E_2)] \tag{20}$$

The intersection of any two subsets E_1 and E_2 of R can be expressed in terms of unions and differences of sets by the following formula:

$$E_1 \cap E_2 = R - (\bar{E}_1 \cup \bar{E}_2) \tag{21}$$

Similarly, we have, for the union,

$$E_1 \cup E_2 = R - (\bar{E}_1 \cap \bar{E}_2) \tag{22}$$

The formulas for *conditional probability* can be readily expressed in set theory notation. For example,

$$P(E_2 \mid E_1) = \frac{P(E_1 \cap E_2)}{P(E_1)} \tag{23}$$

We can rewrite (23) as

$$P(E_1 \cap E_2) = P(E_1) \cdot P(E_2 \mid E_1) \tag{24}$$

If $P(E_2 \mid E_1) = P(E_2)$, then (24) reduces to the case of *independent events* E_1 and E_2, and we have

$$P(E_1 \cap E_2) = P(E_1)P(E_2) \tag{24a}$$

More generally, given a sequence of n events $E_1, E_2, E_3, \ldots, E_n$ such that $P(E_1) > 0$, $P(E_1 \cap E_2) > 0$, $P(E_1 \cap E_2 \cap E_3) > 0$, and so on to $P(E_1 \cap E_2 \cap$

$E_3 \cdots \cap E_n) > 0$, then

(25) $P(E_1 \cap E_2 \cap E_3 \cdots \cap E_n) = P(E_1) \cdot P(E_2 \mid E_1)$

$$\cdot P(E_3 \mid E_1 \cap E_2) \cdots \cdot P(E_n \mid E_1 \cap E_2 \cap E_3 \cdots \cap E_{n-1})$$

In the case of *mutual independence* of E_1, E_2, E_3, ... E_n, (25) reduces to

(25a) $$P(E_1 \cap E_2 \cap E_3 \cdots \cap E_n) = P(E_1)P(E_2)P(E_3) \cdots \cdot P(E_n)$$

III. CALCULUS

Our uses of calculus in the text are concentrated in Chapters 5 and 6 except for a few scattered references in Chapters 4, 7, 9, and 10. We have made explicit use of the concepts of limits, differentiation, and integration and implicit use of the concept of continuity.

We intend to keep the present section brief, elementary, operational, and rather directly oriented toward the particular concepts we have drawn upon in the text.

Functions. If the value of y depends upon the value of x so that to every value of x there corresponds a value of y, then y is said to be a *function of* x. We write this relation as

(1) $$y = f(x)$$

usually read as "y equals f of x." The value of $f(x)$ when $x = a$ is denoted by $f(a)$; a is a numerical value (scalar). To find $f(a)$, we merely substitute a for x in the formula for $f(x)$. Thus, if

$$f(x) = 2x^2 - 4x + 2$$

then

$$f(a) = 2a^2 - 4a + 2$$
$$f(2) = 2(4) - 4(2) + 2 = 8 - 8 + 2 = 2$$
$$f(3) = 2(9) - 4(3) + 2 = 18 - 12 + 2 = 8$$

and so on.

We shall restrict ourselves to variables whose values are real numbers.

All functions may be classed as either *algebraic* or *transcendental*. In the text we have mainly used polynomials such as $y = a + bx + cx^2$ (these are algebraic functions); logarithmic functions (such as $y = a + b \log x$ or $\log y = a + b \log x$); and exponential functions (such as $y = be^{-x^2}$). Logarithmic and exponential functions are classed among the transcendental functions.

If y is a function of x, it is clear that x is also a function of y. When we write $y = f(x)$, we usually think of x as the *independent* variable or *argument* (to which we assign any values we deem appropriate) and y as the *dependent* variable. Conversely, if we write $x = f(y)$, we usually think of y as the independent variable and x as the dependent variable. In general, for mathematical manipulations, either x or y may be chosen as the independent variable. For example, if $y = a + bx$, it is also true that $x = -a/b + (1/b)y$, obtained by dividing the original function through by b and rearranging terms. If $y = f(x)$ is such that to every value of x there corresponds a unique value of y and to every value of y there corresponds a unique value of x, it is sometimes convenient to write $x = f^{-1}(y)$ to indicate that x is an "inverse function" of y. (In set theory terminology, this is the case in which the mapping f: $X \rightarrow Y$ is one-one). If $y = ax^2$, then $x = \sqrt{y/a}$; two different values of x (for example, 2 and -2) will lead to the same value of y (namely, $y = 4a$) and, conversely, a given value of y is consistent with two different values of x. An "inverse function" does not exist in this case. The inverse function notation, $x = f^{-1}(y)$ does *not* stand for $x = 1/f(y)$, a meaningless notation. Nor should the inverse function notation be confused with the notation for the inverse of a matrix.

Limits. When the successive values of a variable x approach nearer and nearer a fixed number a in such a way that the difference $a - x$ becomes and remains numerically less than any preassigned positive number however small, the constant a is called the *limit* of x, and x is said to approach the limit a. We may write this relation as

$$\lim x = a \tag{2}$$

For example, the sequence of numbers 0.9, 0.99, 0.999, ... evidently approaches the limiting value 1. Or, suppose that regular polygons with increasing numbers of sides (an equilateral triangle, a square, a pentagon, a hexagon, and so on) are inscribed successively in a given circle. As the number of sides increases indefinitely, the area of the inscribed polygon approaches the area of the circle as a limit; that is,

$$\lim A_p = A_c$$

where A_p and A_c stand for the areas of the polygon and the circle, respectively.

Frequently, we are interested in observing the behavior of a function $f(x)$ as x approaches a limit. If, as x approaches a, the difference between $f(x)$ and some fixed number k ultimately becomes and remains numerically less than any preassigned positive constant however small, then $f(x)$ is said to approach the limit k, and we write

$$\lim_{x \to a} f(x) = k \tag{3}$$

read as "limit of $f(x)$, as x approaches a, is k."

Unless otherwise specified, it is supposed that the same limit is approached whether x approaches a from the positive or the negative direction. If we wish to consider what happens when x approaches a from the positive side only, we write $\lim_{x \to a^+} f(x)$; from the negative side only, $\lim_{x \to a^-} f(x)$.

If the limits exist for several functions $f(x)$, $g(x)$, $h(x)$, ... at $x = a$, it can be shown that the limit of a combination of the functions is the combination of the several limits. Writing lim for $\lim_{x \to a}$;

$$\lim\{f(x) + g(x)\} = \lim f(x) + \lim g(x) \tag{4}$$

$$\lim\{f(x) - g(x)\} = \lim f(x) - \lim g(x) \tag{5}$$

$$\lim\{f(x) \cdot g(x)\} = \lim f(x) \cdot \lim g(x) \tag{6}$$

$$\lim\left(\frac{f(x)}{g(x)}\right) = \frac{\lim f(x)}{\lim g(x)}, \qquad \text{for } \lim g(x) \neq 0 \tag{7}$$

Equations (4), (5), and (6) can be extended to any number of functions (assuming they are all defined—that is, continuous—on an interval containing a). The operations indicated in (4), (5), and (6) are evidently *commutative*, *associative*, and *distributive*.

A function $f(x)$ is said to be *continuous* at the point $x = a$ if $f(x)$ approaches a definite limit as x approaches a, and if the limit so approached is the value of the function at that point; that is, if

$$\lim_{x \to a} f(x) = f(a) \tag{8}$$

A function $f(x)$ is said to be *continuous in an interval* of values of the argument x if $f(x)$ is continuous at all points of the interval. As x changes from any value a to any other value b, the function $f(x)$, if continuous in the interval $x = a$ to $x = b$, assumes every value intermediate between $f(a)$ and $f(b)$. The graph of the function has no gap or break in the interval $a \leq x \leq b$; it is "continuous" in the ordinary sense of the word.

The most important type of discontinuity is that in which $f(x)$ increases numerically without limit (becomes infinite) as x approaches a. In this case, we write

$$\lim_{x \to a} f(x) = \infty \tag{9}$$

Graphically, such a discontinuity means that the curve $y = f(x)$ approaches nearer and nearer the (vertical) line $x = a$, usually without ever reaching it, while at the same time receding indefinitely from the horizontal x-axis. For example, as x approaches 2, the function

$$y = \frac{1}{x - 2}$$

becomes infinite:

$$\lim_{x \to 2} \left(\frac{1}{x - 2} \right) = \infty$$

The function is a rectangular hyperbola with asymptotes $x = 2$ and $y = 0$.

Often we are interested in observing the behavior of a function as the independent variable becomes infinite. If, when x increases without limit, the difference between $f(x)$ and some fixed number k ultimately becomes and remains numerically less than any preassigned positive constant, however small, we write

$$\lim_{x \to \infty} f(x) = k \tag{10}$$

Graphically this means that the curve $y = f(x)$ approaches nearer and nearer the line $y = k$, usually without ever reaching it, while at the same time receding indefinitely from the y-axis. (We assume in all our references to graphs that $y = f(x)$ is plotted on the vertical axis and x on the horizontal axis.)

The derivative. Given a continuous function

$$y = f(x) \tag{11}$$

if we choose some fixed value of x, the corresponding value of y is given by (11). Now consider another value of x, differing from the first one by an amount which we shall call the *increment* of x and denote by Δx. For this value of x, y will have a new value differing from the original by an amount Δy. We may write

$$y + \Delta y = f(x + \Delta x) \tag{12}$$

where

$$\Delta y = f(x + \Delta x) - f(x)$$

We now form the ratio $\Delta y / \Delta x$ and consider its behavior as Δx approaches zero. Geometrically, $\Delta y / \Delta x$ is the slope of the straight line connecting the following two points on $y = f(x)$: $P = (y, x)$ and $P' = (y + \Delta y, x + \Delta x)$. (See Figure 2.) As Δx approaches zero, Δy also becomes small and the point P' moves closer and closer to P; the line PP' approaches the line PT as a limiting position, where PT is the tangent to the curve $y = f(x)$ at P. Thus $\Delta y / \Delta x$ approaches the slope of the tangent PT as its limit; this limit is called the *derivative of y with respect to x*. This is the fundamental definition of the differential calculus:

The derivative of y with respect to x is the limit of the ratio $\Delta y/\Delta x$ when Δx approaches zero.

$$\frac{dy}{dx} = \lim_{\Delta x \to 0} \frac{\Delta y}{\Delta x} = \lim_{\Delta x \to 0} \frac{f(x + \Delta x) - f(x)}{\Delta x} \tag{13}$$

Other symbols for the derivative are

$$y' \quad \text{or} \quad f'(x) \tag{13.1}$$

$$D_x\, y \quad \text{or} \quad d_x f(x) \tag{13.2}$$

$$\frac{d}{dx}\, y \quad \text{or} \quad \frac{d}{dx} f(x) \tag{13.3}$$

The operation of finding the derivative of a function is called *differentiation.* To obtain the derivative of any function, the general procedure is as follows:

(a) Replace x by $x + \Delta x$ and y by $y + \Delta y$, and write

(i) $y = f(x)$

(ii) $y + \Delta y = f(x + \Delta x)$

(b) By subtraction, eliminate y between (i) and (ii), thus obtaining a formula for Δy in terms of x and Δx:

(iii) $\Delta y = f(x + \Delta x) - f(x)$

(c) By some suitable transformation, throw the right-hand member of (iii) into a form which contains Δx explicitly as a factor.

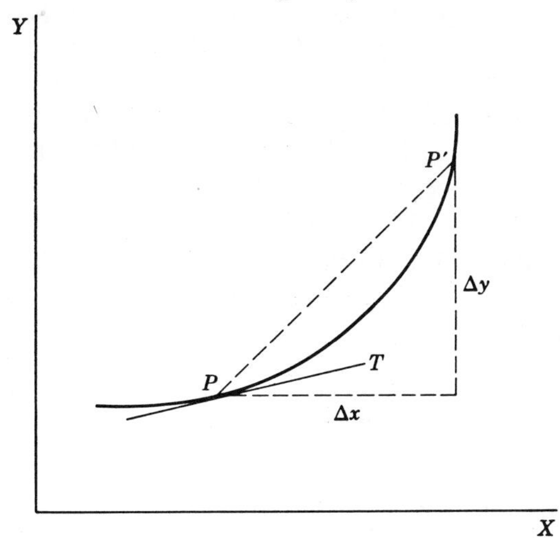

Figure 2

(d) Divide through by Δx:

$$\text{(iv)} \quad \frac{\Delta y}{\Delta x} = \frac{f(x + \Delta x) - f(x)}{\Delta x}$$

(e) Pass to the limit as Δx approaches zero. By definition, the left-hand member becomes dy/dx, while the right-hand member becomes $f'(x)$.

For example, to differentiate a polynomial, say, $y = x^3 - 2x$, we write:

$$\begin{aligned}
& y = x^3 - 2x \\
\text{(i)} \quad & y + \Delta y = (x + \Delta x)^3 - 2(x + \Delta x) \\
\text{(ii)} \quad & \Delta y = (x + \Delta x)^3 - 2(x + \Delta x) - x^3 + 2x \\
\text{(iii)} \quad & = x^3 + 3x^2\Delta x + 3x(\Delta x)^2 + (\Delta x)^3 \\
& \quad - 2x - 2\Delta x - x^3 + 2x \\
& = 3x^2\Delta x + 3x(\Delta x)^2 + (\Delta x)^3 - 2\Delta x \\
\text{(iv)} \quad & \frac{\Delta y}{\Delta x} = 3x^2 + 3x\Delta x + (\Delta x)^2 - 2
\end{aligned}$$

Then, as Δx approaches zero, the term $3x\Delta x$ approaches $3x(0) = 0$ and the term $(\Delta x)^2$ approaches $(0)^2 = 0$, leaving only the terms $3x^2$ and (-2). Hence,

$$(v) \quad \lim_{\Delta x \to 0} \frac{\Delta y}{\Delta x} = \frac{dy}{dx} = y' = 3x^2 - 2$$

We have already noted that the derivative y' of $y = f(x)$ for any given value of x is identical with *the slope of the graph* at that value of x. (Note that the slope of the tangent PT to the curve $y = f(x)$ at the point P is identical with the slope of the graph at P.) The derivative y' may also be interpreted as *the rate of change of y with respect to x.*

We shall write out some of the operational rules for the derivatives of combinations of function in Table 1. Rules 1, 2, and 3 can be extended to any number of functions.[2]

We shall note a few other important cases.

(14) If $y = c$, where c is a constant, Δc is always zero; hence $\Delta y = 0$, and $dy/dx = dc/dx = 0$; that is, *the derivative of a constant is zero.*

(15) If $y = x$,

$$\lim_{\Delta x \to 0} \frac{\Delta y}{\Delta x} = \frac{\Delta x}{\Delta x} = 1$$

or

$$\frac{dy}{dx} = \frac{dx}{dx} = 1$$

[2] Table 1 is reproduced from R. G. D. Allen, *Basic Mathematics*, (New York: St. Martin's Press) 1962, p. 273.

Table 1

Rule	Function	Derivative
1. Sum	$f(x)+g(x)$	$f'(x)+g'(x)$
2. Difference	$f(x)-g(x)$	$f'(x)-g'(x)$
3. Product	$f(x)g(x)$	$f(x)g'(x)+g(x)f'(x)$
4. Quotient	$\dfrac{f(x)}{g(x)}$	$\dfrac{g(x)f'(x)-f(x)g'(x)}{\{g(x)\}^2}$, if $g(x) \neq 0$
5. Composite	$F\{f(x)\}$	$F'(u)f'(x)$ where $u=f(x)$
6. Inverse	$f^{-1}(x)$	$\dfrac{1}{f'(y)}$, if $f'(y) \neq 0$

(16) If $y = cg(x)$, where c is a constant, $dy/dx = cg'(x)$.

This is a special case of Rule 3, with $f(x) = c$.

(17) If $y = c/g(x)$, where c is a constant, $dy/dx = -cg'(x)/\{g(x)\}^2$.

This is a special case of Rule 4 with $f(x) = c$.

(18) If $y = x^n$, where n is a positive integer, then

$$\frac{dy}{dx} = nx^{n-1}$$

In this case, $y + \Delta y = (x + \Delta x)^n$. Expanding the right-hand member by the binomial theorem, we obtain

$$y + \Delta y = x^n + nx^{n-1}\,\Delta x + \frac{n(n-1)}{2!}\,x^{n-2}(\Delta x)^2 + \cdots + (\Delta x)^n$$

Then

$$\Delta y = nx^{n-1}\,\Delta x + \frac{n(n-1)}{2!}\,x^{n-2}(\Delta x)^2 + \cdots + (\Delta x)^n$$

$$\frac{\Delta y}{\Delta x} = nx^{n-1} + \frac{n(n-1)}{2!}\,x^{n-2}\,\Delta x + \cdots + (\Delta x)^{n-1}$$

Then, as Δx approaches zero, all terms in Δx, $(\Delta x)^2, \ldots, (\Delta x)^{n-1}$ approach zero, and

$$\frac{dy}{dx} = \lim_{\Delta x \to 0} = nx^{n-1}$$

(19) If $y = \{g(x)\}^n$, then

$$\frac{dy}{dx} = n\{g(x)\}^{n-1}g'(x)$$

By this means, we can differentiate *any power of any function of x*. Note that n does not have to be an integer. For example, if

(20) $$y = \sqrt{g(x)} = \{g(x)\}^{1/2}$$

$$\frac{dy}{dx} = \frac{g'(x)}{2\sqrt{g(x)}} = \frac{1}{2}\{g(x)\}^{-1/2}g'(x)$$

The derivative of y with respect to x is itself a function of x and may, in turn, be differentiated. The derivative of the first derivative is called the *second derivative*, is written d^2y/dx^2, and is read, "the second derivative of y with respect to x." Further differentiations give d^3y/dx^3, $d^4y/dx^4, \ldots, d^ny/dx$w Other symbols for the second derivatives are y'' and $f''(x)$.

Since y'' is the derivative of y', the second derivative of a function is the *rate of change of the slope* of the graph of $y = f(x)$. First and second derivates have important applications in many economic and statistical problems in which we wish to maximize profits, minimize costs, minimize a sum of squares, maximize utility subject to a budget constraint, and so on.

Maxima and minima. Consider the graph (Figure 3) of a function $y = f(x)$. At point B in Figure 3, the tangent to the curve is horizontal, and $y' = 0$; at point C, the tangent to the curve is also horizontal, and again $y' = 0$. Point B represents a *maximum* point, and point C represents a *minimum* point on $y = f(x)$.

In particular applied problems, it may be clear from the nature of the case that $f(x)$ will have a maximum point (and no minimum point) in a certain interval. Mathematically, however, we can identify a *maximum* point by the fact that *the second derivative is negative*:

(21) $$y'' = \frac{d^2y}{dx^2} < 0$$

As x increases from a little less than x_b to a little more than x_b, the slope of the curve changes *from positive* ($y' > 0$) through zero ($y' = 0$) *to negative* ($y' < 0$). In the algebraic sense in which (say) -2 is less than -1, the slope of the curve, y', is *decreasing* in the neighborhood of B.

Conversely, we can identify a *minimum* point by the fact that the second derivative is positive:

(22) $$y'' = \frac{d^2y}{dx^2} > 0$$

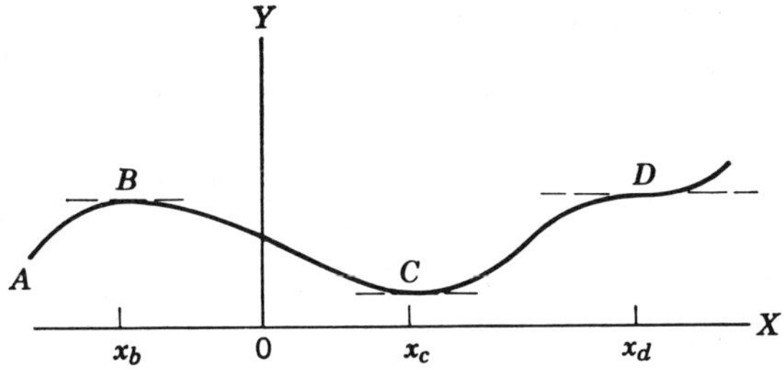

Figure 3.

Point C is a minimum point. As x increases from a little less than x_c to a little more than x_c, the slope of the curve changes *from negative* ($y' < 0$) through zero ($y' = 0$) *to positive* ($y' > 0$). In the algebraic sense, the slope of the curve, y', is *increasing* in the neighborhood of C.

A point such as D is neither a maximum nor a minimum, even though $y' = 0$. The slope of the curve, y', decreases from positive to zero but does not pass through zero and become negative; instead, it immediately becomes positive again. The second derivative, y'', is negative for $x < x_d$ and positive for $x > x_d$ in the neighborhood of D. We shall not elaborate on this case; it does indicate, however, that while $y' = 0$ is a *necessary* condition for a maximum or minimum point on $y = f(x)$, it is not a *sufficient* condition. The *sufficient* condition for a maximum (given $y' = 0$) is that $y'' < 0$; the sufficient condition for a minimum (given $y' = 0$) is that $y'' > 0$.

Derivatives of functions of several variables. The concepts presented so far can be extended to functions of several or many variables. Consider, for example, the function

$$z = f(x, y) \tag{23}$$

Geometrically, this represents a surface in three-dimensional space; for any pair of values of the independent variables x, y (that is, a point in the plane containing all possible values of x and y), the corresponding value of z is the height of the surface above the x, y plane at the point (x, y).

Limits. When x and y approach the respective values x_1 and y_1, the function z is said to approach *a limit* z_1 if the point (x, y, z) of the surface (23) approaches a definite limiting point (x_1, y_1, z_1). We may rewrite equation (3) as

$$\lim_{x \to x_1} f(x) = y_1 \tag{3.1}$$

for the case in which $y = f(x)$. The corresponding definition in the case in which $z = f(x, y)$ is

$$\lim_{\substack{x \to x_1 \\ y \to y_1}} f(x, y) = z_1 \tag{24}$$

A function $f(x, y)$ is said to be *continuous* at the point (x_1, y_1) if

$$\lim_{\substack{x \to x_1 \\ y \to y_1}} f(x, y) = f(x_1, y_1) \tag{25}$$

Similar definitions can be stated for functions of n variables, where n is any positive integer.

We assume here that all functions discussed or implied in our demonstrations are continuous at all points under consideration.

Partial derivatives. If y is kept fixed at some value (say) y_1, the function $z = f(x, y)$ becomes a function of x alone, and the derivative of $f(x, y_1)$ may be found by the rules already described for $y = f(x)$. The derivative of $z = f(x, y_1)$ is called *the partial derivative of z with respect to x*, and is denoted by any one of the symbols

$$\frac{\partial z}{\partial x}, \quad \frac{\partial f}{\partial x}, \quad f_x'(x, y)$$

If x is kept fixed at some value x_1, then $z = f(x_1, y)$ becomes a function of y alone, and we may take the partial derivative of z with respect to y, denoted by any one of the symbols

$$\frac{\partial z}{\partial y}, \quad \frac{\partial f}{\partial y}, \quad f_y'(x, y)$$

The operation of obtaining partial derivatives is called partial differentia, tion, the idea of which can be extended at once to any number of variables, remembering that, in differentiating with respect to any one variable, *all other variables are treated as constants.*

Consider, for example, the function $z = x^2 + 3y^2 + 2t^2 + 3xy + 7xt + 5yt + 76$. Then

$$\frac{\partial z}{\partial x} = 2x + 3y + 7t$$

$$\frac{\partial z}{\partial y} = 6y + 3x + 5t$$

and

$$\frac{\partial z}{\partial t} = 4t + 7x + 5y$$

Given a surface in three dimensions, $z = f(x, y)$, the geometric interpretation of holding y constant at y_1 means that we are cutting the surface with the plane $y = y_1$. In this plane, z is a function only of x; the partial derivative $\partial z/\partial x$ is the slope of the curve of intersection of the surface $z = f(x, y)$ and the plane $y = y_1$.

Note that we can explore this curve of intersection for maximum and minimum points. At such points, $\partial z/\partial x = 0$. If the point is a maximum, *the second-order partial derivative*, denoted by

$$\frac{\partial}{\partial x}\left(\frac{\partial z}{\partial x}\right) = \frac{\partial^2 z}{\partial x^2}$$

is *negative*; that is

$$\frac{\partial^2 z}{\partial x^2} < 0 \tag{26}$$

Conversely, if the point is a *minimum*, the second-order partial derivative is *positive*; that is,

$$\frac{\partial^2 z}{\partial x^2} > 0 \tag{27}$$

Incidentally, the situation just described, in which we explore the curve $z = f(x, y_1)$ in the plane $y = y_1$ for (say) a maximum, illustrates the concept of a *constrained maximum*. For example, the surface $z = f(x, y)$ might have an unconstrained maximum at some point (z_2, x_2, y_2) at which $y_2 \neq y_1$. The highest value of z that can be found in the plane $y = y_1$ may be z_1, which is less than z_2 (z_2 is in the plane $y = y_2$). If the surface is a rounded dome of some sort with the point (z_2, x_2, y_2) at its summit, we find only lower points whichever way we move in either the $\pm x$ or $\pm y$ directions. Hence, at (z_2, x_2, y_2) we have

$$\frac{\partial z}{\partial x} = 0, \qquad \frac{\partial^2 z}{\partial x^2} < 0$$

and also

$$\frac{\partial z}{\partial y} = 0, \qquad \frac{\partial^2 z}{\partial y^2} < 0$$

If z has a constrained maximum, $z_1 < z_2$, within the plane $y = y_1$, then at z_1 we find

$$\frac{\partial z}{\partial x} = 0, \qquad \frac{\partial^2 z}{\partial x^2} < 0 \text{ as before}$$

but $\partial z/\partial y \neq 0$ at the point (z_1, x_1, y_1). If $y_2 > y_1$, the surface slopes upward from y_1 to y_2, and $\partial z/\partial y > 0$. Hence, $\partial z/\partial y > 0$ measures the increment in z associated with a unit increase in y. If z is net profit and y is (say) floor space fixed at the amount y_1, the value of $\partial z/\partial y > 0$ is the marginal cost or "opportunity cost" of limiting floor space to y_1 rather than expanding it to $(y_1 + 1)$.

Our constrained maximum problem may be stated as:

$$\text{Maximize } z = f(x, y)$$

subject to

$$y = y_1$$

The maximization of a function subject to specified constraints is a pervasive theme in the economic theories of the consumer, of the firm, and of national economic policy.

We should note that the necessary and sufficient conditions for an *unconstrained* maximum (or minimum) can be extended at once to functions of n variables, where n is any positive integer. We make use of these conditions in deriving the *normal equations of least-squares regression analysis* (Chapters 9 and 10). Thus we are given an equation form in which y, x_1, x_2 are deviations from their respective sample means,

$$y = b_1 x_1 + b_2 x_2 + z \tag{28}$$

y, x_1, and x_2 are (for computational purposes) given numbers, forming a series of n observations. Our object is to minimize the sum of the squared residuals, namely $\sum_{i=1}^{n} z_i^2$. We rewrite (28) as

$$z = y - (b_1 x_1 + b_2 x_2)$$

square,

$$z^2 = y^2 - 2y(b_1 x_1 + b_2 x_2) + (b_1 x_1 + b_2 x_2)^2$$

and sum,

$$\sum z^2 = \sum y^2 - 2 \sum y\,(b_1 x_1 + b_2 x_2) + \sum (b_1 x_1 + b_2 x_2)^2 \tag{29}$$

where all summations range over $i = 1, 2, \ldots, n$.

We are to choose values of b_1 and b_2 which will make $\sum z^2 = f(b_1, b_2)$ a minimum. Then we must have

$$\frac{\partial \sum z^2}{\partial b_1} = 0, \qquad \frac{\partial^2 \sum z^2}{\partial b_1^{\,2}} > 0$$

$$\frac{\partial \sum z^2}{\partial b_2} = 0, \qquad \frac{\partial^2 \sum z^2}{\partial b_2^{\,2}} > 0$$

Differentiating (29) with respect to b_1 and b_2, respectively, and setting the partial derivatives equal to zero, we obtain

$$\frac{\partial \sum z^2}{\partial b_1} = -2 \sum y x_1 + 2 \sum x_1 (b_1 x_1 + b_2 x_2) = 0 \tag{30.1}$$

and

$$\frac{\partial \sum z^2}{\partial b_2} = -2 \sum y x_2 + 2 \sum x_2 (b_1 x_1 + b_2 x_2) = 0 \tag{30.2}$$

or

$$b_1 \sum x_1{}^2 + b_2 \sum x_1 x_2 = \sum y x_1 \tag{31.1}$$

and

$$b_1 \sum x_1 x_2 + b_2 \sum x_2{}^2 = \sum y x_2 \tag{31.2}$$

which are the required normal equations.

The values of b_1 and b_2 determined from these normal equations will clearly satisfy the *first-order conditions*, or necessary conditions, for $\sum z^2$ to be a minimum. However, these conditions,

$$\frac{\partial \sum z^2}{\partial b_1} = 0, \qquad \frac{\partial \sum z^2}{\partial b_2} = 0$$

would also be found at a point where $\sum z^2$ is a maximum.

We derived the first normal equation by differentiating with respect to b_1, so we may differentiate (30.1) again with respect to b_1:

$$\frac{\partial}{\partial b_1}\left(\frac{\partial \sum z^2}{b_1}\right) = \frac{\partial^2 \sum z^2}{\partial b_1{}^2} = 2 \sum x_1{}^2 > 0 \tag{32.1}$$

Similarly, differentiating (30.2) again with respect to b_2:

$$\frac{\partial}{\partial b_2}\left(\frac{\partial \sum z^2}{\partial b_2}\right) = \frac{\partial^2 \sum z^2}{\partial b_2{}^2} = 2 \sum x_2{}^2 > 0 \tag{32.2}$$

The *squares* of the real numbers x_1 and x_2 are always positive, so the rates of change of the first-order partial derivatives, being sums of squares of real numbers (times the positive real number 2), are always positive; the values of b_1 and b_2 determined by (31.1) and (31.2) are associated with the *minimum* value of $\sum z^2$.

Derivatives of logarithmic and exponential functions. In Chapters 5, 6, 7 and 9 we made some use of logarithmic and exponential functions. The function

$$y = a^x \qquad (a > 0) \tag{33}$$

in which x varies continuously over the set of real numbers and is *not* limited to integers, is called the *exponential function.* To each value of x, there corresponds a unique value of y, and vice versa. In set theory terminology, the mapping $f: x \to y$ is one-one, and an inverse function $x = f^{-1}(y)$ exists, namely, $y = \log_a x$ (read "y equals log x to the base a").

Exponential functions obey the laws of exponents:

$$a^x \cdot a^t = a^{x+t} \tag{34}$$

$$(a^x)^t = a^{xt} \tag{35}$$

If (say) $x = 2$ and $t = 3$, (34) gives us

(34.1) $$a^2 \cdot a^3 = a^{(2+3)} = a^5$$

and (35) yields

(35.1) $$(a^2)^3 = a^2 \cdot a^2 \cdot a^2 = a^{(3 \times 2)} = a^6$$

The laws also hold if (say) $x = 1.56$ and $t = 3.25$; (34) leads to

(34.2) $$a^{1.56} \cdot a^{3.25} = a^{(1.56+3.25)} = a^{4.81}$$

and (35) leads to

(35.2) $$(a^{1.56})^{3.25} = a^{(1.56 \times 3.25)} = a^{5.07}$$

The exponential function is positive for all values of x, including $x < 0$. For example, if $x = -3$,

$$y = a^{-3} = \frac{1}{a^3} > 0$$

We note once again that $a > 0$, as specified in (33).

The inverse of the exponential function is the *logarithm*, defined by the statement that

(36) $$y = \log_a x \quad \text{if} \quad x = a^y \qquad (a > 1)$$

This function is single-valued and continuous for all *positive* values of x. The number a is called the *base* of the system of logarithms. In practical applications only two bases, $a = 10$ and $a = e = 2.71828\ldots$, are widely used.

The definition (36) implies the following facts (given $a > 1$):

(i) Negative numbers have no (real) logarithms,
(ii) Numbers between 0 and 1 have negative logarithms,
(iii) Numbers greater than 1 have positive logarithms,
(iv) As x approaches 0, y becomes indefinitely large and negative,
(v) As x increases indefinitely, y does likewise,
(vi) The logarithm of 1 is 0.

Logarithms to the base 10 are called *common logarithms*. They were often used in computing (and still are when neither electronic nor mechanical computers are at hand) because they possess the great advantage that the "mantissa" or fractional part of the logarithm is independent of the position of the decimal point in the given number (for example, the common logarithms of 2, 20, and 200 are 0.30103, 1.30103, and 2.30103, respectively; a table of "mantissas" includes the value 0.30103 only once and the user puts in front of it the appropriate power of 10).

In applications of calculus, it is more convenient to use so-called *natural logarithms*. Natural logarithms have as their base the number

$$e = 2.71828 \cdots \tag{37}$$

defined by the formula

$$e = \lim_{z \to \infty} \left(1 + \frac{1}{z}\right)^z \tag{38}$$

The natural logarithm of e is 1; that of 10 is 2.30259; that of 100 is 4.60517; and so on. In other words, $\log e = 1$, $e^{2.30259} = 10$, and $e^{4.60517} = 100$. A logarithm to the base 10 can be converted to a natural logarithm by multiplying it by 2.30259. Thus $\log_{10} 100 = 2.00000$; $\log_e 100 = 2(2.30259) = 4.60518$. A natural logarithm can be converted to a common logarithm (a logarithm to the base 10) by *dividing* it by 2.30259 or, what is the same thing, multiplying it by $0.43429 = 1/2.30259$. Thus $\log_e 100 = 4.60517$; $\log_{10} 100 = 4.60517/2.30259 = 2.00000$. More generally, $\log_{10} x = \log_e x/\log_e 10$, and $\log_e x = \log_{10} x/\log_{10} e$.

Derivative of a logarithm. To obtain the derivative of a logarithm to any base $a > 1$, we apply the procedure described earlier:

(i) $y = \log_a x$

(ii) $y + \Delta y = \log_a(x + \Delta x)$

(iii) $\Delta y = \log_a(x + \Delta x) - \log_a x = \log_a\left(\frac{x + \Delta x}{x}\right) = \log_a\left(1 + \frac{\Delta x}{x}\right)$

Then

(iv) $\dfrac{\Delta y}{\Delta x} = \dfrac{1}{\Delta x} \log_a\left(1 + \dfrac{\Delta x}{x}\right)$

Multiplying and dividing the right-hand member by x, we obtain

$$\frac{\Delta y}{\Delta x} = \frac{1}{x} \cdot \frac{x}{\Delta x} \log_a\left(1 + \frac{\Delta x}{x}\right) = \frac{1}{x} \log_a\left(1 + \frac{\Delta x}{x}\right)^{x/\Delta x}$$

Hence

(v) $$\frac{dy}{dx} = \lim_{\Delta x \to 0} \frac{\Delta y}{\Delta x} = \frac{1}{x} \lim_{\Delta x \to 0} \log_a\left(1 + \frac{\Delta x}{x}\right)^{x/\Delta x}$$

$$= \frac{1}{x} \log_a\left[\lim_{\Delta x \to 0}\left(1 + \frac{\Delta x}{x}\right)^{x/\Delta x}\right]$$

Putting

$$\frac{x}{\Delta x} = z$$

we see that the limit in brackets becomes $\lim_{z\to\infty} (1 + 1/z)^z = e$. Hence

(39) $$\frac{d}{dx} \log_a x = \frac{1}{x} \log_a e$$

In case the base a of the system of logarithms is the number e, $\log_a e = \log_e e = 1$, and (39) takes the much more convenient form

(40) $$\frac{d}{dx} \log x = \frac{1}{x}$$

This is the reason for using natural logarithms in applications of calculus.

Suppose now that u is any function of x. Then

(41) $$\frac{d}{dx} \log u = \frac{du/dx}{u}.$$

For example, let

$$y = \log \sqrt{1 + 2x^2} = \log(1 + 2x^2)^{1/2}$$

Then

$$y = \frac{1}{2} \log(1 + 2x^2)$$

$$y' = \frac{1}{2} \cdot \frac{4x}{1 + 2x^2} = \frac{2x}{1 + 2x^2}$$

As an alternative notation, we might write any function of x as $g(x)$. Then (41) becomes

(41.1) $$\frac{d}{dx} \log\{g(x)\} = \frac{g'(x)}{g(x)}$$

Derivative of the exponential function. The derivative of the function $y = a^x$ may be found as follows.

If

(i) $y = a^x$

then

(ii) $\log_a y = x$

Remembering that y is a function of x, say, $g(x)$, we differentiate (ii) as

$$\frac{d}{dx}\log_a g(x) = \frac{g'(x)}{g(x)} \cdot \log_a e = \frac{dx}{dx}$$

Or

$$\text{(iii)} \quad \frac{d}{dx}\log_a y = \frac{1}{y}\frac{dy}{dx}\log_a e = 1$$

Then

$$\text{(iv)} \quad \frac{dy}{dx} = \frac{y}{\log_a e} = y \log_e a$$

from the standard formula for change of base ($\log_b a = 1/\log_a b$), where, in the present case, $b = e$. From (iv) and (i), we have

$$\frac{d}{dx} a^x = a^x \log_e a \tag{42}$$

For the case in which $a = e$, this becomes

$$\frac{d}{dx} e^x = e^x \log_e e = e^x \tag{43}$$

If u is a function of x, the formulas (42) and (43) become, respectively,

$$\frac{d}{dx} a^u = a^u \log_e a \cdot \frac{du}{dx} \tag{44}$$

and

$$\frac{d}{dx} e^u = e^u . \frac{du}{dx} \tag{45}$$

As an alternative notation, write $g(x)$ instead of u; then (45) becomes

$$\frac{d}{dx} e^{g(x)} = e^{g(x)} \cdot g'(x) \tag{45.1}$$

Integrals as "antiderivatives." Suppose we have two functions, $F(x)$ and $f(x)$, related in such a way that $F'(x) = f(x)$. In other words, $f(x)$ is the derivative of $F(x)$. We assume that both $F(x)$ and $f(x)$ are continuous over any intervals relevant to our discussion. The curve $y = F(x)$ has a tangent at x with a slope equal to the height of the curve $y = f(x)$. The function $F(x)$ is sometimes called the *antiderivative* of $f(x)$. In the older, and perhaps the more usual, terminology, $F(x)$ is called the *integral* of $f(x)$.

Recall that the derivative of any constant is zero. This means that two functions which differ only by a constant will have precisely the same derivative; for example,

$$\frac{d}{dx}(x^2 + 1) = 2x; \qquad \frac{d}{dx}(x^2 + 27) = 2x$$

and, in general,

$$\frac{d}{dx}(x^2 + C) = 2x$$

where C is any constant whatsoever. The following tabulation should make the situation perfectly clear:

x	$y = x^2 + 1$	$y = x^2 + 27$	$y = x^2 + C$	$dy/dx = 2x$
−10	101	127	100 + C	−20
−5	26	52	25 + C	−10
0	1	27	0 + C	0
5	26	52	25 + C	10
10	101	127	100 + C	20

The graphs of the first two functions are shown in Figure 4. The graphs are identical *except* that the level of one is higher than that of the other by the same constant amount ($27 - 1 = 26$) for all values of x. The graph of the derivative, $y' = dy/dx = 2x$, is also shown; it is identical for both of the $F(x)$ functions or antiderivatives.

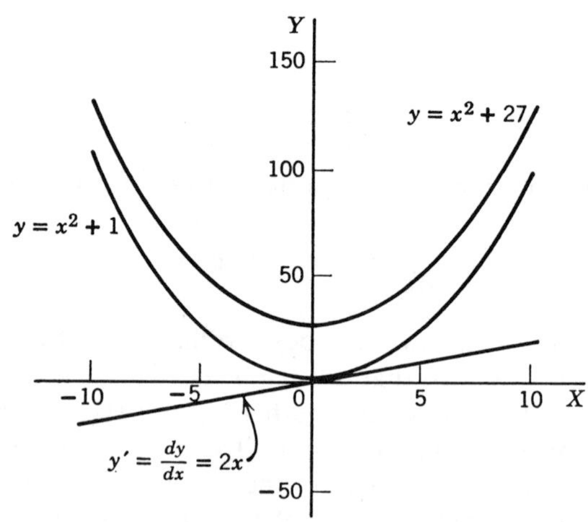

Figure 4

If we are given no information about a function $F(x)$ other than that its derivative is $F'(x) = f(x) = 2x$, we can be sure that $F(x)$ includes the term x^2 and no other terms in x but we know nothing about a possible nonzero constant term. The expression $y = x^2 + C$, where C is an arbitrary constant, is called the *indefinite integral* of $dy/dx = 2x$.

The differential. To define integration operationally, we must first define a differential. If $y = f(x)$ is a function of one variable, x, and if the derivative $f'(x) = dy/dx$ exists [that is, if $f(y)$ is continuous], the *differential* of y is

$$dy = f'(x)\,dx \tag{46}$$

where dx is treated as an independent variable.

Remember that $f'(x)$ is the *slope* of the curve $y = f(x)$. At point P, this slope is equal to the slope of the tangent PR. As x is increased by a small amount $\Delta x = dx$ (represented by the line PQ), $y = f(x)$ is increased by an amount Δy (represented by the line QP'). In contrast, formula (46) indicates that the *differential* of y, namely, dy, is represented by the line QR which is smaller than QP'. However, if dx is extremely small, the tangent PR will approximate the corresponding segment of $y = f(x)$ very closely, and dy will be only slightly smaller than Δy in percentage terms. (See Figure 5.)

In formula (46), dy and dx are both *infinitesimals*. An infinitesimal is a variable whose limit is zero. The variables dy and dx are infinitesimals of the *first order*; $(dy)^2$ and $(dx)^2$ are infinitesimals of the *second* order; $(dy)^3$ and $(dx)^3$ are infinitesimals of the *third* order; and so on. An infinitesimal of the

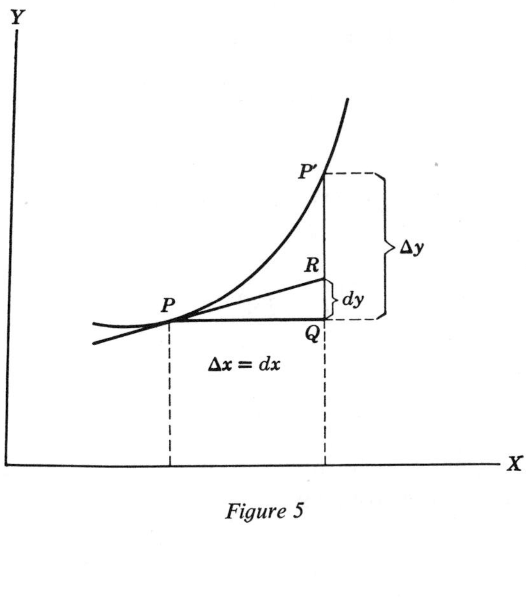

Figure 5

second order approaches zero more rapidly than one of the first order, in the same way that the square of any small quantity approaches zero more rapidly than the quantity itself. For example, if the quantity itself is $1/n$, its square is $1/n^2$; when $n = 100$, the quantity $1/n = 0.01$ and its square $= 0.0001$. More generally, if an infinitesimal v is defined as a function of another infinitesimal u, as in (46), and

$$\lim_{u \to 0} \frac{v}{u} = k \qquad \text{where } k \neq 0 \tag{47}$$

then v and u are said to be *infinitesimal of the same order*. Further, if

$$\lim_{u \to 0} \frac{v}{u} = 0 \tag{48}$$

v is said to be an infinitesimal of *higher order* than u.

Consider the function $y = x^2$, and let x increase by successive increments $dx = 1$. Using the curve-and-tangent graph in Figure 5, the value of dy corresponding to any dx_i will be

$$dy_i = 2x_{i-1}\, dx_i \tag{49}$$

where i denotes the ith successive increment of $dx = 1$ starting from the point $x_0 = x_{1-1} = 0$. Then $x_i = i \cdot dx_i = i \cdot 1 = i$ and $dy_i = 2x_{i-1}$. Note, however, that

$$\Delta y_i = y_i - y_{i-1} = x_i^2 - x_{i-1}^2 \tag{50}$$

Given that $dx_i = x_i - x_{i-1}$, (50) is equivalent to

$$\Delta y_i = (x_{i-1} + dx_i)^2 - x_{i-1}^2$$

or

$$\Delta y_i = x_{i-1}^2 + 2x_{i-1}\, dx_i + (dx_i)^2 - x_{i-1}^2$$

so

$$\Delta y_i = 2x_{i-1}\, dx_i + (dx_i)^2 \tag{51}$$

Then, from (49),

$$\Delta y_i - dy_i = [2x_{i-1}\, dx_i + (dx_i)^2] - 2x_{i-1}\, dx_i = (dx_i)^2$$

The numerical results for the first six successive increments of $dx_i = 1$ (starting from $x_0 = 0$) are as follows:

i	x_{i-1}	dx_i	$x_i = i$	Δy_i	dy_i	$(dx_i)^2$
1	0	1	1	1	0	1
2	1	1	2	3	2	1
3	2	1	3	5	4	1
4	3	1	4	7	6	1
5	4	1	5	9	8	1
6	5	1	6	11	10	1
Selected totals		6		36	30	6

The cumulative sum of the Δy_i's gives the values of $y = f(x) = x^2$, that is, a series of points on the curve. In each successive interval dx_i, $dy_i = \Delta y_i - (dx_i)^2$.

Now, if we let $dx_i = 0.1$ in the neighborhood of (say) $x_{i-1} = 4.0$, we find that $\Delta y_i = (4.1)^2 - (4.0)^2 = 16.81 - 16.00 = 0.81$, $dy_i = 2(4.0)\ (0.1) = 0.80$, and $(dx_i)^2 = (0.1)^2 = 0.01$; the difference between dy_i and Δy_i is reduced to about one part in 80.

Every infinitesimal such as dy, where dy is a function of another infinitesimal dx, can be written in the form

$$dy = dx\,(k + \eta) \tag{52}$$

where $k \neq 0$ and η is infinitesimal. For, dividing by dx and passing to the limit, we have

$$\lim_{dx \to 0} \frac{dy}{dx} = k + \lim_{dx \to 0} \eta \tag{53}$$

Comparing (53) with (47), and noting that $dy = v$ and $dx = u$ in (47), we must have

$$\lim_{dx \to 0} \eta = 0$$

If we apply (52), (53), and (47) to our present illustration, we have $k = 2x$ and $\eta = dx$; that is, $dy = dx(2x + dx) = 2xdx + (dx)^2$ and, when we divide by dx, we obtain

$$\frac{dy}{dx} = 2x + dx$$

Obviously, then,

$$\lim_{dx \to 0} \frac{dy}{dx} = 2x$$

Definition. The differential of any function of one variable, say, $y = f(x)$, is equal to its derivative multiplied by the differential of the independent variable:

(54) $$dy = y'\,dx$$

It follows from this definition that *all the fundamental formulas for derivatives become differential formulas if we merely multiply through by dx.* For instance, the formula for the *differential* of a product is

(55) $$d\{f(x)\cdot g(x)\} = f(x)\,g'(x)\,dx + g(x)\,f'(x)\,dx$$

The corresponding *derivative* was

(56) $$\frac{d}{dx}\{f(x)\cdot g(x)\} = f(x)g'(x) + g(x)f'(x)$$

Consider the function $z = f(x, y)$, a surface in three dimensions, with partial derivatives

$$\frac{\partial z}{\partial x} \quad \text{and} \quad \frac{\partial z}{\partial y}$$

and define the three infinitesimals, dz, dx, and dy. Given any point on the surface, say (z_1, x_1, y_1), a tangent to the surface at this point and lying in the plane $y = y_1$ has the slope $\partial z/\partial x$, while a tangent to the surface and lying in the plane $x = x_1$ has the slope $\partial z/\partial y$. Suppose that both $\partial z/\partial x > 0$ and $\partial z/\partial y > 0$ at (z_1, x_1, y_1); if we move to a neighboring point $(z_1 + dz, x_1 + dx, y_1 + dy)$ along the two tangents, the increase in z, namely, dz, is the sum of the two component movements $\partial z/\partial x \cdot dx$ and $\partial z/\partial y \cdot dy$. The infinitesimal dz is called the *total differential of* z and is written

(57) $$dz = \frac{\partial z}{\partial x}\,dx + \frac{\partial z}{\partial y}\,dy$$

This formula can be extended at once to the case in which z is a function of n variables, where n is any positive integer.

Integration as determining the area under a curve. Consider the case where

(58.1) $$y = F(x) + C = x^2 + C$$

and

(58.2) $$\frac{dy}{dx} = f(x) = 2x$$

Multiplying both sides of (58.2) by dx, we get

(59) $$dy = f(x)\,dx = 2x\,dx$$

Starting from the point ($x = 0$, $y = 0$), let $dx = 1$ and add successive increments dx_i, so that

$$x_i = 0 + dx_1 + dx_2 + \cdots + dx_i$$

when the ith increment has been added. The corresponding increments in y are $dy_i = 2x_i$ (since $dx_i = 1$ for all i):

i	dx_i	$\sum dx_i(=x_i)$	dy_i	dy_{i-1}	$1/2(dy_i + dy_{i-1})$
1	1	1	2	0	1
2	1	2	4	2	3
3	1	3	6	4	5
4	1	4	8	6	7
5	1	5	10	8	9
Selected totals	5		30	20	25

In Figure 6 we indicate that the *area* between the x-axis and the line $dy = 2xdx$ can be approximated by the sum of either of two sets of rectangles. Each rectangle has the same width, $dx_i = 1$. However, the heights of the rectangles which are entirely below the curve are dy_{i-1}, while the heights

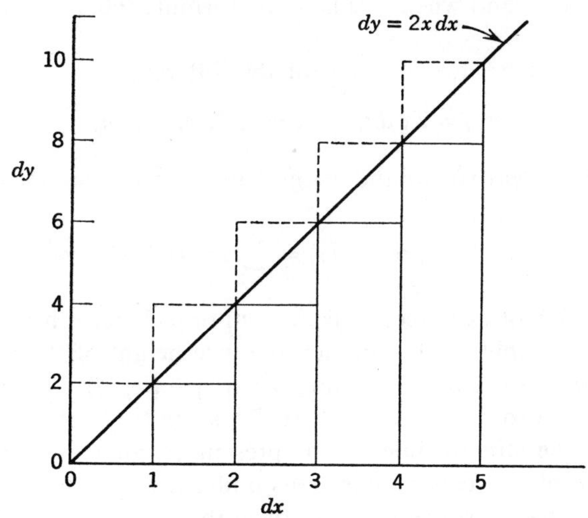

Figure 6

of those which project above the curve are dy_i. The sums of the respective sets of rectangular areas through $i = 5$ are

$$(60.1) \qquad \sum_{i=1}^{5} dy_{i-1}\, dx_i = (0 + 2 + 4 + 6 + 8) = 20$$

and

$$(60.2) \qquad \sum_{i=1}^{5} dy_i \cdot dx_i = (2 + 4 + 6 + 8 + 10) = 30$$

Formula (60.1) underestimates the desired area and formula (60.2) overestimates it. The *average* height of the line within any particular interval dx_i is midway between dy_{i-1} and dy_i. Note that the area of any triangle is equal the the base times half the altitude, and that the area of each small triangle lying between the line and the top of a rectangle is $dx_i\,(dy_i - dy_{i-1})/2 = 1(2/2) = 1$. Then the true area under the line from $i = 0$ through $i = 5$ is

$$(60.3) \qquad \sum_{i=1}^{5} \left(\frac{dy_{i-1} + dy_i}{2}\right) dx_i = (1 + 3 + 5 + 7 + 9) = 25$$

We recall that $y = x^2$ is the *integral* of $dy = 2x\,dx$. If $x_i = \sum_{i=1}^{5} dx_i = 5$, the value of y_i corresponding to it should be 25. We should also be able to calculate y_i as the sum of the areas

$$dx_i\left(\frac{dy_{i-1} + dy_i}{2}\right)$$

for $i = 1, 2, 3, 4, 5$, and when this is done, formula (60.3) indeed gives 25 as the sum.

Hence, starting from the formula for the differential

$$dy_i = 2x\, dx_i \qquad i = 1, 2, 3, \ldots, n$$

we can evidently *approximate* the integral $y_n = x_n{}^2$ (where $x_n = \sum_{i=1}^{n} dx_i$) as

$$(61) \qquad y_n \simeq \sum_{i=1}^{n} dy_i = \sum_{i=1}^{n} 2x_i\, dx_i$$

If we let the width of each rectangle, dx_i, approach zero, it is clear that the height of each rectangle will approach the true height of the line as a limit, and the triangular areas of discrepancy $dx_i\,(dy_i - dy_{i-1})/2$ become infinitesimals of a higher order than dy and dx. This limiting process would apply equally well if the straight line of our present example were replaced by a curve. The sum of the rectangular areas under the curve over any specified interval $a \le x \le b$ can be made to approach the true area under the curve as closely as we like.

This brings us to the fundamental theorem for *definite integrals*. Given a function $f(x)$, continuous in the interval from $x = a$ to $x = b$, divide this interval into n equal parts Δx and form the sum

$$\sum_{i=1}^{n} f(x_i)\,\Delta x,$$

where

$$x_1 = a,\ x_2 = a + \Delta x,\ x_3 = a + 2\Delta x,\ \ldots,\ x_n = a + (n-1)\,\Delta x$$

Then

$$\lim_{n\to\infty} \sum_{i=1}^{n} f(x_i)\,\Delta x = \int_a^b f(x)\,dx \tag{62}$$

Hence, if in any problem an arbitrarily close approximation to the required quantity can be found by adding up terms of the type $f(x)\,\Delta x$ from $x = a$ to $x = b$, that quantity is given *exactly* by the definite integral

$$\int_a^b f(x)\,dx$$

The concept of a definite integral can be extended readily to functions of two or more variables. Thus, consider the *volume* under a surface $z = F(x, y)$ within the limits $a \leq x \leq b$ and $c \leq y \leq d$. The height of the surface corresponding to the point (x_1, y_1) in the xy plane is z_1. (See Figure 7.) Suppose

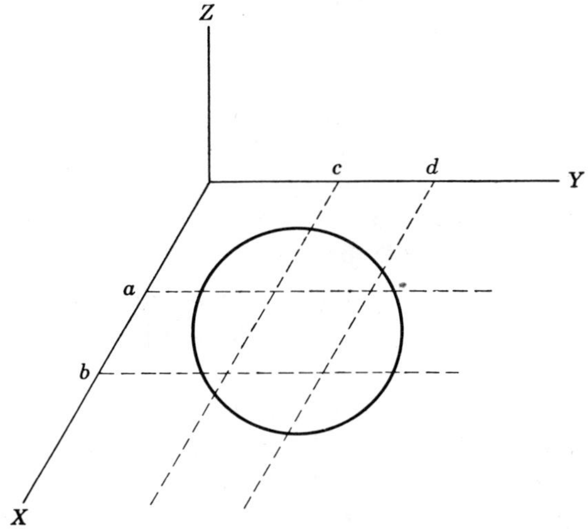

Figure 7

the area shown in the xy plane is the projection of the surface on that plane. Divide the interval a, b into n parts, Δx, by means of lines parallel to the y-axis and the interval c, d into m parts, Δy, by means of lines parallel to the x-axis. By passing through each line of the two sets a plane perpendicular to the xy plane, we produce mn vertical rectangular columns each with the cross-section area $\Delta x \cdot \Delta y$. If the boundary of the required area in the xy-plane includes some curved segments, some of the rectangular columns will extend slightly beyond the curve (and hence beyond the vertical curved surface of the volume we wish to determine). Also, the top of each column may intersect the surface in a curved area; if we cut the column off squarely at a height (say) z_1 where one of its corners touches the surface, the heights of other points on the surface of intersection will be slightly higher than z_1, leaving a small volume of discrepancy.

Approximately, then, the required volume is given by

$$V = \sum_{i=1}^{n} \sum_{j=1}^{m} f(x_i, y_j)\, \Delta y\, \Delta x \tag{63}$$

where $f(x_i, y_j)$ is the altitude of the column and the double summation covers all the full rectangles lying within the required area in the xy-plane.

It is obvious that the error in this approximation may be made arbitrarily small by taking both Δx and Δy sufficiently small. Hence the required volume is given *exactly* by

$$V = \lim_{\substack{\Delta x \to 0 \\ \Delta y \to 0}} \sum_{i=1}^{n} \sum_{j=1}^{m} f(x_i, y_j)\, \Delta y\, \Delta x \tag{64}$$

or

$$V = \int_a^b \int_{y_L}^{y_U} f(x, y)\, dy\, dx \tag{65}$$

This quantity is called a double integral or, more properly, an iterated integral. We could also write (65) as

$$V = \int_a^b \left[\int_{y_L}^{y_U} f(x, y)\, dy \right] dx \tag{65.1}$$

where y_L and y_U are the upper and lower bounds of y *given that* y is a function of x and its values are constrained by that relationship. The integration within the square brackets is carried out first, and *during this integration with repect to y, x remains constant.* The inner differential will disappear in this stage. We then integrate the result with respect to x.

The first or inner limits of integration are, in general, variables (as the limits of y are usually functions of x), but the outer limits are always constants.

We could, of course, integrate first with respect to x and then with respect to y. The same argument as before would lead to the formula

$$V = \int_{c}^{d} \left[\int_{x_L}^{x_U} f(x, y)\, dx \right] dy \tag{65.2}$$

Ordinarily, we do not use square brackets to demarcate the inner integral from the outer one. The expression

$$V = \int_{c}^{d} \int_{x_L}^{x_U} f(x, y)\, dx\, dy \tag{65.3}$$

would be interpreted correctly as implying that the integration with respect to x is to be done first.

We might note in closing that the symbol

$$\int_{a}^{b} f(x)\, dx \tag{66}$$

denotes "the definite integral of $f(x)$ from a to b." Where $F(x)$ is the anti-derivative of $f(x)$, we sometimes use the symbol $F(x)\Big]_a^b$ as meaning $F(b) - F(a)$:

$$\int_{a}^{b} f(x)\, dx = F(x)\Big]_a^b = F(b) - F(a) \tag{67}$$

The constant of integration, C, associated with indefinite integrals, disappears in (67), which may be interpreted as

$$[F(b) + C] - [F(a) + C] = F(b) - F(a)$$

IV. MATRIX ALGEBRA

We shall present and illustrate here those rules and definitions of matrix algebra which are most frequently used in multiple regression analysis.

We assume that all values of the variables and constants with which we are concerned are real numbers. We confine ourselves to estimating equations which are *linear in the constants*, though not necessarily in the variables. For example,

$$Y = a + bX + cX^2 + u$$

is linear in a, b, and c (which are to be estimated) but not in X, which is simply a series of given numbers (observations).

A *matrix* is an ordered set of m times n elements, arranged in an array of m rows and n columns, for given positive integers m and n; for example,

$$X = \begin{bmatrix} x_{11} & x_{12} & x_{13} & \cdots & x_{1n} \\ x_{21} & x_{22} & x_{23} & \cdots & x_{2n} \\ x_{31} & x_{32} & x_{33} & \cdots & x_{3n} \\ \vdots & \vdots & \vdots & & \vdots \\ x_{m1} & x_{m2} & x_{m3} & \cdots & x_{mn} \end{bmatrix}$$

The matrix X is said to be of "order m times n," that is, a matrix with m rows and n columns.

Two special cases of matrices arise (1) when $m = 1$ and (2) when $n = 1$.

When $m = 1$, the matrix is of order 1 times n and is called a *row vector.* For example, the first row of the matrix X, written as $(x_{11}, x_{12}, x_{13}, \cdots x_{1n})$, is a row vector. Any other row of X, taken separately, is also a row vector.

When $n = 1$, the matrix is of order m times 1 and is called a *column vector.* For example, the first column of the matrix X, written as $\{x_{11}, x_{21}, x_{31}, \cdots x_{m1}\}$, is a column vector. If space in books were of no consequence, it would no doubt be preferable to list the elements of a column vector vertically:

$$\begin{bmatrix} x_{11} \\ x_{21} \\ x_{31} \\ \vdots \\ x_{m1} \end{bmatrix}$$

Any other column of X, taken separately, is also a column vector.

Any matrix of order m times n may be regarded as built up from vectors, either from m row vectors each of order 1 times n or from n column vectors each of order m times 1.

Two special matrices of given order are the *zero* (or *null*) matrix and the *unit* (or *identity*) matrix. The zero or null matrix of order m times n is 0_{mn}, consisting of m rows and n columns of elements, all zero. The unit matrix, I, is always square; that is, it has the same number of columns as of rows. The unit matrix of order n times n is

$$I_n = \begin{bmatrix} 1 & 0 & 0 & \cdots & 0 \\ 0 & 1 & 0 & \cdots & 0 \\ 0 & 0 & 1 & \cdots & 0 \\ \vdots & \vdots & \vdots & & \vdots \\ 0 & 0 & 0 & & 1 \end{bmatrix}$$

Thus, the unit matrix has unit elements on its leading diagonal (from upper left to lower right) and zero elements everywhere else. When there is no danger of ambiguity as to the order of the matrix intended, the subscripts may be omitted in writing 0 and I.

Two vectors, each having n scalar elements, can be multiplied to obtain what is called their inner product, also a scalar:

$$\begin{aligned}(y_1, y_2, y_3, \ldots y_n)\{x_1, x_2, x_3, \ldots x_n\} \\ = y_1x_1 + y_2x_2 + y_3x_3 + \cdots + y_nx_n \\ = \sum_{i=1}^{n} y_i x_i\end{aligned}$$

Note that the elements in each vector are *ordered* (that is, they are listed in a specific sequence). In forming their inner product, the first element of the y vector must be multiplied by the first element of the x vector; the second element in the y vector by the second element in the x vector; and so on.

In applied problems, there are logical reasons for pairing specific elements in the y vector with specific elements in the x vector. For example, y_1 could be the weight of Person 1 and x_1 his height; y_2 the weight of Person 2 and x_2 his height; and so on. Or, y_1 could be the observed value of United States personal consumption expenditures in constant dollars in Year 1 and x_1 the observed value of the consumers price index in Year 1; y_2 and x_2 would be the corresponding observations for Year 2, y_3 and x_3 for Year 3, and so on, ending with the observations y_n and x_n for Year n.

Two matrices A of order m times k and B of order k times n are said to be *conformable* and give m times n inner products. We may write

$$AB = \begin{bmatrix} a_{11} & a_{12} & a_{13} & \cdots & a_{1k} \\ a_{21} & a_{22} & a_{23} & \cdots & a_{2k} \\ a_{31} & a_{32} & a_{33} & \cdots & a_{3k} \\ \vdots & \vdots & \vdots & & \vdots \\ a_{m1} & a_{m2} & a_{m3} & \cdots & a_{mk} \end{bmatrix} \begin{bmatrix} b_{11} & b_{12} & b_{13} & \cdots & b_{1n} \\ b_{21} & b_{22} & b_{23} & \cdots & b_{2n} \\ b_{31} & b_{32} & b_{33} & \cdots & b_{3n} \\ \vdots & \vdots & \vdots & & \vdots \\ b_{k1} & b_{k2} & b_{k3} & \cdots & b_{kn} \end{bmatrix}$$

The first inner product in AB is

$$(a_{11}, a_{12}, a_{13}, \ldots a_{1k})\{b_{11}, b_{21}, b_{31}, \ldots b_{k1}\} = \sum_{i=1}^{k} a_{1i} b_{i1}$$

the result of multiplying the first *row* of A by the first *column* of B. If we multiply the second row of A by the first column of B, we obtain the inner product

$$(a_{21}, a_{22}, a_{23}, \ldots a_{2k})\{b_{11}, b_{21}, b_{31}, \ldots b_{k1}\} = \sum_{i=1}^{k} a_{2i} b_{i1}$$

Similarly, the third row of A multiplied by the first column of B yields $\sum_{i=1}^{k} a_{3i} b_{i1}$, and the mth row of A multiplied by the first column of B yields $\sum_{i=1}^{k} a_{mi} b_{i1}$.

This set of m inner products completes the first *column* of the product matrix AB:

$$\begin{bmatrix} \sum_{i=1}^{k} a_{1i} b_{i1} \\ \sum_{i=1}^{k} a_{2i} b_{i1} \\ \sum_{i=1}^{k} a_{3i} b_{i1} \\ \vdots \\ \sum_{i=1}^{k} a_{mi} b_{i1} \end{bmatrix}$$

Since each of these inner products is a scalar, the result is simply a column vector with m elements (order m times 1).

The second column of the product matrix AB consists of m inner products formed by multiplying each row of A by the *second* column of B; for example,

$$(a_{11}, a_{12}, a_{13}, \ldots a_{1k})\{b_{12}, b_{22}, b_{32}, \ldots b_{k2}\} = \sum_{i=1}^{k} a_{1i} b_{i2}$$

The third column of AB consists of terms such as $\sum_{i=1}^{k} a_{1i} b_{i3}$, and the nth column of AB consists of terms such as $\sum_{i=1}^{k} a_{1i} b_{in}$. Specifically,

$$AB = \begin{bmatrix} \sum_{i=1}^{k} a_{1i} b_{i1} & \sum_{i=1}^{k} a_{1i} b_{i2} & \sum_{i=1}^{k} a_{1i} b_{i3} & \cdots & \sum_{i=1}^{k} a_{1i} b_{in} \\ \sum_{i=1}^{k} a_{2i} b_{i1} & \sum_{i=1}^{k} a_{2i} b_{i2} & \sum_{i=1}^{k} a_{2i} b_{i3} & \cdots & \sum_{i=1}^{k} a_{2i} b_{in} \\ \sum_{i=1}^{k} a_{3i} b_{i1} & \sum_{i=1}^{k} a_{3i} b_{i2} & \sum_{i=1}^{k} a_{3i} b_{i3} & \cdots & \sum_{i=1}^{k} a_{3i} b_{in} \\ \vdots & \vdots & \vdots & & \vdots \\ \sum_{i=1}^{k} a_{mi} b_{i1} & \sum_{i=1}^{k} a_{mi} b_{i2} & \sum_{i=1}^{k} a_{mi} b_{i3} & \cdots & \sum_{i=1}^{k} a_{mi} b_{in} \end{bmatrix}$$

The matrix AB could be written much more compactly as

$$AB = \sum_{i=1}^{k} a_{ri} b_{is} = a_{r1} b_{1s} + a_{r2} b_{2s} + \cdots + a_{rk} b_{ks}$$

for $r = 1, 2, 3, \ldots m$ and $s = 1, 2, 3, \ldots n$.

The set of operations just described is called matrix multiplication, or multiplication of matrices. It involves a special set of rules which were designed with an eye on the application of matrices to linear transformations and linear equations.

Only *conformable* matrices can be multiplied. If A is to be conformable with B, A must have as many columns (k) as B has *rows* (k). In the present example, A is of order m times k and B is of order k times n, so A is conformable with B. However, if we reverse the order of B and A, the matrix B of order k times n is *not* conformable with matrix A of order m times k! If we try to calculate the first inner product in this latter case, we find we are trying to multiply

$$(b_{11}, b_{12}, b_{13}, \dots b_{1n}) \quad \text{by} \quad \{a_{11}, a_{21}, a_{31}, \dots a_{m1}\}$$

or an n vector by an m vector, where usually $n \neq m$. No inner product exists in this situation; the commutative law does not hold in matrix multiplication and, in general, $BA \neq AB$.

In the abstract, the rules for matrix multiplication may seem quite tricky. In concrete applications, however, the right way makes sense in terms of the logic of the sequence of operations required to solve the problem at hand, and the wrong way (or ways) is nonsense from an operational point of view.

Matrix multiplication is deliberately designed to agree with successive applications of linear transformations. This is readily seen from the following demonstration.[3]

Write two matrices

$$A = \begin{bmatrix} a_{11} & a_{12} \\ a_{21} & a_{22} \end{bmatrix} \quad \text{and} \quad B = \begin{bmatrix} b_{11} & b_{12} \\ b_{21} & b_{22} \end{bmatrix}$$

both of order 2 times 2. They are conformable and yield the product matrix

$$AB = \begin{bmatrix} c_{11} & c_{12} \\ c_{21} & c_{22} \end{bmatrix}$$

where

$$c_{11} = a_{11}b_{11} + a_{12}b_{21} \qquad c_{12} = a_{11}b_{12} + a_{12}b_{22}$$
$$c_{21} = a_{21}b_{11} + a_{22}b_{21} \qquad c_{22} = a_{21}b_{12} + a_{22}b_{22}$$

Now take the linear transformations

$$z_1 = a_{11}y_1 + a_{12}y_2 \quad \text{and} \quad y_1 = b_{11}x_1 + b_{12}x_2$$
$$z_2 = a_{21}y_1 + a_{22}y_2 \qquad\qquad y_2 = b_{21}x_1 + b_{22}x_2$$

and combine them (in succession) to give

$$z_1 = a_{11}(b_{11}x_1 + b_{12}x_2) + a_{12}(b_{21}x_1 + b_{22}x_2)$$
$$z_2 = a_{21}(b_{11}x_1 + b_{12}x_2) + a_{22}(b_{21}x_1 + b_{22}x_2)$$

[3] Adapted from R. G. D. Allen, *Basic Mathematics*, op. cit., p. 381.

Then

$$z_1 = (a_{11}b_{11} + a_{12}b_{21})x_1 + (a_{11}b_{12} + a_{12}b_{22})x_2$$
$$z_2 = (a_{21}b_{11} + a_{22}b_{21})x_1 + (a_{21}b_{12} + a_{22}b_{22})x_2$$

or

$$z_1 = c_{11}x_1 + c_{12}x_2$$
$$z_2 = c_{21}x_1 + c_{22}x_2$$

We can build upon this example as follows.

A vector is a special case of a matrix. The matrix

$$B = \begin{bmatrix} b_{11} & b_{12} \\ b_{21} & b_{22} \end{bmatrix},$$

of order 2 times 2, is conformable with the column vector

$$x = \begin{bmatrix} x_1 \\ x_2 \end{bmatrix},$$

of order 2 times 1. Then the product

$$Bx = \begin{bmatrix} (b_{11}x_1 + b_{12}x_2) \\ (b_{21}x_1 + b_{22}x_2) \end{bmatrix},$$

of order 2 times 1. We note that the first element in this product is equal to y_1 and the second is equal to y_2. Hence the column vector

$$y = \begin{bmatrix} y_1 \\ y_2 \end{bmatrix},$$

of order 2 times 1, is equal to Bx.

It follows that we can write sets of two or more linear transformations (or linear equations) in vector and matrix notation at two levels of compactness:

$$\begin{bmatrix} y_1 \\ y_2 \end{bmatrix} = \begin{bmatrix} b_{11} & b_{12} \\ b_{21} & b_{22} \end{bmatrix} \begin{bmatrix} x_1 \\ x_2 \end{bmatrix}$$

or

$$y = Bx \qquad \begin{array}{l} y\text{: 2 times 1} \\ x\text{: 2 times 1} \\ B\text{: 2 times 2} \end{array}$$

If we use the second form, $y = Bx$, we should also list the orders of each vector and matrix unless these orders are clear from some earlier statement of the problem at hand.

Similarly, the other transformations may be written as

$$\begin{bmatrix} z_1 \\ z_2 \end{bmatrix} = \begin{bmatrix} a_{11} & a_{12} \\ a_{21} & a_{22} \end{bmatrix} \begin{bmatrix} y_1 \\ y_2 \end{bmatrix}$$

or

$$z = Ay \qquad z, y\text{: 2 times 1}$$
$$A\text{: 2 times 2}$$

We may apply the two transformations successively in either notation. In expanded notation, we are given that

$$\begin{bmatrix} z_1 \\ z_2 \end{bmatrix} = \begin{bmatrix} a_{11} & a_{12} \\ a_{21} & a_{22} \end{bmatrix} \begin{bmatrix} y_1 \\ y_2 \end{bmatrix} \quad \text{and} \quad \begin{bmatrix} y_1 \\ y_2 \end{bmatrix} = \begin{bmatrix} b_{11} & b_{12} \\ b_{21} & b_{22} \end{bmatrix} \begin{bmatrix} x_1 \\ x_2 \end{bmatrix}$$

Substituting for y_1 and y_2, we obtain

$$\begin{bmatrix} z_1 \\ z_2 \end{bmatrix} = \begin{bmatrix} a_{11} & a_{12} \\ a_{21} & a_{22} \end{bmatrix} \begin{bmatrix} b_{11} & b_{12} \\ b_{21} & b_{22} \end{bmatrix} \begin{bmatrix} x_1 \\ x_2 \end{bmatrix}$$

or

$$\begin{bmatrix} z_1 \\ z_2 \end{bmatrix} = \begin{bmatrix} c_{11} & c_{12} \\ c_{21} & c_{22} \end{bmatrix} \begin{bmatrix} x_1 \\ x_2 \end{bmatrix}$$

where $c_{rs} = \sum_{i=1}^{2} a_{ri} b_{is}$, $r = 1, 2$, and $s = 1, 2$.

In compact notation, we may write:

$$\text{Given } z = Ay \qquad z, y, x\text{: 2 times 1}$$
$$y = Bx \qquad A, B\text{: 2 times 2}$$

Then

$$z = ABx$$

or, defining $C = AB$,

$$z = Cx$$

The compact notation is an extremely efficient shorthand for designing and describing linear computations and for many other purposes. The same symbols can represent matrices of any order. The numerical computations involved (say) in successive linear transformations become more voluminous as the orders of the matrices increase. However, standard programs are available in many computing centers for carrying out the required operations on matrices of all orders from 2 times 1 to 100 times 100 or larger.

Three or more matrices can be multiplied in sequence, as in the case $z = ABx$, if they are conformable. For matrices that are conformable, the following properties hold.[4]

Associative: $A(BC) = (AB)C$

Distributive: $A(B + C) = AB + AC$

$(A + B)C = AC + BC$

Zero matrices: $A0_{nn} = 0_{mm} A = 0_{mn}$, where A is of order m times n.

Unit matrices: $AI_n = I_m A = A$, where A is of order m times n.

If $A = [a_{rs}]$ is of order m times n, then the *transpose* of A, $A' = [a_{sr}]$, is of order n times m. That is,

$$A = \begin{bmatrix} a_{11} & a_{12} & a_{13} & \cdots & a_{1n} \\ a_{21} & a_{22} & a_{23} & \cdots & a_{2n} \\ a_{31} & a_{32} & a_{33} & \cdots & a_{3n} \\ \vdots & \vdots & \vdots & & \vdots \\ a_{m1} & a_{m2} & a_{m3} & \cdots & a_{mn} \end{bmatrix}$$

and

$$A' = \begin{bmatrix} a_{11} & a_{21} & a_{31} & \cdots & a_{m1} \\ a_{12} & a_{22} & a_{32} & \cdots & a_{m2} \\ a_{13} & a_{23} & a_{33} & \cdots & a_{m3} \\ \vdots & \vdots & \vdots & & \vdots \\ a_{1n} & a_{2n} & a_{3n} & \cdots & a_{mn} \end{bmatrix}$$

The first row of A becomes the first column of A'; the second row of A becomes the second column of A'; and so on.

We should pause here to define *matrix addition* which also exists only for conformable matrices.

If $A = [a_{rs}]$ and $B = [b_{rs}]$ are the same order m times n, then the sum $A + B$ is the matrix $C = [c_{rs}]$, also of order m times n, where $c_{rs} = a_{rs} + b_{rs}$ for all $r = 1, 2, \ldots m$ and all $s = 1, 2, \ldots n$.

Any matrix can be multiplied by a scalar. Thus, the *scalar product* λA, where $A = [a_{rs}]$ is of order m times n, is the matrix $D = [d_{rs}]$ where $d_{rs} = \lambda a_{rs}$ for all $r = 1, 2, \ldots m$ and all $s = 1, 2, \ldots n$.

Certain rules apply to the set of all m times n matrices. *Sums* are commutative and associative; that is, $A + B = B + A$, and $A + (B + C) = (A + B) + C$. Also, *scalar products* are associative and distributive:

$$\lambda(\mu A) = (\lambda\mu)A; \ \lambda(A + B) = \lambda A + \lambda B;$$
$$(\lambda + \mu)A = \lambda A + \mu A$$

where λ and μ are scalars.

[4] See R. G. D. Allen, *Basic Mathematics*, op. cit., p. 383.

We note that

$$A + 0 = 0 + A = A$$

where 0 is the zero matrix of order m times n; also, that

$$A + (-A) = (-A) + A = 0$$

where $-A = [-a_{rs}]$ and $-a_{rs} = (-1)\, a_{rs}$ for all $r = 1, 2, \ldots m$ and all $s = 1, 2, \ldots n$. In other words, $-A$ is obtained by multiplying A by the scalar -1.

These results together with the definition of a transpose imply some properties of transposes, namely,

$$(A + B)' = A' + B';\ (AB)' = B'A';\ (A')' = A;\ \text{and } I_n' = I_n$$

Some of these are intuitively obvious. Also, we may note that in forming the transpose of I_n' of

$$I_n = \begin{bmatrix} 1 & 0 & 0 & \cdots & 0 \\ 0 & 1 & 0 & \cdots & 0 \\ 0 & 0 & 1 & \cdots & 0 \\ \vdots & \vdots & \vdots & & \\ 0 & 0 & 0 & \cdots & 1 \end{bmatrix}$$

the first row of I_n becomes the first column of I_n'; the second row of I_n becomes the second column of I_n'; and so on. It is clear, however, that the first column of I_n' thus obtained is identical with the first column of I_n; the second column of I_n' is identical with the second column of I_n; and so on. Hence,

$$I_n' = I_n$$

The proposition $(AB)' = B'A'$ justifies a special demonstration. If matrix A is of order m times n, B must be of order n times m (in the present set of propositions, which involve matrices of order m times n and n times m only). Then the product matrix $C = AB$ results from the following multiplication:

$$C = AB = \begin{bmatrix} a_{11} & a_{12} & \cdots & a_{1n} \\ a_{21} & a_{22} & \cdots & a_{2n} \\ \vdots & \vdots & & \vdots \\ a_{m1} & a_{m2} & \cdots & a_{mn} \end{bmatrix} \begin{bmatrix} b_{11} & b_{12} & \cdots & b_{1m} \\ b_{21} & b_{22} & \cdots & b_{2m} \\ \vdots & \vdots & & \vdots \\ b_{n1} & b_{n2} & & b_{nm} \end{bmatrix}$$

where $C = [c_{rs}]$ is of order m times m. The elements in the first *column* of C are

$$\begin{aligned} c_{11} &= (a_{11}, a_{12}, \ldots a_{1n})\, \{b_{11}, b_{21}, \ldots b_{n1}\} \\ c_{21} &= (a_{21}, a_{22}, \ldots a_{2n})\, \{b_{11}, b_{21}, \ldots b_{n1}\} \\ &\vdots \\ c_{m1} &= (a_{m1}, a_{m2}, \ldots a_{mn})\, \{b_{11}, b_{21}, \ldots b_{n1}\} \end{aligned}$$

In the transpose $C' = (AB)'$, these same elements form the first *row*:

$$c_{11}', c_{12}', \ldots c_{1m}'$$

The product matrix $D = (B'A')$ results from the following multiplication

$$D = B'A' = \begin{bmatrix} b_{11} & b_{21} & \cdots & b_{n1} \\ b_{12} & b_{22} & \cdots & b_{n1} \\ \vdots & \vdots & & \vdots \\ b_{1m} & b_{2m} & \cdots & b_{nm} \end{bmatrix} \begin{bmatrix} a_{11} & a_{21} & \cdots & a_{m1} \\ a_{12} & a_{22} & \cdots & a_{m2} \\ \vdots & \vdots & & \vdots \\ a_{1n} & a_{2n} & \cdots & a_{mn} \end{bmatrix}$$

Here B' is of order m times n, A' is of order n times m, and $D = (B'A') = [d_{rs}]$ is of order m times m.

The first *row* of D consists of the following elements:

$$\begin{aligned} d_{11} &= (b_{11}, b_{21}, \ldots b_{n1}) \{a_{11}, a_{12}, \ldots a_{1n}\} \\ d_{12} &= (b_{11}, b_{21}, \ldots b_{n1}) \{a_{21}, a_{22}, \ldots a_{2n}\} \\ &\vdots \\ d_{1m} &= (b_{11}, b_{21}, \ldots b_{n1})\ a_{m1} \quad a_{m2}, \ldots a_{mn}\} \end{aligned}$$

It can be seen that $d_{11} = c_{11}'$; $d_{12} = c_{12}'$; $d_{1m} = c_{1m}'$; and so on. (Note that the inner product $\sum_{i=1}^{n} a_{1i}b_{i1} = \sum_{i=1}^{n} b_{i1}a_{1i}$, from the commutative rule of ordinary algebra that $ab = ba$, where a and b are scalars. This demonstration can be extended to show that every element c_{rs}' of C' is identical with the corresponding element d_{rs} of D, for all $r = 1, 2, \ldots m$ and all $s = 1, 2, \ldots m$. Therefore, $C' = D$, and

$$(AB)' = B'A'$$

It should be noted that this proposition holds for conformable matrices of any orders. For example, if A is of order m times n as before but B is of order n times k, where $k \neq m$, the product matrix $C = AB$ is of order m times k and its transpose $C' = (AB)'$ is of order k times m. The transpose of B, namely, B', is of order k times n and the transpose of A, namely, A', is of order n times m. Hence, the product matrix $D = B'A'$ is of order k times m, the same order as C'. The k times m inner products in each matrix are formed by multiplying pairs of vectors of n elements; there are k such vectors in B (and hence also in B') and m such vectors in A (and hence also in A'). Thus $D = (B'A')$ and $C' = (AB)'$ each contain the same k times m elements, and it can be shown (as we have indicated) that these elements appear in the same order in each matrix.

The same result can be derived in more compact notation as follows.

Given A: m times n and B: n times k with typical elements a_{ri} and b_{is}, $i = 1, 2, \ldots n$, $r = 1, 2, \ldots m$, and $s = 1, 2, \ldots k$. Then

$$(AB) = \sum_{i=1}^{n} a_{ri} b_{is} \qquad \text{for} \quad \begin{aligned} r &= 1, 2, \ldots m \\ s &= 1, 2, \ldots k \end{aligned}$$

The typical elements in B' and A' are b_{si}' and a_{ir}' respectively, with the row and column indexes reversed as compared with B and A. Then

$$(B'A') = \sum_{i=1}^{n} b_{si}' \, a_{ir}' \qquad \text{for} \quad \begin{array}{l} s = 1, 2, \ldots k \\ r = 1, 2, \ldots m \end{array}$$

But $b_{si}' = b_{is}$ and $a_{ir}' = a_{ri}$ for all $s = 1, 2, \ldots k$, all $i = 1, 2, \ldots n$, and all $r = 1, 2, \ldots m$. Hence

$$\sum_{i=1}^{n} b_{si}' \, a_{ir}' = \sum_{i=1}^{n} a_{ri} \, b_{is} \qquad \text{for} \quad \begin{array}{l} s = 1, 2, \ldots k \\ r = 1, 2, \ldots m \end{array}$$

However, the matrix $(B'A')$ has the element $\sum_{i=1}^{n} b_{si}'a_{ir}'$ in Row s and Column r, while the matrix (AB) has the element $\sum_{i=1}^{n} a_{ri} b_{is}$ in Row r and Column s. When we transpose (AB) to obtain $(AB)'$, the element $\sum_{i=1}^{n} a_{ri} b_{is}$ now appears in Row s and Column r of $(AB)'$; we might reflect this fact notationally by writing the transposed element as $\sum_{i=1}^{n} b_{si}' \, a_{ir}'$, which is identical with the element in the same position (Row s and Column r) in $(B'A')$. Q.E.D.

Square matrices, that is, matrices of order n times n where n is any positive integer, are used frequently in multiple regression analysis and in economic models. Square matrices arise particularly in solving sets of "simultaneous" linear equations. We shall therefore begin our discussion of square matrices with a review of the algebraic procedure for solving a set of simultaneous equations. Consider, for example, the equations

$$a_{11}x_1 + a_{12}x_2 = c_1 \tag{1}$$

$$a_{21}x_1 + a_{22}x_2 = c_2 \tag{2}$$

in which a_{11}, a_{21}, a_{12}, a_{22}, c_1, and c_2 are given numbers (scalars) and x_1 and x_2 are variables. This set of equations may be solved by "eliminating" either variable so that the other variable is expressed as a function only of the a's and c's. We may eliminate x_1 as follows:

We multiply each term in (1) by $(-a_{21})$ and each term in (2) by a_{11}, obtaining

$$-a_{21}a_{11}x_1 - a_{21}a_{12}x_2 = -a_{21}c_1 \tag{1.1}$$

$$a_{11}a_{21}x_1 + a_{11}a_{22}x_2 = a_{11}c_2 \tag{2.1}$$

Now we add corresponding terms in (1.1) and (2.1). The coefficients of x_1 cancel out; that is, $(a_{11}a_{21} - a_{21}a_{11})\,x_1 = 0x_1 = 0$. We are left with

$$(a_{11}a_{22} - a_{21}a_{12})x_2 = a_{11}c_2 - a_{21}c_1$$

so

$$x_2 = \frac{a_{11}c_2 - a_{21}c_1}{a_{11}a_{22} - a_{21}a_{12}}$$

Alternatively, we could eliminate x_2 from (1) and (2) as follows.

We multiply each term in (1) by a_{22} and each term in (2) by $(-a_{12})$, obtaining

$$a_{22}a_{11}x_1 + a_{22}a_{12}x_2 = a_{22}c_1 \tag{1.2}$$

$$-a_{12}a_{21}x_1 - a_{12}a_{22}x_2 = -a_{12}c_2 \tag{2.2}$$

We now add corresponding terms in (1.2) and (2.2). The coefficients of x_2 cancel out; that is, $(a_{22}a_{12} - a_{12}a_{22})x_2 = 0x_2 = 0$. We are left with

$$(a_{11}a_{22} - a_{21}a_{12})\,x_1 = a_{22}c_1 - a_{12}c_2$$

so

$$x_1 = \frac{a_{22}c_1 - a_{12}c_2}{a_{11}a_{22} - a_{21}a_{12}}$$

We note that the denominators in the expressions for x_1 and x_2 are identical; they include all four of the a_{ij}'s, which are coefficients of x_1 and x_2, but do not include any (either) of the c_i's. The numerators in the expressions for x_1 and x_2 each include all (both) of the c_i's; however, in the numerator of x_1, the c_i's are multiplied by a_{ij} coefficients with the column index $j = 2$, while in the numerator of x_2, the c_i's are multiplied by a_{ij} coefficients with the column index $j = 1$.

A geometrical interpretation of these results may be helpful. Equations (1) and (2) are the equations of two straight lines in the plane containing all pairs of values of the variables x_1 and x_2. The coefficients a_{ij} determine the *slopes* of these two lines; the coefficients c_i are involved in determining their *levels*. Before drawing the graph of equation (1), we would "normalize" it by dividing each term by the coefficient of x_1; that is,

$$\left(\frac{a_{11}}{a_{11}}\right)x_1 + \left(\frac{a_{12}}{a_{11}}\right)x_2 = \frac{c_1}{a_{11}} \tag{1.3}$$

or

$$x_1 = \frac{c_1}{a_{11}} - \left(\frac{a_{12}}{a_{11}}\right)x_2 \tag{1.4}$$

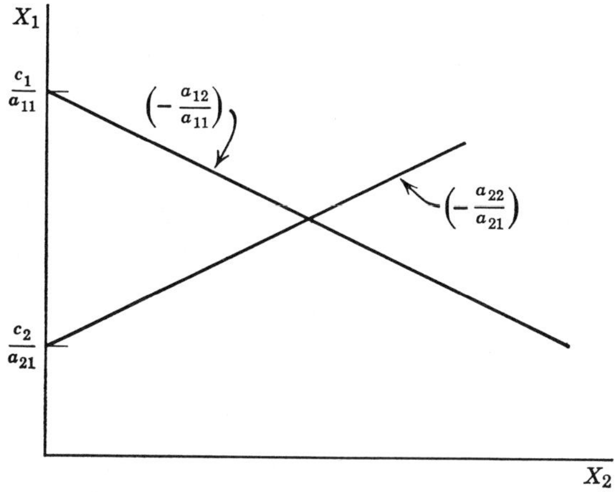

Figure 8

Similarly, we would "normalize" equation (2) by dividing each term by the coefficient of x_1 in that equation; that is,

$$\left(\frac{a_{21}}{a_{21}}\right)x_1 + \left(\frac{a_{22}}{a_{21}}\right)x_2 = \frac{c_2}{a_{21}} \tag{2.3}$$

or

$$x_1 = \frac{c_2}{a_{21}} - \left(\frac{a_{22}}{a_{21}}\right)x_2 \tag{2.4}$$

The normalized equations are shown in Figure 8. Their constant terms or intercepts are (c_1/a_{11}) and (c_2/a_{21}) and their slopes are $(-a_{12}/a_{11})$ and $(-a_{22}/a_{21})$, respectively. These two lines intersect in a point (x_1, x_2), the coordinates of which constitute the "solution" of the pair of equations. This point lies on both lines and satisfies both equations. At this point, x_1 has, of course, the same numerical value in both equations (1.4) and (2.4). Therefore, the right-hand sides of (1.4) and (2.4) at this point are also equal to one another and we may write

$$\frac{c_1}{a_{11}} - \left(\frac{a_{12}}{a_{11}}\right)x_2 = \frac{c_2}{a_{21}} - \left(\frac{a_{22}}{a_{21}}\right)x_2$$

We have eliminated x_1, and the resulting equation includes x_2, the four a_{ij} coefficients, and the two c_i coefficients. To solve for x_2, we rearrange terms to obtain

$$\left(\frac{a_{22}}{a_{21}}\right)x_2 - \left(\frac{a_{12}}{a_{11}}\right)x_2 = \frac{c_2}{a_{21}} - \frac{c_1}{a_{11}}$$

The lowest common denominator for these terms is $a_{11}a_{21}$; placing all terms on this denominator yields

$$\left(\frac{a_{11}a_{22}}{a_{11}a_{21}}\right)x_2 - \left(\frac{a_{21}a_{12}}{a_{11}a_{21}}\right)x_2 = \frac{a_{11}c_2}{a_{11}a_{21}} - \frac{a_{21}c_1}{a_{11}a_{21}}$$

We can now multiply every term by $a_{11}a_{21}$, clearing the common denominator and obtaining

$$a_{11}a_{22}x_2 - a_{21}a_{12}x_2 = a_{11}c_2 - a_{21}c_1$$

from which

$$x_2 = \frac{a_{11}c_2 - a_{21}c_1}{a_{11}a_{22} - a_{21}a_{12}}$$

as before.

We can calculate the numerical value of x_1 at the point of intersection by substituting the value just obtained for x_2 in either equation (1.4) or (2.4). Substituting in (2.4), we obtain

$$x_1 = \frac{c_2}{a_{21}} - \left(\frac{a_{22}}{a_{21}}\right)\left[\frac{a_{11}c_2 - a_{21}c_1}{a_{11}a_{22} - a_{21}a_{12}}\right]$$

To place the two right-hand terms on a common denominator, we multiply both numerator and denominator of (c_2/a_{21}) by $(a_{11}a_{22} - a_{21}a_{12})$, obtaining

$$x_1 = \frac{c_2(a_{11}a_{22} - a_{21}a_{12}) - a_{22}(a_{11}c_2 - a_{21}c_1)}{a_{21}(a_{11}a_{22} - a_{21}a_{12})}$$

Rearranging terms in the numerator, we have

$$x_1 = \frac{(a_{11}a_{22}c_2 - a_{11}a_{22}c_2) + a_{21}(a_{22}c_1 - a_{12}c_2)}{a_{21}(a_{11}a_{22} - a_{21}a_{12})}$$

The first two terms in the numerator cancel; then a_{21} can be eliminated from both numerator and denominator of the remaining terms, leaving

$$x_1 = \frac{a_{22}c_1 - a_{12}c_2}{a_{11}a_{22} - a_{21}a_{12}}$$

also as before.

Some of the special situations that arise in matrix algebra can be clarified by examining their geometric implications and, in fact, the terminology used to describe them is geometric: linear independence, linear dependence, collinearity, and multicollinearity, for example.

Consider the case in which our two linear equations are

(1.5) $$a_{11}x_1 + a_{12}x_2 = c_1$$

(2.5) $$a_{21}x_1 + a_{22}x_2 = c_2$$

where $a_{21} = \lambda a_{11}$; $a_{22} = \lambda a_{12}$, and λ is any scalar. We rewrite (2.5) as

$$(\lambda a_{11})x_1 + (\lambda a_{12})\, x_2 = c_2 \tag{2.5.1}$$

Normalizing (1.5) and (2.5.1) as before, we obtain

$$x_1 = \frac{c_1}{a_{11}} - \left(\frac{a_{12}}{a_{11}}\right)x_2 \tag{1.6}$$

$$x_1 = \frac{c_2}{\lambda a_{11}} - \left(\frac{\lambda a_{12}}{\lambda a_{11}}\right)x_2 = \frac{c_2}{\lambda a_{11}} - \left(\frac{a_{12}}{a_{11}}\right)x_2 \tag{2.6}$$

These are the equations of two parallel lines. They do not intersect; there is no point which lies on both lines, and the pair of equations does not have a unique solution.

If we try to solve (1.5) and (2.5.1) algebraically by eliminating x_1, we first obtain

$$-(\lambda a_{11})\, a_{11}x_1 - (\lambda a_{11})\, a_{12}x_2 = -(\lambda a_{11})c_1 \tag{1.7}$$

$$(a_{11})\lambda a_{11}x_1 + (a_{11})\lambda a_{12}x_2 = a_{11}c_2 \tag{2.7}$$

Adding corresponding terms, we find that the coefficients of x_1 cancel out as before, leaving

$$(a_{11}\lambda a_{12} - \lambda a_{11}a_{12})x_2 = a_{11}c_2 - \lambda a_{11}c_1$$

and

$$x_2 = \frac{a_{11}c_2 - \lambda a_{11}c_1}{\lambda(a_{11}a_{12} - a_{11}a_{12})}$$

The denominator of this expression is zero! The numerator is, in general, a nonzero scalar, say, k, so $x_2 = k/0 = \pm\infty$, depending on whether k is positive or negative.

While the proposition that two parallel lines intersect at infinity is useful in proving certain theorems in projective geometry, two parallel lines in matrix algebra applications usually spell trouble. If we are given two equations in two variables and the equations turn out to be parallel lines, the problem has no solution; there is no point (x_1, x_2) that satisfies both equations.

A linear equation in three variables x_1, x_2, x_3 is the equation of a plane. Two planes which are not parallel to one another intersect in a straight line. A third plane which is not parallel to either of the first two will intersect that line in a point (x_1, x_2, x_3) which is common to all three planes and satisfies all three equations; its coordinates represent the unique solution of three linear equations in three variables. However, if planes 1 and 2 are parallel to one another, they have no line in common; a third plane will intersect planes 1 and 2 separately, and these lines of intersection will be parallel, with no point in common.

Another possibility is that, while planes 1 and 2 intersect in a line, plane 3 might pass through (i.e., contain) this same line, so that again there is no unique solution; *every* point on the line satisfies the equations of all three planes.

We shall show shortly that the case of three equations, two of which represent parallel planes, lead to expressions for x_1, x_2, and x_3 with zero denominators. But we must first digress to mention certain properties of *determinants.*

A square matrix $A = [a_{rs}]$ is an ordered array of n times n scalars (numerical values). The *determinant* of A, which we will denote by $|A| = |a_{rs}|$, is a single scalar to be derived from the n times n scalars of A in a prescribed manner. This manner is expressly designed to facilitate the solutions of sets of n linear equations in n variables.

Notation. If $A = [a_{rs}]$ is a square matrix of order n times n, write A_{rs} for the matrix of order $(n - 1)$ obtained by deleting from A the rth row and the sth column, intersecting in the element a_{rs}. If $|A|$ is the determinant of A, then the *co-factor* of a_{rs} in A is the determinant of A_{rs} with an appropriate sign:

$$A_{rs}^* = (-1)^{r+s}|A_{rs}|$$

where we use the asterisk to distinguish the cofactor from the corresponding determinant.[5]

The rule for evaluating a determinant of order n is given in terms of the cofactors A_{rs}^*, that is, in terms of determinants of order $n - 1$. We proceed by mathematical induction.

Definition. The determinant $|A| = |a_{rs}|$ of order n is obtained from a square matrix $A = [a_{rs}]$ of order n times n. When $n = 1$, $|A| = A = a_{11}$, a single element. The rule for a determinant of order n in terms of determinants of order $n - 1$ is

$$|A| = \sum_{i=1}^{n} a_{i1}A_{i1}^*$$

where A_{i1}^* is the cofactor of a_{i1} $(i = 1, 2, \ldots n)$.

From this definition, determinants of successive orders may be evaluated step by step. The above relation is in terms of the elements of the first column of whatever matrix is being considered, and of the appropriate cofactors of

[5] The page and a half beginning with "**notation**" is adapted from R. G. D. Allen, *Basic Mathematics*, op. cit., pp. 385–386.

these elements:

$\underline{n = 1:}$ $$A = a_{11}$$

and

$$|A| = a_{11}$$

$\underline{n = 2:}$ $$A = \begin{bmatrix} a_{11} & a_{12} \\ a_{21} & a_{22} \end{bmatrix}$$

and

$$|A| = a_{11}\,|a_{22}| - a_{21}\,|a_{12}| = a_{11}a_{22} - a_{21}a_{12}$$

Here the determinant $|A|$ is the "cross-product" $(a_{11}a_{22} - a_{21}a_{12})$ of the four elements of A. We note that this "cross-product" is identical with the denominators we obtained in solving two linear equations for x_1 and x_2:

$\underline{n = 3:}$ $$A = \begin{bmatrix} a_{11} & a_{12} & a_{13} \\ a_{21} & a_{22} & a_{23} \\ a_{31} & a_{32} & a_{33} \end{bmatrix}$$

and

$$|A| = a_{11}\begin{vmatrix} a_{22} & a_{23} \\ a_{32} & a_{33} \end{vmatrix} - a_{21}\begin{vmatrix} a_{12} & a_{13} \\ a_{32} & a_{33} \end{vmatrix} + a_{31}\begin{vmatrix} a_{12} & a_{13} \\ a_{22} & a_{23} \end{vmatrix}$$

Using the "cross-products" for the determinants of second order, this becomes

$$|A| = a_{11}(a_{22}a_{33} - a_{32}a_{23}) - a_{21}(a_{12}a_{33} - a_{32}a_{13}) + a_{31}(a_{12}a_{23} - a_{22}a_{13})$$

or

$$|A| = a_{11}a_{22}a_{33} - a_{11}a_{32}a_{23} + a_{21}a_{32}a_{13} - a_{21}a_{12}a_{33} + a_{31}a_{12}a_{23} - a_{31}a_{22}a_{13}$$

In general, a determinant of order n can be expanded into $n!$ terms of n elements each. In the present case, $n = 3$ and $n! = (3)(2)(1) = 6$.

Actually, a determinant can be expanded in terms of the elements of any column or any row, as long as the appropriate cofactors are used in each case.

Determinants may be used to solve our set of two linear equations as follows.

Our equations are

(1) $$a_{11}x_1 + a_{12}x_2 = c_1$$

(2) $$a_{21}x_1 + a_{22}x_2 = c_2$$

or, in matrix form,

$$\begin{bmatrix} a_{11} & a_{12} \\ a_{21} & a_{22} \end{bmatrix}\begin{bmatrix} x_1 \\ x_2 \end{bmatrix} = \begin{bmatrix} c_1 \\ c_2 \end{bmatrix}$$

The value of x_1 can be obtained as the ratio of two determinants,

$$x_1 = \frac{\begin{vmatrix} c_1 & a_{12} \\ c_2 & a_{22} \end{vmatrix}}{\begin{vmatrix} a_{11} & a_{12} \\ a_{21} & a_{22} \end{vmatrix}} = \frac{a_{22}c_1 - a_{12}c_2}{a_{11}a_{22} - a_{21}a_{12}}$$

The value of x_2 can be obtained as

$$x_2 = \frac{\begin{vmatrix} a_{11} & c_1 \\ a_{21} & c_2 \end{vmatrix}}{\begin{vmatrix} a_{11} & a_{12} \\ a_{21} & a_{22} \end{vmatrix}} = \frac{a_{11}c_2 - a_{21}c_1}{a_{11}a_{22} - a_{21}a_{12}}$$

These are identical, of course, with the values obtained earlier by more laborious methods. The denominator in each case is the determinant $|A|$. In the numerator of x_1, the column vector of constant terms $\{c_1, c_2\}$ is substituted for the first column of $|A|$ which had contained the coefficients of x_1 in the original equations. Similarly, in the numerator of x_2, the column vector of constant terms $\{c_1, c_2\}$ is substituted for the second column of $|A|$, which had contained the coefficients of x_2.

The forms of the original matrix equation and of the solutions for x_1 and x_2 suggest that some process analogous to "matrix division" is going on. For example, it appears that we can display the values of x_1 and x_2 in the matrix notation as follows:

$$\begin{bmatrix} x_1 \\ x_2 \end{bmatrix} = \frac{1}{|A|} \begin{bmatrix} a_{22}c_1 & -a_{12}c_2 \\ -a_{21}c_1 & a_{11}c_2 \end{bmatrix}$$

where the ratio $1/|A|$ is a scalar to be multiplied by each of the four elements of the matrix of order 2 times 2. It appears further that the latter matrix is the result of the following multiplication:

$$\begin{bmatrix} a_{22}c_1 & -a_{12}c_2 \\ -a_{21}c_1 & a_{22}c_2 \end{bmatrix} = \begin{bmatrix} a_{22} & -a_{12} \\ -a_{21} & a_{11} \end{bmatrix} \begin{bmatrix} c_1 \\ c_2 \end{bmatrix}$$

Hence, our solution for x_1 and x_2 may be displayed as

$$\begin{bmatrix} x_1 \\ x_2 \end{bmatrix} = \frac{1}{|A|} \begin{bmatrix} a_{22} & -a_{12} \\ -a_{21} & a_{11} \end{bmatrix} \begin{bmatrix} c_1 \\ c_2 \end{bmatrix}$$

If we compare this solution with the original matrix equation

$$\begin{bmatrix} a_{11} & a_{12} \\ a_{21} & a_{22} \end{bmatrix} \begin{bmatrix} x_1 \\ x_2 \end{bmatrix} = \begin{bmatrix} c_1 \\ c_2 \end{bmatrix}$$

it is evident that (in obtaining the solution) both sides of the original equation have been multiplied by

$$\frac{1}{|A|}\begin{bmatrix} a_{22} & -a_{12} \\ -a_{21} & a_{11} \end{bmatrix}$$

This implies the following identity:

$$\frac{1}{|A|}\begin{bmatrix} a_{22} & -a_{12} \\ -a_{21} & a_{11} \end{bmatrix}\begin{bmatrix} a_{11} & a_{12} \\ a_{21} & a_{22} \end{bmatrix}\begin{bmatrix} x_1 \\ x_2 \end{bmatrix} = \begin{bmatrix} 1 & 0 \\ 0 & 1 \end{bmatrix}\begin{bmatrix} x_1 \\ x_2 \end{bmatrix} = \begin{bmatrix} x_1 \\ x_2 \end{bmatrix}$$

We may disregard the $\{x_1, x_2\}$ vector and focus our attention on the coefficient matrices,

$$\frac{1}{|A|}\begin{bmatrix} a_{22} & -a_{12} \\ -a_{21} & a_{11} \end{bmatrix}\begin{bmatrix} a_{11} & a_{12} \\ a_{21} & a_{22} \end{bmatrix} = \begin{bmatrix} 1 & 0 \\ 0 & 1 \end{bmatrix}$$

The left-hand side is equivalent to

$$\frac{1}{(a_{11}a_{22} - a_{21}a_{12})}\begin{bmatrix} (a_{22}a_{11} - a_{12}a_{21}) & (a_{22}a_{12} - a_{12}a_{22}) \\ (-a_{21}a_{11} + a_{11}a_{21}) & (-a_{21}a_{12} + a_{11}a_{22}) \end{bmatrix}$$

$$= \frac{1}{(a_{11}a_{22} - a_{21}a_{12})}\begin{bmatrix} (a_{11}a_{22} - a_{21}a_{12}) & 0 \\ 0 & (a_{11}a_{22} - a_{21}a_{12}) \end{bmatrix}$$

$$= \begin{bmatrix} 1 & 0 \\ 0 & 1 \end{bmatrix}$$

so the identity holds. Also, the matrix

$$\begin{bmatrix} a_{22} & -a_{12} \\ -a_{21} & a_{11} \end{bmatrix}$$

can be seen to be the transpose of the matrix of cofactors of the original coefficient matrix

$$\begin{bmatrix} a_{11} & a_{12} \\ a_{21} & a_{22} \end{bmatrix}$$

The cofactor of a_{11} is a_{22}; that of a_{21} is $(-a_{12})$; that of a_{12} is $(-a_{21})$; and that of a_{22} is a_{11}. Thus, the matrix in which each original a_{ij} is replaced by its cofactor is

$$\begin{bmatrix} a_{22} & -a_{21} \\ -a_{12} & a_{11} \end{bmatrix}$$

and the transpose of this matrix is

$$\begin{bmatrix} a_{22} & -a_{12} \\ -a_{21} & a_{11} \end{bmatrix}$$

as required by the identity.

Recall that $|A|$ is a scalar and so, of course, is its reciprocal $1/|A|$. To be perfectly clear, however, we shall write out the 2 times 2 matrix

$$\frac{1}{|A|}\begin{bmatrix} a_{22} & -a_{12} \\ -a_{21} & a_{11} \end{bmatrix}$$

as

$$A^{-1} = \begin{bmatrix} \dfrac{a_{22}}{|A|} & \dfrac{-a_{12}}{|A|} \\ \dfrac{-a_{21}}{|A|} & \dfrac{a_{11}}{|A|} \end{bmatrix}$$

where A^{-1} is called the *inverse* of the matrix

$$A = \begin{bmatrix} a_{11} & a_{12} \\ a_{21} & a_{22} \end{bmatrix}$$

The matrix product $A^{-1}A = I$, the identity matrix

$$\begin{bmatrix} 1 & 0 \\ 0 & 1 \end{bmatrix}$$

In compact notation, the original matrix equation can be written as $Ax = c$. To solve for x, we simply multiply both sides of the equation by the inverse of A:

$$A^{-1}Ax = A^{-1}c$$

or

$$Ix = A^{-1}c$$

or simply

$$x = A^{-1}c$$

The concept of an *inverse matrix* is extremely useful both for abstract manipulations of matrix equations (as in proving theorems or outlining generalized computations for sets of equations of any order) and also for obtaining numerical solutions to specific numerical problems. If the original square matrix A in $Ax = c$ is *nonsingular* (that is, if its determinant $|A| \neq 0$), where x and c are of order n by 1 and A is of order n by n, the matrix equation $x = A^{-1}c$ is valid for all integral values of n.

If the original square matrix A is *singular* (that is, if its determinant $|A| = 0$), no inverse A^{-1} exists. If $|A| = 0$ in the two variable case, then

$$A^{-1} = \frac{1}{0}\begin{bmatrix} a_{22} & -a_{12} \\ -a_{21} & a_{11} \end{bmatrix} = \begin{bmatrix} \dfrac{a_{22}}{0} & \dfrac{-a_{12}}{0} \\ \dfrac{-a_{21}}{0} & \dfrac{a_{11}}{0} \end{bmatrix}$$

All elements are indeterminate (infinite), and the original set of equations has no solution. (We note in passing that the case of two parallel lines described earlier,

$$a_{11}x_1 + a_{12}x_2 = c_1$$
$$\lambda a_{11}x_1 + \lambda a_{12}x_2 = c_2$$

leads to the determinant value

$$|A| = (a_{11}\lambda a_{12} - \lambda a_{11}a_{12}) = 0$$

Hence, the matrix A is *singular* if two rows represent the equations of parallel lines; in this case, the coefficients in one row of A are identical with the coefficients of another row except for a scalar multiplier.)

Definition. The square matrix A of order n times n has determinant $|A|$. If A is singular ($|A| = 0$), no inverse exists. If A is nonsingular ($|A| \neq 0$), then

$$A^{-1} = \frac{1}{|A|}[A_{sr}] = \frac{1}{|A|}\begin{bmatrix} |A_{11}| & |A_{21}| & \cdots & |A_{n1}| \\ |A_{12}| & |A_{22}| & \cdots & |A_{n2}| \\ \vdots & \vdots & & \vdots \\ |A_{1n}| & |A_{2n}| & \cdots & |A_{nn}| \end{bmatrix}$$

where $|A_{rs}| = (-1)^{r+s}|A_{rs}|$ is the cofactor of a_{rs} in $|A|$.[6]

Hence, A^{-1}, like A, is a matrix of order n times n. By definition, it is obtained by writing the matrix $[A_{sr}]$ of the n times n cofactors, by transposing it to $[A_{sr}] = [A_{rs}]'$, and by dividing every element by $|A|$. This requires $|A| \neq 0$.

If A is nonsingular, then A^{-1} is the only matrix with the property $AA^{-1} = A^{-1}A = I$. This is the basic property of inverses. Some other properties are as follows:

$$(A^{-1})^{-1} = A; \; (AB)^{-1} = B^{-1}A^{-1}; \qquad \text{and} \qquad (A')^{-1} = (A^{-1})'$$

As a particular case, $I^{-1} = I$; that is, the unit matrix is such that it is the same matrix as its transpose and inverse.

Earlier, we mentioned the case in which two planes in a set of three planes were parallel. This case will shed light on our next topic, the rank of a matrix. We are given the following general set of three linear equations in three variables:

$$(1) \qquad a_{11}x_1 + a_{12}x_2 + a_{13}x_3 = c_1$$

$$(2) \qquad a_{21}x_1 + a_{22}x_2 + a_{23}x_3 = c_2$$

$$(3) \qquad a_{31}x_1 + a_{32}x_2 + a_{33}x_3 = c_3$$

[6] See R. G. D. Allen, *Basic Mathematics*, op. cit., p. 387.

In a particular case, suppose it happens that $a_{21} = \lambda a_{11}$, $a_{22} = \lambda a_{12}$, and $a_{23} = \lambda a_{13}$, where λ is a scalar. Then, in matrix notation, our three equations are

$$\begin{bmatrix} a_{11} & a_{12} & a_{13} \\ \lambda a_{11} & \lambda a_{12} & \lambda a_{13} \\ a_{31} & a_{32} & a_{33} \end{bmatrix} \begin{bmatrix} x_1 \\ x_2 \\ x_3 \end{bmatrix} = \begin{bmatrix} c_1 \\ c_2 \\ c_3 \end{bmatrix}$$

or $Ax = c$. Expanding $|A|$ by the elements of the first column, we obtain

$$\begin{aligned} |A| = a_{11}&(\lambda a_{12} a_{33} - a_{32} \lambda a_{13}) \\ &- \lambda a_{11}(a_{12} a_{33} - a_{32} a_{13}) \\ &+ a_{31}(a_{12} \lambda a_{13} - \lambda a_{12} a_{13}) \end{aligned}$$

or

$$\begin{aligned} |A| = \lambda\,&[a_{11}(a_{12} a_{33} - a_{32} a_{13}) - a_{11}(a_{12} a_{33} - a_{32} a_{13})] \\ &+ \lambda[a_{31}(a_{12} a_{13} - a_{12} a_{13}] = 0 + 0 = 0 \end{aligned}$$

Since $|A| = 0$, A is singular; A^{-1} does not exist; and the set of three equations has no solution.

The rank of a matrix. Suppose we are given a set of m linear equations in n variables.

$$\begin{bmatrix} a_{11} & a_{12} & a_{13} & \cdots & a_{1n} \\ a_{21} & a_{22} & a_{23} & \cdots & a_{2n} \\ a_{31} & a_{32} & a_{32} & \cdots & a_{3n} \\ \vdots & \vdots & \vdots & & \vdots \\ a_{m1} & a_{m2} & a_{m3} & \cdots & a_{mn} \end{bmatrix} \begin{bmatrix} x_1 \\ x_2 \\ x_3 \\ \vdots \\ x_n \end{bmatrix} = \begin{bmatrix} c_1 \\ c_2 \\ c_3 \\ \vdots \\ c_m \end{bmatrix}$$

or $Ax = c$; the matrix A is of order m times n. If $m = n$ and $A_{nn} = A_{mm}$ is nonsingular, we can obtain a unique "solution vector" x containing n elements $\{x_1, x_2, x_3, \ldots x_n\}$. If $m < n$, we can at most determine a unique solution involving some set of m variables and all m equations. If $n < m$, we can at most determine a unique solution involving some set of n equations and all n variables. Obviously, the largest nonsingular square matrix that can be formed from a rectangular matrix of order m times n will at most be a matrix of order m times m or n times n, whichever is of lower order.

Suppose $m < n$, so that the largest possible square matrix is of order m times m. It may happen, however, that two rows k and j are *linearly dependent* (that is, $a_{ji} = \lambda a_{ki}$, where $k \neq j$ and $i = 1, 2, \ldots m$). In the case in which $m = 3$, we have shown that $|A| = 0$ if two rows are linearly dependent. The matrix A of order 3 times 3 has $3! = 6$ submatrices of order 2 times 2. If the determinant of at least one of these submatrices is nonsingular, the *rank* of the matrix A is 2.

Linear dependence may also appear in a more subtle form in which one row is a linear combination of two or more other rows. Consider the following coefficient matrix

$$A = \begin{bmatrix} a_{11} & a_{12} & a_{13} \\ a_{21} & a_{22} & a_{23} \\ (a_{11} + \lambda a_{21}) & (a_{12} + \lambda a_{22}) & (a_{13} + \lambda a_{23}) \end{bmatrix}$$

in which $a_{3i} = a_{1i} + \lambda a_{2i}$ for all $i = 1, 2, 3$. Expanding $|A|$ by the elements of the first column, we obtain

$$\begin{aligned} |A| &= a_{11}(a_{22}a_{13} - a_{12}a_{23}) + a_{11}(a_{22}\lambda a_{23} - \lambda a_{22}a_{23}) \\ &\quad - a_{21}(a_{12}a_{13} - a_{12}a_{13}) - a_{21}(a_{12}\lambda a_{23} - \lambda a_{22}a_{13}) \\ &\quad + a_{11}(a_{12}a_{23} - a_{22}a_{13}) + a_{21}(\lambda a_{12}a_{23} - \lambda a_{22}a_{13}) \end{aligned}$$

Rearranging terms, we obtain

$$\begin{aligned} |A| &= a_{11}(a_{22}a_{13} - a_{12}a_{23}) + a_{11}(0) \\ &\quad - a_{21}(0) \qquad\qquad\qquad - a_{21}(\lambda a_{12}a_{23} - \lambda a_{22}a_{13}) \\ &\quad - a_{11}(a_{22}a_{13} - a_{12}a_{23}) + a_{21}(\lambda a_{12}a_{23} - \lambda a_{22}a_{13}) \end{aligned}$$

The two nonzero terms remaining in the first column cancel; so do the two nonzero terms remaining in the second column. Hence, $|A| = 0$ and matrix A is singular.

Matrix A would still be singular if the coefficients in Row 3 had the formula

$$a_{3i} = \lambda_1 a_{1i} + \lambda_2 a_{2i} \quad \text{for all } i = 1, 2, 3$$

The three terms in the second column of the expansion of $|A|$ would all contain λ_2 instead of λ and would cancel out as before. The three terms in the first column would each be multiplied through by λ_1 and would also cancel out, resulting again in $|A| = 0$. The concept of linear dependence can be extended to the case in which

$$a_{ki} = \lambda_1 a_{1i} + \lambda_2 a_{2i} + \cdots + \lambda_j a_{ji}$$

where j is any positive integer smaller than k. In a set of several equations, multiple cases of linear dependence may occur, with each of two or more rows being linear combinations (or scalar multiples) of still other rows. The term *multicollinearity* is sometimes used to imply all cases of linear dependence, including the case in which one row is a scalar multiple of another (that is, the linear dependence directly involves only two rows).

Hence, the rank of a matrix of order m times n with $m < n$ may be $m - 1$, $m - 2$, or even lower. For example, a determinant $|A|$ of order m times m includes $m!$ cofactors of order $m - 1$. If $|A|$ itself and all of these cofactors are singular, the rank of $|A|$ is less than $m - 1$. If at least one square submatrix of order $m - 2$ in A is nonsingular, the rank of A is $m - 2$.

A *symmetric* matrix is a special kind of square matrix in which $A' = A$. If $A = [a_{rs}]$ for $r, s = 1, 2, \ldots m$, then $a_{sr} = a_{rs}$ for all $r, s = 1, 2, \ldots m$. For $m = 3$, we may write a symmetric matrix as

$$A = \begin{bmatrix} a_{11} & a_{12} & a_{13} \\ a_{12} & a_{22} & a_{23} \\ a_{13} & a_{23} & a_{33} \end{bmatrix}$$

Expanding by the elements of the first column, we obtain

$$\begin{aligned} |A| &= a_{11}(a_{22}a_{33} - a_{23}^2) \\ &\quad - a_{12}(a_{12}a_{33} - a_{23}a_{13}) \\ &\quad + a_{13}(a_{12}a_{23} - a_{22}a_{13}) \end{aligned}$$

Rearranging terms, we get

$$\begin{aligned} |A| &= a_{11}a_{22}a_{33} - a_{11}a_{23}^2 - a_{33}a_{12}^2 + a_{12}a_{13}a_{23} \\ &\quad + a_{12}a_{13}a_{23} - a_{22}a_{13}^2, \end{aligned}$$

or

$$|A| = a_{11}a_{22}a_{33} + 2a_{12}a_{13}a_{23} - a_{33}a_{12}^2 - a_{22}a_{13}^2 - a_{11}a_{23}^2$$

We may note that a matrix of correlation coefficients is symmetric. For $m = 3$, we may write

$$[r_{ij}] = \begin{bmatrix} 1 & r_{12} & r_{13} \\ r_{12} & 1 & r_{23} \\ r_{13} & r_{23} & 1 \end{bmatrix}$$

Comparing $|r_{ij}|$ with $|A|$, we have $r_{ii} = a_{ii}/a_{ii} = 1$ for $i = 1, 2, 3$ and $r_{ij} = a_{ij}$ for $i = 1, 2$ and $j = 2, 3; j > i$. Hence, $|r_{ij}| = 1 + 2r_{12}r_{13}r_{23} - r_{12}^2 - r_{13}^2 - r_{23}^2$.

Multicollinearity (linear dependence), which leads to singular matrices when computing the coefficients of linear equations, also leads to singular matrices of correlation coefficients. In the three variable case, the matrix of sums of squares and cross-products of the deviations of X_1, X_2, and X_3 from their respective means is

$$[\sum x_i x_j] = \begin{bmatrix} \sum x_1^2 & \sum x_1x_2 & \sum x_1x_3 \\ \sum x_1x_2 & \sum x_2^2 & \sum x_2x_3 \\ \sum x_1x_3 & \sum x_2x_3 & \sum x_3^2 \end{bmatrix}$$

Now, suppose $x_3 = \lambda x_2$ where λ is a scalar. Then $[\sum x_i x_j]$ may be written as

$$[\sum x_i x_j] = \begin{bmatrix} \sum x_1^2 & \sum x_1 x_2 & \lambda \sum x_1 x_2 \\ \sum x_1 x_2 & \sum x_2^2 & \lambda \sum x_2^2 \\ \lambda \sum x_1 x_2 & \lambda \sum x_2^2 & \lambda^2 \sum x_2^2 \end{bmatrix}$$

Recall that the $\sum x_i x_j$'s are given numbers, just like the a_{ij}'s of our previous demonstrations. The third row of $[\sum x_i x_j]$ is a scalar multiple of the second row, so $|\sum x_i x_j| = 0$ and $[\sum x_i x_j]$ is singular.

If $x_3 = \lambda x_2$, $\sqrt{\sum x_3^2} = \lambda \sqrt{\sum x_2^2}$. When we convert $[\sum x_i x_j]$ into the corresponding correlation matrix, we divide each $\sum x_i x_j$ by

$$(\sqrt{\sum x_i^2} \cdot \sqrt{\sum x_j^2})$$

In this case, each λ is divided by a λ and the λ^2 in $[\sum x_i x_j]$ is divided by a λ^2, giving the following result:

$$[r_{ij}] = \begin{bmatrix} 1 & r_{12} & r_{12} \\ r_{12} & 1 & 1 \\ r_{12} & 1 & 1 \end{bmatrix}$$

$$|r_{ij}| = 1(1 - 1) - r_{12}(r_{12} - r_{12}) + r_{12}(r_{12} - r_{12}) = 0 + 0 + 0 = 0$$

Hence, the rank of $[r_{ij}]$ is less than 3.

Of the $3! = 6$ cofactors of order 2 times 2 in $[r_{ij}]$, four are equal to $(1 - r^2)$ and two are equal to zero; therefore the rank of $[r_{ij}]$ is 2.

In an applied economic problem, it would be quite unusual to find that (say) $x_3 = \lambda x_2$ exactly. However, it quite frequently happens that $x_3 = \lambda x_2 + \mu_3$, where μ_3 is a random disturbance. Further, x_3 and x_2 are deviations from the means of X_3 and X_2. The original observations on X_3 and X_2 would be related as $X_3 - \bar{X}_3 = \lambda(X_2 - \bar{X}_2) + \mu_3$, or $X_3 = (\bar{X}_3 - \lambda \bar{X}_2) + \lambda X_2 + \mu_3$, and a quick visual inspection of the original paired values of X_3 and X_2 may not reveal how closely they are correlated. We would probably compute $[r_{ij}]$ by a routine method and view the numerical results as

$$[r_{ij}] = \begin{bmatrix} 1 & r_{12} & r_{13} \\ r_{12} & 1 & r_{23} \\ r_{13} & r_{23} & 1 \end{bmatrix}$$

If at this stage we find that r_{23} approaches 1, we shall also find that r_{13} approaches r_{12} and that $|r_{ij}|$ approaches zero. The standard errors of our partial regression coefficients will prove to be large in the multiple regression equation

$$\hat{X}_{1\cdot 23} = a_{1\cdot 23} + b_{12\cdot 3}X_2 + b_{13\cdot 2}X_3$$

and the significance levels of $b_{12\cdot 3}$ and $b_{13\cdot 2}$ will be low.

SELECTED REFERENCES

Allen, R. G. D., *Basic Mathematics*, St. Martin's Press, New York, 1962.

Allen, R. G. D., *Mathematical Analysis for Economists*, St. Martin's Press, New York, 1956.

Allen, R. G. D., *Mathematical Economics*, St. Martin's Press, New York, 1959.

Love, Clyde E., *Differential and Integral Calculus*, The Macmillan Company, New York, 1934.

Wilks, Samuel S., *Mathematical Statistics*, John Wiley & Sons, Inc., New York, 1962.

Index

  Regression, multiple regression model, 384–385